The Macedonian War Machine

In memory of the late
Professor N.G.L. Hammond
DSO, CBE, FBA
(1907–2001)

And to my beloved Lena and parents

'After this I beheld, and lo another, like a leopard, which had upon the back of it four wings of a fowl; the beast had also four heads; and dominion was given to it'.

(*Book of Daniel* 7:6. Verse interpreted by St Jerome as prophesying the rise of the Macedonian Empire.)

'Alexander ... [and his army] ... those leathern-belted demons with dishevelled hair'.

(*Bahman Yasht* 3.34, trans. E.W. West)

The Macedonian War Machine

Neglected aspects of the armies of Philip, Alexander and the Successors (359–281 BC)

DAVID KARUNANITHY

Pen & Sword
MILITARY

First published in Great Britain in 2013 by
PEN & SWORD MILITARY
An imprint of
Pen & Sword Books Ltd
47 Church Street
Barnsley
South Yorkshire
S70 2AS

ISBN 978-1-84884-618-0

A CIP catalogue record for this book is available from the British Library.

Typeset by Concept, Huddersfield, West Yorkshire.
Printed and bound in India by Replika Press Pvt. Ltd.

Pen & Sword Books Ltd incorporates the Imprints of Pen & Sword Aviation, Pen & Sword Family History, Pen & Sword Maritime, Pen & Sword Military, Pen & Sword Discovery, Wharncliffe Local History, Wharncliffe True Crime, Wharncliffe Transport, Pen & Sword Select, Pen & Sword Military Classics, Leo Cooper, The Praetorian Press, Remember When, Seaforth Publishing and Frontline Publishing.

For a complete list of Pen & Sword titles please contact
PEN & SWORD BOOKS LIMITED
47 Church Street, Barnsley, South Yorkshire, S70 2AS, England
E-mail: enquiries@pen-and-sword.co.uk
Website: www.pen-and-sword.co.uk

Contents

List of Illustrations

(1) Plates

(2) Figures

Unless otherwise stated, all figure drawings are by Graham Sumner.

(3) Diagrams

(4) Maps

Preface

My imagination was first fired by Alexander the Great and the ancient Macedonians when just a child and I confess that undertaking a major project of this kind has been a key ambition of mine for a very long time. Producing a work on the Macedonian army took years of gestation and deliberation before I finally entrusted my thoughts to paper. With other commitments, the entire research process took well over a dozen years to complete.

There are many reasons to compel study of this remarkable military organisation. The foremost is that Alexander's army gave birth to the Hellenistic Age. The army and its heirs helped diffuse Greek civilisation 'worldwide', spanning the eastern Mediterranean rim and Middle East, even in pockets as remote as Central Asia and the North-West Frontier of India. It has a faint resemblance to the Roman imperial army, which functioned as a vehicle for the spread and propagation of Latin and Roman civilisation in the opposite direction: the western Mediterranean, west and northwest Europe. In retrospect, we may argue that as a result of the energies unleashed by these two military systems, for centuries a landmass from England to Syria was unified for the first and only time by the ties of a generally homogeneous ('classical' Mediterranean) culture.[1]

The Macedonian army without doubt constitutes a complex subject for enquiry and, like peeling the layers of an onion, is far more involving and labyrinthine than most have been led up to now to suppose. Given its powerful impact on the history of warfare, the institution has become an especial subject for research, much discussed by modern generals and historians, and a topic of ongoing interest in master's and doctoral theses.[2]

During the past 40 years or so around 2,000 books and articles have been produced on Alexander the Great, covering practically every known angle of his life, military or otherwise. His campaigns alone have been more closely probed than for any other personality in the ancient world. Alexander's logistical methods in Asia have even been adopted as appropriate rationale for modern business management strategies.[3] At time of writing, knowledge continues to grow in size and scope with the ongoing publication of new archaeological discoveries, and thousands of yet-to-be-published Babylonian cuneiform tablets from the British Museum may contribute further insights. A multi-faceted study covering Alexander's army, in detail and in its entirety, would require decades of effort and fill several volumes.[4]

What was this army like, and why was it so successful? In the last 100 years rivers of ink have been spent attempting to clarify and answer this deceptively straightforward question. Even though all work on the subject is valuable, and there have been some erudite treatments, endeavours have usually been confined to a well-beaten path. This furrowed trail embraces precise definitions of terminology, qualities of leadership and strategy, the history and characteristics of campaigns, battle mechanics and tactics, troop armament (in particular the cavalry lance and infantry pike), command structure, unit organisation and siege craft. The limited, prosaic nature of what has become an intellectual cul-de-sac for at least Hellenistic armies is recognised by a growing band of academics, but in particular Patrick Baker, Professor of Ancient History at Laval University, Quebec.[5]

It thus appears that for too long the armed forces of Philip II, Alexander and the Successors have been seen through a narrow and sometimes uninspired frame of reference. Undoubtedly, to Alexander and his officers all military aspects were inextricably linked, contributing in sum total to the army's remarkable efficiency and success. As has been achieved in Roman military studies, it is incumbent on us that we nurture a broader field of vision for the Macedonian army so as to appreciate its considerable achievements more clearly and assess their historical impact. There is much to be uncovered, much that remains ignored or only cursorily examined, and much that persists in being beyond the scope of people other than a small community of specialists. All the same, a profusion of hard-to-obtain evidence is available for anyone willing to delve deep enough to acquire it.

With the above in mind, the present work sets out to bring to life the Macedonian army through a range of generally obscure, albeit fascinating, fields of investigation, many of which are less fashionable or have gone largely neglected in research elsewhere. It examines the eventful eighty-year period (359–281 BC) from Philip II's ascent to power to the death of Seleucus, the last of the Successors, but with priority given over to Alexander's army and Macedonian troops. In compiling material, I drew upon suitable supporting evidence from the early-fourth century BC to Hellenistic armies of the third or even second centuries BC.

It is important to stress from the outset that my book is not meant as a light read or introduction to the subject, nor is it intended as a strictly academic work, but occupies a niche in the middle for those who have gained some familiarity with the ancient and modern literature detailing Macedonian warfare. Nonetheless, I hope that a wide audience will find something of appeal to them – from professional historians and archaeologists to students and enthusiastic lay-people who enjoy reading ancient military history for its own sake. My writing is, in effect, an attempt to plug gaps, to target and sketch out the obtuse corners of an otherwise incomplete canvas. In pursuit of this goal my original manuscript was naturally adjusted to fit publication requirements, with sections pruned, sheared or even excised in places. In such a huge subject, there are inevitably blank patches in knowledge that still remain to be painted in by other hands. For example, I discuss training but do not fully address the system of recruitment; I investigate military awards and possible decorations but deliberately leave the question of pay undisturbed.

The consistent aim followed here has been to marshal together in one place as much of the relevant *testimonia* as possible on a specific series of topics, but to let readers draw their own conclusions based on the efficacy of surviving data. Ideally, evidence should be allowed its own 'breathing space' to speak for itself. However, this is rarely achieved in this subject where researchers are detached from the Macedonian epoch by some twenty-four centuries and cut off from contemporary written accounts. This leaves us with only a shattered impression seen through opaque glass. Some might therefore justifiably wish to follow Aristotle's golden mean by which the investigator cultivates a vital in-depth understanding of the source material at hand, while at least attempting to maintain a clear-sighted and objective distance from it – an acutely difficult balance.

To date, many scraps of evidence on the Macedonian army have been put under the forensic microscope of debate and conjecture. The result is that each fragment is usually passed on with its own imposed veneer of scholarly baggage. It can admittedly be helpful in a factual and contextual sense, but personal interpretations can often be based on variables such as the extraneous life experiences of the authority in question and their selective vision. We must also be mindful of the general social, cultural or political (ideological) climate and intellectual trends at the time of writing. Many may yield to inaccurate post-rationalisation, subconsciously or otherwise fitting evidence to correlate with preconceived notions and building a sometimes inflexible and doctrinaire line of argument.

The research found here will hopefully help readers arrive at a more holistic view and should be treated as a supplement to what has already been established in the field of Macedonian warfare. If the reader is left with a lasting picture of the army as an institution that brought about a whole range of revolutionary changes, then I would have accomplished my intended task. Throughout the work there is some emphasis on the outward appearance and paraphernalia of soldiers although with the infusion of fresh insights on evidence. This is combined with material located in rare or out of print monographs, highly specialised articles, conference papers and excavation reports, which are either little known, frequently hard to obtain or in languages other than English. It is true that an abundance of artistic evidence exists for Macedonian and Hellenistic soldiers but these precious fragments have been dispersed in museum collections around the world (Greece, Turkey, Cyprus, Egypt, France, Italy, Germany, Russia, the United Kingdom and the United States) and sadly remain to this day under-scrutinised.[6]

Lastly, it remains to be said that the ideas expressed throughout these pages will by no means remain static. In many cases they will have to be reworked and reinterpreted in light of future discoveries. Even so, being by natural inclination a 'maximalist', I set out from the start to make this study as thorough and as well referenced as possible, ascribing as I do to Thomas Mann's adage that 'the exhaustive is truly interesting'.[7] This really is the only adequate way of presenting the full tenor of the Macedonian military achievement.

Ave atque vale.

Acknowledgements

My wife Lena is chiefly to thank for making this book possible. Without her monumental efforts in formulating an effective editing schedule, my work would have simply remained a dog-eared manuscript gathering dust, hopes and dreams. We worked on the task as a team – I undertook the research while she performed major 'surgery' on my text. I am indebted to her for cutting away the jumble of words, transforming what I wrote into something altogether orderly and readable.

Apart from Lena, I would like to express my appreciation for the late Professor Emeritus N.G.L. Hammond of the University of Cambridge. Professor Hammond was one of the foremost authorities on ancient Macedonia in the English language and his many admirable works have over the years kindled in me a passion for all things 'ancient' and 'Macedonian'. We maintained a written correspondence from August 1997 and he patiently reviewed some of my early drafts and was at hand to encourage me in my slow and laborious progress. His last letter to me was dated 8 March 2001 with an invitation to meet him at his home. Unfortunately, this was not to be as he died several weeks later at the age of 93 after over half a century of phenomenal academic output. I sent my condolences to his family and received a touching response from his wife Margaret. Before his passing, the Professor had most generously provided a short Foreword banged out on a fossilised typewriter and I had requested that this might accompany my book.

Over the years scores of people and institutions have been instrumental in aiding me with the painstaking research for this project and with the use of their facilities.

During my forays into Greece I made contact with individuals who have greatly improved the final outcome. A debt of gratitude must be paid to Georgina Giati for her immense kindness and thoughtful cooperation, Konstantinos Noulas for his inexhaustible generosity and hospitality, Anthi Efstathiou for bringing some rare evidence to my attention and Jordanis Pimenidis for his inspiring enthusiasm.

Other contacts who have assisted and inspired me include Gian Svennevig, George Casstrisios, Evelyn Miller, Philip Greenough, Anthony Dove, Steven Neate, Christopher Webber and many other members of the Society of Ancients. To this roll call must be added Anna Chatzinikolaou, Minor Markle, Roger Scott, Duncan Head, Dean Lush, Ruben Post and Luke Ueda-Sarson, to whom I am grateful for useful points, valued insights and a generous supply of articles. In addition, I would like to recognise Anastasia Maravela and the staff of the Joint

Library of the Hellenic and Roman Societies including the Warburg Institute Library (both of the School of Advanced Study, London).

Heartfelt thanks must go to my parents for always being at hand whenever I needed their support along the long twisting road of study, and to my sister Clare, who transported me around Melbourne in my obsessive hunt for research materials. My editor, Philip Sidnell, should also be singled out for praise, granting me the timely opportunity to have my work published through Pen & Sword Books after years of vacillation. Finally, I must acknowledge Dr Peter Guest of Cardiff University. When we were schoolboys he once teased me about an imaginary book called 'The Macedonian War Machine' – the subtle seedling for this project.

David Karunanithy
St Albans, June 2012

Foreword

'The Macedonian Army' is a very extensive subject. For the Army was almost commensurate with the Macedonian State in the time of Philip and Alexander, and sections of the army continued to be the dominant factor in the subsequent period. That period has been called the 'Hellenistic' period, which implies a decline from the 'Hellenic' period. It should be called correctly the 'Macedonian' period, in which Macedonian commanders and sections of the Macedonian army set up and maintained individual States.

The present study of 'The Macedonian Army' embraces all aspects of the subject. It is based upon comprehensive and meticulous research. It is unique in that it is beautifully and extensively illustrated. Full references are given both to the ancient evidence (literary, archaeological and numismatic) and to modern accounts. It is a great advance as compared with previous treatments of the subject. Indeed it is by any standard a definitive version.

N.G.L. Hammond
Cambridge, April 2000

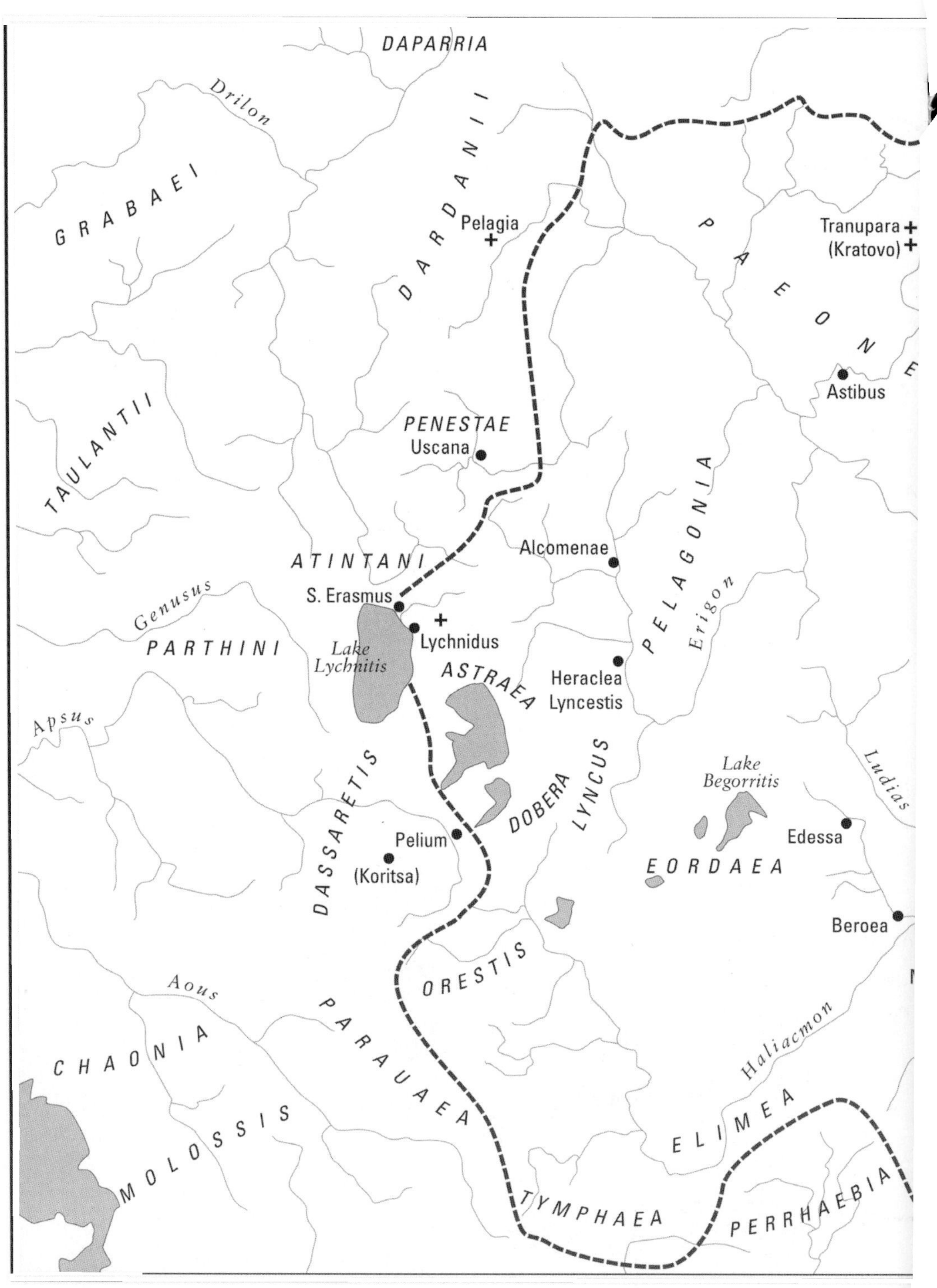

Map 1. The Macedonian kingdom under Philip II.

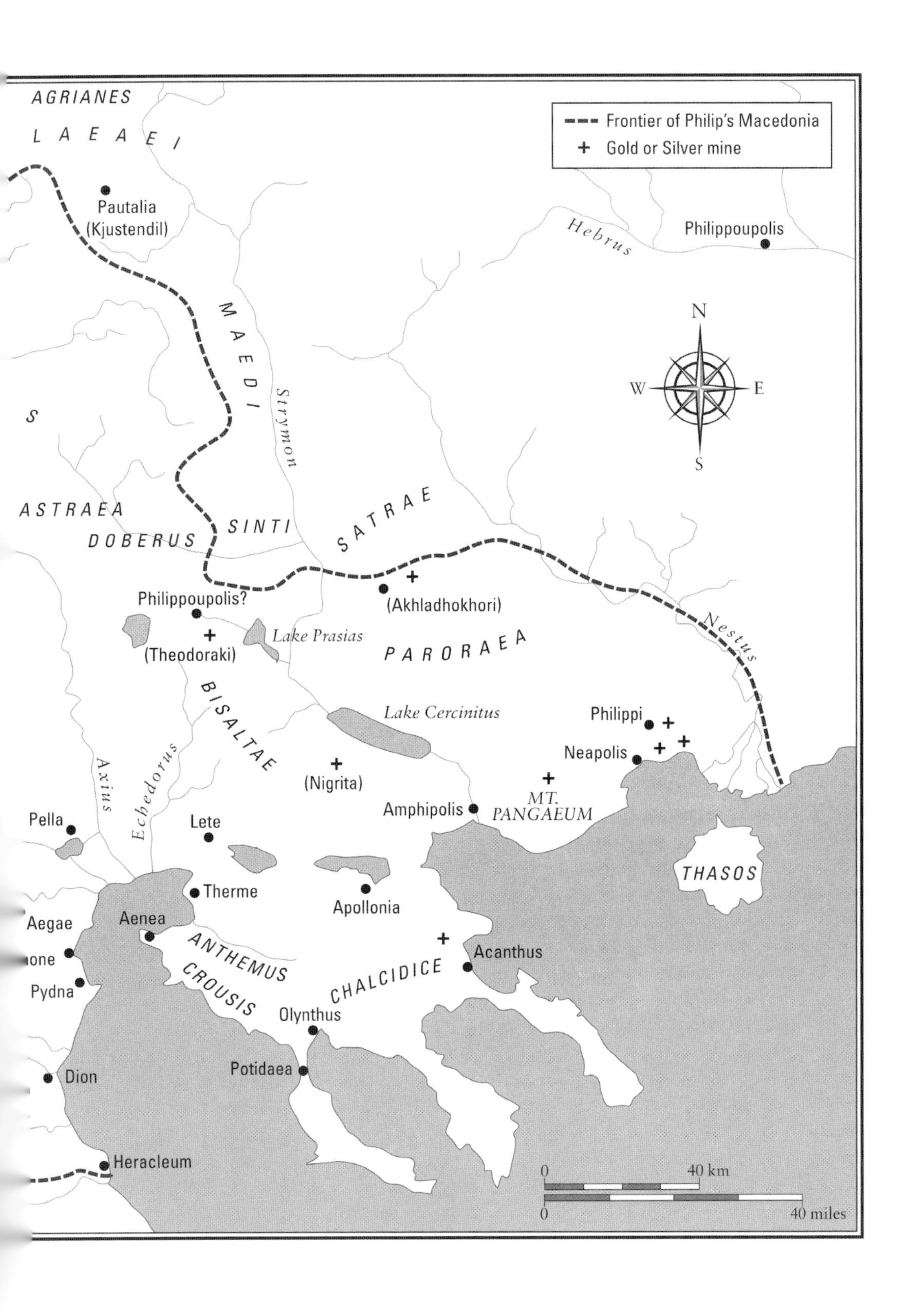

Frontier of Philip's Macedonia
Gold or Silver mine
AGRIANES
LAEAEI
Pautalia
(Kjustendil)
Hebrus
Philippoupolis
N
W
E
S
MAEDI
Strymon
ASTRAEA
DOBERUS
SINTI
SATRAE
Philippoupolis?
(Akhladhokhori)
Nestus
(Theodoraki)
Lake Prasias
PARORAEA
BISALTAE
Lake Cercinitus
Philippi
Neapolis
Axius
Echedorus
(Nigrita)
Amphipolis
MT. PANGAEUM
Pella
Lete
THASOS
Therme
Apollonia
Aegae
Aenea
ANTHEMUS
Acanthus
Pydna
CROUSIS
CHALCIDICE
Olynthus
Potidaea
Dion
Heracleum
0
40 km
0
40 miles

Map 2. Alexander the Great's conquests and empire.

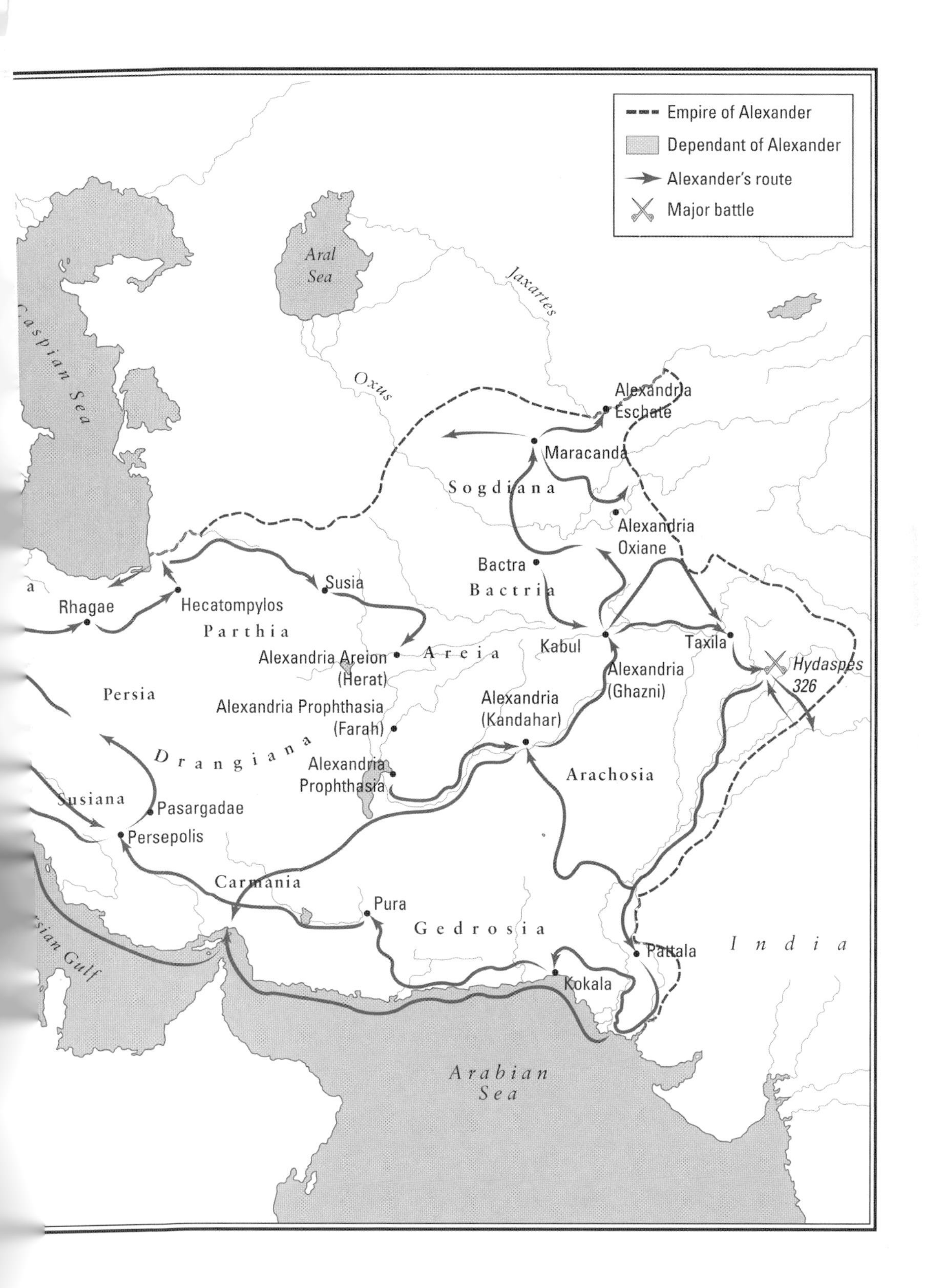

Empire of Alexander
Dependant of Alexander
Alexander's route
Major battle
Aral Sea
Caspian Sea
Jaxartes
Oxus
Alexandria Eschate
Maracanda
Sogdiana
Alexandria Oxiane
Bactra
Bactria
Susia
Rhagae
Hecatompylos
Parthia
Alexandria Areion (Herat)
Areia
Kabul
Taxila
Hydaspes 326
Alexandria (Ghazni)
Persia
Alexandria Prophthasia (Farah)
Alexandria (Kandahar)
Drangiana
Alexandria Prophthasia
Arachosia
Susiana
Pasargadae
Persepolis
Carmania
Pura
Gedrosia
Pattala
India
Kokala
Persian Gulf
Arabian Sea

Map 3. Division of Alexander's empire under the Successors.

N
W
E
S
Ara
Seal
Caspian
Sea
nia
Sogdiana
Bactria
Paropamisadae
Hyrcania
Media
Parthia
Areia
Gandaris
SELEUCUS
Persia
Drangiana
Susiana
Arachosia
Mauryan
Empire
Carmania
Persian Gulf
Gedrosia
Arabian Sea

PART I

ORIGINS AND PERSPECTIVES

The dynamic period of Macedonian conquest in the mid- to late-fourth century BC is understood as one of the pivotal chapters in the history of warfare. It was the Macedonian army and its generals which exploded onto the world with such force, expanding the kingdom's power far beyond the confines of the Balkan Peninsula. In some respects, the army's legacy can be felt even to this day.

Chapter 1

The Macedonian Army's Place in History

The history of Philip II and Alexander the Great is to all intents the story of outstanding men of action leading the Macedonian royal army. Seen with the clarity of modern hindsight, this army is regarded as representing one of the most important leaps in military thinking in the West before Napoleon.[1] At the least, it embodies the pinnacle of evolutionary changes affecting Greek warfare, which grew and gathered momentum from the time of the Peloponnesian War (431–404 BC) to Philip II's accession to the embattled Macedonian throne (359 BC).

The Macedonian army also stands as a watershed – the precursor to a new era of intense activity exemplified by the large well-developed land forces of Hellenistic and Roman times.[2] We can clearly ascertain the extent of this if we momentarily exclude the Macedonians from the historical process. Their exclusion would leave a major gap between the military sophistication of Athenian, Spartan and Theban armies during the fifth and early-fourth centuries BC and that of the Roman Republic from the second and first centuries BC. The intermediate period was dominated by the 'New Model' Macedonian army and its derivatives.[3]

The Rise of Philip's Army State

Notwithstanding that modern comparisons must always be treated with caution, the evidence suggests the meteoric rise of Philip's well-knit state in the mid-fourth century BC to be in broad outline analogous to the march of Prussia in the eighteenth century under Frederick William I and his famous son Frederick the Great.[4] Like Prussia, the newly enlarged kingdom's wealth and natural resources were suborned, first and foremost, to maintaining the armed forces and consequently military strength, security and conquest.[5] The Macedonian *sarissa* (long spear or pike) and concentric pattern shield soon became proud symbols synonymous with the kingdom and its people.

Philip's leadership and state building acumen made Macedonia the culminating point for Greek military inventiveness. Moreover, through the foundation of the army, a ramshackle 'sub-Homeric enclave' on the semi-barbarous fringes of the Hellenic world was transmogrified during his reign into arguably the first nascent 'nation-state' and significant land-empire seen in Europe.[6]

Philip II conceived the image of the Macedonian as a soldier, and it was on this basis that his reinvented and expanded kingdom began its rapid rise to greatness.[7] As noted in a compelling and memorable comment by Ellis, which is just as legitimate now as when written over thirty years ago, 'Philip II reconstructed the Macedonian army not only as a military weapon but also (and I suspect deliberately) as an instrument of social and political unity'.[8] The army was the anvil upon which a distinct Macedonian group identity and consciousness were forged.[9] It was the instrument that amalgamated all groups into an integral whole, while at the same time funnelling an outlet to those in the aristocracy with talent and ambition.

Once Philip had achieved those goals, the army, as the cornerstone to 'national' unification and stability, had to be kept active.[10] In other words, Philip II's kingdom presents us with a paradox: the army was the main pillar of the state but at the same time it had to be continually employed in external wars for unity to be sustained. The result was a policy of expansion and the waging of almost perpetual wars. In a later age Napoleon would lightly remark that Prussia was hatched from a cannon ball;[11] that is, founded on the army and forged through conflict. The recast Macedonian kingdom was little different.

From his inception in power Philip continually strove to systematically nurture his burgeoning might, while testing and honing the military system that underpinned it. The King frequently led the troops in person and it has been said that during his reign he waged no less than twenty-eight campaigns.[12] Some part of the army was in the field every year and, unlike the convention found elsewhere, these activities were all-season affairs.[13] It is thought that in the twenty years after 358 BC Macedonian military strength trebled or even quadrupled in size. Philip II's achievement can be favourably compared to that of Frederick William I, who, during twenty-seven years, a little over doubled the extent of the Prussian army.[14]

The interminable political jostling and machinations of Greek city-states all came to nought in the end. The Macedonians scooped the pool and victory at Chaeronea (338 BC) assured their dominance. By the time of his assassination when only 46 years old (336 BC), the battle-scarred, one-eyed and limping Philip had bequeathed to his son a fully integrated and professionalised force. It was an army tried, experienced and confident at all levels – the finest to be found anywhere in the Greek world and Europe. It can be said that without Philip II's precision instrument, neither Alexander's enormous empire nor the three centuries of the Hellenistic Age would have been realised.

The Macedonian Achilles

Just as Philip II was a brilliant innovator and probably the foremost soldier-king seen in European circles till then, so his heir Alexander III became the greatest conqueror antiquity ever saw. As Appian declared to the Roman readership of his *History* in the second century AD: 'The empire of Alexander was splendid in its magnitude, in its armies . . . and it wanted little of being boundless and unexampled, yet in its shortness of duration it was like a brilliant flash of lightning'.[15]

Brilliant flash indeed – but whilst analysing and appreciating the military genius of Alexander the Great, we must never divorce his exceptional personal talents from the infrastructure so scrupulously prepared by his father.[16] Nor should we forget the superb officer corps inherited and extended by this fierce young pretender to Achilles, some of whom came into their own as state builders during the bloody winnowing which followed his death.[17]

King Philip had successfully harnessed and militarised the manpower available to him, and his son exploited it to the fullest potential. It has been considered that on the eve of the invasion of the Persian Empire by Alexander as many as one in ten Macedonian citizens were serving with the army.[18] The core kingdom, no bigger than modern Denmark, Switzerland or 'North Country' England (Map 1), was destined to become a military recruiting ground catering to the needs and ambitions of royal governments in the East.[19]

It should be said that Alexander's momentous career is a rare example of a single army-commander achieving vast contiguous conquests penetrating from the Aegean Sea to Central Asia, repeated by Tamerlane and his Turco-Mongols, who slashed a swath of expansion from the opposite direction seventeen centuries later. This fact alone should make the army Alexander led worthy of ongoing analysis. In contrast, the Romans attempted, furtively or otherwise, to emulate him on a number of occasions, but failed each time. The Parthians were simply too resilient and for centuries both the Roman and Byzantine empires were destined to wage wars erupting along the volatile Sassanian frontier.

At the head of Philip's army, and in only nine years (334–325 BC), Alexander swept unchecked from the Dardanelles to the Punjab, covering over 30,000km in all forms of climate and terrain. Marching further than any before him, he founded an empire of over 5 million square kilometres (spanning the borders of fifteen countries today),[20] with an estimate of 20–40 million inhabitants, about a quarter of the then world population (Map 2).

With support from the defeated Achaemenid state and its colossal wealth, Alexander continued to make many refinements to the army beyond purely tactical considerations. As will be shown in this book, these include the mass distribution of lavish fabrics and panoplies, improved logistics and medical attention on campaign, and effective alterations to soldiers' benefits and structures of promotion. Both individual and group orders of ranking based on merit were installed, enhancing the concept of a 'Homeric' battle array, motivation and competitive spirit.[21] The eastern venture formed a backdrop to all this, but it was no romantic cavalcade. Some thirteen years of war under Alexander probably witnessed more deaths than Greek battles from Marathon (490 BC) to Chaeronea put together.[22] The Macedonians had made a stock-in-trade of defeating those who dared oppose them on the field.

In the last phase of his life Alexander began to lay remarkable imperial foundations, remoulding his army into a comprehensive, multi-ethnic force of soldiers and reserves trained, armed and organised to Macedonian, and even some Iranian, principles within his newly formed 'Kingdom of Asia'.[23] Literary sources varyingly report that had he lived the King's insatiable appetite may have led him on to absorb the Arabian peninsula, Carthage and the western Mediterranean seaboard. Arrian goes on to express little surprise that had Alexander lived he might have driven remorselessly westwards until even the British Isles were subjugated.[24] Despite Livy's self-assured protestations, Rome would have been swallowed up.[25] Alexander thus seems to have contemplated nothing short of exploring and ruling all inhabited land.[26] The Macedonian army was to be the tool in realising his dream world, quenching his unsatisfied and inexplicable *pothos*, or 'longing'.[27]

The Struggle for Power and its Aftermath

While still only 32, the increasingly paranoid conqueror and his plans for future conquests were abruptly cut short with sudden fever and premature death amid the sultry summer heat in a palace at Babylon (June 323 BC). His parting words were to prove prophetic. On being asked who should inherit the empire, his barely audible reply was 'the best' or 'strongest'.[28]

The King had gained historical immortality, his exploits radiating over the millennia in story and song from Iceland to China. But his officers now unexpectedly found themselves leaderless and rudderless: like a Cyclops blinded of its one eye, to paraphrase Demades of Athens' pithy remarks.[29]

So the funeral games on Alexander's departure became the internecine conflicts of his *Diadochoi* or 'Successors'. This galaxy of larger than life Machiavellian characters was the offspring of fourth-century BC military invention, schooled in warfare both by Alexander and his father.[30] Individually some, like Antigonus Monophthalmus and Eumenes of Cardia, were among the finest commanders in ancient history. Collectively they represent a powerful generation of warlords – professionals with by and large a standard of ability rarely achieved by corresponding groups.[31] Alexander's officers are sometimes compared to the Roman civil war generals of the first century BC, the lieutenants of Genghis Khan, or Napoleon's imperial marshals. They were, as Plutarch says: 'men to whose rapacity neither sea nor mountain nor desert sets a limit, men to whose inordinate desires the boundaries which separate Europe and Asia put no stop'.[32]

Bereft of a stable consensus and with such a preponderance of ruthless and gifted leaders, the carcass of empire was doomed to disintegrate. The Macedonians thus turned upon themselves[33] as Alexander's fragile political edifice was sundered over forty years, with those in the kingdom who were able-bodied and militarily eligible recruited to fuel armed confrontation (Map 3). This was accompanied by the dismemberment of the hitherto unbeaten veteran army, splintered and frittered away between manoeuvring factions and coalitions. The deadly game was played out over a sprawling theatre of operations bounded by Greece, Egypt and the Iranian plateau.

During this time something like an arms race escalated between the contending parties. The great centuries-old treasure-troves of the defunct Persian Empire were squandered to pay for warfare in all its grim diversity. Gigantism was the order of the day with ever bigger or better warships, artillery, siege engines and fortifications, as well as the first widespread use of elephants on battlefields west of India. To cap it all, the almost constant fighting of Philip,

Alexander and their officers triggered another milestone in the intellectual field: a slew of varied military literature.

Within a Greek milieu Macedonia had become an aberration. Philip II had found an internally divided and largely archaic warrior people, which he united and transformed through scientific militarism. Alexander then applied it to carving out a huge empire but ultimately failed in translating this to a workable, long-lasting system of governance. What few chances there were for establishing an *imperium* withered on the vine after his death. It was during the violent convulsions of the 'Wars of the Successors' that the old soldiers who had conquered Persia were gradually pensioned off and dispersed as settlers in the East, their fame passing on into popular myth.

Macedonians now emigrated in their tens of thousands to settle with their families as *katoikoi* (soldier-colonists) in farms and cities spread throughout the wide expanse of conquered lands – particularly Asia Minor, Syria and Egypt.[34] Many of these incomers were destined to form the elite units of the newly formed Seleucid and Ptolemaic militaries.

The fact that the Macedonians were germinated as the ruling *nuclei* of Hellenistic states in Seleucid Asia and Ptolemaic Egypt had profound repercussions. Philip and Alexander's land forces, directly or otherwise, provided the canon for military organisation, weaponry, tactics and specialism across the next few centuries from Sicily to Afghanistan. The Macedonian army archetype remained undisputed until the event of 22 June 168 BC, when at the denouement of Pydna the kingdom went down in defeat and collapse before the maniples of Rome (see Epilogue).

In the second century BC, following Roman defeats inflicted on the home kingdom, there may have been more 'patriotic' Macedonians emigrating to the East. Perhaps an element of the population preferred overseas military service to languishing at home as Rome's subjects. Macedonian emigration possibly accelerated during the first century BC, with growing employment in the Ptolemaic army.[35] The waves of colonists who settled across the Middle East over generations became forever mingled and dispersed within the gene pool, faintly traced to this day.[36]

Chapter 2
Transmission of Military Knowledge

It is an often overlooked fact that many of the basic professional features of modern armies find partial descent from a remote ancestor in antiquity – namely, the Macedonian army (see Diagram 1). Like the Romans, the Macedonians under Philip and Alexander became adepts at experimenting with, absorbing and improving upon the most progressive military thought of their antecedents, contemporaries and enemies. Their activities acted as catalyst for a virtual explosion of intellectual activity on things military – an invaluable resource, even if only a fraction has survived.

The Intellectual Base

The army was Philip II's obsession, pride and joy,[1] and the limited information we have suggests that it was brought into being as an agglomerated formalisation of all practical knowledge on warfare known to him. The King was an astute pupil who took on concepts adopted from a range of eminent figures.

Those personalities who influenced Philip must include ambitious royal forebears such as Alexander I (r. 498–454 BC), Archelaus (r. 413–399 BC) and the more able of the King's immediate predecessors. It is possible that Philip II came at the end of an intermittent sequence of military reforms which had begun in Macedonia half a century before his own time.[2] This was complemented by tactical doctrine drawn from his great mentor Epaminondas (the greatest soldier of the Greek world in his day) and, to a lesser extent, the other prominent Theban generals, Pelopidas and Pammenes. Neither should we ignore the enigmatic Thessalian tyrant Jason of Pherae (d. 370 BC) as another source of inspiration. Ideas were possibly borrowed from the mathematical theorems of Lysis of Tarentum and diverse cultural stimuli. Philip was well versed in the Greek heroic epics and could have made use of a growing genre on Homeric warfare.[3] It is notable that the Athenian playwright Aristophanes, living only decades before Philip's accession, spoke of Homer as a fount of knowledge on army discipline and equipment.[4] The King collected and assimilated much of this large, often disparate, melange of features while combining them with his own ingenious, far-reaching contributions.[5] He may even have been experimenting with different armament and tactics during a local governorship on behalf of his elder brother, Perdiccas III (r. 365–360/359 BC), from as early as 364 BC.[6]

It has gone largely neglected by researchers that the intellectual background and reasoning behind the creation of the new Macedonian army by Philip II was partly based on written military discourse. Writing in the second century BC and drawing on an already established convention, Polybius recommended that to acquire sound generalship leaders must be students of history and should read military treatises.[7] It has been suggested that Philip was acquainted with the writings of Herodotus, Thucydides and other historians.[8] He respected learning and the liberal arts and was known at least for his elegant writing style.[9]

One of the main reasons why military techniques embraced by the Macedonians are often seen today as elusive or problematic is in part due to Philip and his officers having borrowed from literature which is now all but lost. Antipater and Parmenion were the two most trusted generals on Philip's staff and there is evidence that Antipater composed a now lost history on the Illyrian wars of Perdiccas III.[10] Works on tactics, weapon handling and horsemanship were accessible to interested parties in the mid-fourth century BC, ascribed to forgotten authors like Simon of Athens (c. 480 BC) and Democritus of Abdera (c. 360 BC).[11] Aeneas the Tactician offers an excellent exemplar. Internal indications from his extant treatise *How to Survive Under Siege* (c. 350 BC) hint at a number of other titles on warfare written by him in open circulation covering subject areas as diverse as 'tactics', 'naval tactics', 'encampments', 'preparations', 'procurement', 'addresses' and 'siege operations'.[12] Such an assortment implies that there was an eager readership willing to devour studies of this kind.

It is plausible that Philip read histories and military works in private collections as part of his education while residing at Thebes (368–365 BC).[13] The tragedian Euripides is said to have owned a private library as early as the fifth century BC.[14] Given his aspiration to build the best army, Philip would have set out to gather and read as much practically useful literature as possible.

An even better case can be made for the famous Xenophon (d. 354 BC). His surviving books *On Horsemanship*, *Cavalry Commander*, *Memorabilia*, *Anabasis* and *Agesilaus* all, to varying degrees, incorporate excellent advice for up and coming military men. Polybius implies that Philip had read Xenophon's *Anabasis*, on the expedition of the Ten Thousand (401–399 BC), and had studied Agesilaus of Sparta's campaigns in Asia Minor (396–394 BC).[15] Arrian asserts that in a speech before the battle of Issus (333 BC) Alexander bolstered the morale of his troops by comparing their all arms versatility and superiority in cavalry, archers and slingers to the Ten Thousand.[16] This shows that, like his father, he was familiar with the *Anabasis* and had taken heed of the lessons found there.

Xenophon's *Cyropaedia* ('Education of Cyrus'), with extensive advice for making a model army (perhaps with Sparta in mind), is thought by some scholars to have had a bearing on Philip's reforms.[17] As both Scipio Africanus and Julius Caesar were keen readers of this work, there is no reason to think that Philip or Alexander were anything less.[18] In regard to Alexander, Renault speculates that lore digested from the *Cyropaedia* was probably ignored as elementary.[19] However, Xenophon goes into minute detail over many aspects of training, organisation and campaign arrangements in Asia so it is improbable that during his schooling Alexander, or those with responsibility over his education, would have ignored such a valuable resource. Eunapius (c. 400 AD), in the introduction to his *Lives of the Philosophers*, even takes the trouble to praise Xenophon for having inspired Alexander and other great captains.[20]

Plutarch remarks that Alexander was a serious minded youth and an avid reader.[21] That he was reading up on things military from an early age – doubtless with Philip's encouragement – is borne out by Polybius, who states that he studied generalship from boyhood.[22] The *Suda*, a tenth-century Byzantine historical encyclopaedia, describes Marsyas of Pella, brother of Antigonus Monophthalmus, as having written a work called *The Education of Alexander*.[23] If this had survived it would have imparted to us what the future conqueror read.

Alexander may have learnt topics of military importance under his tutor Aristotle.[24] More than Philip, he is known in particular for having had an unusually high regard for the *Iliad*, which he read incessantly (often memorising whole sections). When campaigning in Asia he kept a special 'casket copy' annotated by his celebrated teacher as a sort of travellers' handbook on warfare practice.[25] An anecdote survives purporting to be a dialogue between Philip and Alexander during the *Xandika* festival at Dion, although in all likelihood it represents a later invention. Here father questions son over his fixation with Homer to the detriment of other poets.[26]

Throughout his reign, Philip attracted to his court and service a motley menagerie of military advisers, roistering opportunists and adventurers from across the Balkans and Black Sea coast, Greece and the Aegean world, Rhodes and Crete. Others are documented coming from as far west as Tarentum in southern Italy and also Sicily, to as far east as the Hellespont, Ionia, Caria and Cappadocia – with even a handful of Persian exiles emanating from Asia Minor, Egypt and the Achaemenid court itself.[27] Thus, using the tenets of his own native culture as a basis, Philip's brainchild was essentially syncretic, formed from many places.

The assortment of people gathered by Philip potentially brought with them a wealth of experience in wars waged over the Hellenic and eastern Mediterranean arenas. Some would have an abiding intellectual interest in armies and tactical combat systems or could be itinerant military instructors – a profession that became more pronounced by the early-fourth century BC. Xenophon describes in a dialogue one such called Dionysodorus of Chios, who visited Athens and lectured on generalship, although some were disappointed that his focus was on infantry tactics alone.[28] The mercenary captain Phalinus (c. 400 BC) also professed to give expert advice on drill.[29]

Those invited to Macedonia are best thought of as a 'melting pot', making Philip's court a laboratory within which know-how was disseminated, with experimentalism as the inevitable by-product. A condensed extract from the Homeric commentaries known as the 'Townley scholia' gives a shadowy glimpse into the flourishing scene of exchange witnessed in the Greek world of the early- to mid-fourth century BC, including Philip's circle: 'The tactician Hermolykos says that Lykourgos enacted the synaspismos [locked shields drill order]. Lysander the Lakonian and Epaminondas taught it. Then the Arkadians and the Macedonians learnt it from Charidemos'.[30] This information appears to originate in a lost Hellenistic history or military treatise.

Theopompus suggests that lively debate among the King and his friends was common. In a long-winded tirade he castigated Philip for spending too much time in the company of drinking cronies with whom he deliberated on important matters.[31] That some of Philip's late night conversations were about warfare can be discerned through the filter of the Athenian comic poet and playwright Mnesimachus (c. 345 BC), who amused audiences by parodying stock Macedonian characters speaking of their fellow countrymen banqueting in the following manner:

> Have you any idea what we're like to fight against? Our sort make their dinner of honed-up swords, and swallow blazing torches for a savoury snack. Then, by way of dessert, they bring us, not nuts, but broken arrow-heads and splintered spear-shafts. For pillows we make do with our shields and breastplates; arrows and slings lie strewn under our feet, and we wreathe our brows with catapults.[32]

This colourful fragment hints at the Macedonian interest in combined arms tactics and a general obsession with warfare, viewed through a satirical lens, when even at dinner they reverted to type by talking about their pet preoccupation: military affairs. It implies the extent to which Philip and his subordinates set out to realise the full potential of warrior achievement embodied in the concept of martial *arete* ('virtue' or 'excellence').[33]

In his commentary to historical anecdotes about abusive brawling which arose between men drunk on wine, Athenaeus introduces the same stanza quoted above using a phrase from Xenophon, describing the rowdy scene as a 'workshop of war' – an apt description for Philip's court.[34] We can therefore appreciate the narrative of Justin (epitomising the *Philippic History* of Pompeius Trogus), where Philip II is summarised as a king for whom things military were the leading passion: 'in his view his greatest treasures were the tools of warfare'.[35]

The very fact that so much radical change was established in the Macedonian army over such a surprisingly short burst of time, differentiating it from all other land forces of its day,[36] suggests that Philip was helped by a large number of now nameless aides and specialists. The internal workings of the new army were given birth to by Greek intellect and one cannot help but think this represented the collective cerebral effort of many ruminating and scientifically inclined minds. In this respect, the urge to place in rank order, to clinically analyse and classify, witnessed especially with the Aristotelian approach, was brought to highest fruition. The Macedonians applied single-minded flair in systematically reorganising their army on a level that even the Spartans could scarce outclass.[37] Marsden's investigation into the subject of Macedonian siege apparatus helps bear this out. Technical invention proceeded at such a fast pace in less than twenty years that Marsden was moved to believe that in some sense it anticipated the advance of armaments in the modern industrial age.[38]

Given Antipater's known interest in military history, we can only surmise what part he, Parmenion and younger talented officers within Philip's coterie may have played in helping to realise a broad-based programme of change. After all, Plutarch tells us that there were no less than thirty other sons of leading Macedonians who accompanied Philip as hostages to Thebes.[39] In a warlike society, the young prince was almost certainly not the only one among them with an interest in warfare and it is a genuine pity that none of their names have come down to us.

Philip was thereby from the outset surrounded by a nucleus of like-minded, dedicated followers – the true founders of Alexander's famous army. Polybius rejected the opprobrium some writers had for the King's officers and instead praised their outstanding energy, industry and daring, suggesting that many actively participated in preparing Macedon's military resources.[40] This seems to be further hinted at by the orator Demosthenes in his *Second Olynthiac* (delivered in 349 BC).[41] It is, of course, unlikely that Philip II was solely responsible for *all* that was put into effect, though as head of state he obviously selected and co-ordinated (rather than micro-managed) reform.[42]

Hellenistic Military Works

The dynamic rise to predominance of the Macedonian army in the late-fourth century BC gave rise to a plethora of military writings. These set in train the high watermark of ancient written enquiry into warfare, inclusive of India and China,[43] establishing a convention that continued among Greek speaking authors to the time of the Roman and even Byzantine empires.

The roll call of *known* author-theorists amounts to almost thirty names and incorporates: Daemachus, thought to be a diplomat under the Seleucid king Antiochus I (r. 281–261 BC);[44] Clearchus, sometimes identified as a Peripatetic philosopher from Soli; and the famous historian Polybius. These are followed by such little known worthies as Hermolycus, Apollonius, Bryon, Eupolemus, Evangelus, Stratocles, Hermeas, Nymphodorus, Pausanias, the more familiar Posidonius, and others.[45] Undoubtedly, their varied works were consulted by Hellenistic army technicians.[46] In such manner, methods nurtured by the Greeks, and especially Macedonians, could now be effectively transmitted to succeeding generations. The pursuit of war science hence became a respected field of enquiry in its own right.

The source material originally to hand must have comprised a rich store of knowledge on Philip and Alexander's war machine. It is, however, a disappointing fact that of over 1,100 names of authors covering all subjects which survive to the present, the bulk of Hellenistic prose has been irretrievably lost to us, although a small minority may have been dispersed and diffused among later encyclopaedias and general histories. We can but ponder what these many works contained. The sheer scale of the loss is appreciated from the known output of Greek dramatists in the fifth century BC, with one instructive example reflecting that Aeschylus wrote between seventy and ninety plays, of which a mere seven are intact.[47]

Among the works which have failed to percolate down to our time were writings by the leading officers and personalities of the age of Alexander and the

Successors.[48] It has been estimated that as many as twenty or more individuals who took part in Alexander's expedition wrote about it, implying an intellectual interest in history or military matters.[49] These include accounts of Alexander's wars in Asia by Ptolemy and Nearchus. Other studies incorporate a *strategika* by Demetrius of Phalerum and a *taktika* written by Alexander II of Epirus (died c. 242 BC). An epitome on the corpus of works produced by Aeneas the Tactician in the mid-fourth century BC has even been attributed to Cineas, rhetorician, philosopher and minister to King Pyrrhus.[50]

Arguably the most profound loss of all was a treatise (or treatises) by Pyrrhus himself (319–272 BC) – second cousin of Alexander the Great and father to Alexander II of Epirus. It is feasible that this work was governed by the earlier writings of Aeneas or was partly a critique of them.[51] Pyrrhus was one of the great generals of the ancient world in his own right and his literature may have formed the initial material upon which the extant tactical manuals of Aelian and Arrian were based. An analysis of passages found in Polybius (with their close association to the works of Asclepiodotus, Aelian and Arrian) has drawn Devine to conclude that Polybius was prompted by the writings of Pyrrhus and other third-century BC authors, while establishing a literary tradition underlying the tactical treatises of his Hellenistic and Roman successors.[52]

Pyrrhus' lost writings must have comprised in-depth commentary on many topics, if his interests and qualities as a general given in ancient accounts are any measure.[53] Piecing together information from these, we can propose that his study had tactical guidelines and technical discussion on the disposition of units and formations on the battlefield. Alongside might be sections on intelligence, camps, siege machinery, oratory, propaganda and psychological warfare. Pyrrhus has also been credited by the Roman grammarian Aelius Donatus as inventor of the theoretical teaching of strategy using pawns or pebbles.[54]

Pyrrhus was highly respected by the military men who came after him. In a fanciful anecdote Scipio Africanus is portrayed conversing with Hannibal in the city of Ephesus over the subject of generalship.[55] The Carthaginian chose Pyrrhus as the second most effective commander, Alexander being first. The reasons given according to the different versions of the story that survive include: Pyrrhus' consideration of boldness as the most important hallmark of a military leader;[56] that he was foremost for experience and ability;[57] that he demonstrated virtuosity at marshalling armies and addressing troops;[58] and that he 'had been the first to teach the art of castrametation' (his contribution to camping methods will be discussed in Chapter 12).[59] Livy states that 'no one had chosen his ground or placed his troops more discriminatingly; he possessed also the art of winning men over to him'.[60]

The tail end to the above extract from Livy seems to refer in part to the Pyrrhic use of diplomacy and propaganda, grounded in Philip and Alexander's *modus operandi*, the successful legacy of which academics are still grappling with.[61] Pyrrhic techniques in action are well evinced in his efforts to win over the disgruntled soldiers of his rival, Demetrius Poliorcetes, in 288 BC. This was achieved by sending Epirotes to infiltrate the enemy camp. These men pretended to be Macedonian, which probably means that some wore the characteristic clothing ensemble of *kausia*, *chlamys* and *krepides*, as well as presumably speaking a Macedonian dialect. They were instructed to deliberately speak well of Pyrrhus as a brave warrior-king, popular with his troops and magnanimous to prisoners, undermining Demetrius' all-too-fragile support.[62]

As Pyrrhus never achieved a consistent string of victories, nor built a substantial empire, Hannibal's supposed choice of him as one of the three greatest generals appears oddly inflated. Yet the accolade may derive more from his superb technical grasp, embodied in the detail and subsequent influence of his written discourse on the classical world.[63] The anecdote may hold a kernel of truth in that before his own war with Rome Hannibal could not have failed to have read up extensively on Pyrrhus' battles in Italy.

All this leads to the near certainty that Hannibal closely studied Pyrrhus' lost writings – possibly first introduced by his Roman-hating father Hamilicar Barca as part of his son's education under Greek language tutors such as Sosylus of Sparta. A fragment from a history by Sosylus shows that the writer had a grasp of tactics.[64] Moreover, Hannibal is known to have composed books in Greek. If these had survived, scholars would be left in little doubt that the skills of Pyrrhus, and ultimately the Macedonian combined arms doctrine which he adopted, were of major impact on his own generalship.[65] One Procles of Carthage is noted for having admired Pyrrhus' military abilities, this man being either a native Carthaginian from the third or second centuries BC or a Greek sophist from Roman North Africa in the second century AD. We are told by Pausanias that he rated Alexander higher because of his achievements but believed Pyrrhus was more astute in tactical theory and at devising stratagems.[66]

Pyrrhus' writings therefore appear to have become a major conduit for understanding Hellenistic, and thus Macedonian, warfare. The Roman orator Cicero and his friend L. Papirius Paetus were reading his work in the mid-first century BC.[67] He takes his place as a sort of classical equivalent to the brilliant Habsburg field marshal, thinker and reformer Count Raimondo Montecuccoli (d. 1680), whose assorted titles became required reading among officers across Europe for many years after his death.[68]

Transfer of Military Doctrine in Later Times

Given what we know, it may not have been an admiration of Greek culture alone that spurred the Roman general Lucius Aemilius Paullus in 168 BC to acquire the royal library of the vanquished Macedonian King Perseus, with his sons freely allowed to select whatever books they wanted.[69] This represents one of the first significant influxes of Hellenistic literature into Rome[70] and we can safely assume that as a Macedonian commander Perseus would have stocked his shelves with a range of military titles.

It should be kept in mind that one of Paullus' sons was none other than Scipio Aemilianus, the destroyer of Carthage (146 BC) and Numantia (133 BC). He fought under his father at Pydna when only 17 years old,[71] became a friend and patron of Polybius and other Greek intellectuals[72] and was a highly capable soldier. We can succumb to speculation here and say that he directly benefited from reading matter taken from the Macedonian palace library.[73]

A Latin work on military science, written by Cato the Elder (d. 149 BC), was circulating around the same time, although it may well have owed more to the expertise of Scipio Africanus and Hannibal.[74] Sallust, writing of events occurring in 107 BC, spoke of how Roman officers would gain aptitude chiefly from 'the records of our ancestors and in the military precepts of the Greeks', reminding us once more of the lost book collection salvaged from Pella, the Macedonian capital.[75]

A developed form of combined arms tactical doctrine, with its genesis in the fourth century BC and with Macedonia as its keystone, was transferred over the generations. Philip II learnt from Epaminondas; his knowledge passed on to Alexander and the *Diadochoi*. The principles espoused by them were then captured in writing by Pyrrhus, whose literature was read by Hannibal. The Carthaginian's tactics were in turn studied by progressive philhellenist Roman generals, beginning with Africanus, who decisively reapplied Hannibal's thinking to defeat him.

In reviewing all evidence it can be said with confidence that the Macedonian army of Philip II was the first rationally and scientifically based armed force in the Greek world and Europe. Under Alexander in Asia it became the most effective military organisation seen before Rome, and the Romans almost certainly recognised this fact[76] as they aspired to be the heroic conqueror's foremost heirs and pupils. Alexander's generalship and actions were often the yardstick by which they measured their own achievements.[77]

Polybius, Diodorus, Arrian and Sallust all agree that one of the key factors in determining Roman success in warfare was their ability to adopt what they saw as the best or most useful practices of other peoples.[78] Cicero and Plutarch describe high ranking Roman officers in the first century BC nurturing their understanding through Greek treatises.[79] The study of Greek and Macedonian methods continued from the first to third centuries AD,[80] and was carried on after the fall of Rome by the Byzantines.[81] Given the great impact Greek science and literature had on the Arab world between the eighth and thirteenth centuries, it may be rewarding to investigate the potential effect of Hellenic technical works, and especially the Macedonian legacy, on medieval Islamic armies.

In Europe, the spirit of military thoroughness and professional flair, revolutionised with the army-kingdom of Philip and Alexander and inherited by Imperial Rome through the Hellenistic polities, was ultimately re-discovered through the transmission of classical texts by the emergent nation-states and armies of the Renaissance.[82] The Macedonian and Roman emphasis on training, drill and discipline combined with high morale were disseminated among Swiss, Spanish, Dutch and Swedish armies during the sixteenth – mid-seventeenth centuries. A shield on the Macedonian model was even proposed for the Muscovite army in 1538.[83] That the Macedonian achievement was broadly studied by the military intelligentsia of those days is borne out by the opinion of

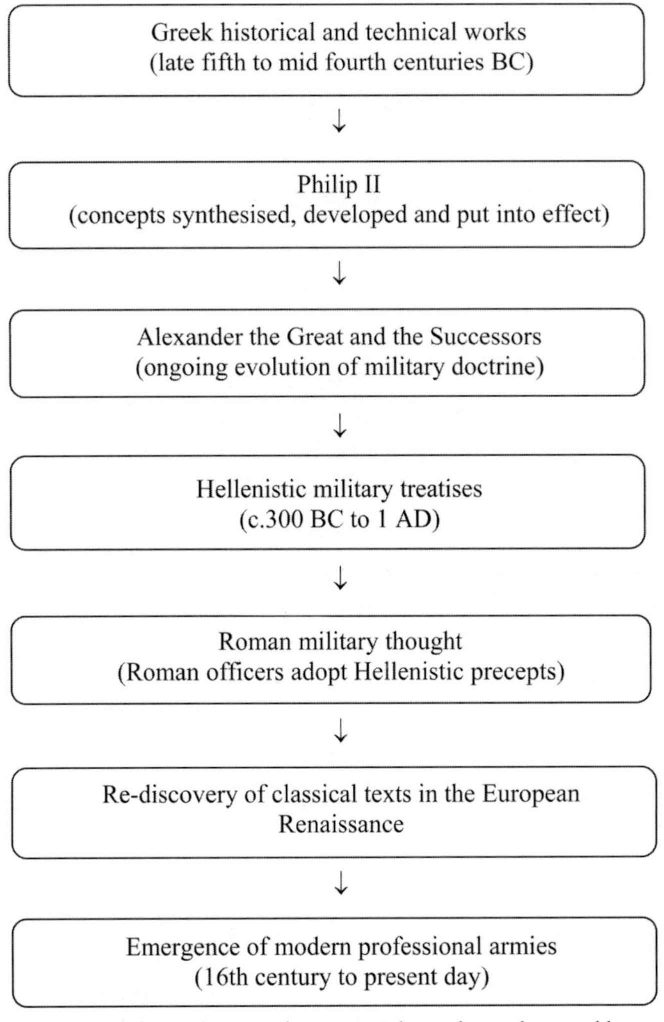

Diagram 1. Transmission of Macedonian military principles to the modern world.

Giovanni Botero, in whose *The Worlde, or A Historicall Description of the Most Famous Kingdomes and Common-weiles therein* (1601) he reminded his readers that: 'the qualities of the weapons, and the order of discipline, are important instruments of … valour. The Macedons achieved great conquests, by means of their pikes, and by disposing of their squadrons; the Romaines, by the means of their darts'.

PART II

PREPARATION

An intellectual base is often essential when devising a powerful military system, but it could only serve as a beginning. Therefore, for the new Macedonian army to be brought into reality it had to be prepared and mobilised for war. This included the necessary training of soldiers, the manufacture and distribution of appropriate arms and armour, and the gathering together of large numbers of quality horses for mounted units.

Chapter 3

Training Soldiers

The installation of formal training served as the foundation for the Macedonian army's battlefield success. Training for war seems to have been conducted through a kingdom-wide programme in Macedonia and it is obviously absurd to imagine it suddenly springing to life fully formed out of a vacuum. Rather, it was only gradually developed over the twenty-three years of Philip's rule, as his policies achieved maturation. It was then transferred by Alexander to Asia and implemented on a far more ambitious scale. The scheme in the East aimed to produce a reservoir of proficient, self-confident soldiers through a course probably combining military with liberal studies.[1]

Origins of Training

Although only circumstantial, we can detect the guiding hand of Philip II behind the training regime found in sources referring to the second half of Alexander's reign.[2] Direct confirmatory evidence for the time being eludes us. Nonetheless, Philip's concern for education is certainly stressed elsewhere,[3] and the title of a work by Marsyas of Pella helps confirm at least elite education as an acceptable subject for study among intellectually inclined Macedonians of the time.[4]

As a backdrop, and on a wider platform, there was a flourishing climate of interest and exchange amongst Greek philosophers as to what defined the ideal socio-political structure and what was the most effective state-sponsored teaching syllabus in the early- to mid-fourth century BC.[5] As already seen in Chapter 2, intellectual study in the time of Philip, Alexander and the Successors often provided inspiration for the Macedonian military leadership's inventiveness.

It has been argued that Plato's aim of a specially trained class of soldiers prescribed in his *Republic* was a precursor to some Hellenistic armies.[6] This cannot be dismissed. The inclusion of Plato in the tactical manual of Aelian, who describes him as one who thought the study of tactics the most necessary of the sciences, suggests that Plato was viewed as a respected authority on such matters.[7] His *Republic*, *Laches* and *Euthydemus* are all lucrative sources for military instruction.

It is likewise illuminating to consult Aristotle, who wrote of the Spartan *agoge*, or training regimen (c. 325 BC), that they were beaten in war and had rivals in education.[8] Given Aristotle's impressive Macedonian contacts and credentials, the most significant of these unspoken and enigmatic 'rivals' was doubtless the Macedonian method.[9] Through the kingdom's conquests it had already proven highly successful at the time Aristotle compiled his work and would not have been far from his mind when forming such passages.[10]

In what manner Macedonian troops were trained can be largely pieced together, albeit tentatively, from the extant literary texts. The following attempted reconstruction therefore incorporates training for the Companion cavalry and *pezhetairoi taxeis* ('Foot Companions' brigades), but with a major emphasis on the latter, i.e. the pike armed infantry units forming the bulk of the army. This includes Asian troops recorded as having been recruited and trained by Alexander.

Where They Trained

In Egypt and Asia under Alexander (from 331 BC or perhaps even earlier) young recruits were taught in the cities.[11] This is compatible with Plato, who details the use of training schools as a primary part of his state model, assessing how there should be 'buildings for public gymnasia as well as schools in three divisions within the city, and also, in three divisions round about the city, training-grounds and race-courses for horses, arranged for archery and other long-distance shooting, and for the teaching and practising of the youth'.[12] Such a description recalls the *stratiotikon logisterion* (war office) with training school of Alexander's Successor, Seleucus I (c. 300 BC), located at Apamea in Syria.[13]

According to epigraphic and archaeological evidence, the *gymnasion* fulfilled a crucial role in army training in the Antigonid period when boys were instructed at sites located in old and new civic settlements across Macedonia.[14] The same may have been true in Alexander's day, but only future discoveries will fully reveal whether this is correct or not. The city *gymnasion* of the classical Greek world certainly served a variety of military functions, including its use as a place wholly devoted to drill and tactical studies during wartime. Three or more extant Ptolemaic papyri signal that uses of this sort continued in the Hellenistic East.[15]

As the Macedonian *gymnasion* was a place for athletic contests during religious festivals in Hellenistic times, Girtzy rightly points out that the institution combined a military education with socio-religious functions.[16] We can thus visualise formal training pursued in a well-equipped city *gymnasion*. In the case of Amphipolis, extant remains reflect a large-scale complex used from the second half of the fourth century BC (Fig. 1), with wrestling grounds and courtyard,

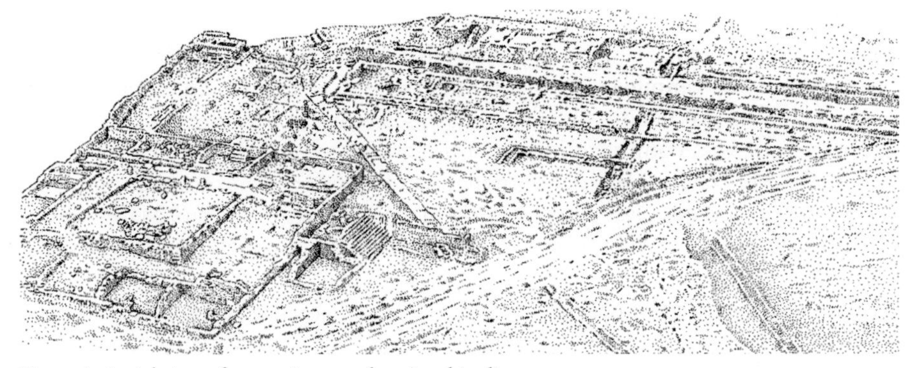

Figure 1. Aerial view of *gymnasion* complex, Amphipolis.

rooms for meetings and physical exercise, running tracks, and areas set aside for banquets and sacrifices.[17]

Based on close comparison with Hellenistic Alexandria, it is likely that Pella was the site of a permanent camp for royal guard units, or detachments of them.[18] Rome of course had its famous *Campus Martius* ('Field of Mars'), used for training and exercise. At the centre of Pella lay the imposing Agora, occupying the heart of the city, and measuring 200m × 182m (70,000 square metres), equal to ten city blocks.[19] It was therefore large enough to be utilised for processions or parades involving guard units stationed in the capital.[20] Manoeuvres could have taken place in the environs immediately outside the city, including the wide plain to the west of it.

Pella was itself originally established on a narrow coastal strip, in an area enclosed by the Axius and Haliacmon rivers, further bounded by the mountain ranges of Bermion and Paikon. In ancient times the vicinity had forests and marshes so troops training there had convenient exposure to many forms of terrain, with most geographic features within a day's march of one another. The surrounding landscape may in time reveal evidence for important military installations.[21]

In Alexander's plans for the Asian provinces, and under authority of the satraps or provincial governors,[22] *epistatai* (supervisors)[23] and *didaskaloi* (teachers) had to select, organise and maintain all trainees, while teaching them 'works of war'.[24] They were fully funded by the King, who also met other costs.[25] This dovetails with Plato's terse advice in which he approves of public teachers receiving state stipends.[26] Some *didaskaloi* were probably hired Greek specialists. Others would have been veterans drawn from the army as a whole, i.e. men who agreed to stay on and continue their service as experts receiving a higher rate of pay.[27] A glimpse into the character and experience expected of such people can be appreciated from Lucian of Samosata who describes a similar teacher (c. 150 AD) as:

> A man of penetration ... with a soldierly spirit ... and some military experience; at the least he must have been in camp, seen troops drilled or manoeuvred, know a little about weapons and military engines, the difference between line and column, cavalry and infantry tactics (with the reasons for them), frontal and flank attacks.[28]

During the first half of the fourth century BC, Democritus of Abdera is noted for allegedly having compiled a treatise on fighting in armour – an inference that the various methods of hand-to-hand weapons training took up a virtual subject on their own.[29] Greek combat masters emerging from the late-fifth century BC were some of the earliest professional instructors of the Western world and the trend intensified in the next century and even more so after the rise of Macedon.[30]

Instructors might resemble *paidotribes* (physical fitness tutors), who were pictured by Lucian wearing *himatia* (gowns) and wielding long forked staffs which served both as an emblem of office and as a means of beating recalcitrant boys. A Hellenistic relief from Messembria, on the Black Sea coast of Bulgaria (Nessebar Museum of History), depicts a number of such boys and civic officials, including generals and magistrates undertaking a sacrifice. Of the six adults shown, two wear Macedonian *kausiai* and also *himatia*, with four in the familiar *kausia* and *chlamys*.

At Opis (324 BC) Alexander is said to have acknowledged how Philip gave his Macedonians *chlamydes*, or cloaks, in exchange for their sheep or goatskins.[31] This alludes to new recruits from the less Hellenised upland regions, for whom ownership of a *chlamys* was a statement of an individual's willingness to trade in a traditionally primitive way of life for active participation in the progressive ideology of Philip's renewed kingdom.[32] The *chlamys* ultimately became a distinct expression of Macedonian national and Hellenistic military dress (see Chapter 6), as Macedonian identity and military aspects had become by that time synonymous.[33]

Greek training specialists are recorded from the late-fourth to second centuries BC teaching a range of skills covering both light and heavy styles of warfare; combat in helmet and corselet with sword, spear and shield; use of the bow, javelin, sling or throwing stones; and operating a catapult or hurling a javelin from horseback. In Hellenistic Macedonia and other polities like Athens, teachers emerge specialising in specific subject areas, in keeping with the more sophisticated and diverse nature of warfare introduced by Philip and Alexander's army.

These professional instructors include the *toxotes* (for archery), *akontistes* (for javelin-throwing), *hoplomachos* (for handling spear and shield) and the *katapeltaphetes* (for teaching the use of catapults).[34] An Athenian inscription (266/265 BC) denotes honorary awards to masters in artillery, archery and javelin-throwing for their careful attention and supervision of pupils.[35] A broad spectrum of combat training is suggested by literary sources describing Alexander's Macedonian infantry and cavalry, proficient in the use of up to three different weapons, as explained further on.

Discipline was rigorously enforced throughout training. We read of how warm water bathing was frowned upon in Philip II's Macedonia,[36] and an extant gymnasiarchal law preserved as a marble *stele* from Beroea (c. 167 BC) even stipulates that those between 18 and 20 years of age (*ephebes*) who were disobedient must be reprimanded with a stick. If not merely intending a moralising convention glorifying the 'virtue' of the West in contrast to an 'Asiatic' lifestyle, Curtius offers a window onto the demands that were required of recruits in Philip and Alexander's time, resembling in some respects Spartan methods: 'Their discipline was subordinated to their poverty. When they were tired, they slept on the ground, when they were hungry, they ate what they scrounged, they slept fewer than the hours of the night.'[37]

Just as indiscipline was punished, so progress, even in the fourth century BC, may have been reviewed at intervals, with honours bestowed upon recruits who won contests proving martial *arete*. The later inscription from Beroea, of which 173 lines survive, regulates in detail the city *gymnasion* in relation to military aspects. It describes age groups between eighteen and thirty competing for prizes (usually shields), which were financed from the revenues of the *gymnasion*.[38] Awards were given to boys who were considered most fit or best conditioned (*euexia*); showed good order or discipline (*eutaxia*); demonstrated love of effort or hard training (*philoponia*); and also to those who won the long foot race (*makros dromos*).[39]

From all this fragmentary evidence it is acceptable to think that among Macedonians the importance of competition and the pursuit of excellence were inculcated early on, with prizes offered as an introduction to the awards, promotions and other inducements which would avail them in their future careers as soldiers. This matches well with Plutarch's rhetorical flourish of the resultant culture in Alexander's army in which there was 'great ambition', 'mutual rivalry of hot youth', 'competition for repute' and 'excellence among his Companions'.[40] Such a fiercely competitive atmosphere is demonstrated in a lively passage from Curtius recording an incident before the battle of the Hydaspes (326 BC) in which a group of rash and daring young soldiers armed only with spears swam out to a well-defended island where they slew many of the opposition but were in turn surrounded and overwhelmed by missiles.[41]

There is now even some emergent artistic evidence for trainees. It consists of six miniature gilded terracotta plaques uncovered from the Eastern Cemetery of Amphipolis (Tomb 130, dated 330–275 BC) (Fig. 2). Each features a naked youth with left arm extended and supporting a small, round shield or *pelte*, the right arm bent at the waist, with hand originally grasping a sword or spear. The figures are either performing a war dance or undertaking *hoplomachia* (fencing with weapons).[42]

Training Characteristics

Formal training lay at the root of the Macedonian army's operational effectiveness. This is well summarised by Lonsdale, who describes the importance of continual and realistic training for the Macedonians and Romans, the aim being that men perform in battle instinctively and with discipline, both as individuals and cohesive units, thereby ensuring optimal command and controllability.[43]

Training curricula would have covered a broad spectrum of subjects and activities, given the versatility of Macedonian soldiers frequently attested in the

Figure 2. Gilded plaques of youths exercising with shields, Amphipolis.

sources. This notion is more acceptable if we consider that the overall period probably began at fourteen and has been estimated, at least for the curricula planned by Alexander in Asia, at about four years – of longer duration than the average modern university degree.[44] A wide number of exercises taught over this time suggest that the scheme was designed to be gradual and systematic, with weapons handling and then complex formation drill built on a basic foundation of physical fitness and literacy. It is possible that this primary basis was practised during the first year, with more vigorous weapons handling and a drill programme requiring greater strength and stamina introduced later on.[45]

Systematised training for Macedonian-derived armies is detected in Polybius' account of Ptolemaic soldiers preparing in the lead up to the battle of Raphia (217 BC):

> Taking the troops in hand they got them into shape by correct military methods. First of all they divided them according to their age and nationalities, and provided them in each case with suitable arms and accoutrements ... they drilled them, accustoming them not only to the word of command, but to the correct manipulation of their weapons. They also held frequent reviews and addressed the men ... [They] inspired them with enthusiasm and eagerness for the coming battle.[46]

These proceedings conform to Diodorus' description of Philip II's first hastily mobilised and reformed army (359–358 BC) in which the King brought the *Makedones*, or citizens, together in assemblies, arousing their battle fervour with eloquent speeches; he then armed them with suitable weapons and undertook manoeuvres and competitive drills.[47]

The model that emerges is that each recruit was first taught to be confident and familiar in the handling of arms, before progression to weapons handling within groups, culminating ultimately in formal mass drill. This methodical approach accords with Philip and Alexander's cornerstone theme of conditioning in which recruits were inured to constant physical exercise before real life warfare.[48] The doctrine espoused by them was later adopted by the Romans and was ultimately rediscovered with the professional armies of the European Renaissance.

Physical Culture

In a society known for its toughness, the Macedonian leadership borrowed features from Greek physical culture for training soldiers.[49] The aim was to practise forms of exercise which cultivated an ability to endure toil.[50]

Plutarch describes Alexander having been averse to athletics and boxing,[51] but this view is skewed as there is incontestable evidence from elsewhere that he sometimes admired athletic achievement. Lindsay Adams lists no less than fifteen separate occasions from Arrian in which the King mounted athletic competitions for the benefit of the army before or after a major campaign, serving as a sort of Macedonian equivalent to an American United Service Organizations show.[52] Philip II practised wrestling,[53] as did some of his son's officers.[54] Strabo intimates the common use of strigils and oil jars in the Macedonian army[55] and such artefacts have been found in graves near Pella, at Pydna and Lefkadia.[56] It is increasingly likely that the Macedonians were aware of the value of Greek boxing and other unarmed combat sports for soldiers.[57] There is mention of a *gymnasion* at Pella,[58] and traces of the impressive athletic complex at Amphipolis shed further light on this training. Literary excerpts, excavations and inscriptions also attest to the use of the *gymnasion* in the Hellenistic period at Macedonian cities such as Lete, Dion, Orestis, Beroea, Styberra, Apollonia, Kalindoia, Cassandreia and Thessaloniki.[59]

All the same, if Alexander sometimes approved of athletic exercise, he disapproved of the professional wrestlers' overindulgence in food. An anecdote survives in which he pours scorn on heavy Persian meals which would only lead to defeat in battle.[60] According to Plutarch, at another time he refused rich delicacies sent to him by Queen Ada of Caria as he is said to have considered 'night marches for his breakfast, and for his dinner his frugal breakfast'.[61] In wider circles, the Companion officer Corrhagus offended the Athenian boxer Dioxippus by scolding him as a clumsy and overfed oaf.[62]

The views of Corrhagus reiterate the earlier creed of Epaminondas, who thought fat men repugnant, declaring that training should encompass athletic exercise and military drill, but with an emphasis on the latter.[63] The Theban general is further remarked upon as interested in physical exercise which was helpful in warfare: namely, agility (*velocitas*) over the strength required of athletes. Although Epaminondas personally trained in running and wrestling, Cornelius Nepos assures us the latter was only followed to the extent of being able to seize and grapple his opponent while still standing; however, he 'devoted his greatest efforts' to weapons training.[64]

Plutarch says the talented Theban cavalry general Pelopidas (a close personal colleague of Epaminondas) likewise upheld 'practice, training, experience, physical toughness and courageous spirit' if an army was to operate at its full potential.[65] Any reading of Alexander's troops on operations – as preserved through competent ancient writers like Arrian – clearly reflects that this doctrine became part and parcel of the Macedonian army. Indeed in this case it may even imply that Philip took directly from Pelopidas himself, who favoured the use of a decisive mounted assault delivered at high speed on the battlefield, resembling the later application of Companion cavalry by Alexander.[66] Philip is said to have become both an associate and lover of Pelopidas during his hostage years in Thebes so he would have debated with him and his friends over military matters.[67]

The physical training ethos of Theban visionaries like Epaminondas, Pelopidas and Pammenes, emphasising stamina, was adopted by their Macedonian protégé and continued well after Alexander's conquests. During the late-third century BC the Achaean general Philopoemen, perhaps influenced by the scepticism of writers in his time, was opposed to an over insistence on athletics and wrestling for soldiers because it was entirely in opposition to a military way of life. Athletes required exact dietary regimes, plenty of sleep and regulated exercise, whereas, as Plutarch puts it, 'the soldier ought to train himself in every variety of change and irregularity, and, above all, to bring himself to endure hunger and loss of sleep without difficulty'.[68]

Stave Fights and Mock Combats

Philip was acutely aware of the critical importance of training over amateur enthusiasm to perfect skill or craft (*techne*) in the handling of arms. Thus, at Chaeronea he saw with satisfaction how the Athenians were eager but in poor condition, in contrast to his Macedonians who were trained and in top fettle.[69]

The foundations of training before progression to using real weapons involved stave fights and the use of dummy weapons in mock single and group sparring. Alexander instituted frequent stave-fighting contests in readiness for the Persian expedition.[70] The use of dummy weapons for drills and mock battles is also affirmed elsewhere. For example, two groups of Macedonian camp followers fought with sticks, stones and clods of earth in a sham battle approved by Alexander in the lead up to Gaugamela (331 BC).[71] Livy even describes how at the *Xandika* festival (182 BC) a combat was arranged in which the army of the Macedonian king Philip V was divided into two. Many wounds were inflicted by the blunt-tipped weapons, making it appear to all intents like an authentic engagement.

The implication from the above is that, on occasion, blunt dummy weapons of wood were used.[72] These were probably not unlike the Roman *rudis*, or wooden training sword, and similar implements are observed among the warrior traditions of the ancient Celts and as far afield as India, China and Japan.[73] The small plain *peltai* of naked youths from the Amphipolis miniature plaques, shown earlier in Fig. 2, even suggests the use of light practice shields.[74]

Mock fights of various sorts appear to have been built into the training schedule on a quite widespread scale. One passage relates to how Philip applied soldiers to exercise under combat conditions, which might imply the use of staged battles as a means of honing fighting technique.[75] The adulterated *Alexander Romance* not implausibly relates how during his childhood years the future conqueror drilled his classmates, organising them into separate teams and having them fight each other.[76] Diodorus remarks on how Alexander busied his troops with tactical exercises: credibly, drills, mock combats or field manoeuvres, prior to the Asian expedition in 334 BC.[77] The benefits of a training regime similar to that of Philip II was fully appreciated by Philopoemen, who in reforming the military organisation of the Achaean League introduced morale lifting drills, competitions and parades for both infantry and cavalry units. This made, in Plutarch's words, 'the dexterity shown by the whole mass in its evolutions to be like that of a single person moved by an impulse from within'.[78]

Earlier in the fourth century BC Plato approved the use of mock fights in armour and combats with anywhere from one to ten contestants per side.[79] Polyaenus preserves a striking anecdote about the Athenian mercenary general Iphicrates, who took this form of training a step further when he contrived a series of mock scenarios covering reinforcements, ambushes, betrayals, desertions, assaults or panics, thereby ensuring his troops were ready for any of these situations should they occur.[80] It is intriguing that Iphicrates held close ties with the Macedonian royal house. He became an adopted son of Philip's father Amyntas III (r. 393–370 BC)[81] and the former met him as a boy when in 368/367 BC the Athenian accepted a request for help from Philip's mother, Queen Eurydice.[82] His reputation was thereafter assured. A son of Iphicrates was part of an Athenian delegation captured by the Macedonians at Damascus in 333 BC. Alexander subsequently kept him 'in attendance and paid him special honour'.[83] Macedonian weapons training exercises probably included proven Iphicratean methods incorporated into the army by Philip, followed by Alexander, and condensed as brief glosses in the narrative of Diodorus.[84]

As with modern armed forces, organised team events were another means of encouraging unity of morale and purpose. Competitive ball games (such as *episkyros*, *phaininda* or *sphairomachia*) seem to have been encouraged as excellent forms of training for soldiers, increasing stamina while refining hand-eye co-ordination.[85] In his medical essay *On Exercises with the Small Ball* Galen (writing in the late-second century AD) praised the use of ball games for military preparation, while showing contempt for wrestling, which he thought only helped weight gain.[86]

An examination of literary accounts for the careers of Macedonian generals confirms that Alexander was fond of exercising with a ball on campaign. Antigonus Monophthalmus promoted soldiers he saw in camp training with a ball while wearing armour, thus simulating battlefield combat. By contrast, he demoted their officers whom he found shirking exercise for getting drunk. Such ball games were followed by the officers of his grandson, Antigonus Gonatas. Hellenistic terracottas and statuettes of young boys and men dressed in the distinctive Macedonian get-up of *kausia*, *chlamys* and *krepides* sometimes even portray them holding a bag or ball (e.g. from Idalion, Cyprus).[87]

Diverse Weapons Handling

For his ideal state Plato recommends a span of topics be adopted for the military education of citizen-soldiers, which match the limited evidence at hand for the training and expertise of Macedonian troops. This education programme was to include 'the use of the bow and all kinds of missiles, light skirmishing and heavy-armed fighting of every description, tactical evolutions, company-marching, camp-formations, and all the details of cavalry training'.[88] Sources reveal that the use and combination of weapons employed by Macedonian soldiers depended solely on the tactical or operational considerations that confronted them at any given time. Infantry weapons handling thus covered not only continual exercises with their *sarissai*, or pikes, but also practice with spears or javelins and swords.

As a consequence, it can be said that the Macedonians were one of the first to introduce the phenomenon of the multipurpose infantryman, making the troops far more tactically flexible than has been generally supposed. A good example is how the *sarissa*, throwing spear and sword could be carried together for staged hand-to-hand combats,[89] and occasionally even for field battles. Among infantry, spears or javelins were sometimes used similarly to peltasts as the principal weapon when deployed for skirmishing, ambushes, executions, forced marches, mountain warfare, sieges or sometimes in battle. Swords were not only made use of as secondary or emergency weapons *in extremis* when infantry were drawn up in a pike phalanx, but, as largely overlooked by researchers, could be utilised in their own right for melee work,[90] after javelins were thrown.[91]

Ueda-Sarson has gone further by making the intriguing hypothesis that small contingents of slingers recorded without ethnic derivation in source texts referring to Alexander could allude to men drawn temporarily from Macedonian infantry units.[92] It should be recalled that as many as 500 slingshots have been retrieved in excavations at Olynthus, dating to Philip II's siege there.[93] Perhaps a minority of *pezhetairoi* were selected as slingers, with a certain, even fixed, number of such specialists conveniently falling out from the ranks of each *taxis* of 1,500 men when required.

As seen from literary extracts, many of the occasions in which slingers are mentioned are for siege work. By way of analogy, Assyrian palace reliefs at Nineveh from the seventh century BC depict at least their regular infantry redeployed as slingers during sieges. It has been assumed that the ability to use sling stones was widespread in Assyrian ranks and later Roman troops, including auxiliaries, were also taught to use the sling.[94] Roman soldiers may even have been trained in the use of hand hurled stones to defend static positions, something approved of in Greek military literature.[95]

Throwing or slinging stones would certainly have been practised in a society like Macedonia and was, for instance, useful among shepherds determined to protect their flocks from wolves and other predators. Those found guilty of treason on Alexander's campaigns were sometimes stoned to death[96] and Antigonus Monophthalmus was in danger of being stoned by a riotous mob of veterans, probably during the conference of Triparadisus (321/320 BC).[97] Evidence from the Hellenistic period suggests that learning to use the sling was incorporated into the Macedonian training model by this time.[98]

Whether slings were used on occasion by discrete groups of Macedonian *pezhetairoi* in Alexander's army or not, as rightly noted again by Ueda-Sarson, the significance of teaching men to use two or more weapons with equal facility was a great achievement. What made it even more remarkable is that these were weapons requiring radically different formations; namely, a pike phalanx or javelins used in loose order.[99]

Emphasis on Formation Drill

Once recruits had achieved a foundation of all round fitness and had learnt to individually wield their weapons with proficiency while strengthening arm and shoulder muscles, they would have then been taught to handle and coordinate them in formed bodies. This was the fulcrum to Macedonian battlefield success.

Polybius quotes a passage on the subject of drill from a lost work by the Athenian Demetrius of Phalerum. This philosopher served with distinction under both Demetrius Poliorcetes and Ptolemy II of Egypt, but in a non-military capacity. In this case he reflects the thought of his times by comparing the conscientious general's accurate arrangement of soldiers and units to a builder setting precise courses of masonry aligned to each other.[100]

Learning varied drill and formation evolutions to perfection must have taken up the longest, most tedious and physically demanding part of training. Nonetheless, the pay off was a standard of drill comparable to that of the Roman legions. Arrian had an obvious intellectual interest in the wars of Alexander and the Successors. Therefore, the stress he places on sound formation training and drill (often inspired by historical cases) is as befitting to Macedonian forces as it is to the Roman armies of his own time:

> The foremost and most important task of the general is to take a disorderly crowd of men and establish them in orderly ranks. I mean that he should assemble and arrange the men in proper ranks and give the multitude a balanced order suitable for battle. For a well-organised army is easier to handle on the march, more secure in camps, and more useful in battle. Indeed we know that in times past, large well-armed armies have been destroyed through lack of order by smaller and less well-equipped forces. For the weaker and less well-armed force conquered the larger by superior discipline.[101]

The importance of training on this scale would not be substantially revived in the Western world until the reforms of Maurice, Prince of Orange, and Gustavus Adolphus, King of Sweden, during the early- to mid-seventeenth century. The standard and sophistication of drill as habituated in Macedonian, Roman and Byzantine armies may have even remained unsurpassed in European circles until the human parade ground automatons of the Prussian army in the eighteenth century.[102] It should be remembered that Frederick the Great was himself inspired by Philip and Alexander's example in having well-drilled infantry.[103]

Manti insists that an exact pace was adopted to preserve rank spacing and cohesion in the Macedonian phalanx formation. A uniform field march step has been proposed, probably measured by *podas.*[104] This could have evolved from the rhythmic movements of the pyrrhic war dance (*pyrriche*) in which participants brandished spears and shields, performing to high-pitched pipe music. It was certainly used for training purposes by Greek hoplites, while in the Hellenistic and Roman periods the term 'pyrrhic' denoted a march step rather than a war dance.[105] Incidentally, the Macedonians enjoyed ancient rhythmic war dances of their own, including the *telesias* and *karpaia*, the former a sword dance in full armour, the latter mimicking a cattle raid.[106]

The depth and complexity of battle drills in the Macedonian army can be appreciated from stealing an oblique glance at the abstract geometrical diagrams and technical annotation found in the extant tactical manual of the theoretician Asclepiodotus (the possible abridgement of a more copious, long vanished *taktika* ascribed to Posidonius of Apamea). It should be noted that some generals distrusted pure scholarly guidance; for instance, Cicero informs us that Hannibal had a low opinion of academic military theorists.[107] Despite the fact that Asclepiodotus lived in the first century BC, his dry academic writing style suggests that his work was still part of an older literary genre. Plutarch tells us that a drill book composed by the otherwise unknown Evangelus, consulted by Philopoemen in the late-third century BC, like Asclepiodotus' manual, displayed tactical schemes and diagrams.[108]

The *Suda* states that formation drill in Philip and Alexander's army was practised exhaustively and, as discussed, included ambiguous armed manoeuvres and combat exercises.[109] Antigonid inscriptions additionally testify to both weapons handling and group formation training being supervised by the *taktikos*. It implies the critical importance of mass drills for the later Macedonian army.[110]

Using the main Alexander literary sources,[111] coupled with the handbooks of Asclepiodotus, Aelian and Arrian as general and very much *cautiously* applied guides, we can conclude that recruits for pike armed units were taught to take up their *sarissai* and smartly form by rank and file for marching and battle, practising the requisite open, intermediate and close order methods for receiving or making a charge.[112] They would have learnt to understand the commands used for military evolutions and react instantly to them, recognising orders given by voice, trumpet or visual signal.

Trainees were instructed together in how to broaden formation frontage by thinning their depth or narrow their frontage by increasing depth. They therefore learnt how to advance, wheel and manoeuvre in line or column. Constant drill would have taught a smooth, self-confident precision and coordination in relation to the soldier as a single cog within the overall machinery of the combat unit.[113] Such complex drill movements and the dexterity required to make them effectual are passed over by Curtius, who describes how before battle with the Persians:

> The Macedonian battle line was fierce and rough looking, and their shields and pikes cover their fixed wedges and close-packed ranks ... [They are] ... intent upon the nod of their commander, they have learnt ... to keep their ranks; what is ordered all obey. How to oppose, make circuits, run to support either wing, to change the order of battle, the soldiers are as well skilled as their leaders.[114]

Given this account, it is not surprising to find Diodorus describing the army (330/329 BC) as having attained an outstanding efficiency, being devoted to Alexander and looking forward to future battles.[115]

The quality of unit training, with its concomitant demands of drill and discipline, is vividly illustrated by Polybius for the campaigns of Philip V. When crossing the River Achelous in the face of hostile Aetolian cavalry (219 BC), the King instructed elite units of infantry to wade into the water and land on the opposite bank in close formation, shield to shield and by unit subdivisions. Finding that the Macedonians remained in excellent order, unaffected by the desultory attack, the Aetolians, for all their high spirits, retreated to the shelter of a nearby town.[116]

The ability of infantry to maintain tight formation integrity when crossing a substantial water obstacle in the face of a mounted enemy with good morale –

and then to force that enemy to withdraw once the terrain feature had been circumvented – is a difficult task to achieve for troops in any era. The Polybian excerpt highlights the quality of Macedonian training in practice, seen here more than 100 years after Alexander's death.

Hunting and Route Marches

Just as the use of formation drills established effectiveness on the battlefield, so hunting was built into the curriculum as a means of sharpening individual and group skills and the objective of teamwork.

After becoming King, Alexander emphasised contests in hunting of different forms,[117] presumably meaning the practical use of sword, spear, javelin and, on occasion, the bow.[118] Such techniques may have been influenced by the doctrines of Xenophon, with whose works both Alexander and his officer circle were well acquainted. In the *Cyropaedia* hunting is recommended as ideal preparation for war, as it inured the body to rising early, enduring temperature changes, jogging, remaining alert and on one's guard from attacks, and showing courage when killing large animals.[119]

As part of their conditioning to fatigue and exercise, recruits also underwent a demanding regime of cross country route marches with full kit and provisions, which is now standard procedure among crack units, to wit the French Foreign Legion, who like the Macedonians in the East historically operated in arid terrain.[120] Philip II is hence described by Polyaenus accustoming them: 'to constant exercise, as well as in peace, as in actual service: so that he would frequently make them take up their arms, and march often 300 stadia [10 Persian parasangs or 59km] carrying helmet, shield, greaves, sarissa and in addition to their arms rations and all gear for day-to-day existence'.[121]

Philip regularly issued orders that a thirty-day supply of grain be taken along on each man's back.[122] Marching could be potentially tough going because Macedonia had a dual climate – from Mediterranean on the coast to continental inland. With its experience of torrid summers and freezing winters, its varied hinterland was made up of plains, mountains, woodlands and rivers. Recruits thus had exposure to most forms of terrain and weather. During these galling endurance tests a continual emphasis on physical effort achieved its purest form. March rates would be quite rapid, tentatively estimated at 4km/h, accelerating to 8km/h during forced marches.[123]

Route march exercises were particularly strenuous, given the King's demand for strict discipline, with men expected to march in line and preserve formation. It was here that recruits were taught the importance of keeping their armour bright and polished; otherwise, as in an extract preserved for Alexander, they could only pretend to be soldiers.[124] Aelian records that Philip was hostile to those who were slack in obeying orders, notably among his aristocratic cadets. At one time he is described whipping Apothonetus for sneaking into an inn for a drink during a march. More extreme is his reputed killing of Archedamus because, when the King had ordered that he stay in armour, the young man had subsequently taken it off.[125]

The King had no truck with those who did not submit to obedience. As officers such high-born youths were expected to lead by example, setting the standard of morale and conduct to their men. Once again, it could be that Philip's severe rigour on such field manoeuvres was directly inspired by the maxims of Iphicrates. An unconfirmed anecdote shows how after making camp this unforgiving Athenian martinet made the sentry rounds in person and, on finding a guard asleep at his post, stabbed him with his spear. When his action was rebuked as cruel he dourly replied that he left him as he found him, suggesting he actually killed the man.[126] Philip's attitude also resembles that of Macedon's capable neighbour, Jason of Pherae, from whose well-run mercenary army the weak or lazy were expelled and those willing to endure toil and danger rewarded.[127]

Riding, Rowing and Manual Tasks

To accustom them to all eventualities, it is possible that at least an element of infantrymen received instruction in horse riding. From the study of an inscription from Amphipolis, it seems that city magistrates in the Antigonid kingdom were responsible for ensuring that men were trained in horsemanship and equestrian exercises. Registers were maintained for young soldiers of the 20- to 30-year-old age group to serve as 'dragoons' on patrols around their respective cities. This epigraphic material is of interest considering the use of mounted infantry in Alexander's army (which will be detailed in Chapter 14).[128]

Neither should we negate the real possibility that some were taught rowing, or that it could be learnt by soldiers out on campaign. Ships were competently manned by hypaspists (royal foot guards, literally 'shield-bearers') and other Macedonian infantry during the siege of Tyre in 332 BC.[129] The training seen there helps support the feasibility of a special naval siege unit envisioned by Murray as having been first developed under Philip and Alexander.[130] As late as 218 BC, during a war with the Achaean League, Philip V's Macedonians underwent constant practice in how to row and their apparent enthusiasm sparked Polybius' observation that they were 'very ready to undertake temporary service at sea'.[131] Disciplined rowing requires team effort, stamina and control of the oar, all of which the Macedonians appear to have excelled in.[132]

Whatever else was taught, the main focus of specialised training was to perfect resourceful manual skills. Macedonian troops did not employ slaves for construction tasks but undertook all activities with their own hands. They were therefore easily converted into a mass labour force undertaking manifold tasks on campaign, setting a precedent for self-sufficiency which would be later adopted by the Roman army to become one of its own more familiar features.

Unlike armies and warrior cultures before, in Macedonia labour was not viewed as a shameful or demeaning business. On the contrary, great pride was taken in hard toil. This was a necessity among a people with a strong egalitarian strain and with many farmers, hunters, pastoralists, miners, woodsmen and lumberjacks.[133] In keeping with their way of life, Livy describes Macedonians as hard living and of austere character.[134] As late as the eighteenth and nineteenth centuries, travellers witnessed the characteristics of initiative and enterprise prevalent among Balkan mountain peoples. The Vlachs of the Pindos range have been praised in particular for their wood and stone crafting skills and Howell describes how he saw the speed and dexterity of modern Vlachs and Sarakatsani in erecting wooden fences, confidently hewing lumber and ramming it into the ground.[135]

Training in woodwork or carpentry was seen as a means of building the spirit of unity and cooperation. Writing in the first or early-second century AD, Dio Chrysostom lists the occupation of carpenter as one of the three most common among the kingdom's inhabitants before Philip II's time.[136] We can expect that in a society covered with small, close-knit settlements, recruits must have already experienced the rudiments of physical work, self-reliance, building and carpentry from older males in their respective family households before they joined up. These abilities were then fully exploited during army service, further honed when out on field exercises or route marches, and especially required in constructing camps. Examples of their woodworking skills are provided by Arrian and Diodorus, who describe Alexander's veterans being ordered to make huts, beds, horse troughs, scaling ladders or up to 100 stakes per man.[137]

Something of the disciplinary procedures employed by the Macedonians to keep soldiers alert, active and attuned to labour are appreciated in an anecdote relating to Iphicrates, who lived by the dictum of never allowing his men to be at a loose end, keeping them busy shifting camp, cutting stakes, digging trenches or repairing equipment.[138] We can describe such activities today as 'make work' tasks.

The resolve of Iphicrates reappears with Philopoemen, who read military handbooks extensively and would have followed the lead of generals like Philip and Alexander. Moreover, he practised what he preached and is known to have personally sought physical exertion by hunting or labouring on his estate. Even as an old man he 'mucked in' with hired help and performed manual chores on the property, clearing land, working in the vineyard, ploughing fields and hewing firewood.[139] The same stress on raw effort is found over two centuries later when Onasander, basing himself on earlier Greek treatises, recommended that generals not permit their men too much relaxation if it led to laziness: 'for this reason the soldiers should never be without occupation'.[140]

Alexander's Future Plans

The Macedonian system of army training was successfully transferred by Alexander to his empire in Egypt and Asia. The population of some of his new-founded cities has been tentatively estimated by Hammond in the vicinity of 10,000 citizens living within an enclosed circuit wall of around 6km, and with the usual political institutions of a magistracy, council and assembly.[141] From these and pre-existing Achaemenid sites, Asian youths from multiple ethnic groups were to be brought up as a military reserve: selected by *epistatai* from each satrapy – chosen for strength, fitness and intelligence – and then taught by *didaskaloi*.[142]

Alexander knew that to achieve stability for his empire it was necessary to gain the trust and assistance of conquered populations. Forcing upon them the yoke of oppression and subservience would only ever be counterproductive. The initial aim during their instruction was for recruits to serve as a pool of hostages guaranteeing the good behaviour of their families, thereby ensuring less likelihood of revolt. By removing the brightest, most promising and energetic young males, isolating them from their home environment and moulding them into pupils training to Macedonian methods, the edge to any future indigenous resistance would be blunted.

The King's long-term master plan was for these youths to attain fluency in spoken and written Greek and to develop into a permanent indoctrinated cadre of Hellenised soldiery fully supporting his new regime in Asia. The use of mercenaries to serve for garrison duties would free up this reservoir of manpower to keep law and order in the satrapies and patrol the frontiers. Only if Alexander had need of reinforcements for his field army would detachments of mercenaries be brought by the satraps to him, with the established Persian communication routes[143] serving as arteries speeding mobilisation.[144]

The overall size and significance of the empire-wide network of training were undeniably breathtaking and more enlightened than the principle followed by the British Army in India, which only began to grant King's Commissions to full-blood Indians in 1918.[145] By contrast, Alexander's intention was to incorporate Asian manpower from the start as he was determined to tap the vast human resources now available to him. It has been cautiously estimated that 120,000 boys were under training from 330–326/325 BC.[146] Ultimately, the system spread to the many colonies established under Alexander's Successors, including Greek cities within the Antigonid, Seleucid and Ptolemaic kingdoms, and some outside their borders.[147] The military education of youths in subsequent cities from Greek Sicily to Afghanistan owes much of its origins to Macedonian precedents – a fact too often ignored.[148]

Young Graduates

Literary sources for Alexander's scheme in Asia describe newly trained soldiers giving a public display of military drill after graduation.[149] Registered on the paymaster lists or muster-rolls, they were provided, at the King's expense, with panoplies stored in state arsenals.[150] At Opis, the 30,000 Asian graduates (hailed the *Epigonoi*, or 'Inheritors', by a delighted Alexander, who had watched their review) are portrayed in literary sources with lavish accoutrements, fresh from the armoury workshops.[151]

As far as the appearance of Alexander's *Epigonoi* is concerned, there may be some disparate evidence for their blending Macedonian with Iranian features.[152] Two terracotta plaques from Campania (southern Italy) suggest this, with one depicting a man on foot in Phrygian helmet, long sleeved tunic and trousers run down by a horseman in Greek apparel. The other wears the same helmet form and has a tube-yoke corselet with two rows of strips or *pteryges* beneath a short sleeved tunic over trousers. He also carries a small, deeply bowled shield.[153] Several Roman terracotta figurines survive in distinctly hooded Iranian costume with shields on which there are central discs with encircling crescents, characterising Macedonian panoply.[154]

More plausible evidence yet exists in the form of a series of tetradrachms from the reign of King Patraos of the Paeones (c. 335–315 BC), which picture a combat on the reverse side. Several versions of this combat survive; one of them has a figure in helmet and trousers with shield displaying a four-pointed sunburst surrounded by four crescents.[155] The lost Kinch tomb mural (Naoussa, late-fourth to mid-third century BC) renders a horseman in a contest with a foot soldier (Fig. 3). The latter figure is this time made up in white hood and trousers with a long-sleeved green tunic and bronze shield on which there is a twelve-rayed sunburst device.[156] A less known fragment is the striking battle scene on a vase from Begram (Afghanistan) which portrays an individual in Hellenistic and Iranian attire (Fig. 4). It consists of a pale blue (even iron) helmet with yellow *chlamys* bordered in red-brown. His short tunic is white trimmed the same colour as his cloak and he wears yellow trousers and light blue shoes. The shield has a pink centre, the radiating lines forming a red-brown coloured sunburst, the inside circle area yellow, the outer rim sky blue.[157]

As explained, there seems to have been a commonly held belief that a total reliance on athletics spoiled the military potential of young recruits with its exact diets and rhythmic movements.[158] This may cast clearer light on the suspicion and jealousy of Macedonian veterans. On viewing this first trained age class of Asians, they scornfully dismissed them as 'war-dancers'; in other words, they might look impressive when square-bashing on the parade ground but would surely wilt in the strain of a real life campaign.[159] Doubtless, many of the *Epigonoi* were keen to

Figure 3 (*left*). Detail of soldier in Iranian dress, Kinch Tomb, Naoussa.

Figure 4 (*right*). Detail of soldier in Iranian dress combined with Hellenistic panoply. Cup or vase, Begram.

disprove this sneering opinion – and, indeed, the veterans were, in essence, paying them a backhanded compliment, reluctantly recognising Alexander's praise and agreeing that their standards of drill did appear to be visually outstanding.[160]

With their training finished, new Macedonian soldiers would have made, like Roman soldiers after them, the customary oath of allegiance.[161] In the presence of the king, this public declaration was apparently spoken out loud in unison and repeated three times, the oath being to defend the royal commander with their lives, never abandon him, and share his enemies and friends. Once the oath was made, the king alone held the prerogative of personally freeing a man from his allegiance and discharging him from the army. Exemption from service could be granted to the ill or wounded, and those physically or mentally impaired.[162] On the other hand, soldiers who broke the oath were condemned as deserters committing an act of religious sacrilege.[163] We read of those who plotted against Alexander's life being shot by javelins or stoned to death.[164] Though it is claimed by one ancient source that Alexander relented on the usual execution of the relatives of traitors, the culture serves to demonstrate the central, even reverential, importance of service and loyalty to the king in Macedonia, with families whose members failed to comply facing humiliation.[165]

The oath was renewed at the *Xandika* festival or at the acclamation or accession of a new king. Such events featured a form of lustration or purification which saw the armed forces, divided into infantry and cavalry columns, marching between the two halves of a female dog which had been ritually cut up by a priest for the event. Its head and forequarters would be thrown down on the right side with the hindquarters and entrails on the left, at the far end of a plain in which the entire army would be led on parade. This peculiar custom may have originated with human sacrificial victims offered up before a campaign in the far distant past. The ceremony was therefore undertaken principally to ensure good fortune in war and signified a form of contact magic, symbolising the unity of the army.[166] The march between the dismembered parts of a dog had a specific precedence, with the king and royal family followed by bodyguard units, the rank-and-file bringing up the rear. The king and cavalry standing under arms would then face the infantry and engage in a mock battle (or 'tourney').[167]

Mobilisation

Their preparations – both physical and psychological – now complete, newly graduated soldiers were ready to mobilise as front-line troops of the prestigious *basilikai dynameis*, or 'King's forces', as or when needed.[168] In the society of a militarised kingdom, training was of course never intended to be static but would have necessitated constant, ongoing practice. To maintain their skills, the *pezhetairoi* were obliged to drill regularly and frequently at their respective localities, and it has been posited that they were mustered for war on a revolving rota.[169]

Despite the fact that figures should be treated with caution, Beloch considers that Macedonia at the time of Alexander and into the third century BC had as many as 80,000 adult male citizens capable of bearing arms. Of these, 30,000–40,000 were consistently available for service in the army, suggesting that almost one half of all men ready for call up were chosen for field duty.[170] His figures concur with the independent statistics of Ellis, who likewise upholds that, of all *Makedones*, 35,000 were under arms in 334 BC, with an additional 50,000–60,000 able-bodied men held as a reserve, and about 50,000–80,000 persons as a 'sub-citizen' class.[171] This would promote the possibility of raising large armies with an extensive strategic reserve without harming the kingdom's economic and agricultural base.[172]

Thomas sets an overall mobilisation pool for Philip II's reign at 160,000–200,000 men, increasing to a general estimate of 240,000–300,000 in Alexander's time. However, these are surely only acceptable as maximum potential figures: namely, what the entire healthy male population of the kingdom *might* have been, rather than solely those recruited into the army.[173] In later times (196–179 BC), there were definite manpower shortages, with Philip V introducing schemes aimed to revive the numbers of *Makedones* available to him.[174] It would also be in the king's interest to help regulate the migration of Macedonians drawn by incentives to join up with Hellenistic armies overseas.[175]

During the Asian expedition, newly trained infantry in Macedonia were probably kept with Antipater's home army and dispatched to Alexander when reinforcements were required, distributed *kata ethne*, or according to territorial *taxeis*.[176] Some have interpreted that Alexander's demands and the ongoing activities of the Successor conflicts were a major burden on the population,[177] but this has been subject to a degree of protracted debate.[178]

What is fairly certain is that kings carefully managed the precious resource of *Makedones* available to them by age sets.[179] This was the case in Antigonid Macedon and hints for it are found scattered in literature referring to Alexander. Even if apocryphal, Justin describes how in selecting troops for the Asian campaign, and in preference to younger soldiers, the King wanted old veterans, 'most of whom had already passed their term of service'.[180] During the siege of Halicarnassus (334 BC), senior age groups were exempt from duties requiring greater fitness.[181] The army reappears elsewhere having been supplemented by drafts from young men,[182] and on one other occasion *neogamoi*, or younger newly married troops, were granted leave to return on furlough to Macedonia.[183] Antigonus Monophthalmus may even have passed an order to billet soldiers, whereby men below the fifty-year category were to be debarred from being lodged with families liable to have young women.[184]

In Antigonid times able-bodied citizens were registered on enlistment catalogues as a form of universal conscription. Fragmentary inscriptions from Amphipolis and Cassandreia reflect that by the time of Philip V remarkable criteria had been developed over who should be called up. These had clauses covering specific conditions and relationships which might prevail across family households. Health or vigour and experience were major prerequisites, and rules set out specifying who could be exempt from service, at what age and circumstance. It is therefore not difficult to imagine that there must have been on the spot physical tests conducted by officials.[185] The preserved text points to something astonishingly thorough, in some respects not unlike conscription legislation circulating in countries today. Errington even considers it to be the most detailed recruitment regulations yet found for an army in the ancient world.[186] With their all too apparent military obsessiveness, Macedonian kings were able to intelligently exploit the male population reserves of the kingdom and maximise their use.

Before a war or campaign, infantry *taxeis* would be marshalled together by age classes from 'the land', or the *chora* (loosely defined as civic territories), and mobilised for drills, parades, mock battles and ceremonial sacrifices.[187] This was intended to keep up training levels and maintain morale and bonds of loyalty to the king. The mustering system in wartime was potentially quite rapid.

Following Alexander's death, his generals took advantage of the pre-existing Achaemenid infrastructure forewarning of invasion, speeding internal communication and mobilisation. To illustrate these preparations, there is mention of lookout posts manned by criers placed within earshot of one another along upland peaks in Persia.[188] The infrastructure was proven so effective for Eumenes in 317 BC that 10,000 archers could on the same day receive the order to mobilise, although some were thirty days' journey time from where this general was located.[189] Perhaps taking his cue from Alexander and the Persians before him, Diodorus says that Antigonus Monophthalmus operated a network of fire-signals and message-bearers.[190] There is also evidence that, like the young conqueror, he preserved and made use of Achaemenid roads.[191]

Turning to the environs of Macedonia itself, the *Suda* refers to later Antigonid kings using beacon signals, with messengers routinely dispatched carrying sealed letters containing orders to mobilise.[192] An oblique reference to the achieved efficiency is captured among remarks of the Roman envoy Titus Sempronius Gracchus, who in 190 BC travelled to Pella via Amphissa in Locris using a chain of horse relays which covered a distance close to 240km in only three days. Livy adds that in Macedon itself 'supplies are prepared for the army on a generous scale, they had bridges at river crossings and roads that were metalled where the passes were difficult'.[193] It is thus not surprising that the kingdom's character of well-knit preparedness under Perseus (r. 179–168 BC) drew Appian to write that it was 'strongly fortified' and that 'its young men were well drilled'.[194]

Although the army, or sections of it, could be mobilised all year round, the month of *Daesius* (May–June) was customarily set aside for not waging war.[195] Times when the national army was brought together for review included the two major festivals in spring and autumn: the aforementioned *Xandika* just after the spring equinox in March, honouring the hero Xanthus (held at a different location each year); and the 'Olympian' festival of Zeus and the Muses in October.[196] A camp would be laid out at each of these gatherings where *Makedones*, arriving *en masse* from all over the kingdom, came into direct contact with their supreme commander.[197] After all, this was often the only time that many saw the king in person, reflecting the strong military flavour of their relationship with him.[198]

Such occasions have been favourably compared (e.g. by Hammond) with the Edinburgh Military Tattoo or, indeed, the Trooping the Colour of the British Army. Parades of this kind must have had powerfully symbolic, morale raising, and propaganda motives, demonstrating to neighbours the ongoing efficiency of Macedonia as a unified military state. It reminds us of the Latin adage *si vis pacem, para bellum* ('if you want peace, prepare for war').

During the *Xandika* the army drilled and exercised daily, its grand pageant preceded by the venerated panoplies of former kings.[199] It was believed that the weapons of past rulers emanated sacral power when the royal forces set out.[200] The Companion cavalry squadrons would execute a series of exercises in full armour and the *telesias* war dance was performed. For his own part, the king took the opportunity to award gifts and prizes. Livy recounts how the entire body was then rallied in a specific way: 'The army in battle order was put through a short movement, though not in a regular manoeuvre, so that the troops should not seem to have merely stood to arms; and Perseus summoned them, in arms as they were, to an assembly.'[201] From a raised platform he would proceed to bombastically inspire the soldiers with speeches 'on the familiar theme of the valiant deeds of their ancestors and the martial renown of the Macedonians'.[202]

Legacy of the Macedonian Training Model

Through continual campaign service the Macedonian army came to be composed of well-trained and motivated soldier-citizens, more at home in the field camp than anywhere else.[203] The *Makedones* under Alexander were trained for every action likely to confront them on operations. As shown earlier, infantry were often taught to perform multipurpose roles, achieved by means of their training in the use of *sarissa*, spear or javelin, and sword. Some sections could have been taught to ride horses and others even deployed as slingers. Manual skills were nurtured and soldiers synchronised as a wiry labour force for sieges and engineering projects (see Chapter 13 for more on the army's technical expertise). The self-contained and economical matrix of training also engendered a sense of Homeric rivalry and the drive to excel in prowess and skill, with individuals and whole units competing for prestige and the acclaim of their peers.

A thoroughgoing approach in shaping a strong army continued as an unbroken tradition in Macedon until its downfall in the mid-second century BC. Florus says how in their wars against Rome the Macedonians were a 'valiant people' with a high repute as soldiers, praised for their persistence and lauded 'Macedonian discipline'.[204] As far as military training was concerned, the goal was rigour on a scale rarely seen in the classical world. This was achieved by amalgamating the best methods practised before, with new forms.

Even though the Roman army did not directly adopt the model introduced by Philip and Alexander, the similarity of doctrine suggests that through the influence of later Hellenistic civilisation (and its literature) they strove to introduce similar characteristics into the training regime of their own soldiers. For the Macedonians, priority was put on all round technical utility and skill. To this we can add tactical versatility and adaptability; personal and group initiative and resourcefulness; and finally, the importance of stamina, unrelenting formation drill and route marches.

Chapter 4

Supplying Arms, Armour and Cloth

Training was of course meaningless without proper functional fighting equipment supplied to troops by fast and effective means. What follows below is an overview of how, between a profusion of raw materials on the one hand and the end product of supplying military equipment and cloth to a demanding army on the other, would be lain an intermediary infrastructure of production, storage and distribution. This analysis begins with its foundation in Macedonia and continues with its growth and eventual zenith under Alexander in Asia.

Natural Resources

From Philip II onward, arms and armour production would have been impossible had Macedonian kings not pursued a deliberate policy of exploiting the natural resources of the state. It would have figured highly in Philip's calculations, encouraging him to steam ahead with his ambitious plans of expansion through forging a single polity with the new institution of an army as its central instrument. Applying Friedrich von Schrotter's observation of Prussia in around 1800 to Macedonia, the kingdom under Philip and Alexander became 'not a country with an army, but an army with a country!'[1] Indeed, natural resources in combination with a flourishing economy were the *sine qua non* for Philip II and his successors' military strength.[2]

Macedonia had the ideal backdrop for developing a powerful military infrastructure, partly because it abounded with mineral wealth facilitating an extensive and self-sustaining manufacturing base. The methods of iron and copper mining and smelting existed from as early as the Bronze Age, allowing for the techniques of making armour and weapons to evolve and improve over centuries. The frequent excavations of sumptuous finds from tombs and pit-burials show the high levels of metallurgical skill achieved in Macedonia by the fourth century BC.[3] From Philip's time, large numbers of mines were operating for the extraction of gold, silver and copper, and for the production of bronze. Gold and silver mines were closed, but iron and copper mines remained open under Roman control in 167 BC.[4]

Many of the mines in the Chalcidice region remained untapped before their working in the fourth century BC and to this day the area produces iron, and small quantities of copper, silver and gold. The same minerals, alongside tin, zinc, lead, lignite, pyrites and molybdenum (used for alloying steel), have been uncovered in other parts of the kingdom, with ancient smelting activities traced in Palaia-Kavala, covering Philippi. Modern maps prepared by archaeologists and museum curators show the wide abundance of minerals, which helped provoke respect for the kingdom from its neighbours until the second century BC.[5]

High quality timber – both coniferous and deciduous – was also bountiful, especially on Mount Bermion, the ranges above the Strymon spanning the Pierian Mountains, the plains of Philippi, Chalcidice, and the Athos peninsula. Theophrastus (c. 370 – c. 285 BC) considered Macedonian lumber to be the best imported into Hellas and one third of all forestland in Greece is still located in the Macedonian districts, with much wooded terrain lying at 200m or more above sea level. Lumberjacks were therefore able to amass impressive stocks of matured timber used by craftsmen for making shafted weapons and warships with masts and oars.[6] The road system founded by Archelaus and expanded under Philip would have helped expedite a more rapid transportation of harvested wood, and ancient roads have in fact been identified on the eastern flank of Bermion, suggesting a link with logging activities there.[7]

As Strabo intimates, Alexander is known to have employed one Gorgus as a *metalleutes* (mining engineer) to investigate the mineral resources of the kingdom of Sopeithes (near modern Lahore, Pakistan), and this man was required to lodge a report.[8] Presumably Alexander acted, or planned to act, on the findings of this before his death. The King followed a similar proactive policy in taking control of mining areas in other parts of his empire. It is alluded to in another passage by Strabo, which describes Alexander sending an agent called Menon accompanied by troops to secure the gold mines of Syspiritis near Kaballa in Armenia.[9]

It seems fairly certain that Philip II would have employed experts to study and survey mining areas across Macedonia, helping to pinpoint and tap the resources required to build his army-state. Philip secured these localities from external attack, improved mining apparatus and, rather than letting his mines out to contract, employed his own managers, indicating that he aimed to maximise efficiency and output in the exploitation of valuable resources under royal jurisdiction.[10] Extant treaties with Athens show that Macedonian kings held a monopoly over all natural wealth inherent in the land: timber (including probably ash and cornel wood), gold, silver, iron and copper. This all helped to ensure that a great quantity of raw materials could be husbanded and channelled to the production of arms and armour on a kingdom-wide scale when required.[11]

Workshops and Arsenals

The royal provision of arms and armour in Macedonia was first established under Archelaus in the late-fifth century BC. He is described furnishing military forces with fighting gear and horses, and probably hiring experienced Thasian, Chalcidian, Amphipolitan and similar craftsmen.[12] As far as is known, Philip II used the system of state provision set up under Archelaus as a foundation for his own production of armaments.

At the onset of his reforms, Philip is passed over by Diodorus having 'equipped the men appropriately with weapons of war': this indicates that even after an interval of 40 years (399–359 BC) – a period often noted for its destructive internal squabbles and foreign invasions – the system established under Archelaus seems to have remained intact.[13] Philip II's methods would have unfolded, developed and expanded as his reign progressed. Errington roundly remarks that the industry of war in Macedonia must have involved many people with the central controlling authority – an infrastructure with tentacles reaching into every corner of the kingdom, urban and rural. All considered, 'royal ordnance' may not have existed to this extent before.[14]

The programme under Philip was in turn inherited by Alexander, and eventually carried on with the overwhelming resources and internal organisation of the conquered Persian colossus as backing support. The rise of the Macedonian army and subsequent Hellenistic Age thus ushered in a new era which underscored a far more concentrated manufacturing base of arms and armour – far greater than had been contemplated previously during the classical age of city-states, under the more localised hegemonies of Athens, Sparta and Thebes.[15]

In which localities weapons and armour were produced in Macedonia is an open question, though the more populated and economically developed cities in the lowland regions are the most likely candidates. This would be especially true of the wealthy former Greek colonies along or near to the coasts whose enterprise had been chiefly responsible for the original exploitation of mines.

Mineral resources had been used by colonists around Pydna from centuries before Philip II's time. One workshop, possibly situated in a Chalcidian colony, has been identified as the manufacturing place for Chalcidian 'type 2' helmets in the sixth century BC, found mainly on Macedonian sites. We know from Herodotus that workshops on Thasos were producing notably elegant silver and gold tableware from as early as c. 480 BC.[16] It is intriguing that in 424 BC, on the adjacent mainland of Amphipolis, the Spartan general Brasidas is described by Diodorus ordering stacks of arms and armour to be made and collected along with supplies of missile weapons and grain.[17]

Craftsmen supplying equipment to the Roman imperial army often set up their workshops near mines, and there are no persuasive reasons to believe that in Macedonia things were any different.[18] Excavations have revealed that an arsenal built in Athens during the early- to mid-third century BC was erected close to metalworking installations.[19] Literary evidence for Athens in the fourth century BC reflects that a single factory concern producing military apparatus would have employed a workforce of as many as 120 slaves.[20] Extracts drawn from the speeches of Demosthenes show that Athenian factories specialised in turning out specific items, whether swords, spears or shields.[21]

Keeping Diodorus' above reference to events in 424 BC in mind, along with the fact that Amphipolis nestled in an area known for its abundant mineral resources and mining activities, it is probable that this city operated as a major hub for arms production in later years. It has been conjectured from the corrupted text of the *Itinerarium Alexandri* that during build-up for the Asian expedition (334 BC) a fleet well-stocked with military stores was moored on the river Strymon off Amphipolis.[22] The case of Brasidas acts as an oblique reference to the town as a well-established armaments centre before Philip II had even become king – possibly dating from the late-sixth century BC, when many of the mines in the Trans-Axius region were opened.[23] Studies of extant late-fourth century BC gorgets from different locations in Macedonia suggest that some of them were made at the same atelier in or around Amphipolis.[24]

Workshops specialising in the manufacture of metal objects of a non-military nature have been uncovered in relative abundance through excavation at Pella, Vergina and at many other places.[25] Stojcev identifies craft centres producing war gear in Gortinia (Republika Makedonija) and the Chalcidice, but without corroborative evidence.[26] Along with Amphipolis, production sites for arms and armour would have been located at Pella, Philippi and major towns of strategic importance, close to or within metal-rich areas. As Pella lies within 40km of Mount Bermion, wood for shields and shafted weapons arrived there by the overland route and by river.[27]

Another later possible example was the city of Cassandreia, founded in the Chalcidice over 316 BC.[28] In 279 BC the tyrant Apollodorus was able to hire a band of Celtic mercenaries there which he provisioned with arms in order that they serve as his personal bodyguard.[29] The passage could allude to workshops or an arsenal for military equipment already existing in the environs of the city before this date, perhaps when Cassander founded the site. This King's other famous foundation, Thessaloniki, is described by Livy accommodating a garrison of 2,000 elite infantry in the reign of Perseus. French archaeologists discovered a fort in Thessaloniki, built into the far right corner of the Hellenistic acropolis.[30] Units of royal troops such as these, stationed in coastal settlements of strategic value, would have required convenient facilities for arms and armour.

Philip II must have arranged for government subsidies to those engaged in war preparation within the kingdom. Given the Macedonian soldiers' pride in owning high quality armour and weapons, and combined with the socio-economic conditions which acted to induce mobility among skilled craftsmen in the fourth century BC, the King would have actively striven to patronise the most eminent armourers from across the Balkans, Thrace and the Greek peninsula, attracting employment with the promise of good wages and standards of living.[31] Dio Chrysostom has captured enemy artisans in Greece being seconded by the victorious side to make panoplies.[32] Philip would have done likewise, deporting to Macedonia metalworkers captured on his numerous campaigns from Thrace to the Aegean world.[33]

King Philip successfully seized Potidaea (356 BC), Methone (354 BC) and Olynthus (348 BC), and Demosthenes alleges that he razed no less than thirty-two Greek cities in the Chalcidice alone.[34] He used the transplantation of peoples as a policy for strengthening and securing the kingdom[35] and such actions could result in artificers being relocated to Pella and other cities to improve arms production. Given his diplomatic leanings and often masterful grasp of human psychology, allied to the appreciation he had of their potential value, it is acceptable to suppose that the King deliberately extended excellent treatment to valued persons of this kind.[36] The often spurious and fantastic legends of Alexander detail him employing Philip's former bronze-smiths, carpenters and other artisans to make helmets, shields, swords, spears and arrows.[37]

If the kingdom led the way in metallurgical techniques by the mid- to late-fourth century BC,[38] introducing a more widespread use of iron helmets to Greece, possibly even pioneering developments such as the brow plate on the Attic helmet, then many first-rate armourers must have been actively involved.[39] After removal of oxidization, some spear and *sarissa* heads from tombs at Vergina show that sophisticated alloys were in circulation. The metal forged with great care had well-made sockets and meticulously worked, shaped points for increased penetration.

Armour recovered from Royal Tomb II, Vergina (the final resting place of either Philip II or Philip III, r. 323–317 BC) was particularly fine and highly tempered.[40] Moreover, an Attic helmet of possible Macedonian origin found near Marvinci (Republika Makedonija) was by report wrought from iron of notable purity.[41] The upper end of one sword blade at the Museum of the Royal Tombs at Vergina (probably 'steeled' by the flames of the funerary pyre) appears to the keen-eyed to be in an almost usable condition even now. Clement of Alexandria actually refers to the nearby Paeonian tribe of the Noropes as famous coppersmiths with the incidental remark that they invented 'purified iron'.[42] It is possible that these artisans were employed too. Distinct decorative elements on products were applied by individual craftsmen to identify themselves as the maker.[43]

Armourers worked to designs which had already been given official consent, with royal inspectors ensuring that a requisite standard was maintained and that orders were promptly fulfilled. Given the imperative of furnishing the lance (*xyston*) to Companion cavalry and *sarissa* for the phalangites, the expert spear-maker would have been prized and well paid, with a prestige status like that of master bowyers in medieval England.[44] Sekunda even contemplates a system analogous to Spain during the sixteenth century and England under Charles I, with an organised system of forestry plantations, tree protection programmes, and weapons shafts produced in the spear-makers' workshop (*doryxeion*), exclusively by artisans who had reached the required levels of proficiency.[45]

Turning briefly aside from Macedonia, the tyrant Dionysius I of Syracuse established what can be seen as the first large-scale military-industrial system in the classical Greek world. Alexander was well read in Sicilian history, having studied Philistus,[46] a faithful adherent of Dionysius, who wrote copiously on Sicily covering everything down to the careers of the tyrant and his son. In addition, Philip's court included ex-patriot Sicilians or men who had fought in Sicily, such as Demaratus of Corinth who had served under the Corinthian general Timoleon (344–337 BC).[47] Given that Philip II was likely well aware of mass armaments manufacture in the West, and that this awareness might have been of some formative influence on his own methods, it is worth considering what the sources impart about Syracusan production.

In 399 BC Dionysius commandeered artisans from cities under his control in Sicily and, according to Diodorus, offered high wages as an incentive for workmen to enter his patronage from Greece, Italy and even Carthaginian North Africa. Having collected the necessary manpower, he organised them into groups according to skill. The aim was to produce 140,000 shields and a similar number of helmets and swords, 'and in addition corselets were made ready, of every design and wrought with utmost art, more than 14,000 in number'.[48] Quality control was improved, with Dionysius offering prizes to craftsmen judged the best.[49] Nor did the centralised production effort perish with him, as details of his son's leadership in 357 BC seem to testify.[50]

Having manufactured new equipment in Macedonia, the next stage would be to have it all checked and secured within appropriate long-term storage facilities. The chief arsenal could have been located in the environs of Pella, built close to the royal treasury just south of the city walls in an elevated citadel surrounded by the swamp of Phakos. In the early-second century BC this site was linked to the city by an easily guarded timber bridge.[51] It is hoped that future archaeological discoveries will help locate traces of the arsenal. Relatively recent excavations in the area have unveiled part of a fortifying wall surrounding two buildings provisionally identified as 'military installations'.[52]

Surviving Alexander legends describe how the original expeditionary force led by Alexander into Asia was accoutred from Philip's armoury.[53] Livy mentions arms being stored in arsenals (*armamentaria*) under Philip V and Perseus, although no further details are given.[54] Recording events for 200 BC, when Roman troops were able to breach the Macedonian-held base of Chalcis, he relates how 'both the royal granaries [*horrea regia*] and the arsenal [*armamentorum*] were burned, with a great store of munitions and artillery'.[55]

Storehouses are similarly mentioned for other Hellenistic kingdoms ruled by Macedonian dynasts. In his rebuilding of Lysimachia (196 BC), the Seleucid king Antiochus III intended that the city's main strategic purpose on the Gallipoli peninsula was to be as a convenient centre of supplies for campaigning in Europe.[56] During war with Rome the city functioned as the principal magazine for military provisioning.[57] This store contained weapons, money and stocks of grain, along with siege engines, probably disassembled in kit form.[58]

Hellenistic archaeological traces for arsenals or state arms factories have been located at Athens, Pergamum (Fig. 5), Memphis (Egypt) and even far away Ai Khanum (in Afghanistan).[59] Of these, the Pergamene structure preserves some intriguing features. Built by Eumenes II (r. 197–159 BC), the facility comprised a series of five parallel magazines of wood set on a stone substructure, the sturdy workmanship being especially made to support heavy loads. Among other artefacts, excavations have yielded 961 beautifully finished andesite catapult balls representing some thirteen calibres. Similar large-scale Hellenistic ammunition dumps have been recovered from sites in Rhodes, Salamis (Cyprus) and Tel Dor (Israel).[60]

Although arsenal buildings within Macedonia await discovery, traces of fortifications, outposts and structures associated with the kingdom's strategic defence, built under Philip II and succeeding kings, have been found at Vrasna, Parembole, Kalyva, St Erasmus, Ditnata, Tsiourba Mandira and Kastania (in Pieria). We can assume that in the fullness of time traces of arms and armour storage facilities will be brought to light.

Figure 5. Remains of a Hellenistic arsenal sited near barrack buildings, north end of the acropolis, Pergamum.

The War Readiness of Macedonia

Some significant physical and artistic evidence survives intact for state made weapons. One such exists in the form of large, bronze arrowheads or catapult bolt-heads (*katapeltai beloi*) for use with non-torsion catapults found on the site of Olynthus and cast with the inscription 'for Philip', which indicates manufacture in a Macedonian workshop (Fig. 6). Produced to set sizes and weights, these objects almost certainly date to Philip II's successful siege.[61]

Sekunda has observed painted bands around head sockets belonging to *sarissai* behind King Darius III of Persia on the Alexander Mosaic (see Plate 3b).[62] The

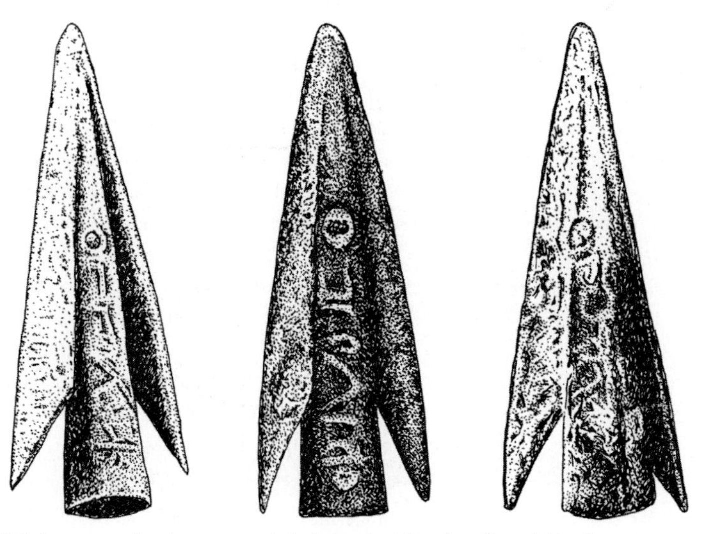

Figure 6. Trilobate arrowheads or catapult bolt-heads, Olynthus (length 7cm).

Mosaic originates from the House of the Faun, Pompeii, dated c. 100 BC, copying a painting from 330–310 BC. These bands compare well to a heavy 38cm cast bronze Macedonian spear butt dated to the late-fourth century BC (Shefton Museum, Newcastle-upon-Tyne) discovered during cleaning in 1977, with similar band markings between which were stamped the letters MAK, an abbreviation for '(Mac)edonian'. This drew the thought-provoking comment that they resemble the familiar WD broad arrow used by the British Ministry of Defence even today to mark armed forces property.[63] Comparable black bands are seen on the socket of a spear carried by a figure on the Hunt Frieze from Royal Tomb II (Plate 9b).[64] Other bands are apparent on a javelin head socket of a possible mounted 'Alexander' from a fragmentary Hellenistic mosaic in Palermo (perhaps based on a Macedonian painting, 325–275 BC) (see Plate 36a).[65]

Bronze Macedonian shield face fragments from Dion (Fig. 7),[66] Vegora (Plate 15),[67] Dodona (Plate 16),[68] Staro Bonce[69] and Palaiokastro,[70] generally dated to the early- to mid-third century BC, often have an inscription on the surface, beginning *Basileus* or 'King', pointing to production at royal expense before distribution to infantry units.[71] An extant limestone model for such a shield has been discovered on the site of Ptolemaic Memphis, now held at the Allard Pierson Museum in Amsterdam (Plate 17). The object seems to have been used as a master-pattern in state manufacture.[72]

The discovery of nine lead tokens (dated c. 250 BC), unearthed in the Athenian Agora in 1971, gives a further unique glimpse into the extent of provision among some Hellenistic polities (Fig. 8). Each piece had been stamped on both sides, with the obverse depicting helmet, corselet, greave, shield or a figure of *Nike* (Victory). By contrast, the reverse has one of three letters: alpha, gamma or delta. The combination of symbols suggests that these were quartermaster chits to be shown in exchange for government-owned stocks. The three letters matched to

Figure 7. Bronze strip from a shield, incorporating the inscription 'King Demetrios'. Sanctuary of Olympian Zeus, Dion, c. 294 BC.

Figure 8. Lead token or quartermaster chit (Athens). Obverse: helmet with wreath; reverse: the letter gamma.

helmet, corselet and greave could equate to three sizes, similar to the modern 'small', 'medium' and 'large'.[73] If Athens was able to supply complete sets of equipment in this fashion, then it is highly likely that from the mid- or late-fourth century BC the Macedonian kingdom, much larger, richer and more aggressively organised for war, did the same – least of all, for elite units.

Literary passages referring to the war readiness of Macedon after Alexander's death help place the artefactual evidence in context, while casting a pertinent light on the type of arms and armour production available. Only with the capacity and backing of a substantial complex of manufacture was Demetrius Poliorcetes able to equip and manage the unusually large force recorded in literary accounts – probably with a degree of exaggeration – which he gradually built during his short reign (294–288 BC).[74] This is particularly outstanding given the near constant obligations and pressures brought to bear upon the military population of Macedonia in the preceding fifty years. It is also a tribute to the durability of the home grown armaments industry, which since the time of Philip II had come under intense use.

It may have been Demetrius' marshalling of men and material that inspired Plutarch's sweeping remark, certainly based on now unknown sources consulted by him, that this king 'was actually thought to be a better general in preparing than in employing a force, for he wished everything to be at hand in abundance for his needs'.[75] But the kingdom was for Demetrius a simple means to an end: both engine and springboard for reconquering the East.

Plutarch describes Philip V over a century later, before his death in 179 BC (and after eighteen years of peace following war with Rome), as having 'filled the fortresses, strongholds and cities of the interior with an abundance of arms, money and men fit for service', with enough equipment to furnish 30,000 soldiers, including a stockpile of 8 million bushels of grain.[76] The organisational intensity can be appreciated even more clearly from Livy's account that the grain supply was sufficient to feed 35,000 troops for 10 years. This was large enough for them not to be dependent on provisions offered by allies or found in enemy territory.[77]

The remarkable reality of an estimate given by Plutarch for the amount of grain at hand suggests that the administration kept a tally on how much was collected. Writing in the third century BC, Philon of Byzantium discusses the subject of military provisioning in detail, with rules governing the selection and inspection of grain, as well as sound methods of storage and preservation.[78] In the increasingly bureaucratic Hellenistic Age there are two inscriptions, preserved from Chalcis and Kynos for the reign of Philip V, with *diagrammata* (royal regulations) on the same subject of supply attested by Philon. These describe with exhaustive rigour the administration of grain, wine and wood stores by officials called *oikonomoi* (finance managers) – the administrative equivalent of *phrourarchoi* (garrison-commanders).[79] No comparable evidence for this degree of intricacy has been found so far for Philip II's and Alexander's time.

Livy helps confirm Plutarch's condensed statements about Philip V's collection of arms, informing us that by the outbreak of war with Rome in 171 BC Perseus had hoarded huge quantities of missiles – which may be reminiscent of the already cited arrowheads or catapult bolts and balls recovered from the site of Philip II's siege at Olynthus. Florus made the evocative comment that under Perseus Macedonia seemed so well prepared, guarded and bristling with arms that only enemies descending from the sky would be able to penetrate and conquer it.[80]

The process of government issued munitions seems to have been better co-ordinated than those in existence elsewhere. Livy writes admiringly of the long-term arrangements with everything efficiently provided for: soldiers equipped with arms drawn from the 'royal state', or apparatus (*regio apparatu*) 'produced in the course of many years as a result of his father's [Philip V] expenditure'.[81]

In terms of soldiers' equipment on campaign, it is striking that the burial mound of Philip II's Macedonians from the battle of Chaeronea yielded evidence for many weapons fragments, but no defensive accoutrements – an indication that in this case armour was salvaged as battleground detritus from the dead and then field modified for the use of fellow soldiers.[82] On occasion, spares were procured from civilian settlements. Philip V's march on Thermum (218 BC) ended with the army bivouacked in the town market-place where they took the best military accoutrements hanging as dedications in the porticoes there and destroyed the remainder – a total of 15,000 weapons or suits of armour (*hopla*).[83] Further available options during an ongoing campaign include individual purchases in arms bazaars or the recycling of battle gear taken from enemy captives.[84]

Equipment was also obtained on a more formal basis. From early Hellenistic history we find the production of large stocks of reserve panoplies in Ptolemaic Egypt,[85] with spare items, such as shields, sometimes distributed among

Macedonian allies for the duration of a war, as was done for a Megalapolitan contingent in the lead up to the battle of Sellasia (222 BC).[86]

Alexander and the Persian Armaments System

Alexander's conquests in Asia saw Achaemenid armaments production surviving intact. There is evidence for Persian naval installations and arsenals established as far west as maritime Cilicia, and there is no reason to reject the notion that Alexander exploited their usefulness for his army as he marched through.[87]

It is important to understand that Persia was far from being a toothless tiger in the face of Macedonian aggression. It integrated an efficient manufacture of arms and armour, which was part of an older Middle Eastern bureaucratic tradition traced to at least the Neo-Assyrian Empire (754–609 BC) where mass produced military gear seems to have been commonplace.[88] An inkling into the impressive Achaemenid infrastructure emerges with Curtius, who attests that Darius III 'at the beginning of his reign ordered that the shape of the Persian sword scabbard [*vaginam acinacis Persicam*] be changed to the Greek model'.[89]

Such remarkably minute modifications to existing designs are again reflected in extracts which describe Darius widely re-arming his troops before Issus and Gaugamela in order to improve tactical potential when opposing the Macedonian army. The order included furnishing many sets of new equipment: 200 scythed chariots, armour for cavalry, and swords and shields to troops who previously had been armed only with javelins. Alongside these, there were longer swords and spears, the latter inspired by the Macedonian *xyston* (implying that the Persians had captured some of these weapons, duplicating their form).[90]

The Persian network of production was therefore both well-organised and flexible enough for tactical weaknesses to be identified and overcome – the appropriate items then being ordered, produced and distributed to troops on a wide basis. Having eventually subjugated Persia, Alexander had no reason to dissolve what was an obviously highly advantageous arrangement – he simply took over managing it and had it serve his own ends.

Reminiscent of King Darius' ordering a change to the standard scabbard pattern, Julius Africanus of Edessa refers to many features of panoply being adjusted and improved under Alexander and what he calls the 'epigones' (probably meaning the *Diadochoi* and succeeding generations of Alexander's Successors).[91] He even describes the mass introduction or promotion of the Spartan helmet for infantry.[92]

That the same primary infrastructure continued long after the demise of the Achaemenids, and well into Hellenistic times, is suggested by metallographic analyses of bronze and iron objects in Asia, which confirm the local origin of their manufacture at that time.[93] We will see later the extent to which Alexander provided panoplies to his soldiers during the Asian campaigns.

Plundered Cloth and Dyes

Clarysse informs us of an army clothing stipend of 10 drachmas per annum recorded in a Petrie papyrus from Ptolemaic Egypt. Cloth was undoubtedly purchased in this case from among small-scale workshops in local marketplaces. Of course, a single undated scrap of Egyptian evidence is of little direct value when attempting to throw light on arrangements among Philip and Alexander's Macedonians a century or two before.[94]

Generally speaking, units of the Macedonian army were issued with dress by kings for parades, state and ceremonial functions, and at the onset of a major venture.[95] Batches of clothing were occasionally provided as a war progressed.[96] However, as with the experience of armies throughout history, soldiers and whole units supplemented their dress with whatever could be bought or forcibly expropriated. A good example of this would be in the event of the formal logistics system breaking down.[97]

Analysing the literary sources, it becomes apparent that under Alexander cloth was obtained as plunder after battle and the seizure of the enemy camp, and the sacking of cities and Achaemenid palace strongholds.[98] Some batches were also supplied by satraps, levied from communities in the empire and subject-ally kings, or even sent out from far away Macedonia (the last at least for Alexander's wardrobe).[99]

Through Alexander's conquests of Persia, an abundance of purple dye in combination with saffron is apparent on clothing and corselets. This is consistent with artistic evidence for this period, as seen on the famous late-fourth century BC Alexander Sarcophagus (Istanbul Archaeological Museum),[100] and a frieze from the Agios Athanasios tomb (found in 1994 near Thessaloniki, dated 325–300 BC).[101] Literary sources also frequently record its distribution in prodigious, exaggerated quantities[102] – a fact recognised by scholars since at least Grote's dissertation in 1913.[103] It is intriguing that warriors in the *Iliad* are described wearing a protective belt, the organic components of which were apparently dyed some red or purple shade.[104] The colour affinity to their own dyed corselets could not have been lost on Alexander and the Macedonians, whose society was often governed by the heroic credo of Homer's epics.[105]

Having gathered excessive amounts of plundered Achaemenid cloth, Alexander fully allowed his army, or individual soldiers, to be clad in purple if they so wished from early on in his reign. The cloth was a suitable reward for the exertions of the eastern campaign, and a potent symbol of wealth. Vivid colours, in combination with the shining glint of silver and gold, and polished iron and bronze equipment, were also appropriate contributors to personal self-esteem and unit pride.[106]

After the battles of Issus and Gaugamela, in Damascus, Babylon, Susa and Persepolis Macedonian soldiers appropriated huge quantities of rich purple and other dyed cloth. Purple-coloured fabrics were first encountered in the aftermath of the battle of the Granicus (334 BC), although much of it was sent back to Queen Olympias in Macedonia.[107] If not merely a literary convention inserted to help dramatise his narrative, Curtius shows Alexander before Issus haranguing his Thracians, encouraging them to see the gold and purple worn by their Persian enemies as abundant booty free for the taking.[108] According to Diodorus, the spoil seized at Issus and Damascus included 'vast numbers of rich dresses from the royal treasure, which they took'.[109] The expensive garments of men and women, some adorned with purple and gold, were found scattered about and eagerly purloined by looters.[110] Likewise, after Gaugamela, the Persian baggage gave up 'no little barbaric dress'[111] or 'costly raiment, since ... the wealth of the entire [Persian] army was concentrated in that spot'.[112]

The occupation of Susa released an even greater stockpile of purple cloth dating to Xerxes' invasion of Greece (c. 480 BC) or before, originally from Hermione in the Peloponnese. Considering that this cloth represented Greek tribute to Persia, what better way of publicising the successful outcome of Alexander's war of retribution than to clothe his troops with the same fabrics that the enemy had originally 'robbed'?[113] It would, moreover, seem more practical to use cloth abundantly to hand than take the logistical trouble of sequestering it from surrounding regions or having it sent out from distant Macedonia. The cloth from Hermione was estimated and described by Plutarch at 5,000 talents weight and it had apparently retained a fresh colour even after 190 years of storage.[114] The material had been treated with honey, which successfully preserved its lustrous texture.[115]

Some thought-provoking calculations can be made based on the information Plutarch supplies. If we accept that he (or his source) was referring to a standard Attic talent, then Alexander had stumbled on an immense hoard approximating to about 130 metric tons of purple cloth. Using the evidence of modern Roman legionary studies as a guide, Lee calculated that a Greek mercenary serving in Asia (c. 400 BC) would have had a tunic of about 1kg and a cloak weighing double, at around 2kg.[116] This is a generally permissible figure for Macedonian soldiers seventy years later. Accordingly, Alexander would have had enough bolts of purple cloth from this *single* source at Susa to potentially outfit more than 40,000 soldiers with one tunic and one cloak each. It is of course unlikely that all the material was destined exclusively for the army. But even if only a fraction was used, Alexander could still have provided many thousands of soldiers with fabric to be made into clothing and applied on some corselets.

It is important to recognise that this windfall was probably not just 'purple' in the broadly accepted sense. In a review of tablets salvaged from the Susa palace (dated around 600–550 BC), Olmstead noted revenue records for textiles covering a wide compass of colours and styles.[117] Among Achaemenid stores there would be cloth sometimes stained with Tyrian purple dye, extracted from murex shellfish, producing many shades from rose-pink through bright red and blue to deep purple, depending on the strength of dye used and its degree of exposure to sunlight.[118] Such colours admirably match up with known evidence for Macedonian and Hellenistic soldiers from the fourth to second centuries BC, which often portrays corselets and tunics in a vibrant range within the red colour spectrum from pink to scarlet through to deep crimson and purple (see Appendix 2 for a catalogue of evidence on cavalry clothing colours).[119] Many of the same colour preferences were inherited ultimately by the Roman army.[120] Pliny preserves an appealing story that when in India Alexander's officers even competed with one another in the colours of their ship ensigns, which again shows an assortment of bright lavishly dyed cloth being readily accessible by this time.[121]

Following on from Susa, the city of Persepolis, according to Curtius, was fated to give up a 'vast amount of clothing', when in their feverish cupidity Alexander's troops grabbed 'the royal robes, as each dragged a part into his possession'.[122] Diodorus remarks that 'many rich dresses gay with sea-purple or with gold embroidery became the prizes of the victors'.[123] Taken literally, this would mean that numbers of soldiers appropriated cloth as part of their private belongings.

After Alexander had passed the Persian Gates (331/330 BC), the cloth garnered by his quartermasters had become so numerous that he freely gave some of it away as surplus to requirements. The generous gifts offered to 800 or 4,000 Greek refugees at this time comprised 10 pieces of clothing each, estimated at 8,000 or 40,000 individual articles of male and female dress.[124] However, much of the overall glut was reserved for the army, which may be construed from Diodorus' account of events in 329 BC, when he tells us that: 'what was distributed to the soldiers, including clothing ... came to 13,000 talents' – a suspiciously expensive sum if the cloth used was relatively plain, unadorned or made with cheap dyes or materials.[125]

Even if a portion of the booty was deliberately destroyed to improve mobility in the Kara Kum desert, some fabrics were still available to the army in the phase following the annexation of Persia to the time of the Indian expedition.[126] In around March of 326 BC Alexander gave a large quantity of Persian textiles as part of a bulk consignment of gifts to the King of Taxila (in the Punjab).[127] Therefore, the army had with it surplus Persian cloth, in part reserved for diplomatic purposes, which was of a suitably expensive quality to be accepted as a royal present.

The troops were still 'clad in Persian dress [*vestem Persicam induti*]' during the ongoing campaign in India in 326 BC.[128] However, Diodorus intimates that the situation soon deteriorated markedly: soldiers were reduced to cladding themselves in re-cut Indian garments, doubtless after the Persian stocks taken along with the army had begun to run out.[129] This reminds us of British officers faced with the same predicament, donning 'Chitrali robes' while on campaign in the Chitral Valley in the 1890s.[130] Indian ambassadors brought gifts, including cloth, which implies that Alexander arranged for clothing supplies to be provided by allied or subject cities and kingdoms in the region.[131]

The Indian cotton cloth is described as *periblemata*, or cloak 'wraps' (which could be similar to those worn by the Nuristani and other peoples of modern Afghanistan and Pakistan), defined as blanket-like, draped over one shoulder or gathered about the waist. These items were first fashioned into recognisable cloaks before being used by Macedonian soldiers.[132] That many of them were white can be deduced from Arrian, who cites Nearchus' description of Indians commonly wearing white linen or cotton clothing.[133]

Indian fabrics may have still seen widespread use during the army's return leg to Persia. In the celebrations that marked the end of the gruelling march through the Makran in 325 BC, baggage wagons were apparently rigged out with items including either white or purple canopies.[134] Curtius describes the vehicles as 'adorned according to the means of each man and hung around with their most beautiful arms'.[135] After the desert ordeal, it seems unlikely that troops would be specially kitted out with hangings for army wagons. It is more credible that they temporarily used their military cloaks as makeshift decorations.

Once back in Persia, Alexander never forgot to ensure that his core army of veterans remained smartly turned out. At Opis he is pictured by Curtius angrily rebuking soldiers for their display of ingratitude in owning purple clothing and quantities of silver and gold.[136] The speech this extract is from is of course largely (or even wholly) fictitious. It may only preserve what words Curtius or his readership *expected* the King to have spoken, but the mention of purple garments cannot be summarily ignored. During this time, Athenaeus states that the King directed the cities in Ionia 'to dispatch purple dye to him' because 'he wanted to dress all his friends in garments dyed with sea-purple'.[137] The sophist Theocritus was so bemused by Alexander's extravagant Ionian directive that he said he now understood the verse in Homer running 'Purple death seized him and a fate overpowering',[138] which may have been made in part to mock the young conqueror's Homeric obsessions.[139] It should not be forgotten that in the years and decades following Alexander's death there is further mention of veteran soldiers continuing to own many expensively dyed bolts of fabric as part of their personal baggage.[140]

All surviving circumstantial evidence presented here leads to the firm conclusion that during the second half of Alexander's reign the army was increasingly inundated with luxurious Persian textiles incorporating a wide range of shades, with an emphasis on true purple but with many other bright colours as well. Although modern commentators seem reluctant to accept that Alexander's imperial army was dressed in costly attire (simply because it was unprecedented), many of his decisions, actions and career aspects as a whole are exceptional in historical annals. More than any other leader, he sought to meet and eclipse challenges, outdo records set in the past and set new standards in his voracious quest for undying fame and *arete*.[141] Macedonian troops under Alexander subsequently came to be so strongly identified with the wearing of lavish cloth that Dio Chrysostom made the facile remark four centuries later that the allure of royal purple and flavoursome Median cuisine was largely to blame for their moral fall from grace.[142]

Alongside purple, Alexander also inherited the Achaemenid penchant for saffron use in clothing.[143] Yellow pigments, which may be representative of saffron or similar, are well evinced on numerous cavalry cloaks and infantry corselets from the Alexander Sarcophagus sculptures.[144] It is again observed for Companion cavalrymen's cloaks dating from the late-fourth to early-third centuries BC on the Agios Athanasios tomb frieze,[145] paintings from the Tomb of Judgement,[146] and a grave *stele* from the Great Tumulus, Vergina.[147] More evidence can be drawn from painted images of soldiers from Ptolemaic Alexandria,[148] the Shatbi Cemetery,[149] Mustafa Pasha Tomb I,[150] a fresco from Pompeii,[151] and a damaged mosaic in Palermo.[152] The fragment of an unguent container for holding saffron was even unearthed as a grave good within the Agios Athanasios tomb.

As a luxury commodity the precious saffron dye is produced by collecting pistils harvested from the autumn *crocus sativus*. In antiquity the chief centre for its production was located in Cilicia, but the substance is still cultivated today on the Iranian plateau, particularly in the provinces of Hamadan and Khorasan.[153] It is especially seen in the bright tribal costume colours of Fars, close to Persepolis in southwest Iran.[154]

The fact that saffron was often seen as a feminine colour identified with the cosmetics of courtesans and indolent Persian aristocrats partly accounts for sources detailing Alexander's soldiers' less than lukewarm reaction to their new attire. Some elements of it were unpalatable to them, making them think that they were dressed more like their defeated enemies, with all the trappings of degenerate pomp and luxury.[155] This sort of contempt is repeated in Virgil's *Aeneid*, which, even if echoing Roman values, describes the Syrian Chloreus (a priest of Cybele) dandified in an outfit of purple, gold and saffron, which was an object of spoil for the warrior-maiden Camilla.[156] Similarly, Numanus compares the soft Trojans to his own sober people, and criticises the former for preferring to wear saffron and purple, taking their ease and dancing, more like women in their deportment than men.[157]

It is significant to add that saffron was sometimes associated with the robe of the god Dionysus, revered by Macedonian kings, the army and society in general – also famed for his mythical wars and exploits in India.[158] It is tempting to surmise from this that Alexander had convenient propaganda motives in mind when having parts of clothing and armour dyed with saffron. It would be especially relevant when preparing for the Indian campaign, imbuing his army with a Dionysiac 'feel' or 'mystique'.[159] Alexander was favourably disposed to comparing himself with Dionysus when he expressed the wish: 'Would that the people of India too may believe me to be a god!'[160] As Arrian puts it, when in India the King thought that the Macedonians would not refuse to join him in outmatching the achievements of this deity.[161] Perhaps Alexander did not personally believe in the veracity of India's Dionysiac connection. He nonetheless exploited it as a useful propaganda tool, part of his brazenly promoted self-image of invincibility.[162]

Distribution of Panoplies in Asia

For what it's worth, the Greek *Alexander Romance* refers to Alexander during the war with Persia writing to satraps in Phrygia, Paphlagonia and other provinces of the western empire ordering them to send him everything in their armouries. More intriguing is another *Romance* extract containing a letter addressed to cities in Persia requesting that every weapon be delivered to storage sites.[163] We can safely assume that there were many thousands of unemployed Achaemenid troops who could act as a source of subversive instability, and the process of disarming them would be vitally important if Alexander were to effectively consolidate his control over Darius' former territories.

The King's care of military supplies during the war with Persia is passed over in more acceptable texts. For one, Arrian describes the case of Arimmas, satrap of Syria, who was removed from office due to his negligence in organising sufficient provisions for the army.[164] Royal provision is also alluded to when Curtius describes Alexander's crossing of the Tigris (331 BC). Troops lost some of their equipment and effects to the fast flowing current and on this occasion the King reacted to their concerns by reassuring them that he would replace everything missing.[165]

Ongoing demands were slightly alleviated by taking advantage of pre-existing Achaemenid depots, including those conveniently sited along the road network.[166] Xenophon describes in his *Anabasis* how the Greek mercenaries came across a handy stockpile of missile weapons, with plenty of animal gut for bowstrings and lead for slingshots.[167] This was particularly useful as the Macedonian army wound its way along the interior lines of Darius' empire, and it is plausible that Alexander's intelligence arm collected information as to where arms dumps were located well in advance.[168]

Some individuals continued to supplement their equipment with captured enemy arms and armour, like the King's ownership of a Persian two-ply layered or quilted linen corselet recovered from Issus and worn at Gaugamela.[169] Such actions conform to the Homeric mindset in which the victor despoiled the vanquished of his panoply. The use of plundered articles could further allow for psychological advantage over the same opponent in a future clash.[170]

Justin describes the Macedonians expressing admiration for the splendour of Persian war gear before combat at Gaugamela.[171] As notably wealthy individuals, the Companions were certainly keen to obtain only the most effective armour and weaponry they could lay their hands on.[172] Archaeological investigation of the Persepolis palace city yielded a number of arrowheads from the armoury halls and treasury rooms (but with only meagre evidence of swords and corselets), and the site shows signs of having been ransacked by Macedonian troops in 330 BC.[173] Access to new, even lavish, weapons from a royal armoury was welcome among weary soldiers who had been marching and fighting with little respite from the Aegean to Persia for some four years, their gear increasingly battered and worn out.

We may observe how at the Hyphasis river in 326 BC Alexander ordered that outsize panoplies be manufactured and left scattered on the ground as items of propaganda, suggesting that the army was accompanied on campaign by some form of mobile armoury with craftsmen serving the King and his immediate bodyguards or elite units.[174] Diodorus refers to an area in Alexander's camp quarters as the 'armoury', sometimes defined as small display cases or chests for different weapons, accessed by pages as part of their royal household duties.[175] Given that significant contingents, including the *pezhetairoi taxeis* and Agrianes, required a steady fund of throwing spears or javelins, it is equally valid that the expedition's armourers had occasion to serve the needs of the whole force.[176]

Apart from the above fragments of evidence, it is possible to *tentatively* trace the course of a lavish and consistent supply of war material from the eve of the Indian expedition until Alexander's death in 323 BC and after. Curtius notes (possibly drawing from the lost work of Cleitarchus) how, before India, Alexander was unsurpassed in wealth and 'not to be outdone in anything', so he 'added silver plates to the shields and golden [or gilded] bits on his horses, and adorned the cuirasses also, some with gold, others with silver'.[177] Justin offers an erroneously transcribed and distorted version of the same event in which 'to match his army's equipment to the glory of the enterprise, he had the harness trappings and the men's arms overlaid with silver, and he called the army 'Argyraspids', after their silver shields'.[178] Presumably the re-fit was paid for, arranged and organised by Harpalus acting as imperial treasurer in Babylon.[179]

Many of the new items rapidly wore out with the campaign waged beyond the Indus. During events in 326 BC Curtius puts into the mouth of the officer Coenus a statement including the complaint that 'already our weapons are dull; already our armour is giving out. ... How many of us have a cuirass? ... Victors over all, we lack everything ... [It] is in war that we have used up the equipment of war.'[180] Diodorus mentions a heavy monsoon lasting months, surely responsible for rusting the metallic components of equipment, similar to the atrocious conditions endured in Burma during World War Two.[181]

Alexander now proceeded to order a second refurbishment. As Diodorus explains, while in the Indus Valley the troops received suits of armour for 25,000 foot soldiers (probably delivered overland from the industrious workshops of Babylon and Persia) escorted by 7,000 infantry.[182] Curtius likewise describes how this specific consignment was to be 'for 25,000 men ... inlaid with gold and silver. These Alexander distributed and ordered the old ones to be burned'.[183]

Of course, the question arises as to what extent Diodorus and Curtius are preserving factual data, and the number 25,000 is certainly questionable as it is significantly in excess of the Macedonian regulars Alexander took with him to India. This makes the reliability of the whole episode suspect.[184] It is possible that the ultimate source used by these writers (perhaps Cleitarchus) had garbled his estimates, or that they were falsified through unreliable transmission.[185] Then again, given the army's hard experience with deteriorating equipage so far, the surplus may only ever have been intended as spare sets.

The number of troops given in the Diodorus excerpt implies that all Macedonian infantry units were to be refurbished with expensive new gear, whatever individual features it happened to contain. To give new pieces of kit to some and not to other units would be unthinkable among prickly proud Macedonian veterans and, more importantly, would only serve to divide the army, thereby undermining morale, when Alexander needed his forces to be united and efficient. We can appreciate the achievement of transporting such a bulk order from McCrindle's shaky calculation that if each (presumably complete) set of panoply weighed 60lbs, then it would have taken a total of 30,000 pack mules to have all supplies safely transferred from their bases in the West.[186]

There are a few other hints and glimpses in the ancient texts for this second mass issue of panoplies worn and used by soldiers throughout the duration of Indian operations. We are told by Curtius that 'for their new arms new enemies constantly appeared'.[187] Diodorus portrays the Companion officer Corrhagus clad in expensive armour, indicating that cavalry units were also lavishly appointed, as noted for the first refurbishment on the eve of the Indian campaign. He further praises the polished brilliance of his gear to the extent that Corrhagus was like a second Ares,[188] which is similar to the Thracian prince Rhesus from a play of the same name, rendered as an Ares lookalike in shining golden armour.[189]

The divine idiom given for Corrhagus is once more traced in other sources. The Sabarcae tribe are depicted gazing in amazement at oncoming Macedonian troops with their gleaming arms, crying out that they were mad to do battle with gods and that the ships carried all-but-invincible foes.[190] Similarly, during temporary anchorage at the Tamerus river, the troops of Nearchus swept through the shallows in the face of 6,000 locals struck with panic by the bright flash of armour and the speed of the attack.[191]

Passages like the above have been naturally viewed more as part of a literary device emphasising the shining appearance of armour in an attempt to impress upon readers exceptional expense and the heroic persona of its wearers.[192] It directly recalls a preoccupation with radiant panoply as witnessed in Homer's epics.[193] The Alexander historian extracts thus reveal more of the psychological impact the King and his Macedonians deliberately aimed or hoped to induce among their opponents when kitted out in such striking paraphernalia. The objective was that they appear like formidable Homeric warriors from Alexander's favourite book the *Iliad.*

The punishing return march across the parched wastes of the Makran saw the deterioration of the second consignment. Many items were lost or discarded through the severe privations endured. Sources refer to a flash flood, the blazing heat, lack of water and difficulty encountered marching through sand dunes while starved and exhausted. With hindsight it was thought by Diodorus that it 'seemed a dreadful thing that they who had excelled all in fighting ability and in the equipment of war should perish ingloriously from lack of food'.[194] After this disastrous ordeal Alexander was moved to arrange for a third delivery of gear, so that old, worn or lost pieces could be replaced with equally elegant sets.[195] It suggests that new gilded or silvered armour was brought overland from localities in Persia.

Alexander had every intention of expanding his distribution of panoplies. In accord with this, the first fully trained age class of 30,000 Asian troops were given expensive sets of armour by the time of the Opis mutiny.[196] It was also planned that the mixed-blood sons of 10,000 Macedonian veterans were to be brought up as soldiers and provided with arms.[197]

Athenaeus (using the lost history by Ephippus) records that when Alexander was at Ecbatana in 324 BC Gorgus, son of Theodotus and a leading citizen of Iasus (in Caria), flattered him by proclaiming that if the King should lay siege to Athens he would provide 10,000 suits of armour and the same number of catapults.[198] Gorgus is described as a *hoplophylax*, interpreted as a 'munitions custodian', 'guardian of arms' or an exceedingly wealthy, independent arms dealer. Two fragmentary epigrams found at Epidaurus help confirm that he held some formal post in the army pertaining to arms and armour.[199] Whatever the case, it suggests that Alexander's administration was more than capable of providing him with impressive quantities of military gear in the closing years of his reign.

In the final analysis, it emerges that Alexander fully exploited the former Achaemenid armaments production system to meet his own far-reaching aspirations. As many as 70,000 or more panoplies were supplied to all armed forces from the King's conquest of Persia to his death (329–323 BC). This estimate signals manufacture theoretically as extensive as that of Dionysius I of Syracuse – itself the largest found in extant texts documenting Greek civilisation before the rise of Macedon.[200] It may even have rivalled or surpassed the mass output of equipment seen with the aggressive kings of the Neo-Assyrian Empire.

Evidence for Lavish Equipment

Arrian contends that Alexander's abilities in arming troops constitute one of his most remarkable strengths as a general.[201] This chimes with a statement in the *Suda* describing the Macedonian conqueror as 'very skilled at marshalling and equipping an army'.[202] Still, the overall impression given by the army's elaborate trappings suggests a growing *hubris*, even megalomania, in Alexander's approach to conquest, witnessed with apparent discomfort by many of his officers and men.[203]

It is undeniable that Alexander had both the opportunity and historical precedence for kitting out his army so lavishly. Opportunity came in the form of finances. Up to the conquest of Persia and after, the King's accumulation of bullion was nothing short of immense. By 330 BC alone it may have stood as high as 200,000 talents, equivalent to over 5,000 tons of silver. This equates to two or nearly three centuries of revenue for the Athenian Empire at its apogee (in the fifth century BC). Now richer than Croesus, this massive fortune instantly made the Macedonian conqueror the most affluent ruler in the world for his time, and he was remembered for exercising generosity with the wealth accrued.[204]

Alexander spent a significant measure of his outlay on the army, which was, after all, the fulcrum to his continual power and success. Some of it would have been used not only to feed, pay and entertain troops, but to furnish them with stores of equipment. Bellinger, Milns and Aperghis all accept that the annual cost for upkeeping the army in Asia may have peaked at the enormous sum of 15,000 talents.[205] Nemeth estimates that the 25,000 new panoplies in India would have cost Alexander a mere 125 to 1,100 talents.[206] In spite of this approximation's validity, and even allowing room for error, it still insinuates that Alexander need only spend a tiny fraction of his financial resources if he desired an ambitious refurbishment programme.[207] His passion for pursuing efficient warfare, as carried on from his father, combined with a taste for heroic display and the bloated wealth amassed by 200 years of Persian Great Kings, spelt one predictable outcome: superbly appointed armed forces.[208]

Historical precedence is confirmed through archaeology. Abundant fragments excavated from Macedonian burial sites prove beyond any doubt that aristocratic warriors favoured gold strip, or appliqué, adornment on arms and armour, steeped in a Homeric tradition for showy gear extending over centuries. One of the most striking collections of finds to show this (incidentally demonstrating how widespread the practice was) exists in the form of a series of pit-graves at Archontiko near Pella. Scores of male burials (dated 550–500 BC) excavated from 2001 onward have brought to light bits of gold foil along with remains of helmets, swords and spears.[209] Comparable examples had already been found (1981/1982) in a rich male grave ('tomb 52') near Sindos, and similar finds continue to be retrieved from many other sites.[210] For instance, Archaic period graves, again at Archontiko, were discovered with five exhibiting purported remnants of silvered bronze shield.[211] Here we are reminded of the silvered shields described by literary sources issued (to the hypaspists) before the Indian campaign.

To a far from comprehensive list of archaeological discoveries down to Alexander's time in the fourth century BC can be added: Grave A at Katerini from which was salvaged sundry gilded silver discs, plaques and appliqués, used as fixtures on a long disintegrated organic corselet;[212] and Tombs A and B at Derveni with silver and gilded sheet, probably from a corselet and shield.[213] A strip of hammered gold decoration bearing repoussé guilloche (embossed patterns) was found from another corselet in the cemetery of Stavroupolis (Thessaloniki).[214] Likewise, an iron cuirass set off with gold attachments, alongside two iron Thraco-Attic helmets (one silvered over and with gold wire), were retrieved from a site at Prodromi, Thesprotia (Plate 20).[215]

The most famous and widely publicised of all the finds is, undoubtedly, the royal corselet from the main chamber of Royal Tomb II, with its gold lion heads, rings, bands and panels.[216] A tomb from Mount Ganos (Bulgaria), thought to belong to a Thracian of royal rank who had taken part in Alexander's conquest of the East, included gold rings and fixtures for a corselet, and fragments of bronze helmet enhanced with gold and silver coloured veneering.[217] The remains of a shield made up from iron sheets, with pieces of leather and gold, were apparently found in a grave in Lefkadia (c. 300 BC).[218] Similar findings will almost certainly accrete with time. An inscription (319–318/304–303 BC) actually mentions a pair of bronze silver-gilt greaves and *pelte epichrysos*, or shields plated or inlaid with gold – all these part of a larger dedication of panoply presented by the high ranking officer Alexander, son of Polyperchon.[219]

Archaeological finds can sometimes be manipulated to fit a preconceived construct. In this case, however, the physical evidence for opulent panoplies is corroborated by scattered mentions in the literary sources. Following Alexander's death, Plutarch portrays the troops of Eumenes of Cardia watching the advance of the army of Antigonus Monophthalmus (317 BC), pinpointing specifically 'the gleam of their golden armour in the sun [which] flashed down from the heights as they marched along in close formation'.[220] Ownership of high quality panoplies among Macedonian soldiers can be adduced from an anecdote preserved by Polyaenus which describes the general Polyperchon, within a decade of Alexander's death, contrasting the rustic garb of Arcadians with expensive Macedonian armour as a means of raising morale.[221] A brief account of Leonnatus recounts the exceptional quality of his armour or weapons.[222] In addition, Demetrius Poliorcetes is pictured flashing a gilded shield while directing his ships against Ptolemaic vessels at Salamis (306 BC).[223]

Further still, an excerpt preserved and quoted by Athenaeus from a lost work written in the third century BC (ascribed to Callixenus of Rhodes) has gold and silver shields placed alternately on the walls of a royal pavilion during celebrations surrounding the accession of Ptolemy II Philadelphus (283 BC).[224] A grand parade to mark the occasion included twenty gold shields and sixty-four sets of golden armour.[225] The latter detail is particularly relevant as it corresponds to an extant Hellenistic military discourse of the first century BC which describes sixty-four men as forming one *tetrarchia* within a Macedonian phalanx.[226] It is therefore feasible that the sixty-four gold inlaid panoplies with some gold embellished shields were for a small elite detachment of foot guards.

The biblical *Book of Maccabees* makes passing reference to gilded arms and armour deposited by Alexander in a temple at Elymais.[227] In the second century AD, the grammarian Julius Pollux listed *chrysaspides* ('gold shields') with *argyraspides* ('silver shields') and *pezhetairoi* as terms suitable for naming Macedonian infantry units, although questions have arisen about the authenticity of this passage.[228] Similarly, the Roman Emperor Severus Alexander allegedly imitated the Macedonian conqueror when re-arming his troops with either silvered or gilded shields during preparations for a campaign against Persia (231–233 AD).[229]

A later text of the *Alexander Romance*, when narrating fabulous events in Egypt, mentions Macedonians in golden body armour gleaming brightly on a sunny day – which brings to mind Plutarch's conceivably metaphorical passage quoted above for the army of Antigonus.[230] An extract survives in the sensational *Alexander's Letter to Aristotle about India* where Macedonian weapons were plated with gold, the army shining like stars.[231] Another version of the *Letter* even refers to a probably fictitious soldier called Zephyrus offering Alexander water from his gold decorated helmet.[232]

Should we dismiss all these late disparate extracts as having no basis in reality whatever? Descriptions of gold could derive from imagery for bronze or perhaps orichalcum. The latter is sometimes defined as an alloy or compound of copper and zinc, yellowish brass in colour, resembling gold when new.[233] Then again, unreliable as these passages are, some might just preserve the vague memory of there genuinely having been at one time precious metals used for enhancing panoplies worn or carried by many troops in the closing years of Alexander's reign.

Panoplies and Imperial Propaganda

It can be said that Alexander continued an already well-established elite Macedonian practice for owning expensively decorated arms and armour. Due to the huge financial and material outlay he inherited from the Persian Empire, this practice was continued on a far grander, indeed unprecedented, scale, and applied to his professional army as a whole or at least to significant contingents within it.

What could have driven Alexander to kit out his army so lavishly? The *Iliad*, lain at the heart of Macedonian aristocratic culture and with its animated accounts of resplendent panoplies, was one definite influence. So were the writings of Xenophon, who approved of soldiers being in their best clothing and armour when about to meet death on the field.[234]

The use of striking panoplies could affect the enemy psychologically before contact in battle was even made. As effused by Onasander: 'The polished spear-points and flashing swords, shining in thick array and reflecting the light of the sun, send ahead a terrible lightning flash of war. If the enemy should also do this, it is necessary to frighten them in turn, but if not, one should frighten them first.'[235] The advance into battle of a highly-disciplined close-formation Macedonian infantry column covered many hundreds of metres: the steady tramp of feet, rhythmic ring and rattle of metal, glistening pikes protruding from the serried ranks like the quills of an angry porcupine, all culminated in a stirring spectacle for onlookers. A display of this kind would undermine the already brittle nerves of opposition forces who knew that they were about to taste the dreadful onslaught of Alexander's soldiers.

Nonetheless, it is safe to say that the distribution of high quality panoplies extended beyond the basic pragmatic purpose of endowing troops with new, vitally needed equipment, and making use of its psychological effect, although this was the primary and immediate goal. A multiplicity of other objectives was intended: raising morale, further consolidating a strong and exclusive sense of corporate identity and solidarity, and even symbolising a new ideology and justification for world conquest.[236]

An illustration of the King's carefully calculated political and propaganda motives can be found in his provision of new gear for the troops after India. It would have helped boost the morale of sullen soldiers recovering from the tender mercies of the Makran and was in keeping with the image of a returning victorious conqueror accompanied by a world beating force.[237] In other words, it acted as an incontrovertible statement that, following years of absence from the main centres of power, Alexander had returned on the scene in command of an army that had lost none of its vigour despite the Makran disaster (with the consequent tarnishing of a reputation for invincibility).

Alexander effectively granted an elevated and privileged status within the 'Companionate' of the King to his Macedonian troops, to be shown in the quality of their armour, regardless of what their individual standing or origins happened to be (see more on Alexander's veterans' social prestige in Chapter 10). In his never ending quest for martial *arete*, he was determined that his army not only be tactically without peer but also excel in visual impact, as stressed in an exuberant (although probably embellished) speech which Curtius attributes to him.[238] The army would thus be transformed from a 'Macedonian' to an 'Alexandrian' institution, stamped with the conqueror's genius. Alexander was, in short, laying the foundations of a new imperial force, spearheading a new 'world order'.[239]

We can summarise the motives for the new panoplies as follows:

- to realise the 'Homeric' appetite Alexander and the Macedonians had for fine flamboyant arms and armour, in pursuit of Homer's heroic ideal. Alexander led an army whose appearance and deeds matched his own prestige, while at the same time recalling, competing or even surpassing the glory of the *Iliad*;

- to gain increased psychological advantage in battle by overawing the enemy with the mass formation flash of gleaming gold and silver, iron and bronze. In the same vein, the linking of Alexander's men to the gods and heroes of Homeric myth referred to above helped spread a belief among their would-be opponents that they were divinely favoured;
- to strengthen the bond between King and *Makedones*, thus increasing ties of loyalty and group unity, and renewing vital concepts of collective kinship (*syngenes*), renown (*kleos*) and reputation (*doxa*);
- to reward the soldiers for their services, honouring them with accoutrements equal to their new status as 'conquerors of Asia', thus reinforcing their military pride. This furthered their popular self-conviction as elite fighters and men who had justified superior quality panoplies through outstanding prowess proven on numerous occasions;
- to continue the opulent Persian custom of distributing lavish equipment serving as a statement to any potential Asian enemies that the new regime was the legitimate heir of the Achaemenids, supreme in wealth and power.

The reaction the soldiers had to their new panoplies, like that of Achilles in the *Iliad*, was doubtless a very positive one.[240] The troops of Peucestas and other eastern satraps that joined the army of Eumenes in 317 BC and were noted for the splendour of their accoutrements helped raise Macedonian morale.[241] Achaean soldiers of the late-third century BC reportedly wore and handled their new equipment with delight, were all the more eager to drill with it and get into action in battle where they could demonstrate fresh courage.[242] We can reasonably assume that some of Alexander's soldiers who wanted to stand out could have personalised their equipment, vying with one another in affixing their own silver or gold loot to armour and weapons, thereby making it unique and even more decorative.[243] Rich personalisation of officers' panoplies is discussed in detail in Chapter 8.

The style of lavish gear first established under Alexander spawned many imitators among the armies of succeeding Macedonian and Hellenistic kings, especially under those rulers who wished, in some way, to emulate his achievements or draw favourable comparisons with him. This includes accounts of units in Ptolemaic (283 BC), Antigonid (168 BC) and Seleucid (c. 200, 164 and 162 BC) service, and the reformed army of the Achaean League under Philopoemen (c. 209 BC).[244] As late as the battle of Chaeronea in 86 BC, the phalanx of the ambitious Pontic king Mithridates VI is described wearing armour finished with gold and silver. As Plutarch says: 'The conspicuousness and splendour of their equipment was not without its effect and use in exacting terror ... the shining brilliance of their arms ... created a frightening play of fire when the army moved and separated, so that the Romans cleaved to their palisade'.[245]

The trend for expensive armour may not have perished with Mithridates. Roman generals like Julius Caesar were ultimately inspired by Alexander when encouraging the use of silver and gold ornamented equipment among their own troops.[246]

Chapter 5

Cavalry Horses

Fine panoplies helped win battles, but without proper quality mounts and effective methods by which they could be procured, the chance of gaining victory would be all but nullified. Philip and Alexander's superb cavalry arm, the *Hetairoi* ('Companions'), were primarily utilised for shock in a well-timed and coordinated tactical assault. Superior horses were hence a major factor when considering the battlefield success of mounted Macedonian troops.

Pasturage and Stables

As with raw materials for producing arms and armour, the kingdom of Macedonia was also favourably endowed with an excellent climate and terrain for breeding a large surplus of cavalry horses. The geography combined alluvial plains with abundant rainfall and mountain ranges providing well-watered alpine summer pastures on upper slopes and basins. Winter pastures are prevalent in Amphaxitis beside the river Axius in the north, and the swampy coastal lowlands west of the Axius including Emathia, Eordaea and Bottiaea. So ideal is the region generally that horses abandoned in the late stages of the German occupation of Greece in the Second World War were able to survive in the wild to the present day beside the Axius and Haliacmon rivers.[1]

Literary evidence suggests that great numbers of mounts were being bred from the dawn of Macedonian history. Hesiod (c. 700 BC) describes how Thyia, daughter of Deucalion, bore Zeus two sons, one being Macedon 'rejoicing in horses'.[2] More specifically, in the *Bacchae* (first performed in 405 BC), Euripides conveys the picturesque canton of Pieria as having many fine steeds.[3] It is tempting to see in this a concentration of studs and state herds during Philip and Alexander's time in Pieria's productive grazing lands and valleys, with areas irrigated by perennial rivers south of Pydna and inland from Dion, including Elimea.[4]

Between the Axius and Strymon rivers were first-rate horse pasturages in Mygdonia and the lands of the Thracian Bisaltae, and, further east, in marshy landscape within the Strymon delta and territory of Amphipolis. The Bisaltic region bred a particular strain of horse later praised for its hardihood and surefootedness in rugged terrain. As enthused by Grattius: 'So too maintenance is easy for horses of the Bisaltae near the Strymon: oh! that they could career along the highlands of Aetna, a sport the Sicilians make their own.'[5]

The kingdom's extensive arable lowlands were capable of producing significant quantities of grain – the staple often required for the breeding of larger, bigger boned horses.[6] State intervention ensured that land was better suited for horses.

Hence the draining of marshland around Philippi in the 340s BC, although ostensibly undertaken for agriculture,[7] was done with the goal of pasturage in mind.[8] Corrigan cites Alexander's reorganisation of Kalindoia as a Macedonian city (in the winter of 335–334 BC) as relevant here because the place-name 'Kalindoia' derives from the Greek verb 'to wallow', identified with the rolling of a horse. Tripoatis, within Kalindoian territory, has been defined as 'triple pasture'. These names therefore refer to the excellence of the area about Lake Bolbe for horse breeding, and Alexander's expected motivation for the reorganisation was to give pasture for cavalry mounts.[9]

Macedonian kings operated a series of stud-farms staffed by horse-breeders and trainers (*hippotrophoi*). The office of horse-keeper to the king was itself as old as the foundation myths of the royal house, and under Philip and Alexander there was probably an official appointed with overall responsibility for the king's herds.[10] Fine chargers from royal stables were identified with a distinct brand which, from coins, appears to have taken the form of a Hermes staff (*kerykeion*) marked horizontally on the rump or shoulder. Horses so branded were not only the sole preserve of royalty but were supplied as mounts to Macedonian cavalry from before Philip II's reign, or could be sold as expensive exports to foreign buyers.[11]

Many horses were entered into prestigious racing events. Among royal competitors was King Archelaus, who won a four-horse chariot race in the 93rd Olympiad (408 BC). Philip II was not only personal winner of the horse race in the 106th Olympiad (356 BC) but entered a team which won another chariot race at the 107th Olympiad (352 BC) and the two-colt chariot race at the 108th Olympiad in 348 BC.[12] Later still, there is mention of a chariot competition being won by a Macedonian horse breeder called Lampus. Writers such as Posidippus of Pella consistently record male and female Macedonian winners for horse and chariot events throughout the late-fourth and third centuries BC.[13]

Large estates were awarded by the King to his Companions as the kingdom expanded, and these incorporated unknown numbers of privately owned stables and horse herds. Theopompus notes how during the reign of Philip II, but prior to 340 BC, 800 Companions 'enjoyed the fruits of a land greater than 10,000 Greeks possessed in the best and wealthiest districts of Greece'.[14]

If we accept the weight of some informed academic opinion, the main site for the Macedonian royal stables and exercise grounds was located in Pella and its environs. The open plains around the city offer exemplary surroundings for horse rearing. The Greek version of the *Alexander Romance* mentions royal stables at Pella and this is confirmed as correct through archaeological discoveries, with apparent stables unearthed to the west of Building III in the palace complex.[15] It is possible that other, as yet unrecognised, stables exist amid traces of buildings excavated on the eastern and western parts of the palace hill.[16]

The fact that model royal capitals could be viewed by the mid-fourth century BC as suitable places for major stables is articulated by Plato in his famous account of the great city-palace infrastructure of Atlantis, which had places set aside for equestrian exercise.[17] Other possible subsidiary localities in Macedonia were at or within the vicinities of Amphipolis, Pydna and Dion, to name but a few. During the third war with Rome in 168 BC, Livy describes Antenor, an officer of Perseus, ordering twenty fine Galatian horses to be sent to Thessaloniki, pointing to the existence of royal stables at or near this city.[18]

There is credible evidence for Macedonian cavalry stables lying well outside the kingdom in Hellenistic times. Antigonus Gonatas ran one such establishment at Sicyon in the Peloponnese, which was an ideal centre for cavalry horses.[19] The city was sited on a plain, 3km from the Corinthian Gulf and 18km west of Corinth. On its seaward side the settlement overlooked fertile flats, the climate tempered with cool breezes offering a valuable location for horse pasturage.[20] The stables operated as a means of providing support to the nearby Macedonian garrison on the vital fortress-outcrop of the Acrocorinthus, granting *de facto* control over Corinth, one of the three strategic 'Fetters of Greece'.[21] The military stables at Sicyon were therefore part of an overall scheme for developing a more tangible presence among protectorate states in the Peloponnese, a region which was an active arena for Macedonian army operations in the third century BC. The stables may have existed before then, as Polyperchon's daughter-in-law Cratesipolis held the town for him between 314 and 308 BC and kept soldiers there. Besides, a Ptolemaic garrison is recorded at Sicyon in 303 BC, and the Achaean League had stables and exercise yards just outside Corinth at the time of Philopoemen.[22]

The seizure of Sicyon by Aratus in 251 BC yielded 500 horses and 400 Syrian captives.[23] These numbers could represent the full complement of horses kept at the Antigonid stables, and mention of 'Syrians' suggests a contingent of mercenaries.[24] It might also refer to grooms hired by the Antigonid kings from Seleucid Syria. If this was the case, then by implication the horses themselves included Seleucid imports. The Seleucids appear to have exported to other kingdoms in the Hellenistic world; for instance, there is evidence to show that the Ptolemies imported horses from their Seleucid neighbours.[25] Strabo recounts how under Seleucus I the city of Apamea in Syria was the centre for the royal stud, which contained more than 30,000 mares and 300 stallions, including colt-breakers.[26] Polybius tells us that the Seleucids kept the satrapy of Media as a major source for numerous cavalry studs, while hoof emblems from numismatic evidence suggest Ecbatana as a chief centre and collections point.[27]

Horse Breeding

Through often successful military campaigns and the large-scale levy of plunder, expanding diplomatic relations, and international commerce, Philip II had the opportunity to attract a wide circle of reputable horse dealers to his court. He strove to increase the importation of foreign strains, building and improving indigenous blood stocks through cross-breeding with southern Greek, Thessalian, Epirote, Illyrian and Scythian horses. Diodorus makes passing reference to a specific type of pedigree Thracian steed, implying that it was integrated into Macedonian horse numbers during the fourth century BC. He explains how in one of the legendary 'Twelve Labours' of Heracles the famously strong and wild horses of Diomedes of Thrace were overcome by the hero. As Diodorus informs us: 'their breed continued down to the reign of Alexander of Macedon'.[28] A lingering descent from breeds used by Macedonian cavalrymen may exist to this day in the guise of the Skyros, Pindos (Fig. 9) and Bosnian ponies.[29]

Late-fourth century BC sources, such as paintings of horses from Phinikas and a bed decoration from the Soteriades Tomb in Dion, 320–250 BC (Fig. 10), show that these mounts were sturdily built. A horse fragment found at Lanuvium, Italy, reveals a strong thick neck (Fig. 11; this fragment may have been part of a larger Roman copy of a statuary group featuring Macedonian cavalry from Alexander's lifetime). A series of silver tetradrachms from Philip's reign presents the King on the reverse mounted on a horse with a large, fine head and well muscled anatomy (Fig. 12). This is in contrast to the slighter examples from Attic vases and sculptures of the fifth century BC, the difference being particularly pronounced in the arching neck, fore and hindquarters. The intention was thus to produce larger, thicker boned horses. These factors were useful to Philip's newly developed battle system, as bigger mounts helped in the application of more weight and force behind the sword or lance of a rider at full gallop, thus increasing the effective impact of a formation charge.[30]

Figure 9. Pindos pony, one of a rare indigenous breed found in Epirus and Thessaly. Stands 13 hands high at the withers (about 130cm).

Figure 10 (*left*). Detail of a lost painting of mounted Persians and Macedonians from a marble *kline*, Soteriades Tomb (Dion).

Figure 11 (*right*). Marble fragment of a horse from a larger monument found at Lanuvium, Italy.

Figure 12. Philip II riding a horse. From the reverse of a silver tetradrachm, 359–354 BC.

It is palpable that Macedonia's long established background in producing cavalry horses of good quality could entail crossbreeding skills being developed there well before Philip II's rule. There are glimpses of this in numismatic evidence. Earliest is from the reign of Alexander I, with a series of octodrachms (475–454 BC), showing a horse on the obverse, which displays features possibly crossed with those of the Persian-Nisaean type. The animals represented on these coins could correspond to some captured by near neighbours of Macedonia during the Persian invasion of Greece (480 BC). Herodotus describes how seven white Nisaean mares were stolen at this time and pastured by 'the Thracians of the hills who dwelt about the headwaters of the Strymon', at or near to the Bisaltae.[31] Since the Persians occupied Thrace for another thirty years, over the next century or so the Nisaeans could potentially have had a considerable genetic input on regional breeds, helping to improve bloodlines.

Ongoing skills in horse breeding for the royal stables are suggested with a preponderance of equestrian images captured on the reverse of extant silver coins (staters, etc.) from the reigns of Perdiccas II, Archelaus, Amyntas III and Perdiccas III.[32] The fact that 400 horsemen provided by Derdas of Elimea defeated a 600-strong force of Olynthian cavalry in 382 BC points in part to a superior breed of horse having seen widespread use by this time.[33]

A vigorous policy of military expansion and the taking control of further excellent, well-watered grazing lands in the east under Philip II, coupled with an interest in land reclamation, created the opportunity to organise ever greater, more ambitious horse breeding programmes.

The 'Spear Won' Scythian Herd

A clear indicator of Philip II's drive to add to domestic reserves, as well as the size and ambition of his breeding project, is seen in events concerning the aftermath of wars waged in the north. In this case, to round off or complete his campaigns in Thrace, Philip launched a pre-emptive strike against King Ateas of Scythia (339 BC).[34]

After inflicting a serious defeat upon the Scythians in the modern region of Dobrudja (near to the Black Sea coast), the victorious Macedonian army retired westward along the line of the river Danube. According to Justin, among the spoil were '20,000 young men and women ... and a vast number of cattle, but no gold or silver. This was the first proof they had of the poverty of Scythia.'[35] Even so, the disappointments were offset by the requisition of some 20,000 fine mares which were to be herded back to Macedonia for breeding purposes.

Philip's plans were, however, dashed – Justin acknowledges how during the return march 'the Triballi met him [in the Danubian plain], and refused to allow him a passage, unless they received a share of the spoil. Hence arose a dispute, and afterwards a battle ... the booty was lost. Thus the Scythian spoil, as if attended with a curse, had almost proved fatal to the Macedonians'.[36] It is more accurate to accept that Philip was able to retain at least a portion of the animals.

Several illuminating threads are deduced from the above passage from Justin. One is the very question of the horses as mares. Hammond discusses this subject at length and suggests that, since it is usual practice to improve breeds of horses through stallions, Philip's otherwise perplexing importation of female horses must have been intended to replace the existing home-bred Macedonian and Thessalian stocks altogether.[37] Hammond raises a prescient point, overlooked by other historians. Nonetheless, he appears to be unacquainted with the following extract from Pliny, which helps place what Justin has to say in proper focus: 'The Scythians prefer mares as chargers, because they can urinate without checking their gallop.'[38] The 20,000 horses obtained by Philip after the otherwise little known victory over Ateas were mainly either seized from dead or captured Scythian horse-warriors or plundered from their camp. That they were mainly female was not due to any intrinsic plan or calculation on Philip's part – it was just that most of the horses the defeated Scythians took with them, or already happened to be mounted on, were mares, which they traditionally preferred as war-chargers over stallions.

Philip's main intention was apparently to have the mares incorporated as the basis for a long-term breeding programme in Macedonia spanning several decades. Another inference taken from the Scythian episode is that the King ordered a roundup of good quality horses as routine policy when on campaign outside Macedonia. This undoubtedly meant a regular inflow of new blood in the form of horses taken over twenty years from Thracian, Illyrian, Epirote and Greek sources further south, principally accessible as spoil after a profitable war.[39]

Of the ancient writers, Pliny describes mares being generally ready for mating from two to three years, continuing to foal yearly up until forty years old.[40] According to modern horse owners and managers, a mare can begin breeding from the age of four, with anything from three to ten foals produced in the course of a lifetime. Given these statistics, through an intensive programme the captured 20,000 Scythian mares could have given birth to a potential 100,000 foals over a ten year or more period. Even allowing for the inevitable losses sustained through natural attrition, with a minority of animals succumbing at birth or before reaching sexual maturation, the figures still demonstrate that a large potential pool of horses were reserved for Macedonian cavalry units. This is particularly true when one remembers that the Scythian intake was representative of only one – albeit a very significant or integral – component of the overall number of mares destined for Macedonia. It is yet again a reminder of the size of the King's horse breeding objectives.

Some Scythian breeds were unusually valuable – a fact reflected in surviving literature. Justin refers to the horses lifted by Philip as *nobilis*, defined in lexicons as noble, excellent or superior, and the same term reappears with the Roman poet Ovid in his description of race horses. Oppian lists the 'Scythian' as one of a range of breeds ideally suited to the hunt.[41] They are also singled out for their powers of endurance when used for stag hunting, and described as swift moving and of an imposing height.[42]

Given the positive characteristics attributed to Scythian horses, it seems that Philip, by setting out to acquire a substantial number of them, was purposefully attempting to develop a superior strain of cavalry mount before and during his projected conquest of Achaemenid territories in the East. Polybius refers to the confidence the King had in his military preparations and the efficiency of his army.[43] Persian horses were certainly renowned for their high quality so Philip aimed to produce a steed equal to that of his future opponents, and which would be capable of tolerating long distance rides in the open plains of Asia Minor and beyond. Considering Alexander's later breakneck pace advances and pursuits, some of which would have required several days of exertion, Macedonian cavalry horses were by this time almost certainly bred and trained first and foremost for control, endurance and agility over brute strength.

There is evidence to show that Alexander was interested in identifying superior strains of domestic animal and, wherever practicable, had them transported to Macedonia in order to improve indigenous stocks and upgrade the home economy. On entering India, he is documented having commandeered 230,000 head of cattle, though these numbers may be conflated or even fabricated. In Arrian's words, he proceeded to 'select the finest ... because he thought them of unusual beauty and size and wished to send them into Macedonia to work the soil'.[44]

Such activities are remarkably similar to those of Philip in the Dobrudja region and imply that, like his father, Alexander had occasion to send back horses as part of an ongoing long-term plan to improve the native breeds. The Macedonian achievement of developing a superior cavalry horse across southeast Europe and the Middle East through the filter of the Hellenistic kingdoms may even have helped determine the types of mount selected ultimately by the Roman army.[45]

Horse Casualty Rates under Alexander, 334–323 BC

The amount of dead or incapacitated horses suffered during the course of the eastern campaigns must have been prodigious. A rare, though generalised, glimpse can be viewed through the mainly convoluted *Itinerarium Alexandri*. In Sogdiana Alexander replenished the number of his horses because many were lame or otherwise unable to be ridden.[46] Although only a rhetorical statement, at the Hyphasis river Coenus exaggerates the plight of soldiers and laments that nobody has a horse.[47] This implies that many of these animals had died or been injured and that difficulties had been encountered when procuring replacements.

Casualties were sustained through battle and on innumerable small scale operations. For example, it is recorded that over 1,000 horses were lost in the pursuit at Gaugamela from injuries and distress, half belonging to the Companion cavalry.[48] Exhaustion, dehydration, heat stroke and breathing in dust must have put many mounts out of action. Mysterious diseases were irksome, not forgetting the hazards involved in feeding horses exotic, unfamiliar foodstuffs. Want of fodder, including lameness and the wearing down of hooves through overexertion on long distance rides in remote and hostile regions, added to the percentage of disabled animals.[49] On one occasion, horses are recorded being lost in a sudden flood. They could also be slaughtered for meat by soldiers when rations became scarce during extreme episodes of hardship.[50]

Nevertheless, the very fact that Alexander regularly and often deliberately drove his cavalry horses so hard suggests that he had the reassurance of knowing that the army kept a fair-sized reserve of spare mounts as standby. The system of obtaining them was well-organised, adaptable and effective enough for adequate numbers to be quickly gathered in when required.[51]

Methods of Acquisition in Asia

With high casualty rates to be expected for a major ongoing campaign of military aggression, Alexander (and his Successors) followed a practical approach in Asia by procuring new and surplus mounts, using every suitable option or expedient. The army administration was flexible enough to assimilate and account for horse stocks arriving from a diverse range of sources, then being organised and distributed to the soldiery. The chief basis for obtaining horses for the remount corral took the form of working pragmatically within the pre-existing Achaemenid model, thereby preserving sensible, socially well-established infrastructures intact.

Acquisition in Asia took many forms: through tribute-tax, stud farms and royal herds, direct purchase, field booty, diplomatic gifts, and – as a last resort – sequestration from within the army.

Although it is rare for anything of direct value to surface in textual sources, there is evidence to show that Alexander was keen to continue levying an annual tribute of horses from subject cities and peoples.[52] Thus, in 333 BC a delegation from Aspendus was instructed to have their city immediately dispatch fifty talents to help pay the army and to surrender the usual mounts they bred as tribute for the Persian king.[53] In other words, the levy was to be paid as a wartime 'contribution', after which the city would be obligated to render regular tribute. The same general policy is repeated further east with the Uxii, who were ordered to pay the same annual tribute as they had under Achaemenid authority, despite the fact that, as Arrian asserts, figures were this time fixed at '100 horses every year with 500 transport animals and 30,000 sheep. For the Uxians had neither money nor arable land; but they were for the most part herdsmen.'[54]

Most Achaemenid satraps and Persians of high military rank owned private studs which could be used to furnish units of cavalry. Alexander certainly gained control of such establishments during his advance. The triumphal entry into Babylon in autumn 331 BC brought with it one of the largest provincial stud farms west of Persia. Tritantaechmes, the satrap of Babylonia, is recorded in former times keeping a huge reserve in which, besides war chargers, there were some 800 stallions and 16,000 brood mares.[55]

An even more invaluable resource existed in the form of the great royal reserves of *Nesaioi*, or Nisaean horses, located in the plains of 'Nisaya', convincingly identified as breeding grounds over 300km southwest of the Caspian Gates and about 140km south of Ecbatana (modern Hamadan), in the Vale of Borigerd.[56] Under the Achaemenid kings, there had been as many as 160,000 Nisaeans, of whom 50,000 or 150,000 were brood mares. By the time Alexander had returned from India, a survey was ordered and numbers found to have dwindled to only 50,000 or 60,000, as most had been rustled by gangs of robbers.[57]

The former Achaemenid system also allowed for the running of much smaller royal herds dotted across the empire. Some of these were still in existence in the period following Alexander's death, as late as 320 BC. Plutarch mentions off-handedly that Eumenes requisitioned horses from one such herd located at Mount Ida (a fertile, upland district in the Troad), taking 'as many horses as he wanted' and sending 'a written statement of the numbers to the overseers [*episkopoi*]'.[58]

The above incident involving Eumenes reveals how royal herds were administered by officials, who kept a regular check on the numbers of animals dispatched for military service.[59] The fact that Eumenes took the trouble to deliver a written slip to the overseers of the horses procured suggests that under normal circumstances officials responsible for herds held records over how many could be supplied at any given time. The whole episode bears a 'natural freshness of truth' about it. That Eumenes quibbled over bureaucratic niceties (given his own secretarial background) explains why Antipater reportedly laughed when he got wind of the event.[60]

A few accounts and receipts lifted out of day-books are preserved as piecemeal papyrological scraps from Ptolemaic Egypt. These hint at an administrative apparatus in place at least for the maintenance of later Hellenistic horse herds and include details of corn rations issued to grooms, along with other expenses (c. 250 BC).[61] A surviving receipt from the horse keeper Lysis from 152 BC acknowledges a delivery of hay for the royal mounts in his charge at Crocodilopolis, within the Arsinoite district.[62]

In 322 BC Eumenes appears to have purchased large numbers of suitable mounts from independent dealers. In this case, he was admired for recruiting and training men, and then distributing horses for a new effective force of Cappadocian cavalry.[63] This is similar to the activities of Seleucus, who while on campaign at Babylon in 312 BC bought up horses from local sellers and then had them distributed to handlers,[64] which meant that each animal was first sent for training to a professional horse breaker, before being assigned to riders.[65]

Many communities hastened to send diplomatic gifts to Alexander as his army passed through their respective territories. Some of these would have taken the form of excellent quality horses – above all, from peoples and satrapies famed for their fine breeds. It may even be possible that Alexander occasionally sent ahead representatives notifying local authorities that he specifically desired horses to be prepared and sent to him. Although most records for these have vanished, there still remain a few disparate references.

Diodorus tells us that, while Alexander was in Egypt, envoys from Cyrene brought a variety of gifts, including 300 chargers.[66] The Cyrenean breed was thought by some ancient writers to be the largest in Libya or North Africa, favoured for long distance riding and possessing all the good points of Greek horses without being hampered by their defects.[67] It has been considered that, after crossing the Euphrates, Alexander deliberately followed a more northerly route in order to facilitate the collection of Armenian horses.[68] Strabo relates how, in his own time (first century BC), Nisaeans were bred there and how, earlier on, the satrap of Armenia 'used to send to the Persian king 20,000 foals every year'.[69] Likewise, the victorious entry into Babylon saw Alexander followed by gifts in the form of herds of horses and cattle.[70] During the return from India, he even received some 1,030 chariots each drawn by four horses as a royal present – suggesting a stock of 4,120 if ever his cavalrymen required fresh mounts in an emergency.[71]

Significant amounts of horses must have fallen into Macedonian hands following victory at the Granicus, and even more so after the battles of Issus and Gaugamela. Curtius mentions gifts given by Alexander to King Ambhi of Taxila comprising 'thirty of his own horses with the same trappings to which they were accustomed when he himself rode them'.[72] This excerpt obviously implies a surplus of mounts available for at least the supreme commander's use at the commencement of the Indian campaign.

Roundup operations could sometimes be organised from reputable horse rearing localities in the army's immediate reach. In 330 BC, following the tremendous exertions of the racing pursuit of Darius which had seen horse casualties sustained through heat exhaustion and inappropriate fodder, a raid was launched against the Mardi, with the intention of lifting as many Mardian horses as possible.[73] It is further briefly mentioned how 1,800 Scythian horses were driven off as booty (329/328 BC).[74] Similar methods are recorded as a gloss when Arrian explains how 'Alexander brought his cavalry to full strength with horses from the vicinity for a good many horses had been lost in the crossing of Mount Caucasus [the Hindu Kush via the Khawak Pass] and on the marches both to and from the Oxus'.[75] Alexander's punitive expedition against the Cossaean mountaineers (southwest of Ecbatana), prosecuted over five weeks in late 324/early 323 BC, was partly undertaken to put a stop to the widespread depredation of nearby Nisaean horses. Alexander intended to recover as many as possible of those that had already been rustled.[76]

On being discharged at Ecbatana (330 BC) but before embarking for the coast, Thessalian cavalrymen first sold their horses – presumably to troopers from standing units, most likely the Companions.[77] However, given that Ecbatana was a major horse collections depot under the Achaemenids and, indeed, later during the Seleucid Empire, it is equally valid that the Thessalians sold their mounts to the administration that Alexander had simply taken over. In any case, having more horses was a welcome development. The situation also affirms that Thessalian mounts were not originally government supplied.

When all other means for obtaining fresh mounts had been exhausted, the army resorted to redistribution of horses within ranks and units under the jurisdiction of Antiphanes, described as the *scriba equitum* (secretary of horses), doubtless supported by a team of undersecretaries. It was considered usual to have mounts sequestered from men with larger than average establishments and redistributed to troopers who expected a replacement (or reimbursement in cash) when their own mounts had died or been otherwise rendered unfit.[78]

Individuals with particularly large numbers of surplus mounts could see them drastically denuded if the demand from other men was high. Amyntas complained that from his original stock of ten, Antiphanes earmarked eight for redistribution to those without any, leaving him with only two.[79] This extract from Curtius concedes that having two mounts was generally considered to be too few and confirms that up to that time it was usual for the cavalry to have more than two horses per rider, perhaps from the large pool captured from defeated Achaemenid troops. The secretary and his clerks had to consult some form of ledger to identify the current horse numbers of each cavalryman and verify which of them could afford to give some up to the central authority. It hints at the substantial paperwork produced by Alexander's army administrators, now almost entirely lost.[80]

Horse Management Procedures

The long standing Macedonian tradition of expertise in horsemanship (which predated the reign of Philip II) was combined with the central importance that cavalry played as an offensive arm in Philip's revolutionary battle system. This, added to the continual efforts the King took to secure better horses to build up an impressive state breeding programme, and the later various means of acquiring suitable chargers in Asia, meant that horses were managed with care and efficiency for the army. Unfortunately, only a cursory amount of evidence has survived to show how this was pursued.

Epigraphic material pertaining to cavalry horses in Antigonid Macedon has been discovered, although poorly preserved. No such comparable evidence exists for Alexander's time, but literary texts do point to records being maintained for horses that had died or were incapacitated. These texts include statistics on incoming horses, showing how and to whom they were assigned, and the numbers of surplus mounts available. A further insight into how thorough and methodical the management process was is offered by fascinating epigraphic material uncovered for late Classical and Hellenistic Athens – historically *far less* of a horse rearing state than Macedon ever was.

In 1965 some 570 or more third-century BC inscribed lead tablets were recovered from a well in what had been the ancient Athenian Kerameikos.[81] This was followed in 1971 by a similar finding of 111 tablets dumped in a well from the Agora: one lot of 26 dating from c. 350 BC, and the remaining 85 from the period 250–225 BC.[82] Each tablet sets out precise notation for cavalry administrative purposes recording the horse-owner's name on the outside. The inside face details horse colour, description of the horse-brand (or 'unmarked' where there was no brand) and an annual monetary 'evaluation' in hundreds of drachmas. A total of twenty-five separate and distinct brand-marks have been collated from the Agora tablets alone, signifying that horses could originate from numerous commercial studs and stables across the Greek world, sold directly or through intermediary dealers.[83] The tablets refer to evaluations and inspections of horses, rather than tests of physical fitness for cavalrymen.[84] In Athens, horse inspections were of two types: checking to see if each had been properly fed and maintained, and assessing performance (comprising several reviews per year) to ensure that each animal was well-trained, obedient and fast.[85]

In the case of the Athenian lead tablets, the evaluations are best judged as monetary appraisals for each horse. The values are estimated from a minimum of 300 drachmas, rising in regular 100 (or occasionally 50) drachma increments to a maximum of 1,200 drachmas (12 minas).[86] Such evaluation provided a means of insuring against the loss of a horse, if it was killed or seriously injured on operations. By annually recording the horse's 'current actual worth' a basis was established for a fair claim and compensation. In this case, the 12-mina level represented the limit of state liability.[87] Sekunda interprets a vague extract from Thucydides as suggesting compensation for horses lost in battle, installed in Macedonia from as early as the reign of Archelaus.[88]

Processes organised similarly to those in Athens were actively followed elsewhere in the Hellenistic world. For example, in Ptolemaic Egypt there is reference to inspectors of horses in the third century BC. Horses supplied from royal establishments were distributed among officers and men who had no mounts of their own.[89] Hellenistic epigraphic sources indicate royal donations to allied or satellite cities in the form of horses for cavalry which patrolled their frontiers. We must presume from this that many horses were likewise supplied to royal armies. The aforementioned Antigonid military regulation, preserved from Cassandreia, describes secretaries involved in documenting which cavalry mounts were in a good or poor condition for service. Horses which passed inspection were to be branded with the Macedonian Hermes staff motif as a mark of approval. Obscure mention is also given to the sum of 1,000 drachmas per horse.[90]

Mounts Favoured by Companion Cavalry, c. 323 BC

It is almost certain that Alexander had issued Nisaean thoroughbreds to his Companion cavalry, or to many individual Companions, before his death in 323 BC. The Nisaean was itself highly praised and celebrated and, as stated earlier, was reared in plains near Hamadan – a region exceptionally good for horse rearing due to its cool climate and plentiful pastures of protein and calcium-rich *Medike poa*, or 'Median grass' (alfalfa or lucerne). Horses bred and grazed there were arguably the best ever produced in the ancient Middle East. As Hyland opines, the mixing of diverse strains coalescing from ancient equestrian cultures in Media, Urartu and Mannea, with centuries of experience in selecting and improving stock, meant that the region was able to produce the finest horses by the time Alexander arrived on the scene to exploit the situation.[91]

Possibly akin to the modern Arabian horse or Akhal-Teke, the Nisaean was of better quality and more reliable than the Thessalian breed – being esteemed for its speed, strength and aesthetic beauty, with a temperament which made it obedient to the bit.[92] It is further described as having a ram-shaped head, heavy shoulders and solid quarters, valued more for its weight and muscular bulk than its height. Coat colours were usually white, grey, black, dun or red-blonde, with a characteristically flaxen-yellow mane and tail.[93]

New breeds of large, strong horses were gradually introduced to Macedonian ranks, as the expedition rolled eastward. From as early as 329 BC an officer called Menedemus is portrayed by Curtius riding an unusually powerful animal with which he was able to careen at full speed into close formations of enemy troops.[94] Bactria was particularly noted as a place abounding in horses, and Alexander exploited its rich reserves, with the town of Bactra-Zariaspa (Zariaspa possibly meaning 'having golden horses') as a roundup point.[95] On leaving behind the ordeal of the Makran Desert, it is further related how the King had fresh horses, representative of east Iranian breeds, transferred to him from the satraps of Carmania, Persia and neighbouring areas.[96]

Following multiple defeats of Persian forces in battle, many Companions took ownership of captured Nisaeans as war booty. The very fact that Alexander ordered a complete survey of Nisaean numbers on his return from India[97] helps demonstrate that he was making preliminary arrangements to remount his cavalrymen with Nisaeans on a more formal basis, even if they did not necessarily have a 'specific combat advantage' over closely related or neighbouring breeds.[98] He even appears to have been planning to organise army units from the mixed-race sons of his Macedonian veterans, the cavalry component equipped with arms and horses provided at state expense.[99]

Figure 13. Rock-cut relief of Alcetas riding a horse, possibly of the Nisaean breed.

Individuals continue to be referred to with expensive mounts at this time. Anecdotally, Pyrrhus is said to have had a dream featuring Alexander mounting a Nisaean before galloping into battle.[100] This reflects that he was popularly expected to have ridden Nisaean thoroughbreds. The Indian ascetic Calanus was transported to his funeral on a Nisaean provided by Alexander, the animal given away, in turn, to one of his favourite students, the officer Lysimachus.[101]

Evidence for a continual use of robust war horses during Alexander's period in Asia is well attested in art. The tomb of the officer Alcetas (d. 319 BC), the brother of Perdiccas, located at Termessus in Pisidia (Turkey) depicts a particularly sturdy looking animal with strong neck and powerful forequarters, hindquarters and rump (Fig. 13).[102] It closely resembles three other thickset, deep-chested horses painted into the left side of the Agios Athanasios tomb frieze (Plate 10a). The same or a related type is ridden by all the Macedonian and Persian figures on the Alexander Sarcophagus and these are actually described by one scholar as Nisaeans.[103]

There was a widespread use of Iranian breeds by Macedonians for years after Alexander. Leonnatus is said to have received Nisaeans (or mounts from Phasis) for himself and his bodyguard in 322 BC.[104] It shows access to royal herds for the use of personal Companions within only eighteen months of Alexander's death. Furthermore, mounts requisitioned by Eumenes in 320 BC came from former Achaemenid herds and were thoroughbreds of presumably Median-Nisaean stock or breeds related to them.[105] A few years later, Antigonus Monophthalmus distributed 1,000 fully caparisoned Median horses to his cavalry, showing that entire units could be furnished with such mounts by the closing decades of the fourth century BC.[106]

PART III

DRESS AND PANOPLIES

The type and design of arms and armour and their tactical combination played a crucial role in determining the individual and collective performance of soldiers on the battlefields of the premodern world. The *xyston* and *sarissa* are the two most well-known examples of Macedonian arms. These have received exhaustive debate and are well served with a plethora of technical studies, with as many as thirty academic papers published so far. To date, analysis has embraced questions as to when and in what context these weapons were first introduced.[1] Problems have been addressed in relation to their size and weight dimensions, properties of composition and manufacture, handling techniques and penetrative power. This is all supported by literary extracts sustained with archaeological finds and even mathematical formulae, backed up with experimental tests utilising facsimiles.[2]

In contrast, what follows here is an investigation into the less studied and sometimes even neglected realms of clothing and armour based on the evidence available for cavalry and infantry, including distinctive officers' features, followed by an examination of emblems used on unit standards, and sword types and their general application in the Macedonian army.

Chapter 6
Cavalry

The outfit of *kausia*, *chlamys* and *krepides*, which became almost the sartorial manifestation of Macedonian military and national identity from the time of Philip and Alexander onward, was no less popular with the cavalry. In terms of armour, different styles of helmet (Phrygian, Attic, Chalcidian, Boeotian) and corselet were adopted by Macedonian cavalrymen. Luxurious horse accoutrements also deserve a special mention. As we have seen earlier in this book, in the closing years of his life Alexander distributed expensive panoply sets among his soldiers, which contributed in no small measure to the stunned reaction of onlookers viewing the advance of the King's cavalry and infantry formations. The Macedonians had an extraordinary pride for the quality of their armour, and the overall appearance of these impressive accoutrements must have been dazzling.

Kausia, Chlamys and Krepides

Of the three clothing elements, the *kausia* is known to have been a uniquely Macedonian form of headgear from as early as c. 500 BC.[1] It is more precisely defined as a one-piece mushroom-shaped cap made from moulded leather (or even lambskin), felt lined with a thick rolled band around the forehead and nape.[2] It has been argued, albeit controversially, that a derivative survives in the guise of the woollen *pakol* or *chitrali* exported from the Chitral Valley within Pakistan (Plate 30). The hat is worn today by the Nuristani of Afghanistan, including the Dards, Kafirs and Pathans, especially in the regions around Peshawar, in Dir, Bajaur and Swat, and as far afield as Karachi.[3]

In an authoritative enquiry, Saatsoglou-Paliadeli identifies the *kausia* from ancient literature as defining specific bodies within the Macedonian royal court and army, posited as high ranking officers, some generals, *somatophylakes* (officer-bodyguards), lead unit of the hypaspists, and page boys.[4] A profusion of evidence from coins, sculptures, reliefs, paintings, mosaics, cameos and above all terracottas (found from Greece to Iraq and Pakistan) indicates that the hat became fashionable among boys and young men throughout the Hellenistic East (see Fig. 14 for one likeness). It was usually worn within the context of primary political, social and civic institutions, such as the army, theatre or *gymnasion*.[5]

We can supplement the list of *kausia* wearers cited by Saatsoglou-Paliadeli with evidence for its use among Macedonian cavalry units from the late-fourth century BC. The most famous visual example comes from the right side of the Agios Athanasios tomb frieze (Plate 10b). Among other painted figures here are four soldiers wearing *kausia*, *chlamys* and *krepides* (three in armour), interpreted putatively as dismounted Companion cavalrymen. A fragmentary relief of a man

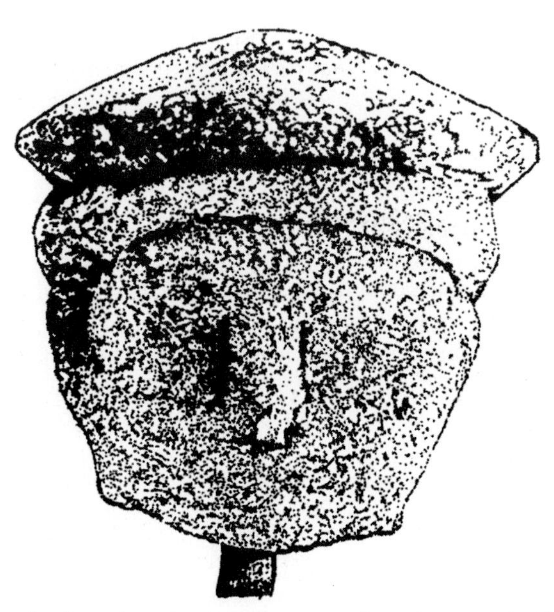

Figure 14. Limestone head wearing a *kausia*, perhaps part of a votive statuette. Cyprus, late-fourth or third century BC.

on horseback in *kausia*, with sword slung upon a baldric, was retrieved from an archaeological site in Republika Makedonija.[6] Images of mounted horsemen wearing *kausiai* and a wide-brimmed travelling hat (*petasos*) are further seen on a painting from Mustafa Pasha Tomb I, with another from a grave *stele* at the Hadra Cemetery (both examples of evidence from Alexandria dated to the third century BC).[7] Lastly, terracotta figurines of mounted males in *kausiai* are a common theme throughout the Hellenistic period.[8]

Saatsoglou-Paliadeli raises the possibility of the right to wear the cap becoming more widespread under Alexander and his Successors.[9] This is suggested in an anecdote referring to Memnon of Rhodes who is described dressing himself and his men in *kausiai* as a *ruse de guerre* to deceive the men of Cyzicus into thinking they were Macedonians led by Calas (335 BC).[10] Pollux describes the *kausia* as a common feature of stage costume for Macedonians in the New Comedy plays of Menander (c. 300 BC).[11] Demetrius Poliorcetes is even portrayed disguising himself as a common soldier with a cloak and *kausia* when escaping on horseback from camp in 288 BC.[12] As he went unnoticed, the implication is that by this time the hat was worn by many mounted Macedonian troops.[13]

The *kausia* was a versatile form of headgear, useful as a helmet substitute in battle and comfortable in both hot and cold weather.[14] The name may derive from the Greek *kausis*, meaning 'heat'. Its ability to warm the head reminds us of the Afghan-Pakistani hat, which is worn rolled up in summer and rolled down to cover the ears in wintertime. Like the modern military beret and Tam'o Shanter, the *kausia* was probably cheap and easy to manufacture in ample numbers. Philon of Byzantium mentions its use to ingeniously convey secret messages, the newly tanned leather surface scribbled upon by the sender using non-visible oak gall ink, later sponged by the recipient, revealing the message with the aid of a copper sulphate solution.[15]

The colour of the *kausia* can be deduced from a range of sources, including Plautus (c. 200 BC), who describes it (Latinized as *causea*) coloured like iron, possibly meaning rusted iron.[16] All six *kausiai* from the frieze and doorway of the Agios Athanasios tomb are rendered pale or cream-white, which indicates the badge of the Companion cavalry at the time the tomb was built. A figure from the Hunt Frieze of Royal Tomb II is shown on foot wearing a cream-coloured *kausia* (Plate 9a). As with the portrayals from Agios Athanasios, he is dressed in *chlamys* and *krepides* and may represent either a royal page or preeminent Companion.[17]

A mounted horseman in tunic and cloak from the Eastern Frieze of the Kazanluk tomb (Bulgaria) wears an unusual form of ochre-coloured headgear which bears resemblance to a *kausia*.[18] A terracotta from Boeotia, dated to the early-third century BC (Museum of Fine Arts, Boston), reveals a youth undergoing military training appearing in a yellow-brown *kausia* with short pink tunic.[19] One terracotta head from Egypt wearing a *kausia* (at the British Museum), also of the third century BC, includes a cream-yellow pigment preserved on the brim.[20] One more, identified as the figurine of a seated Macedonian soldier, has white paint on both *kausia* and *chlamys*.[21] Ambiguous drab hued *kausiai* are also depicted with Macedonian figures from more obscure Alexandrian tombstone paintings.[22]

It is interesting to observe from this ancient evidence that the *kausia* corresponds not only in shape (as noted earlier) but also in colour to the modern Afghan-Pakistani hat, which covers various shades from creamy-white to brown and bright, almost orange, 'rust'.[23]

Beside the distinctive headgear, the ubiquitous Macedonian *chlamys* can be detailed as a fairly short, semi-circular wraparound cloak fastened with a clasp in the usual military fashion at the right shoulder. It offered a practical improvement on earlier apparel used by horsemen in the Greek world and the earliest depiction to date seems to be from a *stele* dated c. 425 BC.[24] The design sometimes incorporated tiny elliptical shapes or weights on each of the two lower corners with different styles of cloak determined by season.[25] This is implied from torso fragments of two statues – purportedly of the Macedonian king Cassander (r.305–297 BC) from museum collections at Piraeus and Chalcis. The Piraeus fragment has a *chlamys* fastened at the right shoulder without any noticeable clasp and falling in heavy folds, implying wool (Fig. 15). In contrast, the fragment from Chalcis appears to be light, betraying thinner linen fabric for warmer weather.[26]

A collection of preserved papyri from Ptolemaic Egypt, dated to the third century BC, affirms the existence of contrasting versions of the *chlamys*. This archive provides an inventory of a government official's clothing, with nine cloaks in summer (linen) and winter (heavy woollen) varieties. Different apparel for summer or winter weather is known for soldiers in the Roman imperial army.[27]

Figure 15. Poorly preserved statue of a Macedonian king, identified as Cassander.

Traditional mantles worn today in Afghanistan and by Vlach and Albanian mountain communities (the *talagan* or *tambari* made from black sheep or goats' hair) have even been suggested as possible descendants of the Macedonian garment.[28]

The final element to complete the cavalryman's ensemble were *krepides*, arguably the most remarkable of the three aspects of dress noted here. They were largely standard wear for the cavalry, and infantry on occasion, as Pliny describes their use for both riding and marching.[29] They are defined as shin or calf high leather boots of open-toed strap work form with hobnailed soles, often worn in conjunction with *pellytra*, or socks, which left the ends of the big and index toes exposed.[30] *Pellytra* (akin to Roman *impilia*) were probably made from felt, useful in preventing chafing from the sharp edges of boot straps.[31]

One of the earliest forms of *krepis* in the Hellenic world is registered on two Cycladic funerary reliefs (470–460 BC).[32] Experiments were being made into the use of more effective military footwear in the years or decades before Philip's developing of the Macedonian army. The influential Iphicrates, described as the son of a shoemaker,[33] is said by Diodorus to have invented light military boots, coined 'Iphicratids' after him, that were easy to untie.[34]

The degree of sophistication in footwear achieved by the time of the late-fourth or early-third century BC is best seen from pieces of an originally gilded bronze equestrian statue, excavated in 1971 from the fill of an ancient public well in the Athenian Agora, and identified as Demetrius Poliorcetes.[35] The single largest fragment comprises a muscular left leg shod with a *pellytron* over which is secured a *krepis* with double-layered contoured sole and attached spur (Plate 28). The details have been wrought with clarity, pointing to a specific style. From a close examination of the find, Morrow was led to believe that it represents a type not illustrated on monuments prior to this statue's date.[36] A similarly stout *krepis* is worn by a bareheaded figure of a possible Demetrius in *chlamys* with corselet, part of an equestrian bronze statuette from the National Archaeological Museum of Florence.[37]

Existing forms of army boot were perfected during the period of Alexander and his Successors. After all, adequate footwear was an essential prerequisite when attempting to conquer the vast interior of the Persian Empire, traversing varied, and often inaccessible, terrain. The fact that the *krepis* (in combination with *kausia* and *chlamys*) was closely associated with Macedonians, evidenced in abundant artistic remains, means that they appreciated its importance and took full advantage of the latest technical improvements.

Macedonian *krepides*, like Roman *caligae*, were made to be robust. Their open, comfortable design and heel netting were more practical in the sweltering East than closed boots because the feet stayed ventilated and water, sand or mud could pour through easily.[38] The flat sole would have obviated the risk of impact hazards sometimes seen with modern heeled footwear.[39] Ancient writers also document the *krepis* with hobnailed soles. For example, Pollux imparts *krepis* soles often studded with hobnails to reduce wear and tear.[40] This feature made them a heavy object of apparel. An incident from Curtius records how Hector, son of Parmenion, was almost drowned in the Nile because of the weight of his water logged clothes and tightly laced *krepides*.[41]

Boot hobnails were in fact uncovered on the Vergina acropolis by Faklaris (1992/1993). This site included a rest room for what was a possible detachment of royal troops, so an inference can be made for these hobnails belonging to military footwear.[42] The reconstruction of a *krepis* sole has been attempted with a hobnail pattern like those of later *caligae* – or even the cleats of modern sports shoes. The layout has an outer border of studs, with two parallel rows near the back of the sole and three more unevenly placed to the front.[43] It seems that the problems of moving over soft or rough ground on foot, with factors of grip, traction and weight distribution, were understood long before the advent of the Roman imperial army.[44]

Applying modern research into *caligae*, Lee accepts that the *krepis* had a comparable life of about 1,000km, and that boots were maintained by rotating the hobnails around the sole for more even comfort. Different forms of *krepis* are shown in fourth century BC sculpture such as the seven or nine loops per side styles.[45] As with corresponding Roman boots, they required a few minutes to firmly secure and lace up.[46] The complexity of lacing is certainly alluded to in surviving sculpture and reminds us of Iphicrates being remembered as one who invented footwear that was tied up more simply.

Paintings and grave *stelai* in general picture *krepides* as mid or dark brown, though occasionally yellowish or pale brown shades are seen. Red pigments are detectable on the footwear of a minority of painted figures from the Alexander Sarcophagus and Tomb of Judgement.[47] Incidentally, the socks were often cream to light grey, yellow or brown (Plate 36b; note the barely visible left foot in a grey sock secured within a brown strap work *krepis*). We can speculate that some sort of meaning was ascribed to the wearing of red boots, in order to denote a certain rank or status, but the colour may only indicate a personal preference.[48]

As with the Agora leg fragment, Macedonian cavalrymen sometimes attached small bronze prick spurs with a strap to the back or heel of their boots.[49] Theophrastus seems to intimate the commonplace use of spurs among horsemen undertaking exercises at Athens (c. 320 BC).[50] A pair of bronze spurs was found in a Macedonian tomb in the area of Charilaou (dated 250–200 BC) and two pairs were recovered from a grave of a warrior in the Agios Athanasios tumulus.[51] A collection of six spurs from the fourth century BC onward are on open display at the Archaeological Museum in Thessaloniki (Plate 29).

There is further evidence that after the conquest of Persia some Macedonian cavalry officers opted to have expensive decoration on their footgear. This is seen with one of Alexander's Companions, namely Hagnon of Teos, who enjoyed wearing *krepides* with silver or gold nails.[52] Even if Hagnon was singled out for comment because he was exceptional, close analysis reveals that the *krepides* worn by all four of the dismounted cavalry figures from the Agios Athanasios frieze have their soles and laces picked out in yellow. A mounted hunting figure from the Alexander Sarcophagus, often identified as either Alexander or a youthful Demetrius Poliorcetes, has been occasionally described in red-brown *krepides* with yellow-coloured soles.[53]

The painted yellow soles from Agios Athanasios resemble funerary items from Iron Age burial sites which show gold work on shoes. One of the most recent is that of a warrior (from about 650 BC) found at Archontiko in 2010. Three graves at Vergina have yielded remains of women's gold or silver sandal soles (500–460 BC).[54] Such evidence brings to mind a detail from Polybius in which the son of a wealthy Ptolemaic official visited Philip V and on his return to Alexandria remarked on the young pages at court displaying notably elegant footwear.[55] Justin even ascribes gold plated *krepides* to the soldiers of the Seleucid king Antiochus VIII Sidetes during his campaign against the Parthians (129 BC).[56] This literary extract is considered of doubtful value, reflecting more the author's aim of unfavourably comparing the effete lifestyle of the Hellenistic East with the new martial vigour of Rome. However, in the light of other literary evidence and artistic and archaeological finds already discussed, the accuracy of the passage needs to be reappraised.

Granting all this, there are as yet no surviving remnants that can be identified on absolute authority as *krepides*. It has been advanced that an ancient form of sheep or goatskin shoe laced up with thongs, worn by Kalash shepherds to protect their feet when journeying through the mountainous Pakistan-Afghanistan borderlands, is a modern descendant.[57] Alternatively, Pottier considers that the *krepis* is preserved in modern times as a tough peasant boot found in rural parts of Greece.[58]

Helmets

The Phrygian helmet was patronised by Macedonian cavalrymen from the mid- to late-fourth century BC, as evident from grave *stele* reliefs and tomb paintings (Pelinna, Apollonia, Naoussa and Vergina).[59] A recently published marble statue of a standing cavalry or infantry officer from Amphipolis dated to the late-fourth to early-third centuries BC (and with head and right arm missing) has a Thracian or Phrygian helmet at its base, behind the left foot (Fig. 16).[60] Alexander, who was fundamentally a cavalry general, is portrayed in a Phrygian headpiece on the reverse of some 'Alexander' medallions as well. A likeness of him in this helmet form is seen on a bronze fraction minted at Naucratis in Egypt (c. 330 BC),[61] and a Roman steelyard weight, possibly of the first century AD.[62]

Figure 16. Damaged statue of a Macedonian soldier in corselet with *pteryges* and helmet at feet.

Although it required some labour to produce, it is easy to see why the Phrygian had such ready appeal in Macedonia. Popular in neighbouring Thrace, it was compact, close-fitting and open faced, combining adequate all round protection without the undue sacrifice of weight, hearing or wide-angled vision.[63]

Phrygian, Attic and hybrid types have been retrieved from sites, most famously Royal Tomb II, followed by Prodromi, Vitsa, Ohrid and near Marvinci.[64] The *stele* of Nikanor, son of Herakleides, found at Gephyra (c. 300 BC) shows a rider in an Attic form helmet (Fig. 17),[65] and another painted grave *stel*e from Vergina presents a dismounted cavalryman leading his horse and wearing a crested Attic type with the skull ridge and neck guard still visible.[66]

Figure 17. Grave *stele* of Nikanor, son of Herakleides. Cavalryman in full panoply with groom behind on foot.

Apart from the Attic form, other styles were adopted, some possibly originating in Thrace. Finds include the discovery of a bronze Chalcidian helmet from Arzos by the Hebrus river and another in a grave from Kitros (350–300 BC), including an Illyrian type recovered from the Kilkis area.[67] A helmet with white horsehair crest discarded under Alexander on the 'Macedonian' side of the Alexander Mosaic appears to be of the Chalcidian variety, and a tiny gold Chalcidian helmet ornament is seen atop the hilt of a sword from the main chamber of Royal Tomb II.[68] A conical bronze type with cheek pieces from Apollonia Museum (Pojani, Albania), dated to the late-fourth century BC, has been tentatively credited as Macedonian, although it more likely belonged to a native Apollonian or Illyrian warrior.[69]

We cannot be certain whether any of these isolated specimens were worn by members of cavalry or infantry, but it is rational to suppose that fully functional equipment was sometimes taken from the dead and recycled by troops as or when needed, allowing for different types on active service.[70] This is best realised with the mostly weapon fragment artefacts retrieved from the Macedonian burial mound at Chaeronea. The lack of defensive elements (helmets, corselets, greaves and shields) could testify to the practicalities of an army on campaign, with these items salvaged and utilised by other soldiers.[71]

One of the most popular helmet forms associated with the Companions was the type labelled 'Boeotian' or 'Boeotian-made', although, as can be seen from the evidence, it is a misguided article of faith to believe that it was ubiquitous among the cavalry.[72] Scattered illustrations of the Boeotian are apparent, with mounted figures from the Alexander Mosaic (Plate 3a; figure to the left of Alexander in a burnished iron (or silvered) helmet; and Plate 3b with two possible examples to the left of Darius, although the artistic rendering is highly impressionistic)[73] and Alexander Sarcophagus (Plates 2a and b), alongside a small marble head from a battle relief, now in the collection of the Metropolitan Museum of Art (New York).[74] A figure in a badly eroded Boeotian helmet is seen gazing back at Alexander on a stone relief from Isernia, Italy, 100 BC (Plate 4),[75] and a yellow-hued example may be carried by a servant for a Macedonian horseman on a tombstone from the Shatbi Cemetery, dated to the late-fourth century BC (Alexandria).[76] The brownish imprint of another Boeotian helmet with plume is distinguished above the horse's rump on a fragmentary monument of a possible Macedonian officer from Larisa.[77] Physical evidence is best embodied with a well preserved bronze sample (dated c. 300 BC), dredged from the bed of the Tigris river in 1854 and eventually deposited in the Ashmolean Museum (Oxford).[78]

Given the Boeotian's widespread use, it may be worthwhile concluding this section by detailing its manufacture and probable appearance.[79] Boeotia had a reputation for making some of the finest quality helmets in Greece.[80] The design was easily and relatively cheaply manufactured from a single thin sheet of bronze (avoiding the inherent weakness of helmet crowns made from separate parts) hammered over a wax mould secured to a former. Its shape was praised by Xenophon as the most suitable available, combining defensive coverage with unobstructed sight.[81]

Soft limestone formers, some of which were used for the making of probable Boeotian forms, were uncovered at Ptolemaic Memphis, now part of the collections in the Allard Pierson Museum, Antikenmuseum (Berlin) and Louvre (Paris).[82] These models have been dated to either 325–300 BC (Dintsis), c. 300 BC (Moorman), 325–275 BC (Ponger, Gagsteiger, Hamiaux) or, least plausibly, the mid-second century BC (Van Essen). One of the artefacts from Allard Pierson is plain in aspect, but another in the Louvre has a finely wrought stylisation of acanthus leaves, palmettes and lotus shapes. All the delicate working is placed below the calotte because any decoration on the calotte itself would weaken the helmet, rendering it inappropriate for combat.[83]

The Macedonians appear to have appreciated the Boeotian's specific advantages. It was high domed and padded to avoid concussion from blows delivered above. Besides, it came well down over the back of the neck, with a front projection shielding the forehead, eyes and face from the sun's glare, the steep trajectory of enemy missiles and hand-held weapons. The design further featured an elaborate series of double folds at either side – perhaps a translation into metal of the laced down brim of the felt, straw or leather *petasos* hat. Such a design feature gave the helmet a unique look, removed the need for actual cheek pieces, and provided effective protection defined by Frazer as the ability to guard against a slash aimed between the collar bone and ear, which could so easily damage or sever the common carotid artery and jugular vein, causing death.[84]

Chin strap holes are absent from many surviving Greek helmets; however, if made of a standard size, and without strap supports, cavalry headgear was liable to fall in front of the eyes during a flat-out charge. This was proven correct through the practical experiences of Robin Lane-Fox during the making of film-director Oliver Stone's *Alexander* movie epic (2004). Dark chin-straps are in fact detected strung through holes in helmet cheek guards worn by Macedonian horsemen on the Alexander Mosaic, presumably in order to circumvent this problem.[85] Punch holes to allow for fastening can be seen on the surface of the Boeotian type from the Ashmolean collection. Some Boeotian helmets may even have featured a strip of leather padding sewn at the rear rim to prevent the metal edge from rubbing against the horseman's back.[86]

To recapitulate, the Boeotian concentrates economy of metal, maximum protection and excellent all-round visibility allied to light weight and ventilation, and could be combined with other helmet designs to create modified variants. This reflects the inventiveness of the armourer's art, as various styles and characteristics were mixed and matched to suit the tastes or needs of different clientele.[87]

Corselets

The Macedonians fully exploited the armourer's increasingly adroit talents and prestige, which during the fourth century BC facilitated refinements in design, a more emphatic use of iron, and bolder, more creative applications of silver and gold.[88] We can speculate that in an age of experiment and change a Companion's expert judgement had a potentially broad range of corselet styles from which to satisfy his requirements. Torso protection could take several basic forms:

- the anatomically moulded iron or bronze 'muscle cuirass', usually with a one or two layered kilt of *pteryges* below the waist;[89]
- the linen or leather tube-yoke type, with a kilt of *pteryges*;[90]
- the iron foundation tube-yoke style, lined outside with linen or leather.

Details of relief sculpture on a wall from the so-called 'Monument of Shields and Breastplates' at Dion reflect the popularity of both the anatomical and tube-yoke forms in Hellenistic Macedonia.[91] The anatomically moulded cuirass was often worn by cavalry units, as suggested by the specimen from Tomb A (Prodromi) and an artistic remnant of a horseman from Apollonia in Albania. To these can be added the rock sculpture of Alcetas, a bronze Alexander statuette from Begram, a funerary *metope* (architectural feature on a frieze) from near the site of Tarentum, and Ptolemaic grave *stelai* and tomb paintings from around Alexandria. However, despite all this, the regnant notion in non-academic circles that the bronze or iron 'muscle cuirass' was the preferred armour type bar none is simply not borne out by the weight of evidence.[92]

The clear majority of artistic finds to date (350–275 BC) indicate that the non-muscled tube-yoke style was the more approved, even archetypal, choice for Companion cavalry. Evidence incorporates images from 'Alexander' medallions, the Alexander Sarcophagus, the Agios Athanasios tomb, Bella Tomb II, and the Tomb of Judgement. A painted likeness of a tube-yoke corselet is also seen in a tomb at Amphipolis and on grave *stelai* from Pelinna and Gephyra.

We can specify even more evidence for the tube-yoke type. The partly damaged *stele* of Demainetos at Vergina portrays a mounted protagonist with sword raised, about to strike an enemy on foot (Fig. 18). The horseman wears a tube-yoke corselet with two unusually short rows of *pteryges* beneath the abdomen.[93] The statue of a soldier from Amphipolis depicts a shoulder piece type with barely seen *pteryges* at armhole and waist (Fig. 16). A similar tube-yoke corselet with shoulder pieces and waist band worn on the *stele* of the dismounted horseman Nikanor, son of Menander, at Pella (340–320 BC) integrates at least two long layers of *pteryges* worn high (Fig. 19), not unlike the painted samplings from the well known Tomb of Lyson and Kallikles, Lefkadia (c. 200 BC).[94]

Later dated finds of the tube-yoke corselet from Italy comprise the second of two funerary *metopes* close to Tarentum, perhaps inspired by portraits of Alexander,[95] and an Alexander battle scene from a funerary urn (Copenhagen, Nationalmuseet).[96] The same form is seen on five marble rider torsos of possible Macedonian Companions found at Lanuvium (see Plates 21a and b for two

Figure 18 (*left*). Grave *stele* of Demainetos.
Figure 19 (*right*). Grave *stele* of Nikanor, son of Menander.

examples),[97] a bronze horseman statuette of Alexander (Naples National Archaeological Museum)[98] and one tentatively identified as Demetrius Poliorcetes from Florence.[99]

There is evidence that some cavalrymen wore tube-yoke corselets with an iron foundation, as witnessed with the famous type from Royal Tomb II made up of nine sections of 5mm thick iron plate, the inside lined with linen cloth, the outside with leather, a few remnants of both preserved at time of excavation. Allegedly, traces of leather *pteryges* adhered to the bottom.[100] The Alexander Mosaic depicts Alexander outfitted with what could be a similar iron core corselet (see Plate 3a).[101] A bronze statuette of an Alexander wearing a corselet from Klagenfurt Landsmuseum (Austria) bears some resemblance to it.[102]

In the Middle East, armour with directly exposed metal features would act as a solar heat collector, making the surface hot to the touch. This made the wearing of such a corselet uncomfortable, inducing profuse perspiration and premature exhaustion if worn during the rigours of an extended combat or pursuit. Thus, corselets with a covering of cloth or leather on the outer side represent an advance over earlier forms. Besides, armour covered in a layer of such material would not have required frequent polishing and rust proof maintenance. Another advantage is that, in comparison with the pure textile format, iron armour was stronger and heavier, offering better defence from enemy weapons.[103]

Interestingly, some troopers may have preferred to divest themselves entirely of body armour due to searing temperatures encountered in the East. On the subject of Alexander's panoply at Gaugamela, Curtius states that the King very rarely wore a corselet,[104] although he happens to mention Alexander in a corselet four times elsewhere in his narrative.[105] The Alexander Sarcophagus shows a mounted Alexander and possible Hephaestion without corselets, and the lost fresco from the Kinch Tomb displays a cavalryman wearing a helmet but without body armour.[106] In addition, one of six Lanuvium torsos appears without armour, and Demetrius Poliorcetes is seen on horseback in a *petasos* and *chlamys*, armed with a *xyston* but without a corselet, from the reverse of staters and tetradrachms minted at Pella, Amphipolis and Chalcis (294–288 BC).[107]

The fact that there is no solid evidence for the iron foundation style in the Greek world prior to the Royal Tomb II find theoretically spells out its point of origin in the Macedonian royal court of the mid-fourth century BC, although this may be a premature hypothesis.[108] The earliest examples of the body armour in the West are Thracian, and Archibald has discussed the possibility that Thracian iron corselets provided the necessary blueprint, adopted and adapted by the Macedonian army.[109] Philip II's campaigns in Thrace and Scythia (342–339 BC) were the likely arena, introducing the King to the effectiveness of Thracian military technology at this time.[110]

Amongst the most remarkable artistic sources so far discovered are the distinctive purple-coloured corselets worn by three probable Companion figures in the banquet scene from the Agios Athanasios tomb frieze (Plate 10b).[111] Purple corselets were occasionally displayed by the wealthiest of Greek warriors centuries before, but surviving evidence suggests that it was only with Alexander's conquest of Persia that it became a more pervasive sight.[112] The form shown on the wall painting is based on the standard linen-leather format, yet they also appear curiously bulky. This could betray an iron foundation – even if contradicted by a passage in Curtius in which Alexander's army shed old sets and heaped them on bonfires, suggesting that entirely linen or leather armour remained commonplace.[113] The appearance of the Agios Athanasios corselets, if wholly of fabric, could be the result of some layering and multi-folding technique, resembling that of the later Byzantines.[114] Alternatively, the armour might have been reinforced or impregnated with kaolin.[115] These are just two examples of methods which could have been known and applied.

Under close analysis two of the Agios Athanasios corselets are pictured with yellow-gold trim, and yellow lines and knobs across the chest. Their virtually identical rich colour and design may offer a unique visual glimpse into the lavish gold and silver decorated armour provided to the Companion cavalry by Alexander in the latter half of his reign, as described in literary sources.[116]

Macedonian corselets included attachments with a diversity of motifs. Excavations of fourth-century BC sites reveal relief work decoration (often silver-gilt), rosettes, palmettes, sunbursts, shields, perched eagles, lion and Gorgon heads, a Nereid on a sea-horse, a crowned face with sphinxes and a head of Heracles wearing the Nemean lion scalp.[117] Although debatable, academics contend that the application of hammered relief work ornamentation for corselets originated in a Thracian metallurgical tradition, as seen with silver-gilt objects from the Golyama Kosmatka tumulus (near Plovdiv), dated to the mid-fifth century BC.[118]

Saddlecloths and Harness Fittings

As well as armour, Macedonian cavalrymen required a comfortable seat when mounted. Saddle pads or cloths were thus used to cushion the backs of horses with prominent backbones. Xenophon is known to have recognised the problem, advising horsemen to select mounts with a 'double back' only.[119] This appears to refer to horses with a recessed backbone and, as Gaebel considers, with adequate muscle at either side 'to keep the human pelvic bone from resting on the horse's spine'.[120]

The advantages of having saddlecloths are in general terms spoken of by Xenophon elsewhere when it is proposed that they serve in part to protect the mount's underbelly and that 'the quilting of the cloth should be such as to give the rider a safer seat and not to gall the horse's back'.[121] Yet, this is tempered by his recommendation that a rider avoid too many cloths and his exaggerated dismissal of the Persians having more coverings upon their horses than on their beds.[122] It is thought that the cloth described by Xenophon could well have taken the form of a contoured seat pad appropriate enough in size to protect the horse's barrel.[123]

Preserved papyri from Ptolemaic Egypt, originally part of a family archive belonging to the Greek cavalry officer Dryton, son of Pamphilos, provide an inventory of arms, armour and tack used by Dryton himself or his son Esthladas. Some of the items listed suggest that padded saddlecloths were promoted in the Hellenistic era. The list includes saddles sewn together, i.e. with a form of filling, possibly wool or linen.[124] Strabo mentions how, when in India, Alexander's Macedonians recognised the usefulness of cotton as padding for horse blankets.[125] Several of the Macedonian mounts from the Alexander Sarcophagus seem to depict an upper shag pile or felt-like saddlecloth with another underneath, maybe used as an alternative to padding. The Dryton papyri refer to sweat-cloths placed beneath the seat to absorb perspiration. Two of these sweat-cloths were round and, as Vandorpe explains, the 'round shape is undoubtedly due to the shape of the saddle proper'.[126]

The Alexander Sarcophagus displays mounted Macedonians with saddlecloths coloured yellow-ochre, with a scalloped blue or purple border.[127] However, under Alexander and his Successors, Companion cavalry cloths were often partially or entirely pigmented in shades of red or purple. A sculptural relief from Athens, purporting to be of a Macedonian with mount, reveals red colouration on the contours of an animal pelt shabraque,[128] whereas a Shatbi Cemetery grave *stele* portrays a square crimson-red or purple coloured type.[129] Similarly, a Persian horseman in a combat scene of Persians and Macedonians on a bed from the Soteriades Tomb displays one as pale rose-purple,[130] while traces of an apparent red tint are seen on the cloth of a cavalry horse from the battle frieze of the Tomb of Judgement.[131] One terracotta from the British Museum (325–275 BC) shows a rider trampling an enemy on foot with red pigment on the seat, the mounted protagonist further clad in a *kausia* with a red patch on the *chlamys*.[132]

The Macedonian horseman on the right side of the mosaic from Palermo (who may represent one of Alexander's staff officers or *Diadochoi*, possibly Craterus) still preserves much of the saddle area intact. It can be described as a square over-cloth on top of a larger undercloth showing a rear dagged pattern. The overcloth appears too damaged to make out, although the edge is still observable as a pink-red (Plate 36b).[133] Literary extracts also refer to red or purple dyed ship sails, wagon canopies, carpets and elephant blankets.[134] It is therefore difficult to reject red or purple horse coverings. Verification for red or purple appears again with a translation of the Dryton cavalryman papyri listing *kases ephippios phoinikous*, interpreted by Vandorpe as a skin, blanket or saddle in the red-purple spectrum.[135] Purple saddlecloths with expensive tack seem to have continued as a hallmark of equestrian wealth in Greece up to the second century AD, with Apuleius recounting Thessalian horses in silver, gold and purple housings.[136]

When exploring the feasibility of uniform horse accoutrements, Sekunda notes that on the Alexander Mosaic the Macedonian king is shown in a green-coloured waist sash, while his horse displays corresponding green trim on its shabraque neck hole (Plate 3a). These matching details suggest some form of designation based on colour, worn as squadron facings on the saddlecloths of Companion cavalry, formulated as green, red or yellow.[137] This is an attractive proposal but it needs expanding. Unless the mosaicist had a limited range of colours to hand, it is probably no coincidence that green is repeated on specific and easily recognisable features on both Alexander and his mount.

The Macedonian army undeniably represents a sophisticated institution into which a system based on colour coding would seem to fit well.[138] Yet, despite its appeal, Sekunda's hypothesis remains unsubstantiated. There is contrary evidence to show that even *if* such a form of squadron identification was used, regulations were relaxed enough to allow many Companions to opt instead for personal saddlecloths, either purchased as luxuries or plundered from defeated wealthy Iranian-speaking peoples. On this point, one of the mounted Macedonian figures from the Alexander Sarcophagus displays an ornate specimen rendered in red with dotted white border, decorative white figures and contours of pale blue (Plate 2a). The original design probably consisted of a lion and two winged griffins. The horse's breastband is noted as light blue with an inner border in yellow-ochre, outer border coloured red-brown.[139] Two other examples of richly decorated saddlecloth from the Alexander Sarcophagus can be seen in Fig. 20 and 21.

Turning our focus to harness fittings, it is clear that after the conquest of Persia war chargers were increasingly caparisoned with lavish accoutrements, in keeping with Alexander's vast material outlay and expenditure. Companion units are described being supplied at this time with extravagant Persian harnesses,[140] silvered horse trappings,[141] or gold (gilded) bits.[142] Although hardly an impeccable source, the apocryphal *Alexander's Letter to Aristotle* mentions gold bridles.[143] It is possible that some individuals retained expensive cavalry furniture procured earlier. For example, gilded bridles, which are obviously reusable, were taken from the Persian camp following victory at Gaugamela and some may have been grabbed at Persepolis.[144] Excluding Alexander's mount, three or four Macedonian horses on the Alexander Mosaic have harnesses fitted with bronze (or gold) disc attachments.[145]

The use of luxurious harness fittings witnessed in literary sources is itself corroborated by appliqué-work offerings dating to the fourth century BC from tombs at Vergina, Derveni and elsewhere in Macedonia, which confirm the

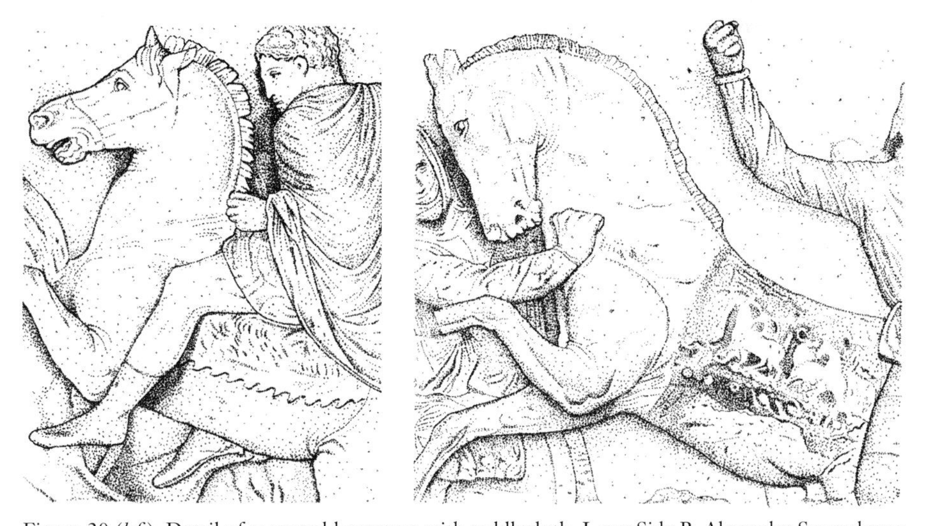

Figure 20 (*left*). Detail of mounted horseman with saddlecloth. Long Side B, Alexander Sarcophagus.
Figure 21 (*right*). Detail of saddlecloth. Short Side B, Alexander Sarcophagus.

wealth, culture and ideology of the horse riding, land owning ruling class.[146] The evidence for such furniture in Alexander's army is congruent with that of later Hellenistic armed forces, which often aped their great predecessor. Accounts of the Seleucid parade at Daphne (166 BC) have some of the Nisaean mounted cavalry in trappings of gold with the rest using silver, while horses of the Companions and royal Friends were all dressed with gold.[147]

The fact that tack often had a uniform appearance can be gauged from details on the Alexander Mosaic, with Macedonian bridles depicted as brown or reddish-brown leather.[148] This compares favourably with the Alexander Sarcophagus, where one mount has a girth painted red, and another red-brown.[149] An extant grave *stele* from the Shatbi Cemetery features a Macedonian cavalryman on a mount clad with matching tan brown or red-purple reins and breastband.[150] Alexander's own reins appear in the Mosaic as white with two thin yellow-gold lines running their full length, similar to the reins of the cavalry officer directly behind him (Plate 3a).[151]

One interesting facet, often overlooked in modern analyses of Companion cavalry, is how the forelock just forward from the poll on at least three horses on the 'Macedonian' side of the Alexander Mosaic seem to be tied up using narrow white ribbons (Plate 3a; one appears over Alexander's right shoulder).[152] These are depicted wound several times around, the hair elegantly forming an ornamental plume in a 'tea-whisk' style. All horses from the Alexander Sarcophagus, the Agios Athanasios frieze and many other equestrian sculptures of Macedonians similarly show them with cropped manes in the Iranian manner.

The 'forelock tie' or 'banded forelock' seems to have served no useful purpose, probably favoured simply for its neat aesthetic look. By contrast, the form of mane referred to in modern equine parlance as 'clipped', 'hogged' or 'roached' may have been undertaken to prevent horse hair getting caught up with the rider's harness, bridle and weapon movements. This was vital if horsed troopers were to attain full and safe manoeuvrability. It could have another intention too; namely, providing a clear field of view for the rider, as is the rule for polo players to this day.[153]

Another hint of eastern influence noticeable on the Sarcophagus is where the mount of a figure identified as either Alexander or Demetrius Poliorcetes is shown with its tail tied up using a brown strip. This was done in order that it did not get entangled with the lance when handled and thrust at the enemy.[154] The same features continue for other equestrian sculptures of Alexander, including a bronze statuette from Herculaneum and another preserved at Klagenfurt.[155] As the forelock plume, hogged mane and tail band are seen on the reverse of some silver tetradrachms from Philip II's reign, Macedonian horsemen were already adopting eastern equestrian decorative styles pre-336 BC.[156]

Royal Pages' and Grooms' Dress

Having investigated aspects of dress and equipment among cavalrymen, we will finally attempt to reconstruct the dress and equipment of the king's page boys (*basilikoi paides*) and cavalry grooms in Macedonian armies.

The royal 'boys' or pages received the king's horses from the grooms, brought them to him when he was ready to mount, and then followed their royal master into battle.[157] They accompanied Alcetas in Pisidia (319 BC) and were deployed with Eumenes' army (317 BC), but it is possible that in these last two examples, rather than being Macedonian, they represent the sons of Pisidian and Cappadocian aristocrats.[158]

Hammond asserts that Macedonian pages first saw combat in battle as opposed to the hunt only in their senior age class, i.e. during their seventeenth year.[159] The upgrade of duties at this age was a sensible evolution from their prior training, distinguishing the older pages from their younger comrades by signifying that they had now graduated to being strong and skilled enough to participate in battle. From an examination of literary extracts,[160] Heckel cogently argues that infantry training was introduced to young nobles from about eighteen or nineteen within a systematic and broad based programme.[161]

It is conceivable that some of the pages were armed. Diodorus describes 'armed soldiers' or 'arms-bearers' (*hoplophoroi*) standing before a unit of Persian royal spearmen (*melophoroi*, or 'apple-bearers') on the scenes decorating Alexander's catafalque. Hammond considers *hoplophoroi* to be an element among the pages.[162] Drawing together all artistic and textual evidence, we can tentatively summarise that in battle senior age group pages were mounted similarly to Companion cavalry, clad in armour and armed with swords and possibly the *xyston.*[163]

Pages in Antigonid times wore expensive, high quality fabric, at least in an official and ceremonial capacity, when attending the king. A Ptolemaic visitor to Philip V's court spoke of young men (*neaniskoi*) who sported the superior quality of dress and footwear.[164] Plutarch adds how at Argos (209 BC) the calculating politician Phayllus persuaded his beautiful wife to dress as a page in *kausia*, *chlamys* and *krepides* in order to slip undetected into the presence of the same monarch.[165] The royal accession parade of Ptolemy II included a contingent of 120 pages in purple tunics, which suggests Macedonian youths sometimes wore lavishly dyed fabrics.[166] Hellenistic terracottas depict boys and young men dressed in garments showing rose or pink pigment on tunics, yellow or pink on cloaks, and black on boots, with Macedonian *kausiai* in ochre, red, blue or turquoise tinctures.[167]

For hunting, and possibly at other times, pages went naked beneath their cloaks. The Hunt Frieze from the facade of Royal Tomb II has two or three figures often identified as pages unclothed but for pale brown mantles.[168] Near naked youths wearing either plain white or white with red bordered cloaks are seen on the Mosaic of the Stag Hunt and Mosaic of the Lion Hunt, both from men's dining halls in villas at Pella (c. 300 BC).[169] Apparent nudity complements Plutarch's passage of pages in 209 BC in which only hat, cloak and boots are characterised.[170] The body tunic was occasionally discarded in order to toughen the pages up as part of their military preparation, which is entirely believable given that some neighbouring peoples to Macedonia allowed boys in certain situations to go naked until they were sixteen.[171]

When reconstructing the appearance of grooms, a number of helpful sources have survived, one of which is a marble horse with groom relief found at Larisa (Fig. 22). It has been proposed as belonging to a larger monument commemorating a Macedonian officer or leading Athenian general of the late-fourth century BC, possibly Leosthenes or Phocion the Good, although it has been cautiously dated elsewhere to the mid-third or second centuries BC. The groom is portrayed as a black African, grasping reins with the left hand while gripping a goad in the right. His get up consists of a short-sleeved tunic, shin high socks and strap work *krepides* with thin form-fitted soles. A red pigment is preserved on the reins or bridle of the horse.[172]

An even more valuable source for the appearance of Macedonian grooms can be found among eight colourfully painted unarmed figures from the banquet scene frieze on the Agios Athanasios tomb (Plate 10a).[173] The three persons on robust horses in this part of the frieze could be friends of the deceased. Of particular significance, however, are the men and boys on foot, all of whom appear to attend those who are mounted. The five footmen all wear different coloured tunics (bright red, orange-red or white) and cloaks of various shades (crimson,

Figure 22. Relief showing a restless war horse being restrained by a groom, c. 320–260 BC/150 BC.

orange-red, bright blue or white) with a purple border. Two of them have wreaths on their heads and all footmen are portrayed carrying vessels or torches.

Several of these figures are acceptable as putative grooms from the manner of their artistic characterisation, especially the man who is holding the reins or harness of a mounted figure's horse (third footman from the left). His appearance is essentially no different from that of a boy-groom shown bareheaded in *chlamys* and tunic from a relief sculpture grave *stele* (Vergina), dated to the late-fourth or early-third centuries BC.[174]

One of the footmen has a narrow black stripe on the right sleeve of his red tunic (stripe not viewable in Plate 10a). The use of vertical stripes on tunics (similar to Roman *clavi*) is an interesting feature found on the clothing of warriors from painted scenes preserved in the Thracian tomb of Alexandrovo (southeast Bulgaria, belonging to the fourth century BC). Kitov suggests that within a Thracian cultural context there was some form of association between the colour of the stripes and the social standing of the wearer, with red signifying higher status and white lower.[175]

Chapter 7
Infantry

Infantry armour will be discussed by focusing on helmets (the metal *pilos* arguably the most prominent) and corselets (including *hemithorakes*). As noted before, Alexander's troops were lavishly re-equipped in the closing years of his reign, which may have been part of the King's extravagant desire to 'recreate' the *Iliad* through the army. Shield emblems will be shown to bear witness to a taste for heroic display, utilising rich colours and striking designs. As we shall see, apart from helmet, corselet and shield, greaves were another important piece of kit which allowed for full protection.

Helmets

Contrary to the view espoused by many researchers, financial constraints under Philip II (or during the early years of his reign) could mean massed pike unit headgear was of a non-metallic composition.[1] This was an ideal cheap alternative: light, disposable and easily replaced, tough, hard wearing, water resistant and less prone to overheating.

The paucity of artefactual evidence for metal helmets suggests that many were made from perishable materials, although it remains speculative until confirmed through archaeology or inscriptions. Even so, Cassius Dio offers a rare glimpse into the characteristics of Macedonian infantry armour, albeit writing five centuries later. He describes the fanatical Alexander imitator and Roman Emperor Caracalla (r. 212–217 AD) issuing his 16,000 reconstructed phalangites with helmets of ox hide, as allegedly worn in Alexander's day.[2]

There is corroborative archaeological evidence for neighbouring Thracian warriors with helmets made from leather from around the fifth and fourth centuries BC.[3] This is extended in the opinion of some academics to military cultures separated in time and space across the ancient world.[4] Curtius appears to hint at the modest equipage of Macedonian troops prior to the conquest of Persia, and the fact that thousands of old panoply sets were burned in India points to headgear made from combustible materials such as leather.[5] Artistic depictions of helmets dated to soon after Alexander's death indicate that metal types became standard at some point during the conqueror's reign – perhaps around the time of the Indian expedition (327–325 BC) when smart expensive panoplies were issued to the army.

As first posited by Juhel, and accepted by Sekunda, there is evidence that the conical metal *pilos* type (in Hellenistic times, the *konos*) saw widespread service or even became regulation wear among pike armed Macedonian infantry in the closing years of Alexander's reign or not long after.[6] An extract from the *Kestoi*[7] of

Julius Africanus (c. 200 AD) mentions how the 'Lacedaemonian helmet', or Spartan *pilos*, was introduced into Macedonian ranks by either Alexander or his Successors as its design promoted an unobstructed field of view to the wearer.[8] If an isolated statement from this late encyclopaedic work by Africanus is applicable, we might conjecture that another reason for the helmet's adoption was the sacrifice of defensive capability[9] for relative speed and simplicity of manufacture in which a flat disc of bronze was hammered out over a conical (elliptical) former.[10]

As Africanus consulted now unknown handbooks and collections of stratagems, what he imparts may be from an authentic source.[11] Moreover, Sparta was without doubt a pioneer for military innovations adopted by other states in the fourth century BC – a factor that could not have been overlooked by the meticulous Macedonians.[12] There is in fact support from elsewhere which helps to strengthen Africanus' claim. The *pilos* appears on some red-figure pottery from Pella dated to the fourth century BC, showing near naked hoplites.[13] More significantly, the grave *stele* of Nikolaos, son of Hadymos, found at Gephyra reveals a figure in a *pilos* (Fig. 23) portraying a likeness more convincingly identified as a fourth- to third-centuries BC *sarissa*-armed Macedonian foot soldier than any other publicised image.[14]

It is possible that a helmet re-equipment programme under Alexander embraced entire army units. As explained in Chapter 4, the Achaemenid polity had the organisational expertise, massive wealth and resources to make this real. Alexander took over management and is disclosed ordering new panoplies to many thousands of troops during his Indian campaign. He was subsequently praised by ancient writers as a general talented in marshalling and equipping field armies.[15] In addition, the *pilos* form ideally lends itself to an organised 'assembly-line' effort, where items had to be routinely and repetitively produced with economic speed and efficiency.[16] If the helmet was introduced under Alexander, then the re-provisioning episode in India was surely the likely time it was done. Numismatic evidence dated from Alexander's time to the third century BC signifies that the *pilos* or *konos* became affiliated with the Macedonian shield (indicating infantry) and, to some extent, overall Macedonian national and military identity (Plates 13 and 14 display some *pilos/konos* helmet samples on coins).[17]

A sound basis exists for this type of helmet becoming typical in the Hellenistic Age. Philon of Byzantium seems to describe the *pilos* as a regular feature of military outfit, while Arrian has the Laconian or Arcadian headpiece as part of the equipment of 'true' heavy infantry.[18] In light of the latter extract it is intriguing to recall an anecdote of unknown derivation preserved by Polyaenus. He refers to Polyperchon within a decade of Alexander's death fighting in the Peloponnese and demonstrably comparing the Arcadian *pilos* and cloak with his own high quality panoply as a means of rousing morale among a crowd of Macedonian soldiers.[19] It is possible to imagine that Polyperchon was ridiculing Arcadian headgear in comparison to the expensive metal *pilos* worn by himself and his troops.

Figure 23. Grave *stele* of Nikolaos, son of Hadymos.

The Amphipolis military code (c. 200 BC) stipulates that phalangites be equipped with the *konos* form,[20] and two surviving inscriptions from Delos record the dedication to Apollo of gold wreathed *konoi* belonging to Philip V's sons, Perseus and Demetrius.[21] Artistically, evidence for phalangites wearing the *konos* helmet in the second century BC is seen on the *stele* of Zoilos, son of Ischolaos, held at Skopje Museum,[22] and a bronze belt plaque fragment from Pergamum, which purports to represent Seleucid pikemen in a scene from the battle of Magnesia (190 BC) (see Fig. 31).[23]

It is, however, disingenuous to deny important evidence for Macedonians, including foot soldiers of the time of Alexander and his Successors, wearing other helmet types, such as Phrygian or Attic forms. Pottery from the late-fourth century BC found in Pydna (South Cemetery, Tomb 52) reveals combatants in both *pilos* and Phrygian helmets.[24] Small gilded clay plaques from a pit-grave in Amphipolis,[25] a marble head from Archontiko (250 BC),[26] and a likeness on an amphora sealing from inside Pella,[27] all confirm the use of Phrygian and Attic types. So do painted or sculptured foot figures from the Alexander Sarcophagus[28] and Agios Athanasios tomb.[29]

Of a total of ten helmeted footmen on the Alexander Sarcophagus and Agios Athanasios frieze, virtually all appear wearing the Phrygian style, with not one identified with certainty in a *pilos*. The discrepancy here is readily explained (though not proven) by supposing that these particular images are of hypaspists or elite guardsmen, arrayed in different helmets to distinguish them from the *pezhetairoi*.[30] In contrast, the relief sculptures of Macedonian foot soldiers in identifiable helmets from the Belevi Mausoleum (early–mid third century BC) mainly depict the Attic form (see Plates 22a and b for two examples). Head suggests that these figures could represent 'Seleucid guard infantry' as some of the helmets they wear are crested.[31]

By way of a brief summary, it remains potentially convincing that the *pilos* became standard for phalangites as part of Alexander's re-equipment provision in the second half of his reign. However, physical evidence is scant. The only Macedonian find of a *pilos* helmet which may be of the fourth century BC is that of a bronze sample discovered in the northern end of Amphipolis (west of the acropolis) (Fig. 24). The object has two holes on each side for riveting now lost cheek pieces, no apparent crest fixture, and a diagonal crack from brim to peak, which might have been caused by weapon impact.[32] Until more are found, we can assume that it was only one of several popular helmets patronised by the infantry, whatever the particular allocation (if any) was. The greater number of Attic, Phrygian and Chalcidian forms recovered from random gravesites (detailed earlier in Chapter 6) offer further support that a range of typologies were adopted. At this stage any further refinement is impossible.

Before we leave discussion on infantry helmets, it is worthwhile noting that their outer surfaces were often painted.[33] From numerous helmets on the Alexander Sarcophagus it seems that sky blue was a preferred colour badge associated with portions of Macedonian infantry (Plate 1a; the helmet of the far left figure shows a faded bluish pigment) – reminiscent of United Nations peace

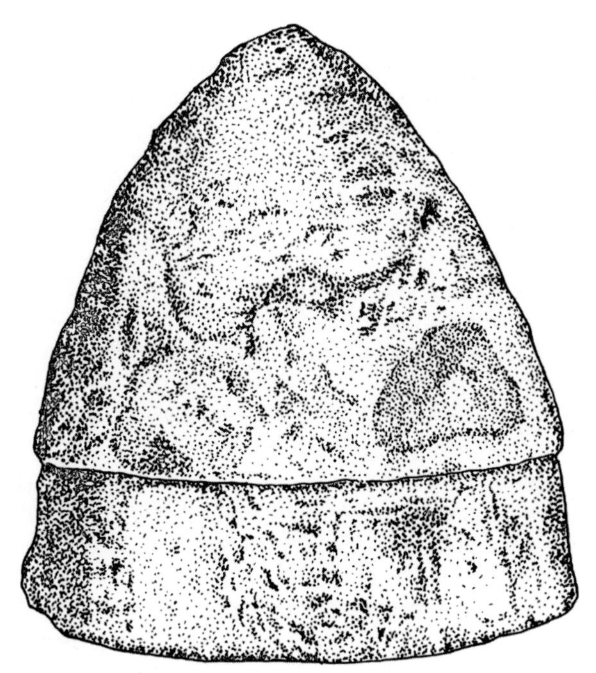

Figure 24. *Pilos* helmet found in Amphipolis.

keeping troops.[34] A little known vase with gold work (350–300 BC) from the Amphipolis Museum renders two Macedonians on foot battling Amazons wearing blue-coloured yellow-rimmed headgear (Plates 33a and b), which tally with those displayed by the Sarcophagus sculptures.

Painting of this sort offered some advantages. As with medieval European helmets, it helped to prevent rust and preserved the surface from adverse weather. There would be no need for constant cleaning – a tedious chore for soldiers marching and fighting in the often dusty conditions of the East. In the open air, a thick coat of paint would have cut down glare, discouraging the surface from overheating and becoming untouchable on torrid summer days.

It cannot be ruled out, of course, that in many instances blue paint was used to signify iron. After all, the colour blue could be exhibitive of iron in earlier military societies. Academics agree that this was the case in Assyrian art, with blue on helmets from wall paintings at Til-Barsip (eighth century BC).[35] When we examine Macedonian evidence, four of the obviously iron spearheads from the Agios Athanasios paintings are shown in blue. In addition, blue colouring has been applied to indicate sword and spear blades on the friezes of the Kazanluk tomb, and iron cuirasses on paintings of Ptolemaic cavalrymen from Mustafa Pasha Tomb I and several grave *stelai* from the Hadra and Shatbi Cemeteries.[36] Most interesting of all is an intact Babylonian cuneiform tablet describing Macedonian troops under Ptolemy III (246–245 BC) in iron panoplies, suggesting that some Hellenistic states had contingents of Macedonian-style infantry mass equipped in iron helmets and corselets by the mid-third century BC.[37] If blue as iron is tenable, it would make the Macedonians some of the keenest promoters of the iron helmet since the Neo-Assyrian army, primarily for officers and choice units of the rank-and-file.

It is equally plausible that some blue paintwork was meant to represent a silvered surface. Snodgrass, concurring with Winter, upholds the blue Sarcophagus helmets as corresponding to a 'silvery' colour rather than the usual 'peacock' blue associated with iron or steel.[38] Closer scrutiny of Winter's published colour plates could confirm this as correct for some of the headgear worn by the infantryman sculptures.[39]

Corselets

The Alexander Sarcophagus and a miscellany of literature and artwork of the fourth- to second-centuries BC evoke a mid thigh length tunic (*chiton*) of wool or linen as proper wear. Over the tunic was donned a protective corselet.

The evidence for the form of corselet worn by *pezhetairoi* in the thirty years spanning Philip II's accession to Alexander's invasion of India is sporadic and sometimes contradictory. This might suggest that various experiments were being carried out in relation to mobility and tactics to decide whether defensive armour of this sort should be worn or not and, if so, by which ranks. If corselets were utilised, then debate would have arisen as to what was the preferred type. Questions such as these were possibly linked to the variable of financial income and military expenditure.[40] There is a consequent lack of consensus in opinion among academics regarding the extent to which corselets were worn by the *pezhetairoi*. It may be that only a percentage of phalangites (file and half-file leaders) were ever intended to wear them.[41] A relevant historical comparison can be furnished in the guise of heavily and lightly armoured varieties of pikeman deployed by Spanish armies in the sixteenth century. Yet literary evidence for the Macedonians is still too sketchy or unreliable to allow for dogmatic assertions.[42]

The dubious nature of documentary evidence inevitably propels us toward the artistic and archaeological remains. But these are equally ambiguous. The Nikolaos *stele* from Gephyra has been interpreted as a file-leader or file-closer in a metal muscled cuirass, which incidentally could support an anecdote of tenuous value preserved by Polyaenus.[43] A fighting Greek in a corselet appears on the right of the Mosaic of the Amazonomachy (House of the Abduction of Helen, Pella, c. 300 BC).[44] Gilt clay reliefs of three 'hoplites' found in a grave from Stavroupolis (325–300 BC) show these warriors with obscure crested helmets and corselets (one with a waist belt), and most of the infantry figures from the Belevi Mausoleum wear corselets as well.[45]

Of nine clothed Macedonian foot soldiers intended with weapons on the Alexander Sarcophagus, seven wear corselets, one is too far damaged to make out, and one on Pediment B is without defensive elements of this kind, dressed only in a tunic gathered at the left shoulder.[46] This last figure reminds us of three painted infantrymen with shields from the Agios Athanasios tomb.[47] Two are shown without corselets and the third has his torso hidden by a shield (Plate 10b). A

Figure 25. Horseman and footman combat. Reverse of a silver tetradrachm.

possible Macedonian is shown on foot without a corselet, this time on the reverse of a tetradrachm from the reign of King Patraos of the Paeones (Fig. 25).[48] The Amphipolis vase exhibits two soldiers on foot wearing only tunics pinned at the shoulder (Plates 33a and b), and a 'Sack of Troy' fresco from the House of Menander (Pompeii) depicts another man in strikingly similar tunic colours (Plate 34), claimed as an Antigonid Macedonian 'White-Shield' by Sekunda.[49]

Such a jumble of largely independent shreds of evidence – some reliable, others less so – might refer to different units, ranks, circumstances, individual preferences or even artistic intentions. It is still fair to assume that components of Macedonian infantry had occasion to wear torso armour, whatever the specific arrangements might be. To avoid tar pits of controversy the researcher should await definitive artistic and epigraphic material before drawing a more refined conclusion.

Corselets worn by Macedonian foot soldiers under Alexander appear at least to have been often constructed from combustible substances with no apparent iron foundation, inferred from Curtius, who describes how in India old rotten sets were to be gathered and burned.[50] This insinuates something organic, made from either linen or leather. Both materials were used concurrently for corselets in the Greek world of the mid- to late-fourth century BC, as construed from Aeneas the Tactician and a fragmentary Delos inventory list (342–340 BC).[51]

On the side of linen, Cassius Dio has Caracalla issuing each of his revived 'Alexandrian' phalangites with a three-ply linen corselet.[52] The writer's dry exactitude may suggest that he had access to a genuine Roman document outlining supplies of equipment. Moreover, the fact that other aspects of panoply described by Cassius Dio have been verified as accurate using independent evidence adds weight to his account as an authentic description for corselets worn by Macedonian infantry in the fourth century BC.[53] Foot figures from the Alexander Sarcophagus are in the main some form of elite soldier, guardsmen or hypaspists, with corselets manifested in the typical Greek format consisting of tube and yoke with lace down shoulder pieces and a kilt of *pteryges* (Plates 1a–b).[54] As rightly thought by Everson,[55] in agreement with Lush,[56] the Sarcophagus samples seem more suggestive of leather. This can also be said of eight of the corselet wearing infantry sculptures from Belevi.[57] The discovery of a warrior's tomb of early though unspecified date at Pydna has led to an unconfirmed preliminary report even claiming evidence for a leather corselet trimmed with bronze.[58]

Sekunda observes that a number of footmen in corselets on the Alexander Sarcophagus display small sky blue segments painted onto the surface, especially at shoulder pieces or waist bands, and colouring to sword baldrics.[59] As with blue painted helmets, this may have been a distinction common among Macedonian infantry units. The colour of the sword baldric is intriguing considering that in a Roman painting of an 'Alexander' from Pompeii the conqueror is presented heroically nude with a straight sword hanging upon a thin blue thong.[60]

On the subject of *hemithorakes*, or 'half-cuirasses', Polyaenus mentions how an element of Macedonian phalangites, perhaps file-leaders only, wore this style of armour, ostensibly following the first attempt to penetrate the Persian Gates:[61]

> Alexander gave half-breastplates to soldiers instead of breastplates so that if they stood firm they would be safe because their front parts were covered, but if they fled they would be unable to protect their backs. Accordingly, no one fled lest he be without armour, but they always stood firm and conquered.[62]

The reasons given for the adoption of the *hemithorax* may be apocryphal, and in fact the underlying veracity of the passage has been seriously questioned.[63] Even if this feature of panoply was ever embraced by Alexander, it would be done as an experiment in manoeuvrability and to promote better comfort and ventilation when fighting in a humid environment.

The exact appearance of the problematic *hemithorax* has been open to conjecture for some time. Keeping the originally leather covered iron corselet from Royal Tomb II in mind, and citing both the Alexander Mosaic and sculptures from the Artemesion at Magnesia (200–150 BC), Everson has made the suggestion that what designated a cuirass from a half-cuirass was the amount of metal used.[64] Yet this hypothesis misinterprets the criteria from Polyaenus where the *hemithorax* is clearly delineated protecting the front only, leaving the back exposed.

A bearded figure depicted on one of the pediments from the Alexander Sarcophagus (Plate 1b; figure to the right) is thought to represent Alexander's half-brother Arrhidaeus (later Philip III).[65] His yellow-coloured (for bronze or gilded) armour is sculpted as a 'muscle cuirass' but its lack of *pteryges* and omission of shoulder pieces may denote that the front plate alone is depicted. If correct, then the artist could well have intended a *hemithorax*. Lumpkin proposes another possible image of this mysterious armour on the victory frieze from the Sanctuary of Athena at Pergamum (first quarter of the second century BC), the object appearing, once more, as the front plate of a muscle moulded cuirass.[66] The Nikolaos *stele* shows a Macedonian phalangite clearly wearing a muscled cuirass (Fig. 23).[67] In view of the artistic evidence from the Sarcophagus and Pergamum, we might have the likeness of a *hemithorax* on this *stele*, provided that the back plate was absent.

Pollux mentions the *hemithorax* as Thessalian in origin, invented by the tyrant Jason of Pherae,[68] theoretically part of a developing trend for lighter gear within the cavalry.[69] Evidence for the type of armour used can be gleaned from a series of coins minted by Alexander (d. 367 BC), Jason's successor, which show an armed horseman in Boeotian helmet and metal muscled cuirass with *pteryges*.[70] Jason was certainly known to have set in train a raft of military reforms[71] and the earliest extracts to mention the *hemithorax* refer to Thebes in the early- to mid-fourth century BC, which cultivated close diplomatic ties with Jason's regime.[72] His military leadership could well have had some bearing on Philip II and his model army.[73] Given that Thessaly was an immediate neighbour, the armour style would have found its way to Macedonia directly or via Philip's Theban borrowings, although this remains speculative.[74]

Shield Devices

The characteristically convex Macedonian infantry shield lacked the offset rim associated with that of the Greek hoplite and was smaller, usually about 60–70cm in diameter.[75] However, some larger, more deeply bowled shields are carried by *sarissa*-armed foot soldiers in art.[76] Depictions from wall paintings and actual fragments have yielded diameters rounded as 72, 74, 75 or up to 80cm.[77]

Shield faces often had concentric patterns with crescents and sun motifs and several creditable theories have been put forward about their origin and meaning. The sunburst motif may have had some astrological significance associated with the power and protection from the Sun-god, with crescents evocative of the moon or orbit of planets.[78] We should also take into account extracts from Diodorus, Curtius, Plutarch and a fragment from a late Hellenistic or Roman poet which help confirm Alexander's identification with Helios and his use of solar imagery, symbolic of universal dominion in the East.[79]

Liampi traces the sun emblem to the reign of Archelaus, supporting her argument with a small decorative shield found at Olympia which displays a central studded circle surrounded by ten double crescents (425–375 BC). This contention implies the use of crescentic theme shields within close proximity to Macedonia well before Philip II's reign.[80] Liampi's thesis for the motifs introduced under Archelaus has been accepted by other academics such as Stella Miller.[81] A faded wall painting from a tomb at Katerini has two shields which show concentric patterning. Both have preserved paint pigments described as pale yellow with a plain central disc in a deeper yellow hue.[82] The fact that this tomb has been dated to the earlier decades of the fourth century BC suggests that shields with this ornamental form were known to Macedonian soldiers around fifty years before Alexander's Asian conquests.

An extant shield monument from Beroea gives further clues to the appearance of shields. It has been posited that the site was built to commemorate the bloodless overthrow of Demetrius Poliorcetes and the accession of Pyrrhus as King of Macedonia and integrates a series of shields on its north elevation.[83] Interestingly, the position of the five smaller shields at the centre is believed by Markle to convey five different units of pike armed infantry in the Macedonian army. We can deduce that the absence of any visible relief decoration on them means that originally they had been painted with devices. Another imposing monument discovered at Archontiko close to Pella has a course of marble displaying round shields. They are once more smooth and unadorned, suggesting that the fabric was painted.[84]

A disc at the centre of shields was a popular feature and could display a number of devices. Anonymous bronze coinage from the time of Alexander and his immediate heirs, showing the motif of a Macedonian shield, has a central device in the form of an eight-, twelve- or sixteen-rayed sunburst, or alternatively a torch, spearhead, double-headed axe, horizontal thunderbolt, Gorgon's head, club or head of Heracles or staff of Hermes.[85] Plates 13 and 14 provide two examples of such central devices: the obverse of the coin in Plate 13 has a head of Heracles in lion skin cap at the centre with five double crescents and pellet patterns around it; the obverse of the coin in Plate 14 displays a Hermes staff at the centre with six double crescents.

Kings, deities or mythical personalities were all popular central disc themes. The Alexander Sarcophagus has three bronze examples of shields with the heads or busts of royal personages or male and female deities (the face of one of these shields can be partially viewed in Plate 1b).[86] This same style may have been prevalent in the army as late as the third century BC, seen with the excavation of fourteen round miniature terracotta shields in the Tomb of the Erotes, arguably the resting place of a high ranking dignitary from the Macedonian garrison town of Eretria.[87]

From reproductions kindly supplied by the staff of the Museum of Fine Arts (Boston), the Tomb of the Erotes miniatures can be grouped into three main central device themes: the bust of Alexander (or of one of the *Dioskouroi*, twin star-crowned gods); a possible bust of Alexander-Helios; and a winged Gorgon's head. The bust of Alexander motif can be seen in Plates 18a and b with a central device showing a young male head with white face and pale yellow hair (flanked on each side by an originally yellow eight-pointed sunburst in Plate 18b). Plate 18c demonstrates the second type (Alexander-Helios) where the head is surrounded by a yellow halo of sun rays. The third central device theme variation can be observed in Plate 18d displaying a Gorgon's white face, yellow hair and wings upon a sky blue cape of scales (*aegis*). Although not easily discerned in the photographs, the shield rims appear to be pale yellow and shield face backgrounds light mauve or purple, terracotta red, pale pink or sometimes sky blue.

Of the above categories, the one with a possible bust of Alexander-Helios is particularly relevant to the subject of the armies of Alexander and the Successors because the same design appears on a terracotta shield decoration from a couch found in a burial at Pydna dated to the end of the fourth century BC.[88]

The central disc was often rendered in some shade of red or purple. Indeed, the application of a central purple disc on shields became a trademark of Macedonian panoply, adding to a sense of military distinctiveness and strong *esprit de corps*. A figure from one of the Sarcophagus pediments wields a shield with a particularly large purple disc, and a vase from the Amphipolis Museum shows a soldier handling a shield with an indistinct sky blue blazon within a bright red disc surrounded by a broad white band.[89] For another example of a shield with a central red disc (and eagle theme), see the description of a young man representing a symbolic portrait of Ares or a deified Alexander from Bella Tomb II in Appendix 1.

Surprisingly overlooked by academics is the appearance of two possible Macedonian bronze shields from the Alexander Mosaic. These can be partially seen within a stand of *sarissai* to either side of Darius' chariot (Plate 3b). The shield to Darius' left has a human face within a purple disc encircled by a thin red band. On the other hand, the shield to Darius' right is almost entirely obscured but seems to have a white centre surrounded by a red border.[90] A relief battle scene from a clay 'Italo-Megarian' cup (found in southern Umbria, 125–75 BC) may derive from the same original early Hellenistic masterpiece as the Alexander Mosaic. Here a foot soldier wears a helmet, his body covered by a round shield rendered with a concentric band design, comparable to the painted specimens listed above.[91]

One of two stucco relief shields from the facade of Royal Tomb III at Vergina (the possible burial place of Alexander's posthumous son Alexander IV, r. 323–310 BC) has a faded garland of leaves in red comprising the outer band, with a large purple central disc within which is a Gorgon's head with dark tangled hair.[92] Given the indisputable Homeric flavour implicit in Macedonian warrior society, it is intriguing to recall the shield of Agamemnon described by Homer, with a roundel of *cyanus* at its centre upon which was emblazoned a wide-eyed and grimacing Gorgon's head.[93] *Cyanus* can be defined as a dark blue enamel or glass paste resembling *lapis lazuli* and is not far removed in colour from the purple disc so often seen on Macedonian shields.[94] Curiously, the ceremonial shield found in Royal Tomb II had transparent glass inlays as a decorative element.[95]

The overwhelming weight of evidence in the artistic corpus attests to the use of vibrant, even garish, polychromatic colours applied on shields. This was because the intention was to overwhelm and intimidate the enemy with a striking and varied mass of colour.

The Agios Athanasios tomb frieze serves as an excellent example of this enthusiasm for brilliant colour combinations. Out of all the figures on the frieze, three foot soldiers are shown with shields (Plate 10b).[96] Two of these shields have purple central discs and the third a pale grey-purple or mauve one, all of which are encircled by broad bands of different colours (blue, cream or red). The central disc devices appear in the form of a yellow-gold sunburst or a vertical winged thunderbolt. The bands around them have sets of crescents in matching yellow-gold, some separated by miniature thunderbolts.

It can be observed that one of these three shields has a bright red outer rim. This particular shield is carried by a figure wearing a purple-coloured (or perhaps iron) Phrygian helmet with white side feathers, denoting an officer. Therefore it is possible that the red rim suggests his field rank. A similar red rimmed bronze shield carried by a Greek or Macedonian footman appears on an ancient Italian vase with battle imagery possibly deriving from the same early Hellenistic painting as the Alexander Mosaic.[97] It is tempting to think that the colour of the outer rim had some deliberate visual recognition purpose when troops were in formation.

The back wall within the Agios Athanasios tomb is adorned with traces of a large blue and red coloured shield, while the doorway is flanked on each side by an almost life-sized and spear-armed human effigy in mourning with a painted shield directly above each (Plates 11a and b). As both these persons wear *kausiai*, it is likely that they represent dismounted cavalrymen.[98] Of the two shields, the one to the left (Plate 11a) has a broad ochre or gold outer rim and a pink-purple field upon which is a Gorgon's head with brown hair and pale skin. The shield to the right (Plate 11b) has a blue outer rim and a bright red field upon which is a horizontally set thunderbolt with white wings and ochre lightning strokes.[99]

If the shield face designs seen with Macedonians and described here are acceptable as authentic unit devices, some of their artistic features may be symbolically significant, conveying a specific standing or prestige. Alexander is known to have relentlessly driven the competitive ethos in his army, differentiating between those who had fought better and those who had not, either as individuals or groups.[100] It would not be surprising if this was shown in some way through their shield designs. The typical use of coloured concentric bands (often red, blue, white or left plain bronze) between the central disc and outer rim of Macedonian shields, and the inclusion of embossed pellets, sunbursts, crescents and thunderbolts may have been far from arbitrary. It is fascinating to theorise that their number, arrangement and interrelatedness were meant to signify distinct units or even subdivisions within a particular unit.[101]

Rossi applies the above reasoning to the shield badges of Roman legionaries and auxiliaries sculpted on Trajan's Column, with units displaying an encircling laurel wreath emblem on the shield face to signify being awarded the status of *torquata* (or 'wreathed') for valour in battle.[102] The same hypothesis is effectively borrowed by Sekunda to explain the wreath iconography on a cavalry shield painted on a wall from the Antigonid Lyson and Kallikles tomb.[103] Anson recently suggested the sunburst emblem as a distinction only seen on shields of hoplite equipped Macedonian *asthetairoi*.[104] Studies of late Antigonid Macedonian shield attributes continue apace with Juhel's analysis of the thunderbolt on a shield from the *stele* of Amyntas.[105]

Evidence for Macedonian shield designs can be possibly sought even further afield. In Italy, Roman murals from the extant remains of villas in Neapolitan towns (second century BC to 79 AD) were influenced by artwork executed in Macedonia or the Hellenistic world.[106] The potentially rich, though little explored, vein of visual material from Pompeii, Herculaneum and Stabiae offer examples of purple disc shields, though some may only be funerary *tondi*, or circular paintings.[107] The Roman Villa of Poppaea at Oplontis (near Pompeii), which was probably owned by the Emperor Nero and occupied by his second wife Poppaea Sabina, includes a fresco depicting a shield with a dark purple disc within which is an idealised face. In the same villa, hanging high along a marble colonnade, can be found a row of three striking bronze shields, each one with an eight-pointed sunburst (a single pellet between each ray) within a central purple disc (Plate 12).[108]

Turning finally to shield interiors, all indications are that their leather surface was dyed, with a variety of colours employed. The Alexander Sarcophagus has three in apparent red, dark red or red-brown, with one on Pediment A in the same general shade but with an indeterminate pale blue or purple rim (Plate 1a).[109] The extreme bottom left-hand corner of the Alexander Mosaic displays a shield interior described as black, suggestive of a deep or shadowed blue or purple tint. By contrast, the rim is coloured pale pink-red or mid-purple.[110] This shield may be Macedonian[111] or, more likely, that of a Greek mercenary fighting in Persian service.[112] One of the two Macedonian foot soldiers from the Amphipolis vase presents the inside of his shield as bright pink-orange (Plate 33a). Ptolemaic murals (dated to the second century BC or earlier) from a building dedicated to Alexander in the site of Kom Madi near Cairo have soldiers marching in step; four of them display identical blue shield interiors rimmed in some shade of red, two pale yellowish with red or blue rims, and one dull red with a matching rim.[113]

Greaves and Bare Feet

The full-length bronze greaves (*knemides*) from Royal Tomb II were regarded by Andronikos as the least remarkable items of panoply found there.[114] Greaves were nonetheless considered important enough pieces of kit to be included in the initial ordinance for Philip II's phalangites, which apparently omitted such a notable element of armour as the corselet.[115]

In conventional terms, greaves were worn to protect their wearers from steep trajectory missiles penetrating a hoplite phalanx and doing damage to the human shinbone. There were different reasons why they were chosen for Macedonian infantry.[116] With the addition of greaves to helmet and shield, the leading ranks of Alexander's pike formations were fully protected almost from head to foot. Moreover, we should take into account that enemies would be faced with the formidable physical and psychological challenge of an advancing wall of gleaming metal behind a dense forest of *sarissa* shafts so that the collective effect must have been overwhelming.[117] There is no coincidence between this effect and glorified descriptions of massed Homeric heroes in the *Iliad*.[118]

Extant Macedonian greaves cover the leg from the top of the knee to the instep.[119] In the Greek world, they were usually hammered out by the armourer from thin bronze sheets, often deftly moulded to imitate the muscularity of the lower leg and calf, ensuring a more comfortable fit, the design opening behind the knee for articulation. Yet despite moulded musculature having been the norm before, it seems to have been less so for the Macedonian army during the mid- to late-fourth century BC. Everson perceptively concludes that those seen on the Alexander Sarcophagus[120] and retrieved from Vergina and a tomb at Korinos (Plate 23) are purely functional and have less anatomical detail than usual.[121]

Greaves from Macedonian gravesites at Derveni (Plate 24), Sevasti and royal tombs at Vergina have been measured at around 1mm thick at the edge, with some inner parts at 5mm. Most finds of bronze greaves from the classical hoplite era weigh only 0.5–1kg, and the practical experiments of reenactors demonstrate that wearers could run, climb and even dance in them without notable hindrance.[122] In the case of those worn by Macedonians shown on the Sarcophagus, a single garter strap (for a closer fit and weight support) was located just below the knee. The imprint of similar straps was found on gilded greaves from Royal Tomb III, noted for their apparent palm frond decoration at the ankles (Fig. 26).[123] Red lining, probably implying soft leather, was often turned over and stitched at the greaves' edges to prevent chafing.[124] Padding helped to make them tolerable to wear for extended periods. Of four sets from Royal Tomb II, a gilded bronze pair has scraps of leather with preserved thread on the upper edges, indicating where it was sewn to the metal.[125] The swelling at the kneecap could allow for thicker padding to be inserted from behind at this vital point.[126]

Those *pezhetairoi* who wore extensive body armour were required to put on their greaves first before donning the corselet simply because it would have been

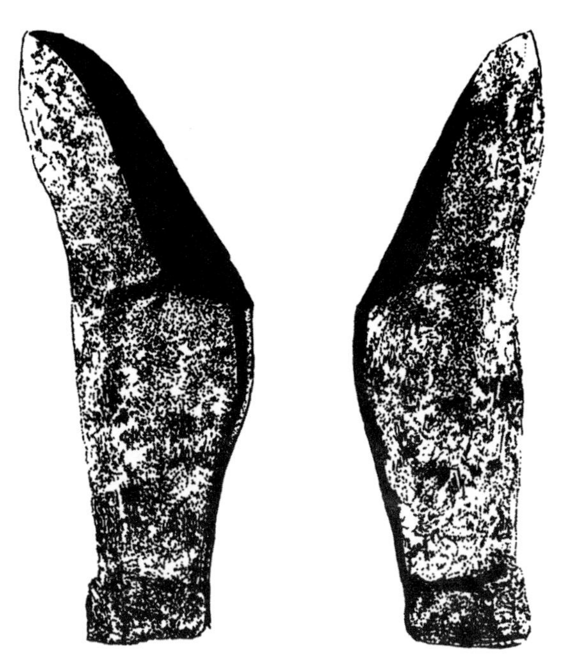

Figure 26. Gilded bronze greaves from Royal Tomb III.

difficult to bend over and undertake this task in stiff and bulky chest armour.[127] It is unlikely that each greave was custom fitted for common rankers. There is evidence for state issued armour tokens in the Hellenistic world denoting the existence of set sizes – a reference to 'small', 'medium' and 'large' has been mooted.[128] A soldier would therefore have had to bend and tweak his own pair until satisfied, attempting to avoid the discomfort of the greaves cutting into the ankles or top of the feet.[129]

It seems plausible that even after having achieved a good fit, if sweaty skin was always rubbing against the inside surface when on the march, occasionally friction over time gave rise to irritating skin abrasions or infections. Hanson adds that because of the constant movement of the leg, the greaves of Greek hoplites became distorted in shape, causing the nuisance of frequently having to rebend them.[130] These tedious drawbacks would have encouraged men to abandon their greaves on campaign, although some soldiers might stow them away with their personal baggage.[131] Many centuries later in the *Strategikon*, attributed to the sixth century AD Byzantine Emperor Maurice, the use of heavy boots and greaves is disapproved of for marching infantry.[132] In Hellenistic times problems of comfort seem to be hinted at in a speech given by Polybius for Philopoemen who recommends that 'a man starting for a review or campaign should in putting on his greaves take more care to see that they fit well and look shiny than he does about his shoes and boots'.[133]

It is possible the rear ranks of a pike phalanx had occasion to discard greaves altogether. The view that they were worn less often by Macedonian infantry in the late-fourth century BC is supported by evidence from the Alexander Sarcophagus. From a total of seven armour clad figures on foot identified as Greek or Macedonian, only two wear greaves. From other features of dress, these two figures were intended as line officers, possibly file or half-file leaders.

Regarding infantry footwear, Xenophon describes Greek mercenaries with boots the straps of which froze to feet during long marches in terrain with deep snow cover. He refers to soldiers having to make do with crudely made moccasins stitched from freshly flayed hide when their old shoes wore away.[134] Besides this, hundreds of 'bone buttons', probable eyelets for 'running sandals', were retrieved from the Greek mass burial at Chaeronea, which may be the resting place of the Theban Sacred Band. It reveals that these men literally died with their boots on.[135]

Although the evidence from Chaeronea proves that some infantry wore footwear during combat, surviving art usually testifies to barefoot Macedonian infantrymen.[136] After all, there were definite practical reasons for boots to be left off. In essence, the phalanx formation relied on its speed and physical impact to come to grips with a startled enemy on the battlefield, and modern reenactors have discovered that bare feet give improved physical control and balance in combat. A preserved pair of Etruscan sandals weighs 0.67kg.[137] It is clear that had the *pezhetairoi* worn *krepides* in combination with greaves, leg movements would have over time become laboured, with mobility impaired. Footwear was hence unnecessary *impedimenta* during a speedy charge or engagement.[138]

Xenophon opined that soldiers who were frequently on campaign had 'feet case-hardened by constant training, and, when tramping over rough ground, must differ from the uninitiated to the sound man from the lame'.[139] Analyses of Roman gladiator remains unearthed from grave plots in Ephesus and examined by forensic anthropologists at the University of Vienna show a significantly higher than normal bone density in the feet, indicating strength gained through intense training and fighting without footwear.[140] Incidentally, to this day going barefoot is sometimes approved as an effective form of infantry training. Modern long distance runners and ramblers report a number of health benefits including stronger tendons and muscles in the ankles, feet and legs, more traction in muddy ground and nimble movements.

It is, however, rash to suppose that the *pezhetairoi* never wore footwear under *any* circumstance whatever. We should recall Pliny, who said *krepides* were used for marching.[141] Also, four out of sixteen Macedonian infantry sculptures on the Alexander Sarcophagus whose feet can be viewed wear *krepides* in the action scenes.[142] Of the overall total, six are heroically nude, so it is no surprise to see them with bare feet.[143] On balance, we are justified in proposing that Alexander's infantry had occasion to wear *krepides* for lengthy marches, but would often take them off for improved agility in battle. This practice was not unknown in antiquity, as evidence from New Kingdom Egypt suggests that soldiers wore sandals on the march, but shed them before combat.[144]

Chapter 8

Officers

Having examined select aspects of panoplies relating to cavalry and infantry units, we shall now study those features which distinguished kings, generals and officers. Clearly, many features of Macedonian officers' attire were calculated for maximum propaganda effect – be it the overall magnificence of their panoplies, purple and gold clothing (following Persian fashions), distinctive ornaments and attachments on their helmets and corselets, emblems on the standards of the units they led, or the use of panther skins as saddlecloths. The heroic feel the officers and their men wanted to project is also evident in the use of often decorated swords, plied with deadly effect in combat.

A Taste for Rich Panoplies

Generals and subordinate officers favoured ostentatious panoplies. It was a trend that would have become more marked after the seizure and dispersal of up to two centuries of hoarded Achaemenid wealth during and after Alexander's time. It was an opportunity which allowed them to give full vent to their personal vanity and military triumphalism. Still, no matter how outstanding their equipment was, Alexander himself could be recognised from among those who attended him for the magnificence of his own gear.[1] His armour and weapons were identified by veterans as belonging to him even years after his death.[2] Historically, Alexander's armour may have become so recognisable that some Roman generals centuries later, such as Germanicus (d. 19 AD), could have purposefully copied elements of it.[3]

Speaking more generally, evidence suggests that officers could be distinguished on the battlefield by some characteristic features in their dress and weapons. Speaking in his own defence during a trial in 330 BC, Amyntas, son of Andromenes, who was in usual command of a *pezhetairoi taxis*, requested that the apparel of an *armiger* (literally an arms or armour bearer) be restored to him.[4] More intriguing are details emerging from the duel between Eumenes and Neoptolemus. Both men are described by Diodorus having picked out one another by their horses and other 'insignia'.[5] The original noun for 'insignia' was *episema*, usually meaning a device on a shield. However, as both men are narrated on horseback it is unlikely that their alleged marks of recognition were on this type of armour. Shields do not crop up in surviving accounts of the fight and such equipment was not strongly associated with mounted Macedonians at this time. On the other hand, the combatants are described by Plutarch wearing helmets and corselets.[6] We might therefore suggest that they were able to recognise each other as *individuals* by distinct features on their helmet, corselet, weaponry or costume, announcing who they actually were, rather than simply what their rank was.[7]

The sophistication of panoplies for the highest ranking officers required them to be wrought by leading artisans who combined the most up-to-date technical improvements with artistry flaunted with exquisite flair.[8] It is possible that, as with kings, a leading man could own several sets.[9] The elite especially made use of a flourishing international market, patronising the finest talent from around the Greek world and eastern Mediterranean rim. The best example of this is Plutarch's account of the master armourer Zoilus of Cyprus, who presented Demetrius Poliorcetes with two iron corselets during the siege of Rhodes (305–304 BC).[10] Both were about 18kg in weight and in a test to prove their worth one of them withstood a catapult bolt shot at a distance of 20 paces, apparently with nothing but a scratch to show for it. The other set was granted to the officer Alcimus the Epirote, a man distinguished for his powerful physique.[11] Such equipment must have been quite thick and stands comparison with plate armour from the early European Renaissance.[12]

Cypriote armour had a long established reputation for excellence by the fourth century BC. In the *Iliad* Agamemnon was sent a gift of a splendid *cyanus* adorned corselet from King Cinyras,[13] and some bronze figurines (c. 700 BC) found at Salamis, Cyprus, have scale corselets inlaid with blue paste, which may be an attempt to recreate *cyanus.*[14] A small remnant of iron armour (c. 470 BC) composed from many splints was discovered by a team of Swedish archaeologists digging at Idalion during the 1930s.[15] This belonged to a corselet made up from many thousands of splint pieces overlapping one another two or even four times. The evidence attests to a meticulously wrought work of art, produced by patient artificers of exceptional calibre.

Demetrius' armour was probably constructed differently but it is possible that, by obtaining his corselet in the way he did, his intention was to draw a parallel with Agamemnon and his opulent corselet. As with Demetrius, Agamemnon's armour was worn during an epic siege. Demetrius could have ordered his own set for propaganda purposes in order to emulate the legendary Homeric king.

Officers' Dress

As far as the clothing of commanders is concerned, several literary extracts seem to point to a unique cloak worn by Alexander, but little can be gleaned about its exact colour and configuration. Plutarch (following the lost narratives of Chares or Callisthenes) has the King in a long antique cloak (*epiporpoma*) made by Helicon, gifted by the citizens of Rhodes. This was said to be more ornate than any part of his already ornamented armour, and he habitually wore it in battle at least up to and including Gaugamela.[16] Its appearance is possibly witnessed with the figure of Alexander on the reverse of large better preserved 'Alexander' medallions.[17]

Curtius informs us how at the Hydaspes an officer called Attalus was instructed to wear royal robes (*vestis regia*) to fool Porus into believing that Alexander was on the opposite bank and not intending to cross.[18] Furthermore, after recovering from a terrible wound received at a town of the Malli in early 325 BC, the King was informed in his tent of many troops who were clamouring to see that he was well, so he took his cloak and went out to greet them.[19] This incident proves once again that Alexander's rich military gear, and specifically his cloak, were familiar enough for a throng of soldiers to instantly distinguish him at a distance and from among those around him.

Although Hammond assumes white as the colour of choice for Alexander's battle cloak, basing himself on the decidedly non-military white *himation* worn by Philip II at Aegae (336 BC), purple and gold offers as good, if not a better, candidate.[20] Demetrius Poliorcetes and Pyrrhus were both keen Alexander imitators and are documented with fabulous purple and gold cloaks either in or out of battle.[21] The Roman emperor Gaius Caligula bridged the Bay of Baiae in Italy wearing an outfit consisting of a corselet purportedly worn by Alexander and a mantle which was either cloth of gold or purple silk decorated with gold embroidery and Indian gems.[22] The mosaic from Palermo reproduces 'Alexander' hunting in a white tunic and purple *chlamys* with one or possibly two central vertical yellow stripes (Plate 36a).[23] Versions of the *Alexander Romance* even convey the King being sent a purple- and gold-threaded cloak as a royal present.[24] Richly dyed articles of apparel resurface with a surviving Sibylline Oracle which pictures Alexander in poetic form invading Asia with a purple cloak over his shoulders.[25] Nor should we overlook the long-sleeved faded purple-grey (originally deep purple) tunic and cloak worn by the young Macedonian king on the Alexander Mosaic (Plate 3a).[26]

For the sake of our argument, it is helpful to bear in mind that Alexander dressed in purple even before the more famous and controversial introduction of Persian features into his regalia. There is reference to him being sent from Macedonia a consignment of clothes (*Macedonicas vestes*) and a large quantity of purple material.[27] Despite the looming possibility of literary invention, he is portrayed by Curtius saying of the same stuff that 'these clothes I am wearing are not merely a gift from my sisters, but also their handiwork'.[28] This seems to complement Athenaeus, who drawing on Ephippus informs us how Alexander wore purple for everyday use.[29]

After the annexation of Persia Alexander seems to have extended the privilege of wearing purple to more of his elite officers.[30] In literature, the satrap Orsines is noted by Curtius travelling from Pasargadae with gifts of 'purple vestments' (*vestesque purpureae*) intended only for the King and his 'Friends'.[31] Cicero conveys officers in purple as part of the immediate entourage of Hellenistic Macedonian dynasts, as does Livy.[32]

Of the different officers in the Macedonian army, the elite *somatophylakes* were some of the closest to the royal presence.[33] They originally consisted of seven, later eight,[34] personal bodyguards or adjutants, charged with special missions and command appointments.[35] Leonnatus was promoted to the post of *somatophylax* when a vacancy appeared with the death of Arybbas in Egypt (332/331 BC).[36] Curtius refers to him as a 'wearer of purple'.[37] Heckel maintains that at the time of the battle of Issus Hephaestion was one of the *somatophylakes* as well.[38]

In the days after Issus, Alexander and Hephaestion had an audience with Sisygambis, the captive Persian Queen Mother, who mistook Hephaestion for the King as he was taller and more handsome.[39] In light of all the evidence for Alexander's dress, Hephaestion was most likely purple clad, as both of them are described resembling each other so that Sisygambis was unable to single out the King due to any intrinsic sign or aspect on his clothing. A mounted figure from the Alexander Sarcophagus in long sleeved purple tunic and under-tunic (Plate 2a) has been widely (though tentatively) accepted by scholars as an idealised portrait of Hephaestion.[40] The colours of this figure accord well, in some respects, with the purple pigmented costume of 'Alexander' from the same monument.[41]

A likely portrait of a preeminent officer, perhaps even a *somatophylax*, can be seen in the Hunt Frieze facade from Royal Tomb II, which shows a spear-armed man on foot poised to strike a wounded lion (Plate 9b). He is deliberately located between two mounted royal figures painted in purple tunic and cloak (either Philip II with Alexander or Philip III with Alexander IV) and both his *kausia* and *chlamys* display a faded purple tint.[42]

Purple-dyed *kausiai* (composed of a finer grade of soft leather) and *chlamydes* were bestowed on the king's authority to denote a privileged status.[43] In 320 BC Eumenes was empowered to distribute such items to each of his 1,000 *hegemonikoi*, or officer-bodyguards.[44] Moreover, Onesicritus presumably had to wear purple *kausia* and *chlamys* (with *krepides*) when assigned to interview Indian ascetics.[45]

The almost completely destroyed figure of an officer from the Alexander Mosaic running alongside Alexander is shown wearing a purple *kausia* with a faded red cloak. This figure has been variously identified either as an officer of hypaspists, an officer of Agrianes or a leading personality such as Ptolemy, Peucestas, Seleucus or Lysimachus.[46] The reverse of coins from the reign of King Patraos of Paeonia depict an armed man with *kausia*, who is similar in appearance to the Mosaic soldier (Fig. 25). In addition, a Roman wall painting from the villa of P. Fannius Synistor at Boscoreale near Pompeii incorporates a figure wearing a purple *kausia* and a long sleeved tunic, leaning on a *sarissa* with a Macedonian shield at the feet, speculated to be either Achilles or Alexander the Great, Alexander IV, Antigonus Monophthalmus, Antigonus Gonatas, Pyrrhus, Cleopatra or a female personification of Macedon (Plate 31).[47]

Apart from the Alexander Mosaic footman (wearing a red cloak, as discussed above), cloaks coloured red emerge from a number of artistic sources said to feature infantry officers. Two fresco compositions on the subject of 'Jason and Pelias' found in the House of Jason and the House of the Gilded Cupids in Pompeii (Roman copies of a possible single late-fourth century BC Hellenistic work) portray 'Jason' in a crimson cloak with pale blue band (Plate 32; far right figure), identified by Sekunda as suitable for a senior officer of Macedonian infantry.[48] Similarly, a painted Ptolemaic grave *stele* from Alexandria, showing a Macedonian soldier on foot (320–275 BC, Louvre Museum), has a *chlamys* which photographic reproductions reveal as red with a wide border in light blue.[49] The same red and blue colours occur on the cloak of a Macedonian soldier from the so-called 'Stele of the Warriors', Pagasai-Demetrias (250–200 BC).[50]

Disparate strands of evidence imply that, similar to their commander, certain high-ranking officers in Alexander's army wore purple and gold cloaks. Whether these references are to *somatophylakes*, other prominent individuals, or both, is unclear. In one instance Curtius tells us that Hephaestion was given orders to invest Abdalonymus with the title of King of Sidon, with a 'purple and gold embroidered robe' and other distinctive royal insignia, elevating him to 'Companion' status.[51] A statement in Justin reveals how after the conquest of Persia the King's 'friends' were instructed to wear 'long gold and purple robes',[52] which could suggest the distribution of Persian court garments,[53] the term 'gold' maybe conveying exaggeration or misidentification for saffron.[54]

Curtius describes how leading officers were required to don Persian clothes, but no further details are given.[55] Some stray lines from the lost *Histories* of Duris of Samos (c. 300 BC) render Alexander's older generation officer Polyperchon clad in saffron during a drunken revel.[56] Such articles may have been worn as much for hunting and war as for ceremonial occasions and drinking parties. The sculpture of a naked and youthful warrior slaying a deer from Long Side B on the Alexander Sarcophagus has a purple *chlamys* bordered in yellow.[57]

It seems that a similar fashion style prevailed among bodyguards later in Hellenistic times. An extract from a work on Alexandria by Callixenus specifies the pavilion of Ptolemy II festooned with cloth of gold tunics and cloaks (probably embroidered with gold), set off with portraits of the King and mythological scenes.[58] These resemble purple and gold cloaks showing comparable subjects, worn by Companions, Friends and other armed groups at the Seleucid Daphne parade.[59] In the *Aeneid* the Syrian priest Chloreus is pictured in battle clad in a gold decorated purple cloak with saffron-dyed tunic, suggestive of the stereotypical clothing expected of those in close attendance around eastern kings.[60]

Archaeology may help to confirm sumptuous attire of this kind. At Agios Mamas (on the coast south of Olynthus) a series of grave sarcophagi was found dating to the early Hellenistic period, with artefacts including a gold-enhanced iron spear point and what has been quoted in one recent report as purple and gold garments (although perhaps only remnants of funerary cloth).[61]

Finally, it is telling that Craterus (an indefatigable paragon of military virtue and conservatism), after his return to Macedonia, was not averse to wearing splendid dress resembling that of Alexander himself, although he had been initially opposed to the introduction of Persian fashions in the King's lifetime.[62] After victory at Crannon (322 BC), Craterus was even seen by attendant Greek envoys sitting in state upon a golden couch and wearing a purple cloak.[63] He was so popular with veterans that, as told by Plutarch, 'if they could only see his cap [*kausia*] and hear his voice, they would come to him with a rush, arms and all'.[64] During battle near the Hellespont, Craterus fell from his horse and was killed by a Thracian or group of Paphlagonian soldiers, who failed to recognise who they had slain as his (presumably purple) *kausia* had been removed.[65]

Officers' Helmets and Helmet Fixtures

Along with purple clothing, such items as polished iron or silvered helmets, helmet crests, plumes and other ornaments could all serve as distinctions.

A silvered or polished iron helmet was commonplace among elite units, officers and generals. According to Plutarch, at Gaugamela Alexander wore a helmet wrought from fine iron that shone like polished silver, fashioned by Theophilus, a renowned armourer.[66] Helmets described as iron covered in silver are also mentioned in a sanctuary inventory from Delos (279 BC).[67]

Actual iron specimens from archaeological sites dating from the mid- to late-fourth century BC and the Hellenistic Age include the stoutly made Attic form with raised 'griffin' beak from the main chamber of Royal Tomb II. This looks similar to a helmet worn by King Priam pleading before Achilles from a wall painting (House of Ottavio Quartio, Pompeii).[68] An obscure fragment of an iron helmet calotte from another tomb at Vergina has been found, and a well-preserved type resembling the Tomb II find discovered near Marvinci.[69] It is plausible that Roman copies of earlier Hellenistic paintings preserve evidence for the use of iron headgear in Macedonian armies. One such copy is a vividly presented silver coloured crested helmet shown off by a young warrior from a 'Sack of Troy' fresco in Pompeii.[70]

The silvered iron Thraco-Attic headpiece from Tomb A (Prodromi) (Plate 20; helmet to the left) incorporates a rectangular device on both sides of the crest ridge, probably to accommodate feathers. A plainer iron sample from Tomb B (Plate 20; helmet to the right) has stylised relief ears behind the cheek guards and may likewise have displayed feathers or a horsehair plume. Of the two finds, the latter has been conjectured as belonging to a subordinate of the man who wore the former helmet.[71]

Tall helmet plumes were often part of a leading Macedonian soldier's regalia. Plutarch informs us that at the Granicus Alexander stood out from among his acolytes for the brilliance of his crest and high white feathers fixed on either side of his helmet,[72] although this headgear was damaged in the same battle.[73] The appearance of the helmet is supported by mounted and dismounted images of him on better preserved 'Alexander' medallions, which characterise the King in a Phrygian type topped off with a distinctive raised crest and two unusually large (perhaps feather) plumes.[74] Several extant Hellenistic onyx cameos of Alexander also depict him in ornate helmets, sometimes with a medial crest or side plumes.[75] The 'Craterus' bust from the Villa of the Papyri (Herculaneum) is defined as a Roman copy of a Hellenistic original portraying an unidentified successor of Alexander and includes a rounded peak which could have served to fix a plume.[76]

Apart from plumes, horns are another feature that turns up on the helmets of Macedonian kings. Pyrrhus is described with a helmet that was familiar to onlookers for its 'lofty' or 'towering' crest and goat's horns.[77] Seleucus (or even Alexander himself) is shown in an Attic form adorned with a mottled panther skin and the ears and horns of a bull on the obverse of a silver tetradrachm struck at Susa or Persepolis (303–300 BC).[78] A coin minted at Ecbatana from the same period has a mounted Alexander or Seleucus in horned helmet with billowing *chlamys* and lance levelled for the charge.[79] Horn fixtures (imitating a goat or bull) may in like manner have been fixed on helmets worn by Antigonus Gonatas[80] and Philip V.[81] This was for dramatic show and a way of declaring superhuman potency, courage and strength or, in the case of goat horns, as a means of invoking supernatural *paneia* from the god Pan – that is, the terrible power to 'panic' enemies.[82] A bronze appliqué found in Italy of a man wearing a Pan faced helmet with decorative ears and horns and the decorative border along the spine of the cockscomb offers a striking example (Fig. 27).

It is tempting to consider that the attributes attached to Alexander's helmets were not only chosen for practical purposes but were inspired by epic literature. In the *Iliad* Agamemnon is extolled for wearing a helmet with four white crests.[83] It is possible that the striking headgear worn by Alexander at the outset of the eastern *anabasis* was a specially commissioned object of propaganda, symbolic of the young King's newly inherited political and military status as 'Captain-Generalissimo' of a united Hellas. He may have wanted intentionally to draw

Figure 27. Hollow-cast bronze appliqué, identified as Antigonus Gonatas. Found near Grosseto, Italy.

favourable comparisons between his invasion of the Persian Empire and the great Agamemnon, 'lord of men', who led the Achaean host to Troy in ages past.

That Alexander was not averse to wearing armour intimating parallels with heroes is well seen with the apparent gold or bronze Nemean lion scalp worn by a mounted likeness of him on the Alexander Sarcophagus.[84] Portraits of the King wearing a lion skin cap persist in sculpture and numismatic evidence.[85] The imagery of extant coins indicate that he appropriated the Persian griffin emblem, allegorical of irresistible divine power,[86] and this actually turns up on helmet crests worn by him on two Roman weights, one preserved in a collection from the University of Delaware, the other from Pompeii.[87] The griffin emblem on helmets is observed on other evidence. Philip V wears a comparable helmet with griffin head-crest on a series of coins from his reign.[88] The same characteristic design is noted on a surviving combat scene from a battle of gods and giants frieze at Pergamum[89] and is found on a huge gilded bronze Phrygian 'show helmet' dredged from Lake Nemi in Italy.[90] From what we know, the griffin was to endure throughout the Hellenistic and Roman eras as a corselet emblem.[91]

Although Alexander is shown in action bareheaded on the Alexander Mosaic, it has been hypothesised that an unremarkable bronze Chalcidian type with white horsehair crest, only partially viewed on the ground in a damaged area directly below the King's horse, is the elusive royal helmet. It might have been portrayed here as having fallen from Alexander's head due to the sudden jolt experienced in transfixing the Persian horseman with a sharp thrust of his *xyston*. Alternatively, it was only intended by the artist as an aid to dramatising the general turmoil of a fast and furious engagement.[92]

Three mounted Macedonian figures in helmets with white plumes can be seen on the Alexander Mosaic (see Plates 3a and b for two of them). The plumes suggest that they represent different ranks of officer – perhaps a *somatophylax* and two line officers.[93] Their position near the apex of the assault column (as recommended in Hellenistic military manuals), supports the argument in favour of them being cavalry officers.[94] Oddly overlooked by researchers is the additional image of a Macedonian cavalryman barely seen on the deceptively less important, almost completely effaced far left of the artefact (Fig. 28).[95] Although executed in a highly impressionistic style, part of a gleaming iron or silvered cheek guard (and possible scale shoulder piece) can still just be made out.

By contrast, none of the horsemen on the Alexander Sarcophagus displays a helmet plume, despite there being several crested and feathered types painted into the flat background surface.[96] It is possible that in this case the artist excluded delicate crests from the combative figures due to their potentially brittle construction.

A number of features on helmets displayed by Macedonian infantry officers are worth relating here. As previously discussed in this book, infantry units from the last years of Alexander's reign onwards may have commonly worn the *pilos* helmet. Therefore coins depicting a *pilos* form could offer an insight into infantry officer headgear. These helmets include elements such as medial or even transverse crests, side plumes or an encircling wreath (see the reverse of two coins in

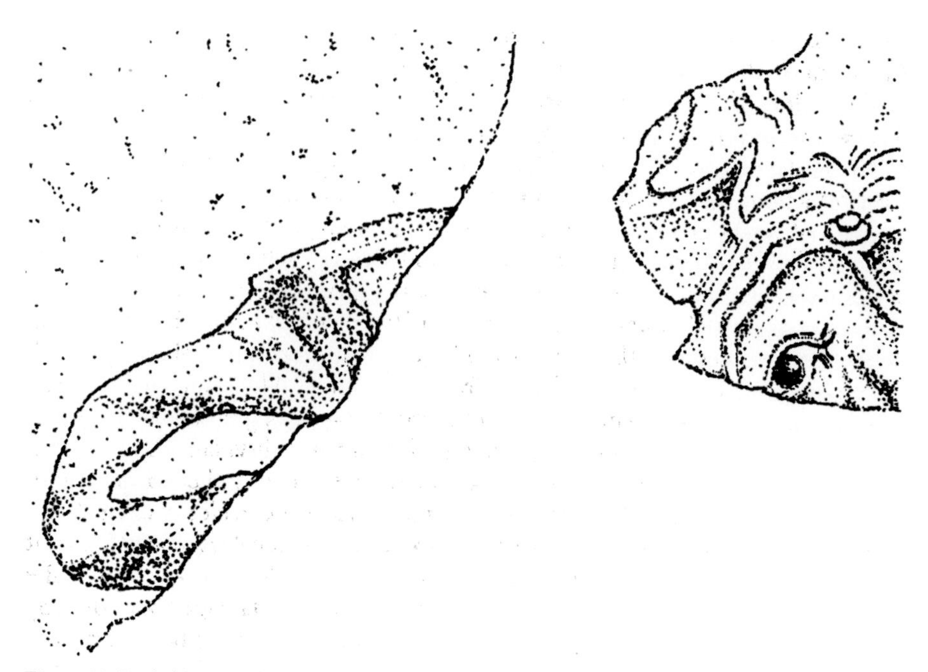

Figure 28. Probable Macedonian Companion on the far left of the Alexander Mosaic.

Plates 13 and 14 with a visible encircling wreath in Plate 13, cheek guards and transverse crest).[97] The distinctive transverse style seems to have been designed to be recognised from a distance, as with the crests of Roman centurions.[98]

Several foot figures from the Alexander Sarcophagus display small holes on their helmets presumably used for fixing miniature side plumes or feathers, which have long since vanished.[99] One of the soldiers on Pediment A (Plate 1b; standing figure on the left) is shown wearing a Phrygian helmet with holes for side plumes and a gilt spine. A purple-coloured helmet with a gilt spine is also seen lying abandoned on the far-right side of Pediment B.[100]

Surviving specimens give further evidence for plume attachments. The bronze Phrygian type from Vitsa (Plate 19; although with missing cheek guards and slight damage located along the ridge) has an appliqué flame palmette riveted on the peak of the cockscomb, above which are two small spools for the inclusion of a probable horsehair plume. Two other spools are soldered at either temple for the attachment of plumes or feathers.[101] A closely related Phrygian helmet (350– 325 BC), with repoussé palmettes on its peak and tubes for fixing plumes, was put up for auction by Christie's New York in 2004 (having formerly belonged to a private collection in Europe). Comparable palmettes are seen on a Phrygian type from the Thracian site of Kovachevitsa (Bulgaria).[102]

The helmet worn by one of the soldiers in the banquet scene from the Agios Athanasios frieze (Plate 10b; fourth figure from the left) closely resembles the Vitsa find in that it appears to be Phrygian with a medial crest and side feathers.

Close analysis reveals that this figure is wearing a finger ring, which may suggest that he is an officer. Plutarch actually describes rings worn by Greek cavalry and infantry officers under Timoleon in Sicily (c. 340 BC).[103] We can establish that this man is an infantryman because of his shield.

The surface of the helmet worn by the above soldier in the Agios Athanasios frieze can be detailed as red with spots of bright blue and faded purple.[104] In general, the way numerous bands, loops and dabs of colour were determined and brought together with other ornaments and fixtures on different sections of helmets was far from random. Rather, they were possibly meant to denote some form of individual 'heraldry' or defined ranks within the army's organisational hierarchy. One specimen from an infantryman on the Alexander Sarcophagus has a sky blue (maybe iron) helmet and a nape protector in contrasting purple.[105] This last detail suggests a specific rank (that of file-leader?), viewable to soldiers behind when in formation. It might be equivalent to unit numbers or rank markings seen on helmets of British, American, Soviet and German troops during the Second World War.[106]

Bright colours of different shades within the blue-purple-pink-red spectrum (and occasionally white) on crests and plumage were similarly chosen given the fact that they made infantry officers far more visible to their men, although they were possibly also inspired by Homer's epics.[107] Further, their high mounted position gave the impression of more height while ensuring that soldiers in ranks further back in the fighting body would be able to follow their leaders through the dust and tumult of battle, thereby aiding tighter cohesion in movement.[108] The bronze belt fragment from Pergamum depicts a front rank phalangite in *konos*, with a phalangite directly behind (though in the same rank) wearing a *konos* with a crest (Fig. 31).[109]

Among cavalry, those in charge could similarly be recognised by troopers thanks to their conspicuous white plumes. This is well evinced by the theorist Asclepiodotus, who likened the cavalry wedge, or *embolon*, to massed birds in flight 'because all the members fix their eyes on the squadron-leader [*ilarch*] at their point, like a flock of cranes which are flying in formation'.[110]

The wearing of helmets with specific features as an emblem of command continued for officers into Hellenistic times and is typical of soldiers in Greek New Comedy plays.[111] One such play dated to the late-fourth century BC portrays an infantry officer at Corinth in command of 1,000 men wearing a feathered helmet as a badge of his authority.[112] More intriguing is a distasteful incident from the Aetolian seizure of the Achaean town of Pallene (241 BC) in which officers termed *hegemones* and *lochagoi* used their helmets as a means of arranging which captives belonged to them.[113] In this instance one of the leaders of a picked corps seized a woman and 'set his three-crested helmet upon her head'.[114]

Wreaths, Finger Rings and Bracelets

After the victory at the Hydaspes Alexander is described rewarding each of his leading officers with a crown of gold.[115] Arrian records more comprehensively how it was later at Susa that he 'gave presents ... varying in proportion to the honour that rank conferred or to conspicuous courage displayed in dangers. He decorated with gold crowns [from *stephane*] those distinguished for bravery'.[116]

The recipients are then specified as Peucestas and Leonnatus (for saving the King's life and defeating the Oreitans), Nearchus and Onesicritus (for their successful voyage back from India and nautical duties), Hephaestion, and the *somatophylakes*.[117] The precise tone of these extracts suggests that Arrian's source had access to, and was perhaps even paraphrasing, genuine documents from the army administration.[118] The decoration may be thought of as a 'Distinguished Services Order' granted to officers or bodyguards of or above a certain rank who had shown exemplary conduct.[119] Decorations of this sort were granted by the King in person before the assembled troops in a ceremony not unlike those of modern armed forces.

As viewed by Roisman, the granting of crowns was a reminder of discipline, confirming the rank order and hierarchy of the army, with Alexander at the top, as the ultimate judge, source, and controller of gifts. It has been further propounded that crown awards only came into vogue or became more prevalent in the later phases of the Asian campaigns as a means of raising morale in an exhausted and increasingly disaffected force.[120] Alexander certainly had access to Persian gold in these years to allow for such extravagant distinctions.

The literary extracts are supported by visual evidence. The cavalry officer from the Alexander Mosaic prominently positioned directly behind Alexander wears a gold wreath around the calotte of his silvered and plumed helmet (Plate 3a). Similarly, the Alexander Sarcophagus has a horseman in a plain yellow (for a bronze or gilded) headpiece encircled by a white or silver laurel wreath (Plate 2b).[121] His deliberate location on Long Side A, at the exact opposite end to the mounted likeness of Alexander (both turned in, fighting Persian foes in the middle), suggests a man of unusually high pedigree, often identified as Perdiccas, Parmenion or Antigonus Monophthalmus.[122]

Early Hellenistic kings are sometimes portrayed with wreathed helmets, recalling their combat role heading mounted bodyguard units. A seal claimed by Lyall as having an idealised portrait of Lysimachus shows him in an apparent wreathed helmet;[123] and a coin of the obscure Sophytes, a Greek ruler or mercenary leader in Bactria (c. 305–294 BC), sets his portrait on the obverse in an Attic type with encircling laurel wreath.[124]

The marble 'Papyri Pyrrhus' bust (Villa of the Papyri) was identified in 1891 as a study of Pyrrhus in an unusual Celtic or Italic influenced helmet form,[125] with oak wreath and cheek guards.[126] It is interesting to add that the oak wreath was a long established emblem of the Epirote state.[127] Plutarch recounts how in 288 BC the Macedonians abandoned the unpopular Demetrius and went over to Pyrrhus at Beroea. There, in celebration, they put on oak garlands to imitate the soldiers who accompanied the Epirote king.[128] This incident therefore seems to allude to a group of cavalry officer-bodyguards in wreathed helmets.[129] Plutarch writes that, in his last battle at Argos (272 BC), Pyrrhus took off 'the crown with which his helmet was distinguished, and gave it to one of his companions'.[130]

The silver wreath seen on the Alexander Sarcophagus in contrast to gold ones mentioned in texts may intimate distinctions worn for different achievements or ranks. Both Arrian and Curtius refer to awards being issued relative to an individual's bearing in the military hierarchy.[131] Arrian suggests that gold crowns were distributed to a far greater range of veteran troops, although the authenticity of this is in question.[132] A surviving agreement recorded between the Pergamene king Eumenes I (r.263–241 BC) and mercenaries at Philetaera and Attaleai describes, *inter alia*, provision for soldiers adorned with wreaths of white poplar as a possible distinction of valour. The Amphipolis regulations from the time of Philip V add that in the later Macedonian army those men who have been awarded a crown should be given a double share of booty.[133] An award for military valour (*stephanos aristeios*) is seen as late as a first century BC inscription from Aphrodisias (Asia Minor).[134]

Finger rings were another distinct item and gold or iron specimens have occasionally turned up in Macedonian and Epirote warrior burials.[135] They were worn as tokens of rank and as a means of sealing and authorising documents when dealing with unit paperwork.[136] Courtiers in Hellenistic kingdoms were sometimes awarded with the gift of a royal gemstone ring, the gesture being one in which personal loyalty was symbolically secured through benefaction (*euergesia*).[137]

As said before, Plutarch mentions cavalry and infantry officers serving under Timoleon in Sicily who were recognised by them each wearing a ring. Philotas, one time commander of the Companion cavalry, is described with a ring he used for correspondence, certifying to leaders with distinctive seals.[138] In Plautus' comedy *Pseudolus* a Macedonian officer is portrayed this time having a letter delivered with his image impressed in wax.[139] This resembles an extract from *Curculio* which portrays the braggart Therapontigonus having a personal ring showing an image of him armed with sword and shield cleaving an elephant in half.[140] It is possible that for security purposes officers were expected to memorise the ring emblems of their fellows and even those of notable enemies.[141]

The display of bracelets was yet another means of emphasising high status. A twisted gold 'bracelet' appears to be worn on each wrist by the wreath-adorned horseman sculpted at the opposite end to Alexander on the Alexander Sarcophagus (Plate 2b), although it may represent the pronounced or decorative end of yellow-coloured sleeve cuffs.[142] If genuine bracelets, it is unlikely that these items are mere personal adornments, as they are absent from all other Macedonians on this monument.

The uniqueness of the Sarcophagus bracelets imparts that they were worn as a special exclusive honour, reminding us of Achaemenid royal gift-giving practices which included the bestowal of rich bracelets, torcs and necklaces.[143] Aspects of this custom were adopted in turn by the Seleucids, probably through Alexander.[144] In the late Hellenistic Age there is a rare inscription detailing the granting of a silver bracelet by Archelaus (a general of Mithridates VI of Pontus) to one of his soldiers called Apollonius the Syrian as a distinction for courageous service rendered at Sulla's siege of Piraeus (86 BC).[145] Bracelets in the form of *armilllae* were later worn as military decorations in the Roman army.[146]

Other Insignia

Among other insignia, the use of gorgets, Gorgon's head attachments, Persian field sashes and panther pelts for horses deserve particular attention.

For those wealthy enough to be heavy cavalrymen in the Greek world, Xenophon advocated that: 'since the neck is one of the vital parts, we hold that a covering should be available for it also, standing up from the breastplate itself and shaped to the neck. For this will serve as an ornament, and at the same time, if properly made, will cover the rider's face, when he pleases, as high as the nose'.[147]

The equipage outlined here is a far more extravagant feature than anything discovered so far. An apparent gorget agreeing with this excerpt appears on a Paeonian (wrongly identified as a Thessalian) coin of King Patraos, dated to the late-fourth century BC. Here a rider wears a bizarre specimen which comes all the way up to the mouth reminding us of a medieval European gorget with bevor. However, we do not know how accurately the original subject is reproduced in photographic illustrations of this coin. As far as evidence goes, it can be cautiously set aside as dubious.[148]

Despite the fact that samples of *peritrachelion* (gorget, or throat protector) have been recovered from wealthy gravesites in northern Greece, it is still a generally neglected part of Macedonian armour studies in the English language.[149] This is especially surprising given that literary extracts demonstrate the intrinsic importance of neck protection.[150] The aristocracy may have been influenced in part by Xenophon's advice, or were emulating the conventions of the neighbouring Thracian and Scythian warrior-elites.[151]

The only literary reference to a Macedonian *peritrachelion* is found in Plutarch's heroic rendition of Alexander before Gaugamela in which the King's armour included one made from highly polished iron set with precious stones.[152] It is described being attached to the helmet, unique in surviving accounts of Greek or Macedonian panoply, hinting at a breakthrough in armour design.[153] An inscription actually records Alexander's wife Roxane having made a dedication of gold *peritrachelia* to the sanctuary of Athena Polias on the Athenian Acropolis (327–316 BC), although these were female necklaces, devoid of any military function.[154]

Artistic evidence may give us a better understanding. There are unfortunately only a few representations of Hellenistic armour with gorgets, partly due to the fact that in art the area about the neck is often obscured by a cloak. An unusual object from a trophy of arms shown in relief on the Temple of Athena at Pergamum has been identified by Faklaris as one possibility, although the majority of academics consider it to be a *kausia*.[155] A series of corselet reliefs from a Hellenistic frieze at Side (Turkey) are rendered with greater clarity, each with a field sash at the midriff and a low encircling neck fixture.[156]

Some academics claim that larger specimens of 'Alexander' medallions display the King wearing a neck guard, although this is far from certain.[157] A Hellenistic onyx cameo (Vienna Kuntshistorisches Museum) displays a likeness of either Alexander and Olympias or Ptolemy II with his sister-queen Arsinoe II. The cameo has a crested Attic helmet, with thunderbolt on the cheek guard, serpent on the calotte and a profile head of Ammon on the nape. Faklaris identifies what he believes to be a *peritrachelion* at the throat, with a layered or 'corrugated' pattern.[158]

When examining the gorgets discovered so far, two main typologies emerge. One is a smaller, built up style fitted or moulded to the neck area alone. The other is a much larger flatter lunate-shaped 'pectoral' which covers the zone around the neck, parts of the upper chest and shoulders. Faklaris has managed to document a number of finds of both types from Macedonia, briefly listed below.[159]

Tomb B (Derveni) yielded a gorget composed from eighteen rows of overlapping gilded bronze scales on a leather foundation with a thong for fastening to the back of the neck (Fig. 29).[160] In the main chamber of Royal Tomb II there was a contrasting low cut iron specimen at the top of the corselet.[161] A gorget-pectoral found in Royal Tomb III was constructed from iron plates with bronze, leather and some preserved traces of gilding.[162] Similarly, a gorget of lunate shape was retrieved from an unplundered burial (Tomb 1, Pydna), dated to the second half of the fourth century BC, wrought from iron with two layers of leather on the back. The front side is overlain with a thin sheet of silver-gilt finely worked using intricate motifs arranged in concentric bands (Plate 25).[163] One grave at Katerini belonging to the reign of Amyntas III gave up several other fragments of the same iron type, ornamented in gold and silver. Their original form would have integrated a hinged device to link the terminals at the nape of the neck.[164] This last find from Katerini is of particular interest as it helps to

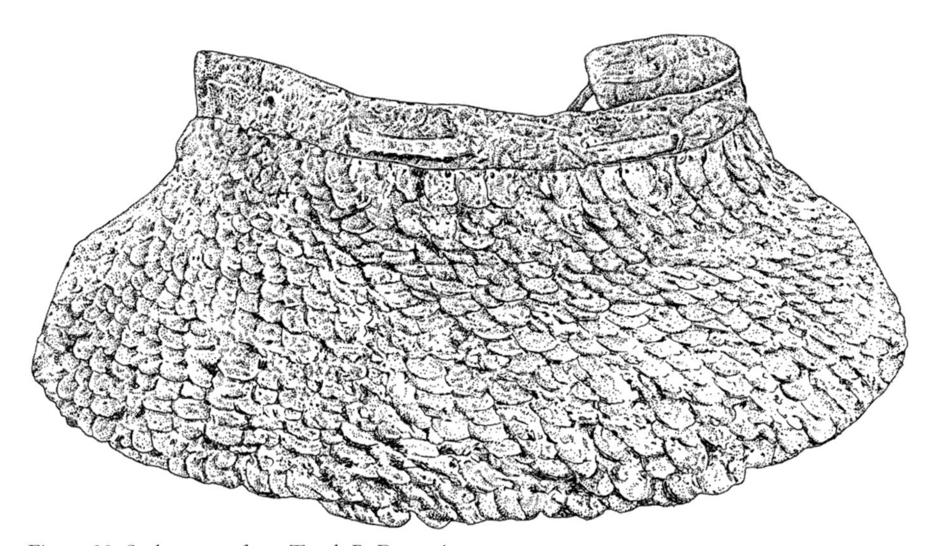

Figure 29. Scale gorget from Tomb B, Derveni.

confirm that lunate gorgets were patronised by aristocratic Macedonian cavalry-men before Philip II's time.[165]

In addition to the above finds, the antechamber to Royal Tomb II yielded another gorget made from leather and an iron sheet, enhanced with relief work and silver-gilt coating. This appears to be of Thracian workmanship, received either as a diplomatic gift or taken as spoil.[166]

Given the nature and quantity of the evidence, it is possible that gorgets were only ever luxury items occasionally worn for battle and parades by generals or high-ranking cavalry officers. This largely concurs with Faklaris, who reckons that such a device was not only a serviceable item but was also the insignia of leading personalities, kings and Companions.[167] It would have therefore served not only as a practical mode of defence, but as a means of signifying status, not far removed from the more formalised wearing of gorgets as an indicator of rank in eighteenth and early-nineteenth century European armies.

A large lunate-shaped gorget would usually obscure the top of a corselet's chest plate; however, if it was not fitted, cavalry officers could opt instead for a corselet with a Medusa or Gorgon's head. This would be located on the upper chest as an *apotropaion*, i.e. a symbolic amulet protecting the prestigious armour wearer from the 'evil eye', while in turn striking fear in the enemy.[168] It was doubtless inspired by the mythical arrow proof *aegis* worn by Athena.[169] The most famous example of a Gorgon's head is found on the corselet worn by Alexander on the Alexander Mosaic (Plate 3a). Macedonian generals and officers made the wearing of Gorgon heads on corselets fashionable in the Hellenistic world,[170] the trend continuing into the Roman Empire.[171]

With the exception of the example worn by Alexander on the famous Mosaic, Gorgon heads worn by Macedonian officers on wall paintings show them to be generally yellow-coloured, probably to imitate the golden hue (and divine protective brilliance) of Athena's cape.[172] One of the best examples is seen on the corselet of the central figure above the door lintel to Bella Tomb II (Fig. 50).[173] Another example is a Pompeian fresco showing a Hellenistic general or even Attalus I of Pergamum (r. 241–197 BC) in ornate armour with Gorgon's head on the chest, green tunic, purple *himation* and golden weapons.[174]

A probable infantry officer from Pediment A of the Sarcophagus (Plate 1a; standing figure) has a richly purple-coloured corselet emblazoned with a yellow-gold Gorgon or Silenus head. A fragmentary Hellenistic sculpture of a torso from Cos has a similar corselet with thunderbolts on the shoulder pieces, a Gorgon's head set within a scaly cape on the upper chest and a lion attacking a bovine or deer drawn in relief on the lower end.[175]

The yellow pigment on Gorgon heads represents gilt appliqué-work.[176] Two gold Medusa heads, originally worn as adornments to a corselet, were found in Royal Tomb II, and fairly large silver-gilt discs used on armour were excavated from Macedonian burials of the fourth century BC.[177] A plain miniature shield was uncovered in a male pit-grave at Sevasti and two others with intricate circle and pellet decoration were found in Tomb A (Derveni). A rational location for these objects would be as a central ornament on the chest plate of a perishable cloth or leather corselet, providing modest reinforcement to this area without the handicap of too much surplus weight.[178]

Bound below the chest, across the abdomen of an officer's corselet, and often in conjunction with the Gorgon's head, was the field sash, otherwise known as a Persian 'belt' or 'girdle'. Pekridou and Stupka provide two of the most helpful studies on the emergence and significance of this feature of apparel, citing textual references and samples from art.[179]

The 'Persian girdle' is first recorded by Diodorus having been adopted by Alexander as part of his hybrid royal costume combining Macedonian and Achaemenid elements.[180] It is briefly alluded to by Plutarch, who describes how during the fateful dinner party at Maracanda (328 BC) Cleitus sneered that Alexander should not have invited free born men to sup at his table but rather cowering barbarians eager to prostrate themselves before his Persian belt and tunic.[181]

As far as corroborating artistic evidence is concerned, the appearance of the officer's sash cannot as yet be *definitely* traced to the period before the Macedonian army's arrival in Asia.[182] However, sashes are seen on at least four of the mounted Lanuvium torsos (Plate 21a displays one). If these sculptures are faithful Roman copies of a bronze statuary group commissioned by Alexander to commemorate the twenty-five Companions who fell at the Granicus (as alleged by some scholars), then they could indicate that the sash was worn by horsemen at the advent of the Asian campaign, and perhaps even earlier.[183] The young conqueror on the Alexander Mosaic wears a conspicuous example in green, edged and fringed in yellow (hinting at gold trim) and black, and tied at the front (Plate 3a).[184] In the ancient world knots were sometimes considered imbued with magic potency, the Mosaic showing a reef or Heraclean knot, possibly adopted as an emblem of propaganda, evocative of kinship with the gods.[185]

The relief from Isernia renders a helmeted Alexander on the far left with shoulder piece corselet, discernible *pteryges* and encircling belt or waist sash (Plate 4).[186] This sash or belt reappears with representations of him in armour from bronze equestrian statuettes now at the British Museum and a figurine at the National Archaeological Museum in Sofia.[187] Another sash is seen worn by the mounted figure of Alcetas, from the Tomb at Termessus (see Fig. 13), and it can be further observed with Ptolemaic grave *stelai* from Alexandria dating to the third century BC and depicting likely cavalry officers.[188]

In literary texts Eumenes has his hands bound with his own sash when the treacherous Silver Shields delivered him up to Antigonus in the aftermath of the battle of Gabiene (316 BC).[189] This testifies to an object made from stout wool or leather.[190] Eumenes' arrest also suggests that the wearing of a sash, at least in its inceptive Achaemenid context, served as a token of obligation, tying royal retainers into a close bond of allegiance and obedience to the Great King.[191]

This obligation is possibly exemplified in an incident from 333 BC when the Greek mercenary Charidemus (serving as military advisor to Darius III) denounced the Persians as cowards. Diodorus informs us that 'this offended the king, and as his wrath blinded him ... he seized Charidemus by the girdle according to the custom of the Persians, turned him over to the attendants, and ordered him to be put to death'.[192] Charidemus might have initially received a sash to wear in token of his duty as an employee of the Persian monarch. Being seized by it meant its wearer had destroyed the link he had with his royal master and would be executed.[193]

Alexander altered the symbolism of the sash when worn by officers in his army from one of absolute submission to the King to one of service to a new imperial ideal. Pekridou rightly assumes this ideal to be the strength and security of the new empire.[194] Alexander could therefore have worn it as a means of honouring Achaemenid tradition and the Persian ruling class while at the same time changing its connotation to one more acceptable to his Macedonians. The sash worn by him was extended to officers not only as a badge of rank but also as a sign of solidarity, putting stress on faithful service to his aims and ambitions. The field sash continued as part of standard military gear for kings, officers and some bodyguard units in the Hellenistic kingdoms before being ultimately inherited by the Emperors and armies of Imperial Rome.[195]

To round off discussion on insignia, we should mention the use of the spotted panther skin saddlecloth. As with the mounts of colourful hussar and cuirassier regiments in the Napoleonic wars, this thick pelt would be split open from behind the mask to the middle of the spine for it to accommodate the horse's head and neck.[196] Still keeping its mask, paws and tail intact, it offered a comfortable, padded seat for riders.

A sample of a panther skin can be viewed on Alexander's charger in the Alexander Mosaic (Plate 3a). The Umbrian 'Italo-Megarian' cup with the same battle scene has a Macedonian horseman on the far left in a plumed helmet (with peak and cheek guards) and corselet, sitting astride a mount with pelt thrown over the horse's back. Nearer the middle of the action is Alexander himself wearing a corselet, *chlamys* and *krepides*. He is seated on a skin shabraque, with the animal's mask and some spot patterning still visible.[197] A Roman equestrian figurine preserved in Klagensfurt may also show Alexander on a pantherskin[198] and, closer to his own lifetime, the obverse of some specimens of 'Alexander' medallion manifest the conqueror's horse draped in yet another pelt.[199]

A series of tetradrachms with an equestrian portrait of Philip II on the reverse indicate that mottled panther skins were used by Macedonian personalities before the Persian campaign.[200] This is reminiscent of a mounted Demetrius Poliorcetes on coins minted at Pella, Amphipolis and Chalcis which have him on a coverlet with a pebble effect, suggestive of panther hide.[201] Other visual evidence incorporates a wall painting of a mounted cavalryman with lion or panther skin from the Kinch Tomb and a sculpture relief from Athens purportedly of a leading Athenian or Macedonian officer (Fig. 22). [202] The latter evidence shows the faded outline of a feline covering on the back of the mount.[203] The sculpture resembles a Hellenistic grave relief from a museum collection at Basel, Switzerland, with a Macedonian standing beside a steed clad in some form of feline, perhaps panther, skin coverlet (Fig. 30). A similar coverlet is seen with a seal preserved at Altino,

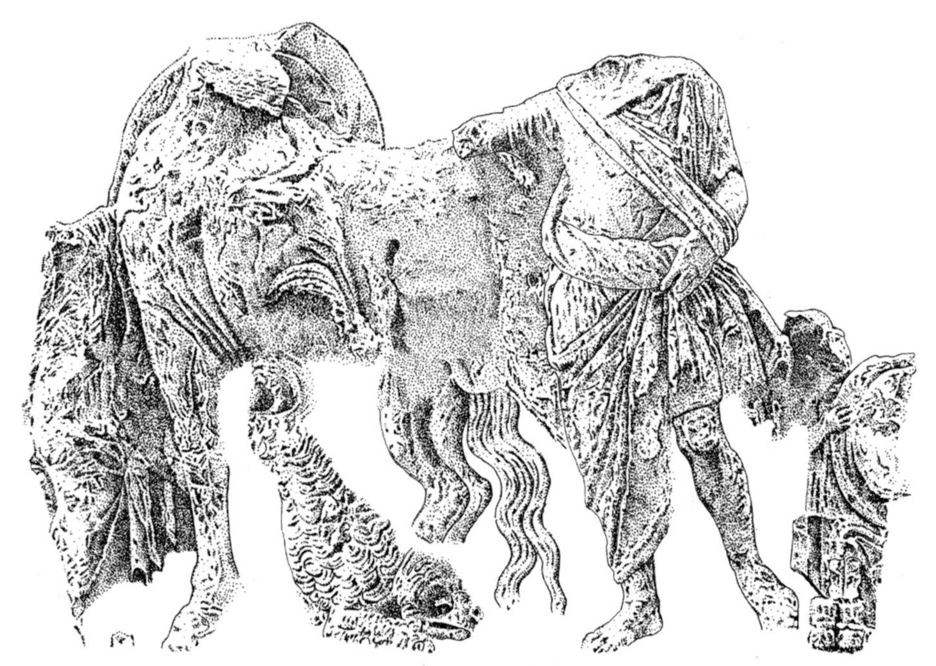

Figure 30. Fragmentary relief of a Macedonian standing beside his horse, 320/150 BC.

Italy, showing a bareheaded Alexander in a corselet inset with a Gorgon's head, mounted on a spotted saddlecloth with a single paw.[204]

Xenophon confirms the use of panther skins for horses as Persian in origin, so it is conceivable that after Alexander's conquest they became even more popular among Macedonian cavalry officers than before.[205] During the Asian campaigns the skins of some unidentified exotic animal were once observed in camp by Nearchus, who described their appearance as being akin to leopards.[206] Alexander is confirmed elsewhere as having received lion and tiger pelts and the skins of huge lizards as diplomatic curios and presents.[207]

The Macedonian preference for panther pelts over those of other equally accessible animals, such as lions, tigers or bears, seems to imply a cultural determinant. Anderson considers how panther skins were portrayed in Greek art as the saddlecloths of mythical figures, Amazons and Scythian Arimaspians. They are narrated draped over the shoulders of some Homeric heroes.[208] As seen earlier with saffron's possible Dionysiac connections, panthers could also sometimes be associated with Dionysus as the companions of the god, pulling his chariot or serving as his mount. The deity is a common motif in fourth century BC Greek vase painting and is often represented as an adolescent wearing a *nebris* (panther or fawn skin). The god's popularity among the Macedonian aristocracy is confirmed by the find of a mosaic in Pella. This was excavated in the men's dining room of a private residence (325–300 BC) and has a naked Dionysus holding a wand in his raised left hand while riding a panther, holding on with his right.[209]

From all evidence accrued, the use of panther skins might have been one facet in an overall tendency, or even a deliberately crafted policy, to glorify Macedonian soldiers in their very dress and equipage, identifying them with heroic Dionysiac attributes. Similar animal pelt saddlecloths were worn on the mounts of Napoleonic hussars for the same dashing reasons. After all, a panther is a predator of great speed, strength and ferocity – so must be the man who sits upon its valuable coat.[210]

Emblems on Unit Standards

In a military context, a *semeion* is defined as a visible sign, token, standard or signal made with flags.[211] We will outline the use and functions of the *semeion* by the Macedonians in Chapter 12 and in this part of the book focus on the appearance of standards.

Officers may have extended their individualist sense of military pride and identity to the battle standard emblems of the units they led. The standard design adopted by Alexander was almost certainly Achaemenid in inspiration, as seen with the *vexillum* type held by a Persian horseman on the Alexander Mosaic showing combat at either Issus or Gaugamela. Here the object consists of a stiff dark red felt cloth attached to a crossbar.[212] This form continued to be used in Persia well into the Hellenistic era, as observed on images from the coins of the 'Fratadara', minor post-Achaemenid rulers of the third to second centuries BC.[213] Heavy red felt had superior durability to wool or linen, as it would never go limp even on a windless day, the emblem upon it remaining at all times clear to onlookers.[214] The fact that the Macedonians borrowed the Persian style is evident from a standard portrayed on the Pergamene belt fragment possibly featuring a combat from the battle of Magnesia (Fig. 31) The object in this case can be described as a rigid square cloth attached to a cross-staff on a shaft about the height of a man[215] with a relatively simple and unadorned form.[216] A reconstruction of a unit standard based on the design from the belt fragment can be seen in Plate 30.

Figure 31. Detail from a now lost bronze belt fragment from Pergamum.

Two apparent Hellenistic battle standards are seen on a frieze from the Sanctuary of Athena at Pergamum. Both are T-shaped without cloth hangings but with floral and vegetal shapes and curlicues along a crossbar. It is possible that all decoration was rendered in metal, presumably gilded hollow bronze because solid gold or bronze would be too cumbersome to carry easily.[217] A Ptolemaic funerary epigram refers to one Ptolemaios who had apparently served with a Macedonian unit as a standard-bearer (*semeiophoros*) and fought with bravery and distinction as a spearman 'with his standard-bearer's staff'. This detail suggests that his standard shaft was tipped with a spear point at the top or a sharp metal ferrule at the bottom.[218]

When reconstructing the appearance of emblems on standards, the images prominently displayed as artistic motifs on Hephaestion's funeral pyre may offer clues as to what the device on his *semeion* looked like.[219] From Diodorus' description we know that his pyre was decorated with red banners (alluding to a man of action and battle), sirens, gold wreaths, centaurs, alternating bulls and lions, and 'perched eagles with outspread wings looking downward, while about their bases were serpents looking up at the eagles'.[220] Significantly, this last motif surfaces in literature as a well-known divine omen of victory.[221] The Tomb of Alcetas in Termessus includes on its northeast face a canopy surmounted by just such an eagle with snake device (Fig. 32), suggesting its popularity as a symbol among high ranking members of the Macedonian army.[222]

There is even less material to work with when researching the appearance of emblems on other Macedonian standards. Royal and civic coin issues (of the fifth to third centuries BC), with their common iconography of divinities, heroes and animals, meaningful to the Macedonian kings and people, offer a fair inkling as to what they *might* have been.

The she-goat, an important feature in the cultural history of the Macedonians, has been postulated as an emblem used on some standards. Justin relates how the first king, Caranus, having founded Aegae (literally 'Goat City'), used goats as war mascots to commemorate the event.[223] However, a goat only appears on coins minted under kings from the fifth century BC, so it is less likely to have been used on standards as they were introduced from the late-fourth century BC at the earliest.[224]

Emblems frequently shown on numismatic evidence from Philip II's time and later are listed as sunbursts (confirmed as a device on the standard from the Pergamene belt), a standing or cantering horse, lion or tripod. Other images include a club of Heracles; eagle on its own, devouring a serpent or with a

Figure 32. Rock-cut relief of an eagle with a snake in its talons. Tomb of Alcetas.

thunderbolt in its talons; a thunderbolt on its own; Gorgon's head; Athena with palm frond and wreath; and Zeus enthroned with sceptre and eagle.[225]

Some Macedonian standard insignia may have copied the personal devices used by unit officers for their seal rings. This helped avoid the confusion that would sometimes arise if an officer's seal motif was not congruent with the emblem displayed on the standard of the unit which he happened to lead. As mentioned earlier, some officers appear to have been familiar with the appearance of seals used by colleagues and enemy leaders. It is known that the anchor and lion were personal devices closely associated with Seleucus and Lysimachus.[226] Lane-Fox also proffers a lion-griffin attacking a stag as the insignia of Antipater.[227] It is plausible that the same images were displayed on their seal-rings and, by implication, might have been duplicated on the standards of units they led during their military careers after Alexander.

Having dealt with emblems, let us consider how the men who carried standards dressed. Unfortunately, there are only a few disparate threads of Hellenistic provenance hinting at the general appearance of some supernumeraries. At the accession parade of Ptolemy II in Alexandria, a trumpeter (*salpinktes*) is described attired in a purple cloak and white strap work boots, which may only be suitable wear for a major royal celebration.[228] The Athanassakeion Archaeological Museum (Volos, Greece) preserves a grave *stele* from Demetrias for one Antigenes, son of Sotimos. The young man seems to have been a Macedonian army trumpeter, described as having fallen near Phthiotic Thebes, fatally wounded in the head, his body pierced with spears while fighting against the Aetolians (possibly in 217 BC). Antigenes is shown here in a sunhat with yellow pigmented tunic and the faint outline of a cloak.[229]

Standard-bearers probably appeared little different to soldiers in their unit. One grave *stele* from the Hadra Cemetery in Egypt, dating to the third century BC, depicts a dismounted Macedonian cavalry signaller from the Ptolemaic army, with banner comprising a small ribbon attached to a thin shaft cradled in his left arm (Fig. 33).[230] He is bareheaded in white tunic, light brown cloak, boots and a white corselet with a red or brown fill at the shoulder yoke and waist band.[231] A shield is viewable behind the left leg, perceived as white with red lines radiating from the centre – a twelve-rayed or sixteen-rayed sunburst.

Figure 33. Ptolemaic grave *stele* featuring a soldier with possible signalling flag.

Chapter 9

Swords

To complete the study of battle dress and equipment, we shall finally examine sword types and their use in the Macedonian army. Although most of the evidence for swords applies to officers, it must be borne in mind that it generally pertains to the army as a whole (both cavalry and infantry units).

Sword Types

In accordance with Greek military convention, straight and sabre bladed swords were the norm in the armies of Philip, Alexander and the Successors. Remains of both types have been retrieved from warrior burials at the Sindos-Thessaloniki cemetery dated as early as the fifth or sixth centuries BC,[1] and fragments have been found in the Macedonian grave at Chaeronea.[2]

Of the two forms used, the straight configuration was probably the more popular and can be defined as 60–70cm long with a cruciform hilt and a leaf-shaped double-edged blade incorporating a mid-rib and tang, forged from a slab of iron. Within a Macedonian context the hilt assembly consisted of animal bone (sometimes ivory), precious woods, gold and occasionally silver decorative elements and fittings, the component parts often made from moulds.[3]

As Sekunda well relates, a swelling toward the sword tip, as far away from the hand grip as possible, weighed the weapon's centre of gravity to the point thereby maximising downward strokes. Some sculptures from the Alexander Sarcophagus are empty handed where once were held miniature metal swords in postures suggesting that both hacking and stabbing techniques were practised.[4] Foot figures executing slashing blows are apparent among the reliefs of the Belevi Mausoleum (Plate 22a).[5]

Plutarch's heroic sketch of Alexander's armament at Gaugamela portrays him with a *xiphos* (meaning simply a 'sword', referring to the straight type) which was 'a gift from the King of Citium [in Cyprus] ... a marvel of lightness and tempering'.[6] The straight form is abundantly confirmed through artistic evidence from the Alexander Mosaic, among figures on the Alexander Sarcophagus, and depictions from Bella Tomb II and Agios Athanasios, the Mosaic of the Lion Hunt and Mosaic of the Amazonomachy. Other less well known painted representations are the interior of a lid from a sarcophagus (Traghilos) and from burials (Pydna and Tanagra).[7] The weapon is further seen on images of Alexander from the reverse of some of the more clearly preserved large 'Alexander' medallions[8] and on the reverse of a series of bronze coins minted at Mylasa in Caria by the general Eupolemus (314/313 BC).[9] Pieces of the bronze equestrian statue excavated in the Athenian Agora (believed to be Demetrius Poliorcetes)

Figure 34. Sword excavated at Veria (length 63cm).

include an oversized (88cm) solid cast sheathed sword, in its original state splendidly plated with gold (Plate 27).[10]

Archaeological evidence abounds for the Macedonian straight sword (dating to the fourth century BC). Some of the better examples come from Grave 2 (Veria) (Fig. 34) and a cemetery at Aiane,[11] with others retrieved from Tombs B and Delta (Derveni).[12] Most famous of all is the object from the main chamber of Royal Tomb II,[13] as well as two excellently preserved iron blades recovered in the funeral pyre above (with hilt for a third) (Fig. 35),[14] and several others from elsewhere in the Vergina Cemetery.[15] Of these finds, the sword from Veria is noted for its gold winged Victory stamped on the grip.[16] Many wealthy officers may have personalised their weapons in this way. Aelian describes the sword used to assassinate Philip II as having a chariot carved upon it; and a close-up of the sword hilt of one of three dismounted cavalrymen on the right of the Agios Athanasios frieze reveals dabs of yellow on its cream-coloured handle, implying some kind of ornament.[17]

Evidence for swords has been found at sites in or near Thessaloniki and Kypseli and continues to be unearthed almost during every excavation season. Relatively recent deposits include samples from a tomb at Pydna (2001/2002) and from the grave of a soldier at the cemetery of Paliomelissa (2004). The former still has traces of a wooden scabbard on the blade, with ivory scabbard mouth and chape left relatively intact (Plate 26).[18] In addition, a sword from Koukkos in Pieria was discovered in 2009 (68cm long, Tomb 5, 325–315 BC), which still awaits

Figure 35. Two sword blades recovered from the pyre above Royal Tomb II (length 51cm).

publication. During a visit to archaeological laboratories, the author of this book handled fragments of its bone cross guard and chape and observed a striking gold Gorgon's head on the top of its hilt.

Neither should we fail to consider a remarkably rich source for the straight blade type associated with Greek-Macedonian armies at the small Zoroastrian temple of Takhti-Sangin on the Oxus river (modern Amu Darya). Fifteen years of meticulous excavation by the Southern Tajikistan Archaeological Expedition recovered artefacts including grips, bone scabbard mouths (Fig. 37) and twenty-three chapes dating from the fifth and especially fourth to second centuries BC.[19]

In contrast, the alternative and less ubiquitous sabre form (identifiable as the *machaira* or *kopis*)[20] may be described having a generally raptor head-and-neck hilt and a 40–60cm long blade with curved inside cutting edge.[21] Its distinct shape, with centre of gravity brought forward due to the blade having been widened and weighted one third of the way to the tip, made for increased kinetic efficiency and the use of the drawing cut.[22]

This style of weapon was, according to Herodotus, used by the semi-mythic King Perdiccas I and his brothers, which implies that Macedonia's ruling elite were well acquainted with the sword long before Alexander's time.[23] Its use is corroborated by finds in the Sindos-Thessaloniki excavations. The slashing blade form may have been part of an established tradition for this type of sword dating back to the early civilisations of Pharaonic Egypt and Mesopotamia. Oakshott commends the Ghurkha *kukri* as an ultimate descendant of the Macedonian implement, introduced into northern India through Alexander's campaigns.[24]

Xenophon recommended that horsemen adopt the *machaira* because it was more effective than the straight sword at delivering a hefty blow from above or when mounted.[25] A surviving Hippocratic medical treatise (c. 400 BC) reports wounds delivered from an elevated position to be more devastating than those that were not.[26] It therefore comes as no surprise to find that it sometimes served members of the Companion cavalry, although in Alexander's army it was used by some infantry as well.[27] A specimen from a Macedonian tomb near Gomphi (reported by *Ethnos* in 2000) had an unusually long blade, probably in order to achieve greater reach from horseback.

Despite Alexander being shown with the hilt of a straight sword on the Alexander Mosaic, he is once described by Diodorus leading the way on foot while armed with a *machaira*.[28] The curved weapon is seen in surviving Macedonian art of the fourth and third centuries BC, held by one of two near naked figures on the Pella Mosaic of the Lion Hunt[29] and painted on the wall of a tomb (Eretria).[30] An eagle head hilt suggestive of this sword type projects from a collection of panoply in the rock cut tomb of Alcetas at Termessus,[31] while a virtually identical solid cast bronze hilt was found at Dodona as part of a general's votive statue (232/230–200 BC).[32] Another ornamental hilt discovered at the same site (and dated 250–200 BC) has the shape of a lion or panther head with a winged thunderbolt at the base.[33] A sword with a handle shaped like a swan's head was excavated at the necropolis of Sindos-Thessaloniki and one more with a horse head-and-neck hilt recovered from Marvinci.[34]

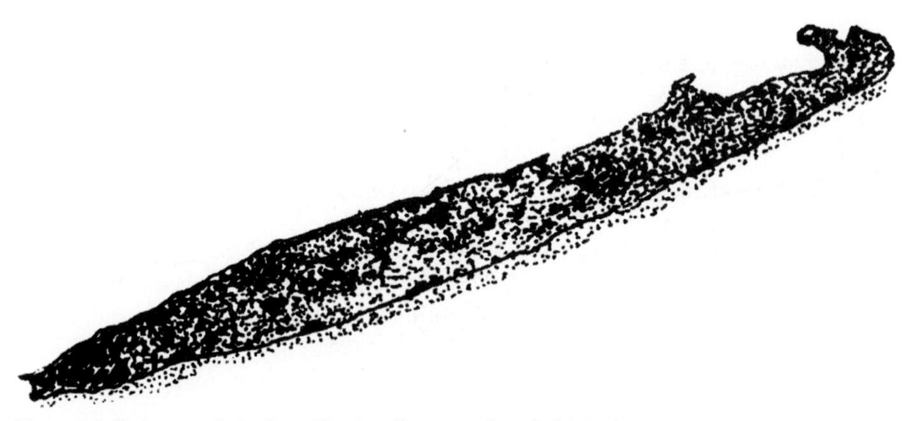

Figure 36. *Kopis* or *machaira* from Vergina Cemetery (length 51.5cm).

A sabre type found in the Chalcidice and dated to the fourth century BC is preserved at the Museum of Polygyros.[35] Vergina has yielded several others, dated 350–300 BC. One on open display at the site museum has an iron core hilt (with wood on both sides and iron strips), although the blade seems badly corroded (Fig. 36).[36] Another example with crane head hilt, from Tomb A (Prodromi), is fashioned entirely from iron with a fine inside cutting edge. Choremis has remarked on the object's light weight, ideal for executing swift blows without tiring of the wrist.[37] Of special relevance are finds from the prior noted Zoroastrian temple of Takhti-Sangin. This site, which provides an abundance of evidence for functional, ceremonial and miniature votive curved swords, can be supplemented by similar fragments found in Dilberjin and in Ai Khanum, both in Afghanistan.[38]

A Note on Sheaths

Turning our attention to sheaths, close analysis of the ceremonial sword from Royal Tomb II reveals streaks of a decomposed wooden scabbard upon the oxidized blade with an ivory chape.[39] The Takhti-Sangin haul similarly encompassed isolated pieces of wood from possible scabbards. It would appear from surviving art, painting and mosaics of the fourth and third centuries BC that the outer surface of the Macedonian sword sheath was made of leather and was commonly red, brown or occasionally ochre-yellow.[40] Evidence for the suspension system for attaching the baldric has been found with the Veria sword (Fig. 34; incidentally, wood debris was found adhering to the hilt and blade, indicating traces of the original scabbard) and at Pydna where the author examined ring and bead attachments at the archaeological storage site.

Other finds from the Takhti-Sangin dig elicit a uniquely vivid insight into their real life use and functionality. The 'aperture', or mouth, was secured by being either glued or riveted on (Fig. 37). Wear and tear was witnessed with many, often in the form of slanted and vertical scratches made by the tip of the blade when the weapon was repeatedly drawn and sheathed. In some, the upper part of

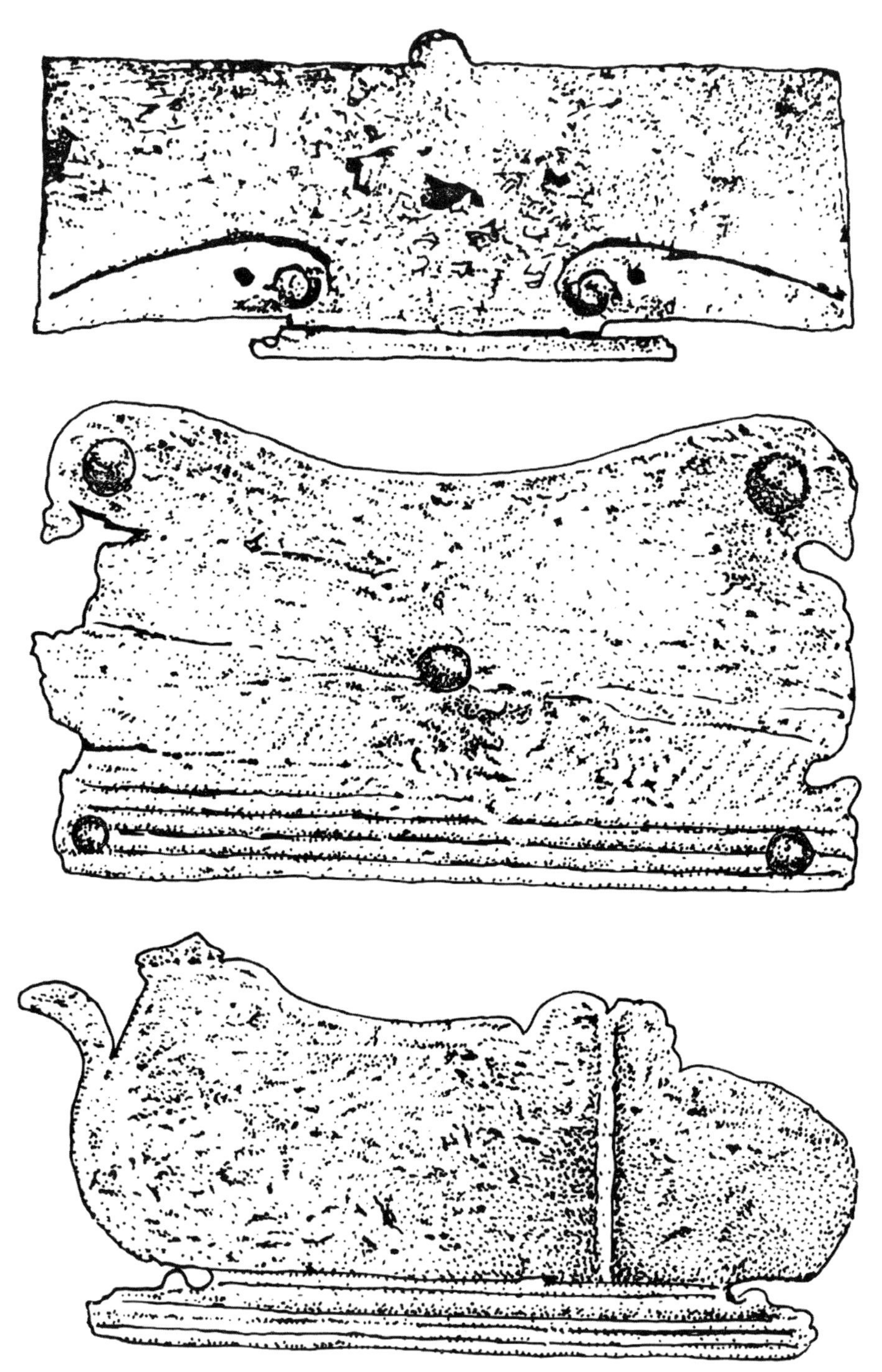

Figure 37. Examples of *xiphos* and *machaira* 'apertures' or scabbard mouths from Takhti-Sangin.

the aperture slits had been widened by the process. Small chisel marks were identified signifying adjustments made over time to scabbards to fit their swords and repairs were occasionally seen in the form of small bronze or iron pins. This is a reminder that Alexander's army was accompanied by armourers who helped maintain weapons. In addition, the 'scabbard cheeks' were joined together with 'rows of bronze rivets along each of the sides'.[41]

Sword Use in Combat

Using evidence from literary sources we shall now examine how these swords were applied in combat. Xenophon considered the learning of swordsmanship an extension of instinctive defence, as natural to man as it would be to animals.[42] He was presumably endorsing the old school of 'natural' training over the more scientific approach to swordplay which began to develop during the early-fourth century BC and in his own time, witnessed in particular with the emergence of specialist trainers in *hoplomachia.*[43]

Within a Macedonian orbit, one of the best indicators of consummate swordsmanship is the duel between Pyrrhus and Pantauchus (289 BC),[44] in which the former deliberately aimed wounds at the least protected, most vulnerable zones of his adversary's body; namely, the gaps between helmet and shield, and shield and greaves.[45] Amberger, a leading authority on the application of edged weapons, gives some valuable insights into the degree of skill required for such contests when analysing a similar engagement described in one of Lucian's *Dialogues.* Here the muscles to the back of the thigh were targeted in order to temporarily weaken the leg.[46] An analogous technique appears to have been applied by Eumenes in his fight with Neoptolemus at the Hellespont (321 BC), although in the process he received wounds to the arm and thighs. The back of the knee includes the *popliteal fossa*, bounded by hamstring tendons. Eumenes may have had knowledge of the vulnerable points of human anatomy, his objective being to cut vital tendons, forcing Neoptolemus to topple to the ground, his leg rendered useless.[47]

Eumenes' action helps to put into perspective Alexander's officers being highly proficient in the use of their swords. They were imbued with the King's Homeric code which included a thirst for glory (*kudos*), pursuit of personal courage (*andreia*), love of war (*philopolemos*) and honour (*philotimia*).[48] A classic case is the contest between Erigyius of Mytilene and the Persian Satibarzanes (328 BC), with the former proclaiming that by dint of victory or death he would demonstrate the quality of Alexander's men. Erigyius proceeded to skewer a lance through his opponent's gullet before severing the unfortunate man's head as a trophy, much to his commander's approval.[49] Pyrrhus of Epirus, especially, was said to recall something of the martial ardour and dash of Alexander[50] and this may even be why he was given the sobriquet *Aetos* ('Eagle') by his men.[51] Eagles hunt by swooping down upon their prey, and a leader of eagle-like bearing was thought to possess not only a powerful physique but also a potent spirit.[52]

In his account of Rome's second war with Macedonia, Livy described the horrified Macedonian response to the power of the Roman adopted 'Spanish sword' (*gladius hispaniensis*). In a cavalry skirmish (200 BC), the soldiers of Philip V, used to arrows, javelins and spears when fighting with Greeks and Illyrians, are portrayed struck with terror of confronting the Romans:

> When they had seen bodies chopped to pieces by the Spanish sword, arms torn away shoulder and all, or heads separated from bodies, with the necks completely severed, or vitals laid open, and the other fearful wounds, [they] realised in a general panic with what weapons and what men they had to fight. Fear seized the King as well, who had never met the Romans in ordered combat.[53]

Given their methods of formation warfare, we can be sure that the Macedonians *usually* encountered the characteristic puncture wounds associated with shafted weapons, just as Livy reports. Nevertheless, even if the reaction to the Roman sword is true in relation to tactics prevalent in the later Macedonian army (silently accepted by many researchers, but disputable in itself as the extract it comes from may be rhetorically charged),[54] references clearly show that sword types used by Macedonian officers and cavalrymen under Alexander and the Successors were equally devastating to Livy's energetic take on the *gladius hispaniensis*. Evidence for this, described below, is scattered throughout the literary corpus.

Cleitus is recorded shearing off the arm and shoulder of the Persian Spithridates with a single blow of his *kopis* at the Granicus.[55] At the siege of Tyre Alexander smote defenders with his *machaira*,[56] while at Gaza he severed the hand of a would-be Arab assassin with one swift stroke of his sword (*gladio*).[57] Ariston (*ilarch* of the Paeonian cavalry) used his sword to decapitate Satropates in an incident near the Tigris[58] and *kopides* looking like 'sickles' were utilised to slash at elephant trunks at the Hydaspes, the aggravated beasts becoming exhausted by their wounds.[59] During the Indian expedition Ptolemy is also pictured driving a weapon through the thigh of an indigenous chief.[60]

One of the most telling examples of the cutting capacity of such weapons is from a surely exaggerated extract referring to that lethal arch-swordsman, King Pyrrhus. While engaging a Mamertine champion (276 BC) Plutarch alleges that he inflicted such a blow with his *xiphos* that 'with the might of his arm and the excellent temper of the steel, it cleaved its way down through, so that at one instant the parts of the sundered body fell to either side'.[61]

Heroic hyperbole aside, some of these harrowing excerpts marry up well with the ghastly though compelling evidence of battlefield archaeology. The gravesite of the Theban Sacred Band at Chaeronea reveals skeletal fragments with multiple cut marks on shinbones (implying close range fighting with edged weapons, as seen from excavations on medieval battle sites). Several skulls affirm traces of sword cuts possibly inflicted by Macedonian cavalrymen. The most graphic example of all (known as 'Gamma 16') is disclosed by Ma as a slashing blow delivered backhanded by a horseman to one on foot, directed at the left part of the victim's face but literally slicing it off from the forehead down. Similar devastating wounds are recorded during the British cavalry brigade assault upon d'Erlon's First Corps at Waterloo.[62]

The injuries wrought by the Greek-Macedonian sabre pattern have been amply proven by a number of weaponry experts. In one experiment Mike Loades used a facsimile against a large side of pork which had been left to harden in the sun for six hours. Applying full force to his sword stroke, the blade made a 15cm gash, slicing through skin and flesh to a depth of 10–12.5cm. This reminds us of Cleitus' feat of chopping through the shoulder of an opponent at the Granicus. Similarly, Jim Hrisoulas forged a reproduction which demonstrated its unequivocal effectiveness as a cutting instrument, penetrating a modern army helmet, with the edges caving in around the cut.[63] It comes as no surprise, then, that some of the head wounds examined from fragmentary bone evidence at Chaeronea appear to have been administered by an identical or at least similar weapon.

PART IV

THE MEN

Alexander's veterans excelled in military performance, combining weapons skill with the experience of unremitting service under an exceptionally driven and inspirational king. Here we will attempt to study those veterans and their achievements not through the all-too-familiar theme of battle tactics, but by examining instead their social ties and the soldiers' daily life on campaign. Such an examination yields fascinating insights, offering a unique glimpse into the lives of men for whom soldiering became a natural – or indeed the only – way of life. Adopting the terminology of modern military psychologists, veterans had a magnified combination of task cohesion and social cohesion (with the corollary of emotional bonds and ties) between mess mates who had fought beside each other in the same ranks for many years.[1] Decades of warfare also ensured effective accumulation of medical lore, while learning directly from indigenous inhabitants probably helped hone survival skills in harsh environments.

Chapter 10

Veterans and their Families

For a veteran who had seen prolonged service with Philip and Alexander, of vital importance were his bonds with the king, the incentives and awards offered, and the pattern of interpersonal relationships developed with comrades and family dependants. Here we may conveniently detail these various facets through the lens of the most notorious veterans of all, the Macedonian Silver Shields.

The Silver Shields

The most feared and famous of all Macedonian veteran infantry in Asia were the 3,000 Silver Shields (*Argyraspides*), who represented a *corps d'elite*. Although Lock has argued to the contrary, it now seems near certain that they were none other than Philip and Alexander's old hypaspist foot guard under a new name.[1] The origins of the title 'Silver Shields' can be tentatively traced to 327 BC when, prior to the Indian campaign, the army was lavishly appointed with gold and silver ornamented armour which included partly or wholly silver plated or veneered shields.[2] With his exchequer bloated with Persian bullion, it would be well within Alexander's budgetary means to have thousands of shields lightly embellished or even heavily plated with true silver.[3]

The overall effect of sunshine on rows of resplendent silver finished shields would be unforgettable, not dissimilar to that of countless bright mirrors *en masse*, their lambent gleam generating a great wave of solar heat (a literal 'heat of battle').[4] There may even be a unique scrap of painted evidence to suggest what these striking shields looked like. The purple *kausia* wearing figure from a villa at Boscoreale includes the portrayal of a Macedonian shield (Plate 31), which some academics identify as 'pale' or 'silvery', notwithstanding that it is more commonly interpreted as a bronze-gold colour.[5]

The Silver Shields next emerge in accounts of guardsmen on duty in the royal tented pavilion (325 BC).[6] Arrian adds how at the Opis mutiny Alexander had formed a unit of Persian Silver Shields, suggesting that a parallel Macedonian unit was already well established by this time.[7] Their final campaigns after Alexander's death were under the generalship of Eumenes (318–316 BC) in Persia.

It is tempting to consider (though with no direct basis in evidence so far) that by the time Eumenes led them the independent minded Silver Shields corps were extravagantly eclectic and non-uniform. They probably appeared wearing vestiges of their original lavish garb and panoply from the time of Alexander, supplemented with whatever was cleaned up along the way, tailored to individual taste. The one aspect that would be entirely uniform was of course the famous shields from which the 'Silver Shields' took their regimental title and corporate identity. They also served as an elegant badge of honour and professional pride. As fragments of Macedonian shields have evidence of an inscription on the outer surface, proclaiming under which king's authority they were made, this would have had a particular resonance for the Silver Shields whose equipment was produced with the consent of the great conqueror Alexander. The men who bore them were therefore unabashedly advertised to younger or less seasoned troops as soldiers of *his* renowned campaigns.

The last hurrah of the Macedonian Silver Shields as a cohesive fighting body was witnessed, after years of distinguished service, at the battle of Gabiene in 316 BC. In this engagement, Antigonus ordered his mobile Median and Tarentine cavalry contingents around one flank of Eumenes' position. Skilfully manipulating dust clouds as a 'fog of war', they advanced under cover well to the rear where they seized all of the Silver Shields' baggage train.[8] The Silver Shields in turn anxiously responded by agreeing to betray Eumenes, delivering him up to Antigonus in exchange for their families and dearly won worldly possessions.

Antigonus was fully aware of the potentially detrimental influence such a duplicitous unit of soldiers would have if allowed to remain intact in his own army. On taking Eumenes into custody, he reneged on the deal and ordered that the treacherous and intractable vagabonds be disbanded once and for all. Their commander Antigenes was duly manhandled, thrown in a pit and burnt alive, and 1,000 of the worst troublemakers who had supported Eumenes' betrayal were banished to frontier service under Sibyrtius (the satrap of Arachosia).[9] This was conceivably part of the defence against the rising power of Chandragupta Maurya in the Indus Valley. However, Antigonus privately informed the governor to ensure their extinction by deploying them in penny packets for near impossible or suicidal missions.[10] Bosworth posits that there was already serious action in the vicinity of the Bolan Pass and Kirthar Range, with the Silver Shields engaged in attacking several targets simultaneously.[11]

Of the remaining 2,000 self-proclaimed 'conquerors of the world', some were settled in secure and secluded sites, with Billows identifying one as a colony at Rhagae in Media.[12] But there may be a postscript to these events. Most of the Silver Shields were exiled to regions which would later fall under the sway of Seleucus, and the fact remains that of all the Hellenistic states only the Seleucid Empire appears to have revived the military institution of infantry guards under the same name. This suggests that some of the original members survived in exile to eventually rejoin the army of Seleucus. Bosworth hypothesises that ex-Silver Shields were absorbed from a military settlement at Carrhae (312 BC). Their deserved reputation as the most competent infantry to be found anywhere, with literally decades of unremitting combat experience under their belts, would have certainly served their new master well.[13]

If Seleucus ever welcomed these irascible hardheads, there were probably only a few hundred of them left and all well past fighting age. Such gnarled veterans were thus more usefully employed as instructors or supervisors. Given their notoriety, we can imagine that many were eager to work with Seleucus, playing an active part in Antigonus' downfall – exacting retribution for his destruction of their regiment. His defeat and death when fighting Seleucus at Ipsus (301 BC) must have been gratifying for those few who had lived long enough to see it.[14] After their exit from the world stage, and with the benefit of hindsight, the Silver Shields can be justifiably viewed as some of the most tenacious exponents of pike warfare that history records.[15]

Old Soldiers Never Die

By the time of the turbulent Successor Wars, the Silver Shields had some remarkably old men numbered in their ranks. Passages deriving from a range of sometimes independent literary sources describe the advanced age of these 'athletes of war', with Plutarch asserting that in 317 BC they were in their sixties (born as early as 387 BC).[16] The existent narratives of Diodorus, Plutarch and others unanimously attest to the seniority of the Silver Shields in the last but one decade of the fourth century BC. Some of these writers were using the now lost history of Hieronymus of Cardia (a possible cousin of Eumenes) – recognised as an eyewitness who saw and probably even interviewed army officers. Hieronymus had no reason to falsify the age of veteran troops as this was politically 'neutral' information.[17]

To dismiss outright what ancient authors say simply because it flies in the face of our modern preconceptions leads to a distorted view of the past. Indeed it is wrong to assume that if current Western society rarely endorses older soldiers in service, then the same naturally applies to cultures far removed in time and place from our own. Provision in the Manchu Empire (1629) called for Mongol militiamen to serve until they were seventy-three. Even as late as the nineteenth century Zulu age-set system, drafts of married men in their sixties and seventies were still serving in regiments against the British (1879).

A distinguished roll call of eminent Macedonians – Parmenion, Antipater, Antigonus Monophthalmus, Seleucus, Ptolemy, Lysimachus and Antigonus Gonatas – shows that they were all physically and mentally alert, vigorous and capable military leaders when well past their 'prime' or, in most cases, even over seventy years old. Parmenion (born c. 400 BC) was in the thick of the fray commanding the left of the Macedonian line at the Granicus, Issus and Gaugamela. A white-haired Erigyius accepted a mounted duel with Satibarzanes and won the ensuing contest while probably in his fifties.[18] Antigonus Monophthalmus in particular fought and died in battle and on foot as an octogenarian and a story preserved by Frontinus suggests a proportion of his troops were more than fifty years old.[19] Lysimachus made a similar defiant stand at Corupedium (281 BC) when he was between seventy and eighty years old[20] and Antigenes is recognised as having led the Silver Shields when over sixty.[21]

We therefore cannot deny the proposition that some Macedonians continued to soldier on into their autumn years. Macedonia was (and is) a land of harsh climate where the inhabitants had to be tough to survive. Unlike the industrial age, here lived a people free from the unsanitary and debilitating effects of pollution, poor diet and urban overcrowding.

Alexander is praised by Diodorus for having originally invaded Asia with most of his soldiers made up of 'battle-seasoned veterans of Philip's campaigns'.[22] Such a force had been carefully readied so that, as elaborated by Justin, the young King had veterans who had fought not only under his father but even under kings as far back as Perdiccas III or Alexander II (c. 370 BC).[23] These were 'not soldiers, but masters in war ... None, on the field of battle, thought of flight, but everyone of victory; none trusted in his feet, but everyone in his arms.'[24] Older soldiers are mentioned by Diodorus as early as Alexander's siege of Halicarnassus, where their redoubtable self-confidence, valour, 'pride and war experience' served to inspire younger troops.[25] Alexander proceeded, in the words of Frontinus, to 'conquer the world, in the face of innumerable forces of enemies, by means of 40,000 men long accustomed to discipline under his father Philip'.[26]

A tireless capacity for enduring hardship, peerless standards of training and a cocksure self image as the best fighters all helped the Silver Shields win a formidable name. However, it is important to stress that they were by no means the only group of veterans active during the Successor age, nor the only ones to be regarded as first-rate troops.[27] As adduced by Hammond, rude health in the soldiers of both Upper and Lower Macedonia was vital for military achievement – a fact glimpsed on numerous occasions in source texts outlining their perseverance on campaign.[28]

What Manner of Men

The physical idiosyncrasies of Alexander's Silver Shields and other stalwarts can be tolerably collated from literary extracts. A fragment written by Theopompus recounts infantry guardsmen under Philip II as the tallest and strongest Macedonians.[29] Selection criteria from elsewhere generally demonstrates that men could be chosen for entry into the hypaspists from all over the kingdom, as long as they met the requisite qualities of martial excellence, had notable bravery citations and possessed superior size, physique or vigour.[30] The tallest of all seem to have been picked out for the lead unit.[31] In a militaristic society such as Macedonia, we can well imagine the pride and social status of families who saw one of their members elevated to such an exclusive post.

Alexander continued to apply similar criteria in Asia. During the eastern expedition, the King showed little reluctance in granting on-the-spot entry to Iranians of exceptional merit. At one time he was impressed by the conduct of thirty Sogdian prisoners described by Curtius as men of unusual strength who remained cheerful in the face of execution. Alexander rescinded the order and retained four of them in his 'bodyguard' (*custodes corporis*). They subsequently 'yielded to none of the Macedonians for affection to the king'.[32]

Excavations from the Theban burial site at Chaeronea gave up femur and separate skeletal remains, believed to be of two members of the Sacred Band, giving heights of around 1.79m for both. During his hostage years, the impressionable Philip would have witnessed these famous troops drilling, so it is plausible that on becoming King of Macedon he aspired to select the largest candidates for his own crack infantry units.[33] The average height of soldiers was probably around 1.70m.[34] Bone measurement reveals this as the stature of the male occupant in Royal Tomb II.[35] As with any human population group, a few men were conspicuously tall even by today's standards. One skeleton, beside which lay an iron sword, was discovered in a burial (Tomb 10, Pydna) dated to the last quarter of the fourth century BC and calculated at 1.96m.[36]

Alexander's veterans assuredly looked a grizzled, rough and ready crew. Living an active outdoor life in the sun, wind and rain would have made them lean and leathery, as brown and hard as a nut. Curtius takes account of their grey hair and marred faces,[37] and on reaching the Hyphasis they are figuratively portrayed as having been 'drained of blood, pierced by so many wounds, rotted by so many scars'.[38] The often erroneous *Itinerarium Alexandri* has the troops grumbling of their many injuries, while longing to take their ease back home.[39] Justin, when writing of soldiers, remarks how 'one pointed to his hoary hairs, another to his wounds, another to his body worn out with age, another to his person disfigured with scars saying that they were the only men who had endured intermittent service under two kings, Philip and Alexander'.[40]

Excerpts from Plutarch and Diodorus help build an even more vivid sketch of this tough-as-nails fraternity. Eumenes had a 'pleasant face, not like that of a war-worn veteran, but delicate and youthful',[41] whereas Antigenes was a 'stout fighting man whose body was covered with wounds'.[42] Some certainly displayed significant bodily trauma. Antigenes (or Atarrhias) was left bereft of an eye due to a catapult bolt, stoically endured at the siege of Perinthus (340 BC).[43] Demetrius Poliorcetes narrowly avoided his good looks being ruined when one such bolt pierced his jaw or mouth during an attack on Thebes (291/290 BC).[44] Philip II is catalogued by Demosthenes having an eye wound, broken collar bone, and maimed arm and thigh.[45] Some skull reconstructions of the King from Royal Tomb II sway from a horrendous crater running diagonally across where the right eye should be to a far less repellent, even 'nice', version showing a flap of skin with no serious disfigurement.[46]

Antigonus Monophthalmus was sensitive enough about his own lack of an eye to have his portrait painted in profile, depicting the good side only, but we can imagine that his body was riddled with scars.[47] Alexander's seventeen-year action-packed career (340–323 BC) was far shorter than many of the men he led, yet during this time he sustained multiple injuries to the head, neck, shoulder, breast, legs and feet.[48] Still, Didymus Chalcenterus believed that after having taken ten serious wounds in his lifetime he escaped the scarring of his father.[49] All of this is in some sense a tribute to the skills of the medical component which accompanied Alexander on his campaigns (of which more later).[50]

Strabo implies that many Macedonians had a distinct hair cut, which Durham interprets as something resembling the traditional head shaves seen among Albanian mountain clans. A close cropped head would aid comfort when wearing a metal helmet in humid environments, but there is not the slightest evidence of it in artistic sources.[51] Alexander had also issued a directive that facial hair be shaved off for practical combat purposes; however, given Macedonian conservatism and remembering that many Silver Shields were of the older generation, a proportion would have preferred to retain their short, curling beards.[52] Antigonus is said to have dismissed Ptolemy's victory over his son Demetrius at Gaza (312 BC), pointing out that the troops he had defeated were 'beardless youths' and he 'must now fight with men'.[53] This seems to imply that many older veterans were still bearded at this time. Their overall visual impact may have been akin to those of elderly Sarakatsan shepherds in modern northern Greece, detailed by Fermor as hale, weather-beaten and dignified patriarchs.[54]

Wealth and Social Prestige

It does not require Holmesian detective work to realise that Alexander's veterans could be comfortably well off. After years of living the life of wanderers and countless military operations came personal wealth and socio-economic prominence. Polyaenus explains how the property of the Silver Shields under Eumenes consisted of 'slaves, gold, silver and everything else they had acquired from their campaigning with Alexander'.[55] This is augmented by a surviving passage from Menander's *The Fisherman* which has a mercenary of the same wars congratulating himself for the quality of his spoils: 'We're living high, and I don't mean moderately: we have gold from Cyinda; purple robes from Persia lie in piles; we have in our house, gentlemen, embossed vessels, drinking cups, and other silverware, and masks of high relief, goat-stag drinking horns and wide-eared vessels'.[56]

Words similar to Menander are repeated under the contemporaneous writings of Theophrastus, who paints a man of fraudulent character: 'On a journey he is apt to put one over on a companion by relating how he campaigned with Alexander, and how Alexander felt about him, and how many jewel-studded goblets he got, and arguing that the craftsmen in Asia are better than those in Europe (he says all this even though he's never been out of town).'[57] In *The Shield*, possibly referring to the actual enrolment of troops by Eumenes around Lycia (318 BC), Menander portrays a slave character recalling an operation by the river Xanthus in which troops left camp to loot and sell their ill-gotten gains.[58] The slave's master, the mercenary Cleostratus, is said to have won a fortune in gold coins, ornate cups and a crowd of slaves.[59]

The wealth of Alexander's soldiers was often the result of prolonged freebooting reaching as far back as Issus,[60] in the aftermath of which Plutarch notes that 'the Macedonians, having once tasted the treasure and luxury of the barbarians, hunted for the Persian wealth with all the ardour of hounds on the scent'.[61] The haul was so copious that at times it was simply too much to be transported easily.[62] The wonders of the Oxus Treasure (found in 1877), with its collection of Achaemenid gold and silver objects, jugs, bracelets, plaques, statuettes and a scabbard, offer a tiny glimpse into the bumper harvest so exuberantly reaped. We can therefore well imagine Alexander's eastern venture as one of the greatest plundering expeditions known.[63]

The seizure of riches raised the spending habits of Macedonian veterans dramatically.[64] Individual consumption was so enormous that in 324 BC Alexander paid off the debts of his soldiers to the tune of nearly 10,000 or even 20,000 talents.[65] The troops took the trouble to flaunt their *nouveaux riche* affluence and social prestige, bringing to mind the boisterous bravura of pike-armed *landsknechts* or the swagger and dash of Spanish *conquistadores* in the sixteenth century.[66] Certainly, the much vaunted Silver Shields were keen to invest generous sums of money in the quality of their equipage and no doubt many arms-dealers and craftsmen were equally happy to satisfy their requirements. This meant an accent on gold and silver for helmets, swords and corselet housings, set off with plenty of luxuriant bright or purple cloth. Besides, decorating arms and armour made for more easily portable wealth, especially relevant to veterans, whose only home had become the marching camp.[67]

When veterans went home to Macedonia in their thousands, performing Odysseus to Alexander's Achilles, the King was quick to realise that their new-found treasure had transformed them into men of respect.[68] He took pains to specify in a letter to Antipater that the old soldiers should be treated with honour in their hometowns – awarded with foremost seats at public contests and theatres and allowed to wear garlands.[69] Alexander was possibly influenced in this regard by Iphicrates, who at festivals and assemblies is said to have granted front row seats to troops who had distinguished themselves.[70]

A decree describing the privilege of *proedria*, or front row seats, at athletic and dramatic contests survives from the precinct and sanctuary of Olympian Zeus at Dion (late-fourth century BC).[71] Similar honours are preserved in extant decrees from other Hellenistic cities incorporating the privilege of front row seating for various performances. The individual so honoured was elevated by virtue of the services he had rendered above the regular citizen, on a par with the elected community representatives.[72] We can thus envision well-to-do Macedonian veterans in their twilight years regaling fellow citizens at such public events with rollicking anecdotes of adventures in exotic lands.

Literary extracts mentioning the respect and prosperity conferred upon returning soldiers are confirmed by epigraphic and archaeological fragments. One is an inscription originally deposited in the courtyard of the Skripu monastery at Orchomenos, now lost. This documents soldiers in an allied squadron of Orchomenian cavalry returning from Ecbatana to their homes after seeing service with Alexander in Asia. A portion of the bounty received by the troopers was apparently spent on a votive offering to Zeus (c. 329 BC).[73] The magnificent Tomb of Judgement, with its sculpted scenes of Macedonians fighting Persians, strongly implies that it was financed by (and housed) a wealthy officer who had served on Alexander's expedition. There is further evidence of a statue at Amphipolis with dedicatory verse inscription on the base citing 'Antigonos, son of Kallas', who claimed to have been one of Alexander's Companions.[74]

Many elaborate household and funerary furnishings discovered in Macedonian tombs of the second half of the fourth century BC serve as testament to veteran officers of eastern campaigns having gained a high-tone taste for the comforts of life as seen and enjoyed by upper classes in Asia.[75] The owner of the impressive 'Derveni Krater' and forty-three other metal vessels from Tomb B (Derveni) has been theorised as a leading Thessalian cavalryman who had fought under

Alexander in the East.[76] It is inevitable that an element among those veterans who stayed longest in eastern lands became to some extent acculturated to indigenous habits and customs.[77] The same phenomenon was witnessed among early groups of Italian, Portuguese and English military renegades in India, a process referred to as 'chutnification' in one of Salman Rushdie's novels.

Promotions and Medals

Although the institution of the army was far from being a full blown meritocracy, able and ambitious rank-and-file veterans could aspire to greater esteem through promotion to lower cadres of officer. One such is the gruff Bolon, numbered among the Macedonian officers (*duces*) and described by Curtius as 'valiant in deeds of arms, but unacquainted with the arts of peace and with civil manners, an old soldier, who had risen from a humble rank to the position which he then held'.[78]

Following the victory speech at Issus, Alexander announced citations for all those who had stood out in the battle and, after studying the necessary paperwork, granted promotions to higher rank or honour (*axioma*) – some as replacement *somatophylakes* and others to fill military and administrative posts.[79] We are told by Diodorus that following Gaugamela he 'scrutinised closely the reports of good conduct and promoted many to an even higher responsibility'.[80] The method of choosing higher-ranking officers usually followed a set procedure. A royal-appointed panel of judges presided over a public gathering of Macedonian soldiers selecting candidates based on their deeds of valour. The troops signalled approval or rejection in Homeric style by rowdy shouts and clashing weapons on shields.[81] Curtius recounts that it was through such competitive selection that a first prize appointment was awarded at Sittacene to 'old Atarrhias, who before Halicarnassus, when the battle was abandoned by the younger men, had been chiefly instrumental in arousing them to action'.[82]

Lendon correctly points out that Alexander had a tendency for 'explicit rankings of human excellence and an old-fashioned, Homeric conception of excellence that combined social status and performance'.[83] Basing himself on the usual Greek custom of prizes for first through to third place in athletic contests (with Achilles granting rewards from first to fifth for chariot races in the *Iliad*), Alexander applied this practice to his army and then extended it by placing his officers in order of achievement from first to eighth.[84] It was all part of the King's unceasing efforts to instil a spirit of competition. Such gestures helped inspire those who failed to enter the honour roll and kindle rivalry between those who already had.[85]

Some ancient writers may have therefore upheld the armed forces of Alexander as an ideal template when making recommendations for an efficient structure of promotion. Onasander writes that a general was to select the bravest private soldiers to 'receive the lesser commands', with officers who had won distinction on the field taking 'higher commands, since these rewards strengthen the self-esteem of those who have deserved well, and encourage others who desire similar rewards'.[86]

Decorations for merit were another means by which a soldier could be distinguished as a veteran or someone esteemed by others. Such items have been part of warfare since at least the armies of Pharaonic Egypt embodied with the 'golden fly of valour', assorted collars and bracelets.[87] As far as soldiers in Alexander's army were concerned, we have already observed in Chapter 8 that wreaths on helmets were used to denote officers of high distinction, sometimes gifted as awards for courage.

Philip II's provision of generous pay bonuses for his men reveals that inducements imparted in sources for Alexander's army existed in his father's day. According to Diodorus, at the siege of Perinthus 'the rewards of victory challenged the daring of the contestants, for the Macedonians hoped ... to be rewarded by Philip with gifts, the hope of profit steeling them against danger'.[88] The Macedonian scheme potentially had some influence on the later, far more evolved system of Roman decorations.[89] Awards granted for merit by Alexander to Persians are preserved in early versions of the questionable *Alexander Romance*, with its mention of gold or silver drinking bowls, and, in the case of an equestrian or 'war-horse' contest, a 'goblet and a Persian robe'. However, 'if the winner prefers rewards in the Persian style, he shall have a golden crown ... a plain Persian robe and a golden belt and two cups weighing 170 staters'.[90]

Physical evidence has come to light over the last century with the discovery of silver 'Alexander' (or 'Elephant') medallions from the period of Alexander's conquests in Asia. They comprise a larger 'decadrachm' series depicting Alexander on a stallion attacking an Indian war-elephant on the obverse and a standing Alexander on the reverse (Plate 35), and a smaller 'tetradrachm' series of two different sets, with either a bowman (Fig. 38) or a chariot (Fig. 39) on the obverse and an elephant on the reverse. These pieces are devoid of lettering and poorly executed with haphazard hammering, anomalies and inconsistencies in size,

Figure 38. 'Alexander' medallion of the 'tetradrachm' series. Obverse: Indian archer drawing bow. Reverse: Indian elephant.

Figure 39. 'Alexander' medallion of the 'tetradrachm' series. Obverse: Indian chariot with charioteer and warrior with bow. Reverse: mahout and warrior on Indian elephant.

weight and shape implying that they were not intended as conventional coinage.[91] In 1992 the Mir Zakah hoard from Afghanistan gave up a much smaller coin-like artefact suggesting a previously unknown gold series. This features Alexander on the obverse in elephant scalp, and an elephant on the reverse.

Professor Frank Holt (University of Houston, Texas) has gone further than any other scholar in studying the inexplicable nature of these intriguing objects, assessing their background and function. His conclusions point to them serving as non-circulating military awards or, more precisely, commemorative medals of the Indian campaign bestowed between the summer of 326 BC and 324 BC.[92] It was a subtle, calculated reminder to occasionally near mutinous troops how much they ultimately depended on the King's power and generosity when far removed from their Macedonian homeland.

It seems possible that most surviving specimens artistically convey stages in the battle of the Hydaspes, with Alexander as god-general. Through the medallions, Alexander's intention was to retell the event in pictures with an emphasis on his own supernatural powers of leadership. As Holt makes clear, there was no need for inscriptions because the recipients were Macedonian soldiers who had personally experienced the battle and therefore knew all the events firsthand.[93] It can be hypothesised that different ranks or units were issued with specific silver or gold medals. In line with this, Olbrycht has made an interesting suggestion that the smaller 'tetradrachm' series may convey an Iranian contribution in the Indian campaign.[94] Perhaps these medallions were even issued to Iranian units which had taken part in it.

The identification of the medallions as military awards adds an intriguing dimension to our understanding of severance bonuses (of one talent each) paid by Alexander to the Thessalian cavalry and other allied contingents in 330 BC and again to discharged Macedonian veterans in 324 BC. Rather than plain bullion, these could have been specially minted issues of 'medals' – equivalent in currency to a talent – awarded at the termination of service.[95] Holt's conclusions to date provide convincing answers to questions about the nature of these enigmatic artefacts. If correct, they would be the earliest recognised exemplars of campaign medals in European history, bearing silent witness to Alexander's innovations.[96]

Militant Masters of War

The distinctions conferred on veterans for their military prowess were more than well founded. A remarkable, albeit little known, example of their combat skill and resourcefulness can be appreciated from events surrounding eight pugnacious officers, quoted as Attalus, his brother Polemon, Docimus, Antipater, Philotas and three others. These gentlemen were captured following the victory of Antigonus Monophthalmus at the battle of Cretopolis in 319 BC and were imprisoned in a near impregnable fortress called *Leontos-Kephalai* ('Lion's Head'). This remote site has been identified as the glowering outcrop of Kara Hissar in west-central Anatolia and described through personal observation by Ramsay as a natural rock fortress around 150–180m high, confined and narrow with almost sheer sides.[97]

Having spent two years in prison guarded by a contingent of 400 men, the 8 officers contrived to make their escape. They first managed to persuade or cajole some of their jailers to set them free, armed themselves, and attacked the guards around midnight. During this process they captured the hapless Xenopeithes, the garrison-commander, who was hurled down the mountainside to his death. They subsequently overwhelmed and cut down the remaining guards, some joining their leader among rocks at the bottom of the mountain, and in the chaos set fire to buildings. Diodorus admits that this was all achieved because 'they excelled in daring and dexterity, thanks to their service with Alexander'.[98]

The outcome of such daredevilry was that those guards who had stood by and took no part in the pandemonium now threw in their lot with the desperados and their followers, increasing their number to fifty. However, the health of Attalus had suffered during the prolonged incarceration and the eight officers fell to quarrelling as to whether they should stay put in the fortress (which was well stocked with food provisions) or try their luck by dispersing into the surrounding terrain.

Meanwhile, a relief force of 500 foot and 400 cavalry, along with a native levy of over 3,000, were assembled from nearby strongpoints, with orders to retake the location. Docimus now betrayed his comrades by sending out a secret message to Stratonice, the wife of Antigonus Monophthalmus, who was residing in Phrygia at that time. With her connivance, he attempted to slip out of the besieged site accompanied by one attendant, but was soon captured. He was consequently arrested and his companion employed to lead large numbers of the Antigonid troops across the besieged position, successfully occupying one of its crags. This only served to fuel resistance, even though the now seven officers and their followers were outnumbered by about eighty to one. Diodorus informs us of their resolve which saw them hold ground 'keeping up the fight day after day' for sixteen months before they were finally 'taken by assault'.[99] It is clear from this that their ability to organise and put into practice efficient defence measures were second to none. The fate of the survivors is unknown – presumably the seven officers died fighting or were recaptured and executed.

The whole incident, as racy as a boy's own adventure story, with all the pathos and theatricality of the Kurosawa film *Seven Samurai*, serves to accentuate the fact that, as stated by Bosworth, constant warfare schooled Alexander's veterans into 'the most adept and proficient fighters seen in the ancient world'.[100]

Comparable extracts for the notorious reputation of the Silver Shields under Eumenes are diffused throughout the written sources. Justin describes them as having never known defeat, 'radiant with the glory of so many victories'.[101] In Diodorus' history they are referred to as undefeated troops before the battle at Paraitacene (317 BC), whose fame caused the enemy no little trepidation.[102] The veterans' skill and experience gained over many years coupled with outstanding hardihood meant that opposing troops were reluctant to meet the old rogues head on. So much so, in fact, that 'they had become, so to speak, the spearhead of the whole army'.[103]

Sanguine carnage was repeated at Gabiene, where in the surge of battle:

> The Silver Shields in close order fell heavily upon their adversaries [the Antigonid battleline], killing some of them in hand to hand fighting and forcing the others to flee. They were not to be checked in their charge and engaged the entire opposing phalanx, showing themselves so superior in skill and strength that of their own men they lost not one, but of those who opposed them they slew over 5,000 and routed the entire force of foot-soldiers, whose numbers were many times their own.[104]

This bulldozer advance is particularly astonishing because those who had the misfortune to face it were themselves capable and committed troops. The account suggests that the hesitant enemy formation lost its confidence and simply burst apart under pressure, with most killed in the rout. The Silver Shields were thus a living testament to Napoleon's dictum that in war morale is to the material as three to one.

The outcome of Gabiene shows that veterans were highly adept at the tricks of their grim trade. During a melee there would be close defensive and offensive collaboration between neighbours in the ranks to either side, in front or behind. Analyses of sixteenth-century German training manuals indicate that muscular power alone resulted in slow and unreliable pike thrusts. The Silver Shields would have developed a much closer feel for the weapon and its various capacities, with proper footwork and fast reflexes when pushing, feinting, tripping or jabbing a less experienced opponent. Veterans would be quick to probe for gaps, exploiting weak spots then striking home with deadly, well-timed precision and force.[105]

While basking in the full glory of being 'unconquered' (*aniketos*), and in their own minds 'unconquerable', the Silver Shields were equally infamous for being an obstinate lot, raucously convening like a militant and egalitarian trade union.[106] Justin narrates *in extenso* that they thought only Alexander worthy of devotion while all other commanders were at best their own equals. Eumenes thus set out to win them over through flattery, describing them as 'patrons' or 'companions' whose valour remained undiminished, superior in prowess to even Dionysus or Heracles. In Eumenes' supposed words, 'Alexander had become great' and 'had attained divine honours and immortal glory' thanks to them, and Eumenes 'begged them to receive him, not so much in the character of a general, as in that of a fellow soldier'.[107]

Acknowledged to be in a class of their own,[108] and with their lofty self-regard as the 'best' and most battle hardened, the Silver Shields were surely resented by other Alexander veterans eager for a fair share in the conqueror's victory laurels and legacy.[109] Cornelius Nepos refers to the strong rebellious streak in Macedonian veterans of the Successor age, comparing them to corresponding Roman troops of his own time (the Roman Civil Wars, c. 35 BC), stating that 'there is danger that our soldiers may do what the Macedonians did, and ruin everything by their licence and lawlessness'.[110]

It is tempting to consider that the Greek New Comedy caricature of the 'braggart soldier' owes some of its origins to the widespread reputation of the Silver Shields or soldiers like them.[111] As no personal memoirs survive, fragments from these plays may preserve beguiling, though lampooned, insights into the lifestyle, aspirations and prejudices of the men who served with Alexander in Asia.

The concept of the *miles gloriosus*, or 'braggart soldier', resurfaces in Roman comedy with the surviving plays of Plautus, which are assumed to be adaptations of lost New Comedy pieces. It is fascinating to imagine that the nature of Macedonian veterans may have subsequently fed the prototype for strutting characters of the Tudor stage such as Ralph Roister Doister and others. There is also visual evidence for such larger than life characters. A sketch of a now lost fresco of a 'braggart soldier' from Pompeii renders a man perhaps originally intended as a Macedonian. A damaged terracotta statuette of a similar theatrical figure is found among the collections of the British Museum.[112]

Families

Alexander's father prohibited women and musicians from appearing in camp grounds due to his concern for internal discipline.[113] Nonetheless, by Alexander's time we read of tent quarters set aside for Evius the flautist, though the playing of flutes and all kinds of music were temporarily banned on the death of Hephaestion, Evius' patron.[114] There are also many references to large numbers of female captives during the Asian campaigns, testifying to Philip II's directive against women having been relaxed under his son and the *Diadochoi*.[115] During the course of their roamings many of Alexander's officers and men naturally developed liaisons with Asian women, whether as high class courtesans, common prostitutes or captives who became wives. It is possible that large numbers of these perished during the arduous Makran desert march.[116] Even so, Justin relates that by 316 BC the Silver Shields had 2,000 or more female camp followers with them.[117]

In the events following Alexander's death the Silver Shields resembled a rootless warrior tribe – a close-knit community on the move wandering from country to country, with the overnight camp becoming their only home. After long years of service Macedonian veterans were accompanied on their adventures by trains of dependants. In the *Miles Gloriosus* Pyrgopolynices is described with a band consisting of his 'parasite', three slaves and a concubine called Philocomasium, whereas in another play by Plautus the following of the mercenary Stratophanes comprised a servant, an orderly and two Syrian slave girls.[118] Many soldiers probably developed strong emotional bonds with their women companions.[119] Stratophanes spoke of how he bought his mistress a purple cloak from Phrygia, Arabian incense and Pontic balsam, which intimates cosmopolitan wealth and ready access to luxury products.[120]

According to Droysen's estimate, as many as one in six Macedonians in the invasion force of 334 BC were newly married.[121] Although this ratio has been challenged, it appears that many Silver Shields had adult sons, grandsons and nephews of Macedonian descent serving in other units. It is almost certain that later on, in the years after Alexander's death, some of these could even be half-brothers, with either Asian mothers in Asia or Macedonian mothers residing in Europe. In the lead up to Gabiene Justin pictures the Silver Shields reluctant to make war on their offspring.[122] Diodorus tells us that before the battle a mounted emissary from Eumenes approached and further addressed the troops of Antigonus in the following stolid manner:

> 'Wicked men, are you sinning against your fathers, who conquered the whole world under Philip and Alexander?' and added that in a little while they would see that these veterans were worthy both of the king's and of their own past battles … When this proclamation had been delivered ... there arose from the soldiers of Antigonus angry cries to the effect that they were being forced to fight against their kinsfolk and their elders.[123]

When the Silver Shields accordingly rumbled across the battleground, their violent onrush was preceded by a barrage of expletives as they bellowed: 'It is against your fathers that you sin, you worthless wretches!'[124] The connotation of male relatives on opposing sides is unmistakably correct and witnessed throughout history in countless civil wars all over the world.

As Curtius attests, Alexander was recognised for being well practiced in the manipulation of his troops.[125] The King's often masterful grasp of individual and group psychology was applied to full effect to keep them on side.[126] Thus, the policies pursued included a scheme encouraging relations between his men and native partners. The question of relationships with local women had become an important issue for soldiers in Hellenistic times.[127] Alexander provided wedding feasts and presents at his own expense, registered the names of all those who married an Asian spouse and arranged for a monthly ration to be doled out to women to feed their families.[128]

A humorous window into domestic life is given by Plautus when the mistress of Stratophanes nags him for offering her a mere gold mina from his purse for their child's upkeep. She remonstrates in no uncertain terms that both she and the baby need food, as does the nurse, along with wine, oil, coal, wood, linen, pillows, cradle and bedding: 'Sons of military men can't be reared on carrots, not they.'[129]

From as early as 334 BC the King granted the parents and children of soldiers who had died in combat exemption from all land and property taxes.[130] A royal letter fragment identified as belonging either to the reign of Philip II or Philip V has a grant of rewards, with tax and other exemptions, to a possible military unit which had served with distinction.[131]

Both Arrian and Diodorus acknowledge that thousands of sons by Asian mothers had been born by 324 BC and orphans were granted the right to draw their dead fathers' pay on an ongoing basis – something akin to an early form of child welfare.[132] If Justin can be trusted, soldiers were even paid by the number of sons they had.[133] In *Miles Gloriosus* we find Pyrgopolynices praised for siring only 'sheer' or 'pure' warriors.[134] This reminds us of Alexander's plan to finance the education of all Eurasian boys as part of an expanded programme, with paid teachers combining liberal studies with an emphasis on military training 'in the Macedonian manner'.[135] On completion, graduates were to follow their fathers into soldiering and be equipped at state expense with arms and horses, depending on whether they entered the cavalry or infantry.[136] A nebulous promise was even made that they be re-united with their retired fathers, suggesting that half-Asian soldiery were in time to be transferred to Europe.[137]

In a limited sense, the projected military incorporation of these boys has more recent peers. In the eighteenth century, Anglo-Indian orphans of British soldiers could be educated in England before seeing service in the British East India Company. We also find Napoleon organising a corps of *Pupilles* in his Imperial Guards establishment, which was a unit specifically recruited from the sons of soldiers who had died in action.[138]

Evidently, Alexander was thinking ahead as far as the social order underpinning his new regime in Asia was concerned. Justin provides a long-winded account of the soldiers' children and considers that the King had several believable aims in mind. For the short term, his deliberate concern for the wellbeing of their families acted as a means of encouraging those Macedonian veterans remaining in Asia to stay there, providing the King with immediate access to their outstanding capabilities. As Justin puts it:

> [Alexander] permitted his soldiers also, if they formed a connection with any of the female captives, to marry them; thinking that they would feel less desire to return to their country, when they had some appearance of a house and a home in the camp, and that the fatigues of war would be relieved by the agreeable society of their wives.[139]

A longer term objective for Alexander was to prepare the mixed descent sons as a useful source of manpower for future wars, thus easing the demand for soldiers from Macedon. This brings to mind the much later Portuguese use of *Topasses* in the sixteenth and seventeenth century. These people were the so-called 'Black

Portuguese' of Asia, their numbers necessitated by the perennial shortage of full-blood Portuguese in the eastern colonies.

Alexander's ultimate goal was far sighted and highly ambitious in that he aimed to rear and organise the Eurasians as a new pool of troops which would operate as a guarantor of security: a counterpoise to the Macedonians and trained Asian *Epigonoi*. Unlike the other two, this third force would not owe loyalty to any particular country or cultural creed. Like Ottoman Janissaries, they would be obedient to nothing and no one before the will of the King – the brave standard-bearers for Alexander's propaganda and ideology of world conquest. Justin describes them forming an irresistible army, brought up in camp from birth, knowing no life other than that of being on the road with their fathers.[140] It correlates with Plato's ideal in which he envisions children of his model military caste taken along as 'observers of war', the experience forming the basis for their future training.[141]

It seems therefore that Alexander was planning to expand his empire much further by leading a deracinated, supra-national army in size and scope far beyond the original concepts of Philip II. By 323 BC these plans were still in their infancy, and we are left in stygian dark as to what the final outcome might have been like. Any definite vision which Alexander had largely died with him. In effect, this was a case of a royal army being cut off in mid trajectory on the way to becoming a full-fledged imperial one.

It is possible that the armies of the succeeding Seleucid Empire offer a stylised dilution of what Alexander was striving to achieve. Yet we can advance that the Macedonian conqueror would have expended considerably greater efforts in attaining social cohesion and fusion. Perhaps the King was hoping to follow a limited policy of heredity with the promotion of the mixed blood children of his Macedonians for one generation or so while his wars of consolidation in Asia were in progress. At the same time, his education scheme covering all satrapies would take root and develop. The 'Kingdom of Asia' was to be eventually defended, policed and administered by Macedonians, Greeks and large numbers of Hellenised Asians selected on merit. The objective was to foster a new strategy, overriding age-old ethnic differences and allegiances, with the central unifying theme of worshiping Alexander as a living god for court, army and state.[142]

Despite all the hopes Alexander had for the half-Asian progeny of his Macedonian veterans, they appear to go unmentioned in historical events after the King's death. However, there are at least references to pike armed infantry called *pantodapoi* ('men of all countries') serving in the Successor Wars (321–316 BC), which seem to be Alexander's Asian *Epigonoi* under a different terminological guise.[143] It is interesting to note how the description of them as 'Inheritors' no longer applied. It reflects the general abandonment of Alexander's racial integration plans by the generals who succeeded him.

Along with the *Epigonoi*, we can assume that the mixed blood sons similarly found employment with the Middle East based warlords, and ended up among the military fief holders settled in Asia by Antigonus Monophthalmus, Seleucus and others.[144] Still generally in their twenties, some would have almost certainly served on both sides at the battle of Ipsus in 301 BC. When sources mention army units without any ethnic derivation or alternatively refer to them as 'Macedonians' domiciled in Asia (rather than being traced from Europe to Asia), it is tempting to think such troops had in their ranks not insignificant numbers of Eurasians – half-Macedonian by birth, brought up in Asia according to royal requirements. Alexander aimed to train them as soldiers and, if we accept Justin, some of the Successors followed through with this policy.[145]

The King's Largesse

Philip, Alexander and the *Diadochoi* all recognised that their power base rested primarily upon the loyalty offered to them by their *Makedones*. Ideally, it was therefore of paramount importance to ensure that this indispensable population core was used sparingly in war and was well catered for. Policies were installed by later Antigonid kings to encourage population growth for better recruitment to the army.[146] Extant sanctuary dedications in Macedonia show that girls undertook physical training so that, like in Sparta, there was an interest in encouraging the birth of healthy children for military service.[147]

Unlike with many modern armies, the Macedonian soldier-citizen and his family were given extensive state-sanctioned support. Reviewing the period of the late-fourth century BC and the age of the Successors, free education was provided by kings, and soldiers could receive rent incomes from land allotments assigned to them.[148] They were paid salaries, with the opportunity of obtaining higher rates in the form of double, triple or quadruple pay, back pay, and a system of bonuses offering up to six months or more payment in advance before fighting a crucial engagement.[149] There was extra consideration when entering a new service: accumulated booty, promotions, possible medals and an occasional financial benefit on retiring.[150] Alexander also discharged his soldiers' debts.[151] There is some Seleucid evidence for Macedonian military colonists in Asia even being granted one year's tax exemption when establishing themselves in new settlements, with state help extending to food supplies, the building of homes, shops and fortresses.[152]

Beyond the cash imperative, generals would arrange revels, banquets and athletic games for their men as a relaxation reward after or between the hardships of a gruelling campaign.[153] A newly married soldier could sometimes be granted leave. If he should fall ill or be wounded, he could be withdrawn from operations and receive treatment. Alternatively, if he should die in battle, he and his deceased comrades were granted a funeral with full military rites. This funeral might include a parade of the whole army in battle array signifying their respect with a mass roar of the war-cry and a clash of shields.

Honorary state funerals of this kind were a regular and expected, if not Homeric, aspect of Macedonian warrior custom. Sometimes separate pyres were constructed for officers and men (indicative of social divisions in the military). There would be sports contests, victory sacrifices and feasting, the dead cremated and interred with their arms, a burial tumulus raised over the ashes. Those killed in battle were occasionally even immortalised with a memorial erected in their hometown.[154] The immediate impression one is left with is of many cities in the kingdom boasting outstanding monuments celebrating the Macedonian fighting tradition, victories and the glorified dead,[155] with an accent on shield and weapons imagery.[156] Commemoratives of this sort appear as well to have been extended to Macedonian colonies in Asia.[157]

The aforesaid details are in stark contrast to the miserable conditions prevailing in European armies as little as two centuries ago, where soldiers were inadequately paid, their families frequently reduced to penury and neglect. During the Napoleonic Wars burials were often appalling undignified affairs, and wounded men would rarely receive financial recompense. Alexander's inspiring arrangements can also be favourably set against those in the early Roman imperial army where marriage was forbidden to soldiers in service. Indeed, it was not until the principate of the Emperor Hadrian that the illegitimate offspring of troops were granted the right to inherit their father's property under certain conditions.[158]

Within Macedonia, and in comparison to other military cultures, the king's active concern for the wellbeing of his men kept them faithful and helped encourage a view of army service as an honourable enterprise. The *Makedones* considered their rulers as paternal soldier-kings, benefactors and the first of equals, exercising a just or 'kindly' policy toward them and their families, strengthening the close interdependence between the troops and their monarch.[159] Macedonian soldiers often felt an inborn *veneratio*, or veneration, for the royal house and took great comfort and pride in their privileged status as members of an admired elite, an exclusive king's 'Companionate'.[160]

Medical Back-Up

As there is no decisive evidence to the contrary, the conventional opinion is that doctors (*iatroi*) in the Macedonian army were utilised in a 'private enterprise' capacity, unlike the allegedly institutionalised arrangements cultivated by the Assyrians and Romans.[161] Ruffin has challenged this view and instead argues that as Alexander's army in Asia was a self-contained force operating independently and at a great distance from its homeland, medical care was integrated into its constituent units, similar to what is generally agreed for Roman military organisation.[162]

Philip II's introduction of medical services was certainly an excellent means of demonstrating concern for the health of Macedonians at all social levels. With the increased ability to heal, or order that wounds and ailments be cured, the King could have furthered an aura of magic and moral power over his grateful subjects and soldiers.[163] It is possible that a series of other factors conspired to bring about the introduction of medical experts. This included a tradition going back 100 years before Philip II's reign when members of the royal house acted as patrons to reputable doctors, particularly the famous Asclepiads of Cos, hired almost as consistently as surgeons from the Beaton family employed by medieval Scottish kings.[164]

Nascent medical care existed among some of the more progressive Greek militaries of the early-fourth century BC. Xenophon describes medical men in the Spartan army and in another discourse approved physicians being regularly used as part of a well-run infrastructure, envisaging that states elect boards of health.[165] The employment of doctors by city-states was apparent during the Hellenistic Age, which in turn gave them a high profile as benefactors in these communities.[166]

At around the same time as the Spartans, the enterprising Jason, tyrant of Thessaly, anticipated later progress by providing his mercenaries with what Xenophon calls 'attendance when they are sick, and honour at their funerals'.[167] Philip was partly inspired by the example of these neighbours and precursors. The famous doctor Hippocrates, patronised by Philip's ancestor Perdiccas II (r.454–413 BC), wrote that those who wanted to pursue medicine should follow an army. The idea of such practitioners gaining experience in a military environment would be even more acute in Macedonia under Philip II.[168]

The general scarcity of evidence as to how medical staff were organised in Macedonian forces, coupled with their occasional intriguing mention in the ancient texts, only serve to fuel further enquiry among scholars, such as Salazar pondering who paid them, how much and in what manner.[169] The sources suggest that experts of this kind were well-salaried. Philip agreed to pay one a generous daily allowance when being treated for a collar bone broken in battle,[170] and the doctor Philip of Acarnania (who began his career under Philip II) was awarded with expensive gifts after successfully ministering to Alexander.[171]

King Philip took a particular interest in the latest treatments.[172] He is known to have corresponded with and followed the activities of the so-called 'Zeus physician' Menecrates of Syracuse, who wrote a treatise on medicine. He may even have invited the latter to his court.[173] Moreover, the King received breakthrough treatment at the hands of Critobulus of Cos for the eye wound suffered at the siege of Methone.[174] Critobulus re-emerges treating Alexander, and this same man seems to have been appointed as one of the *trierarchs* (captain of a *trireme* galley) on the Indian expedition.[175]

It is thought that Philip's eye operation was attempted with the aid of a new device. Attributed to Diocles of Carystus (a possible younger contemporary of Aristotle), this is described as the *kyathiskos*, or 'spoon of Diocles', although its use in removing arrowheads is strongly doubted in academic circles.[176] All the same, Diocles is often regarded as the first systematic writer of anatomy, with *On Bandages*, *On Treatment*, and *On the Equipment of a Surgery* being only a selection of the many lost titles accredited to him, whether accurately or not. A letter from Diocles survives advising on how to identify and cure various diseases.[177] Hippocrates, son of Dracon, wrote a medical work and is recorded treating Alexander's wife Roxane. Hippocrates' father served Alexander during his lifetime.[178] The father–son relationship of Dracon and Hippocrates suggests that some doctors were assisted by understudies.

When identifying medical experts who accompanied the Macedonian army in Asia, a small professional nucleus is implied with a total of eight in the literary sources, some of whom may have helped treat the wounded at Issus, Bactra-Zariaspa and Opis. To the names of physicians so far noted we can add those of Alexippus, Androcydes, Thessalus, Pausanias and Glaucias.[179] Nonetheless, as only a small portion of all written material survives, we should not assume that these represent the entire medical component on the expedition.[180] Alexander was personally attended to by more than a handful of physicians and the few names left to us are of those who happen to have been captured in literature treating the King and his officers. When invading the Persian Empire, the army would have been accompanied by the best doctors from across Macedonia and from renowned medical schools in the Aegean world; it is almost certain that a number of unnamed Asian specialists joined the campaign further east.[181]

Soldiers as Medics

Medicines were seen as highly portable and vitally important and Macedonian soldiers, like British and Commonwealth forces in Burma during the Second World War, appear to have carried them on the march.[182]

Lee remarks on the low ratio of medical professionals to incapacitated in the Ten Thousand, where eight doctors would have had to treat a hundred sick or wounded men. Whether Alexander's far larger army was faced with a similar situation or not, common rankers must have relied upon their fellow soldiers, squad servants and family groups for immediate attention.[183] The army may have even operated a system where each man was his own medic, or each infantry *dekas* (with *dekas* meaning a ten-man squad under Philip, sixteen in the time of Alexander) had one man acting in the capacity of a 'team medic'.[184]

A certain ratio among the troops had knowledge of rudimentary treatments.[185] Homerus of Byzantium is said to relate a successful tracheotomy performed by Alexander on a suffocating soldier using the tip of his sword.[186] In the Balkans, the tradition of coming to the medical aid of comrades goes back centuries and in 1809 John Hobhouse noted the rude proficiency of Albanian *klephts* in healing wounds and setting bones.[187]

Historical armies campaigning in Asia suffered just as much from extremes in climate, disease, and tainted food and water, as from injuries in battle. The same problem afflicted Alexander's men.[188] Yet some of the scientists who were brought along collected field data on therapies for the ailing and maimed. Thoroughgoing investigation is alluded to in the surviving *Enquiry into Plants* and *On the Causes of Plants* by Theophrastus, which discuss new flora encountered during the eastern venture.[189] From Aristobulus and Onesicritus, Strabo recounts the examination of 'herbs and roots both curative and poisonous' by Alexander's men in India.[190]

One of the medicinal plants used in India was *rauwolfia serpentina* (a snake venom antidote), sometimes taken as a hot water infusion.[191] A curative balm for watering eyes (*epiphora*) or inflamed eyelids (*ptilosis*) was used, attributed by Aetius of Amida in the sixth century to Alexander himself.[192] Another remedy applied in the East was a resin extracted from the root of the now extinct plant of the fennel family called *silphium* (or its substitute from the same genus *asafoetida*)

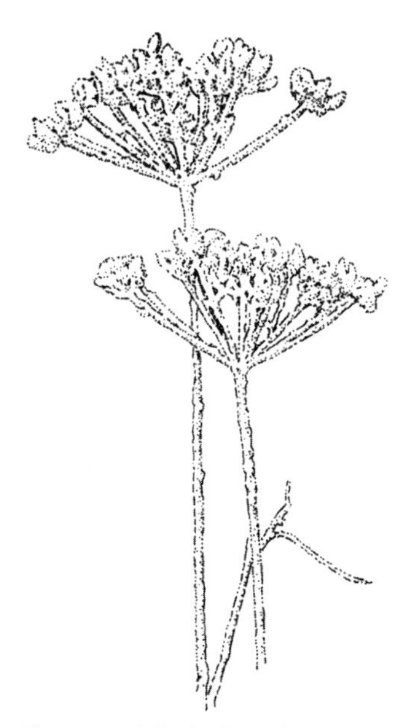

Figure 40. The herb *asafoetida*, grown in the spring and effectively applied for medical treatments.

(Fig. 40). This plant was not only effective in treating infected cuts and swellings but was also good for generic fevers, aches and pains. In addition, it was used to tenderise meat rations for ease of digestion. This last benefit would be crucial when troops in extreme environments resorted to eating the raw flesh of their baggage animals to survive.[193]

It is almost certain that there was a fair amount of resourceful improvisation in the medical sphere when marching and fighting at length in unfamiliar lands, but regrettably evidence for this has mostly failed to turn up in extant accounts. On one occasion, Alexander even forbade his men to eat mangoes, which had caused abdominal pain and dysentery.[194] At another time, oil was found to be an effective remedy for a contagious skin infection suffered by many troops.[195] Oil was also applied in other situations; for example, if cold or blizzard conditions were expected, soldiers were issued with a measure of olive oil or an adequate substitute. The rationale here was that a greasy layer of fat and oil would act as an insulator.[196] Pedanius Dioscorides (c. 60 AD) writes approvingly in his *On the Materials of Medicine* that oil of this kind helped to keep the body warm and supple, guarding it from chills.[197] Curtius mentions troops rubbing themselves down with sesame oil purchased from local inhabitants.[198]

That some ointments or salves used by Alexander's troops could be effective in avoiding skin damage is corroborated by the experiences of those journeying through Bukhara (part of ancient Sogdiana) in the early-twentieth century. Among other hazards and discomforts, the high evaporation of moisture caused skin to chap and develop sores, and exposed areas were treated with oil, which doubled as a mosquito repellent.[199]

The King gave advice on a hellebore treatment for Craterus and personally applied a cure to a debilitated Ptolemy (suffering the effects of a poisoned arrow wound) – perhaps in reality a treatment commended to him by Hindu doctors.[200] As stated by Diodorus, 'now that the value of the remedy had been demonstrated, all the other wounded received the same ... and became well'.[201] At this point we should recall Plutarch, who says that Aristotle's tutorship instilled a lifelong medical interest in Alexander, so that he was attracted to medical theory and could prescribe regimens and cures for his friends when they were sick.[202] The King's health awareness would have filtered down to his officers and troops. Alexander's interests may have inspired the drug and poison antidote pursuits of later kings, Mithridates VI of Pontus being the prime and most notorious exemplar.[203]

Medical Curiosity

Alexander's medical experts sometimes achieved advances in treatment. The herbal skills of Greek doctors joined by those of Persian and Indian specialists (the latter with knowledge of ayurvedic medicine) offered new pharmacological therapies; it has even been suggested that surprisingly sophisticated face and nose surgery was performed, including rhinoplasty.[204] Citing Nearchus, a team of Indians is authenticated by Arrian: 'Alexander had collected and kept by him Indians most skilled in medicine, and had it announced in camp that anyone bitten by a snake was to go to the royal tent. The same men were physicians for other diseases and injuries as well.'[205]

Alexander's known sojourn in Taxila provided the ideal venue for a cross-cultural exchange in expertise of this kind, absorbed into the army. Scholars point to evidence that Taxila was a famous centre for military thought, education and training in India during the late-fourth century BC. It is hard to imagine that Alexander was not aware of this.[206] The imprint Greek medical men have had in the region is inferred from the hakims of Kabul in Afghanistan, who claim a romantic descent from Alexander's expedition doctors and follow 'Greek medicine' (*Tibb-e-Unani*).[207]

Of particular note is the *Suda's* reference to a specialist called Aristogenes, who wrote twenty-four works addressed to Antigonus Gonatas. Titles compiled include *On Diet*, *On Strength*, an epitome of remedies and, most interestingly of all, a work called *On Bites*.[208] This reminds us of Alexander's Hindu doctors treating victims with a snake venom antidote. All these studies were of use to soldiers, and the last one named could partially have been the result of knowledge shared with Indian experts.

Due to the flourishing climate of scientific curiosity established by the late-third century BC, Philon of Byzantium wrote with confidence that for mercenaries Hellenistic cities should ensure that:

> There must also be on the spot excellent doctors, skilled in healing wounds and extracting missiles, equipped with the appropriate medicines and instruments, and provided by the city with ointments, honey, bandages, and lint, not only to prevent the wounded from dying but also to render them, having rapidly recovered their health, useful in subsequent encounters, being ready to court danger in the knowledge that they had been healed and well looked after.[209]

Medical instruments from the mid- to late-fourth century BC have in fact been excavated on the site of Olynthus and again at Aiane (Fig. 41).[210]

Versions of the *Alexander Romance* describe doctors in some form of distinct apparel, although this is a retrospective insertion or accretion, possibly for the dress of personnel in the Roman army.

Medical Routines

Can we reconstruct something of the medical component routines in Alexander's army? The answer is yes, albeit imperfectly. The treatment of incapacitated

1a (*above left*) and b (*above right*). Details of the Alexander Sarcophagus, Pediment A: combatants at left corner and centre.

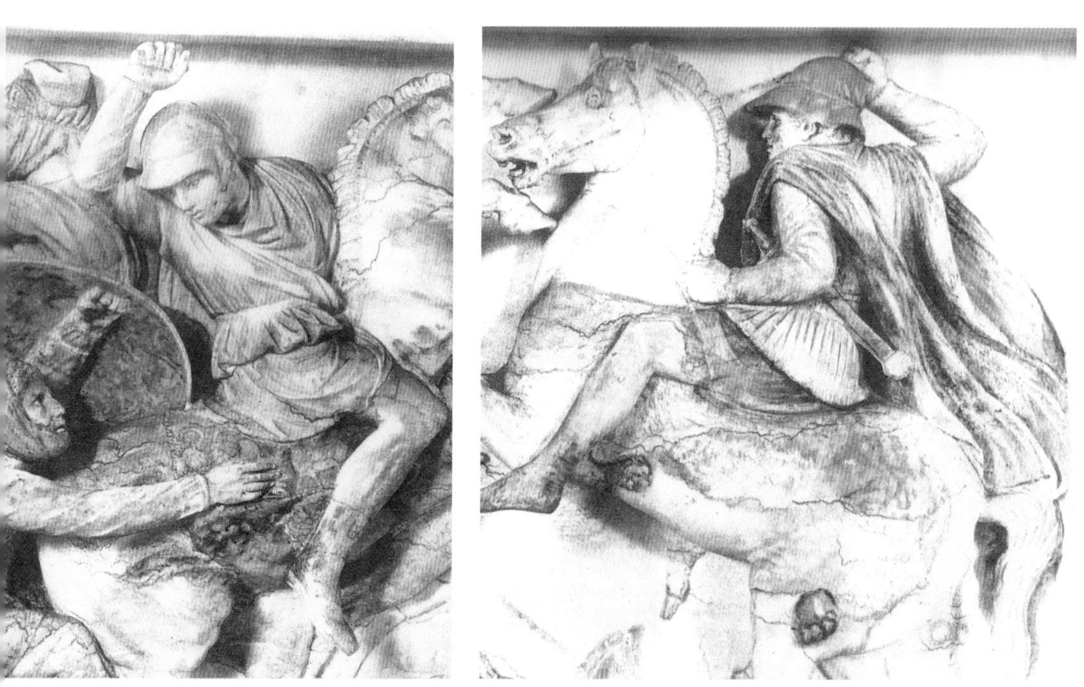

2a (*above left*) and b (*above right*). Details of the Alexander Sarcophagus, Long Side A. Mounted figures at centre and right end of the composition.

3a (*top*) and b (*bottom*). Details of the Alexander Mosaic. Alexander and Darius with surrounding action.

4. Relief from Isernia. Although worn and weathered, an Alexander battle scene can still be traced on the surface.

5. Painted frieze comprising three figures, located above the door lintel from Bella Tomb II.

6. Frieze showing figures in armed combat, Tomb of Judgement.

7. Detail of the battle frieze (far right figures), Tomb of Judgement.

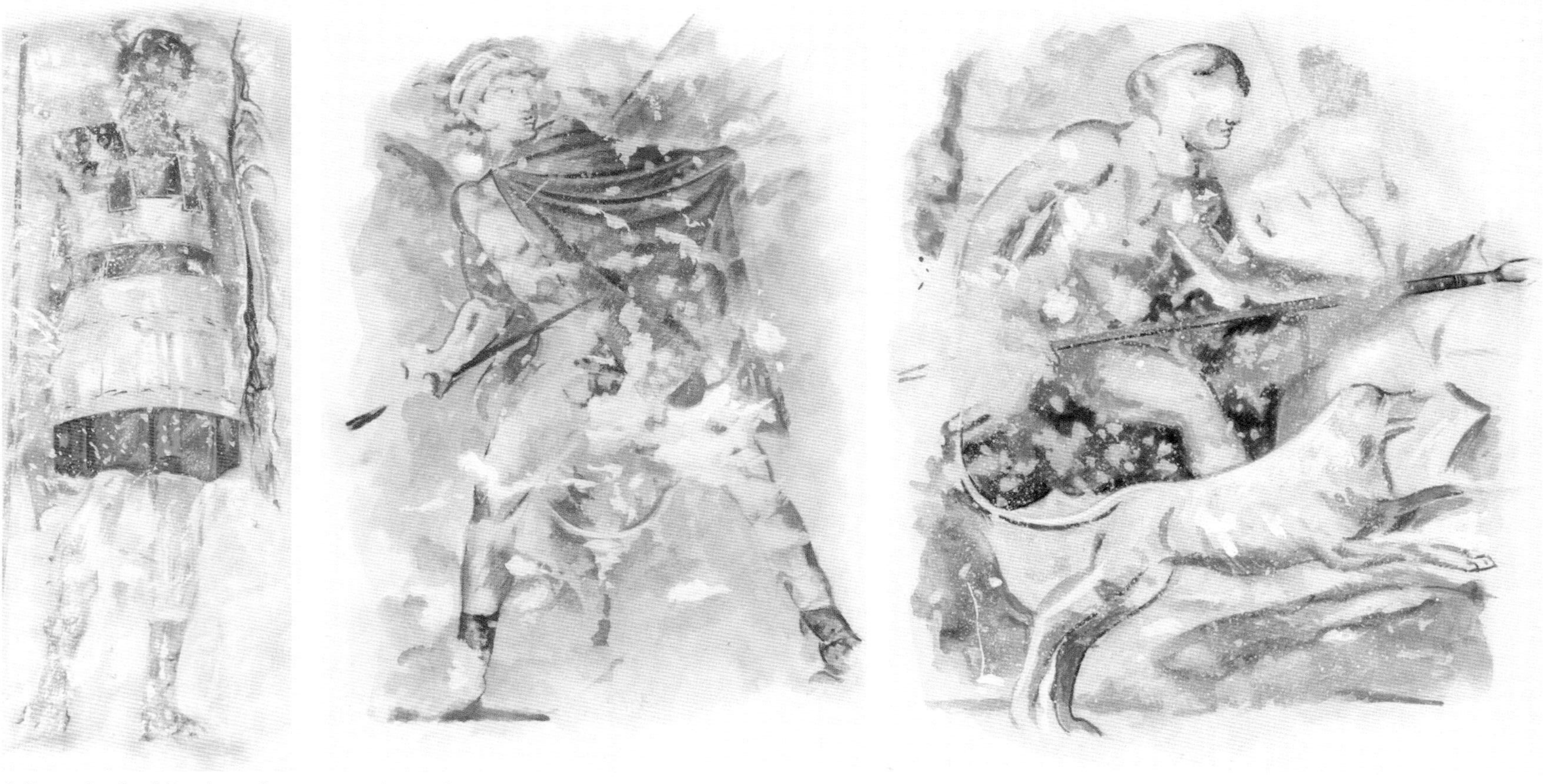

8. Portrait of soldier from far left intercolumniation panel, Tomb of Judgement.

9a (*above left*) and b (*above right*). Hunt Frieze, Royal Tomb II: second and fifth figures from the right.

10a (*top*) and b (*bottom*). Painted frieze above the lintel to the doorway, Agios Athanasios tomb: left side of the banquet scene portraying guests with servants and right side of the banquet scene portraying soldiers.

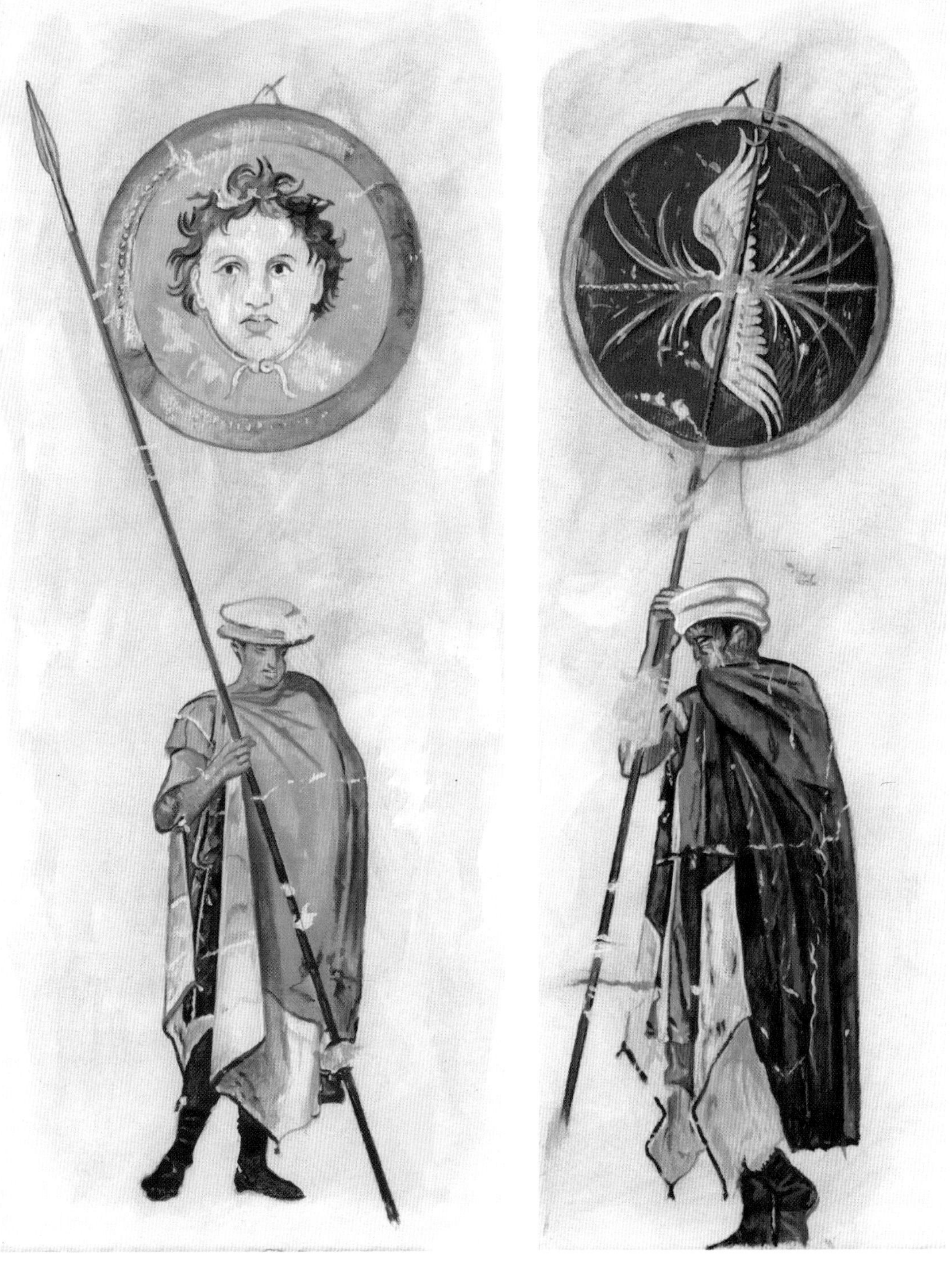

11a (*above left*) and b (*above right*). Armed young men from the left and right side of the entrance, Agios Athanasios tomb.

12. Detail of possible Macedonian shields from a fresco, 90–25 BC, Villa of Poppaea.

13. Bronze coin struck after Alexander's lifetime. Uncertain mint from western Asia Minor, c. 323–310 BC.

14. Bronze coin struck after Alexander's lifetime. Sardes mint, c. 323–319 BC.

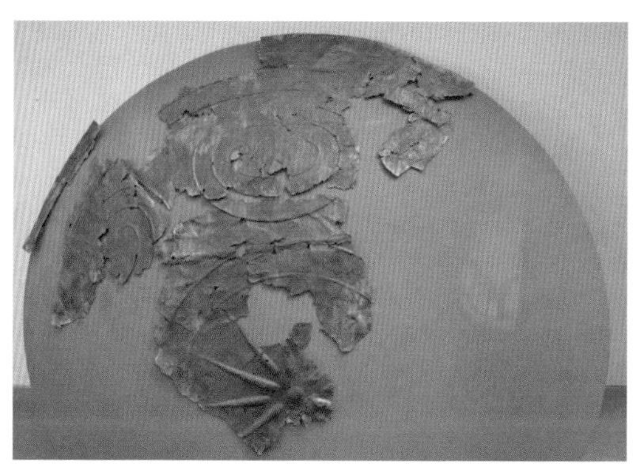

15. Shield face fragment found around Vegora (ancient Lyncus), which may date to the reign of Antigonus II Gonatas (r. 276–239 BC).

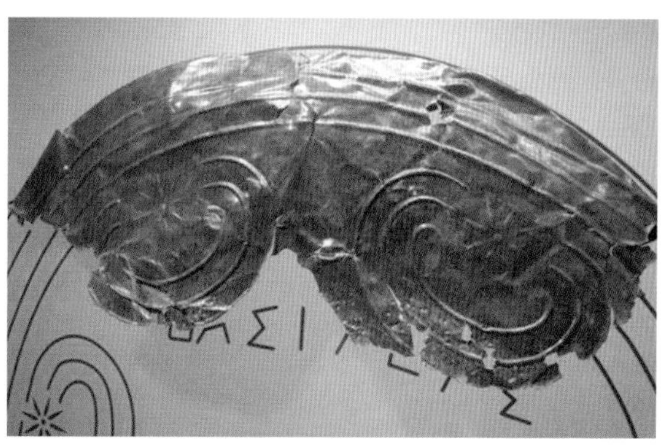

16. Shield face fragment from Dodona, which may have been part of a dedication made by Pyrrhus after a victory over Antigonus Gonatas in 274 BC. Note similarity to patterns on shield fragment from Vegora (Plate 15).

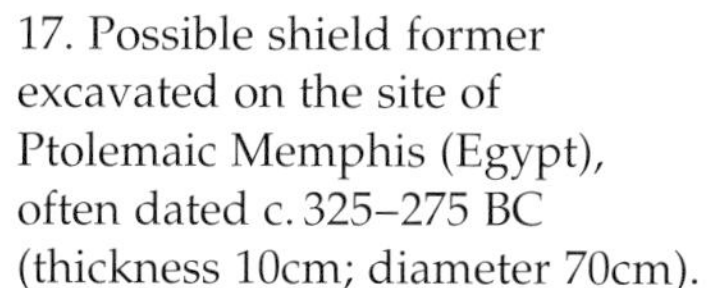

17. Possible shield former excavated on the site of Ptolemaic Memphis (Egypt), often dated c. 325–275 BC (thickness 10cm; diameter 70cm).

18a (*left*), b (*below left*), c (*below centre*) and d (*below right*). Miniature painted terracotta shields from the Tomb of the Erotes, third century BC.

19. Bronze Phrygian helmet, 350–300 BC (height 31.5cm, diameter 27cm), found in House H at Vitsa.

20. Silvered iron and plain iron Thraco-Attic helmet forms with an iron 'muscle cuirass' showing gold nipples and roundels. Found in Tombs A and B at Prodromi, Thesprotia, 350–325 BC.

21a (*above left*) and b (*above right*). Marble torso fragments from Lanuvium, Italy.

22a (*above left*) and b (*above right*). Relief sculptures featuring Hellenistic soldiers doing battle with centaurs. From the Belevi Mausoleum, possibly intended for either Lysimachus (d.281 BC) or the Seleucid king Antiochus II (d.246 BC).

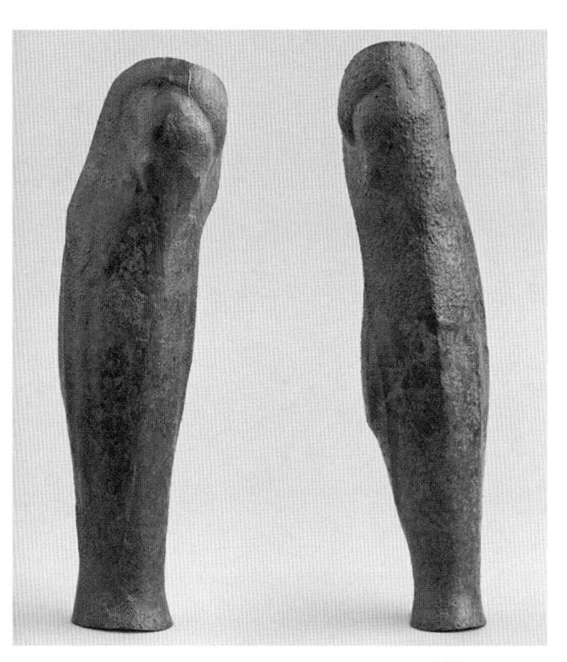

23. Pair of bronze greaves (height 42cm, width 12cm). Recovered from 'Toumbes', tumulus B, Korinos, c. 350–300 BC.

24. Pair of fully restored bronze greaves from Tomb A, Derveni, c. 350–300 BC. Intact or fragmentary evidence for greaves were also found in Tombs B and Delta.

25. Lunate gorget from Tomb 1, Pydna, 350–325 BC.

26. Sword (length 75cm, width 12cm). From a tomb located in the North Cemetery, Pydna, 350–300 BC.

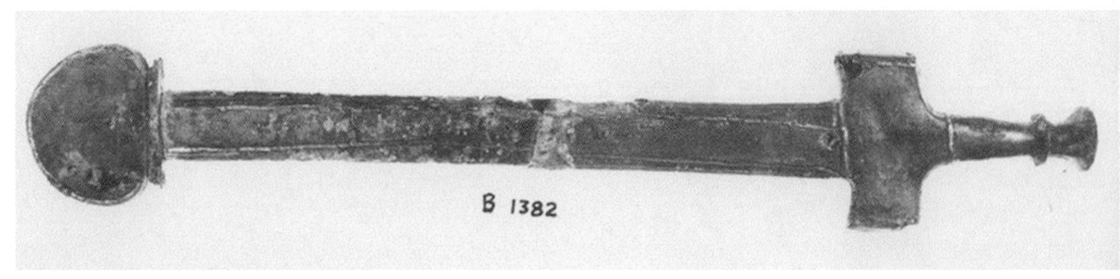

27. Sword from a gilded bronze statue at Athens, tentatively identified as a likeness of Demetrius Poliorcetes. Dated to the end of the fourth century BC.

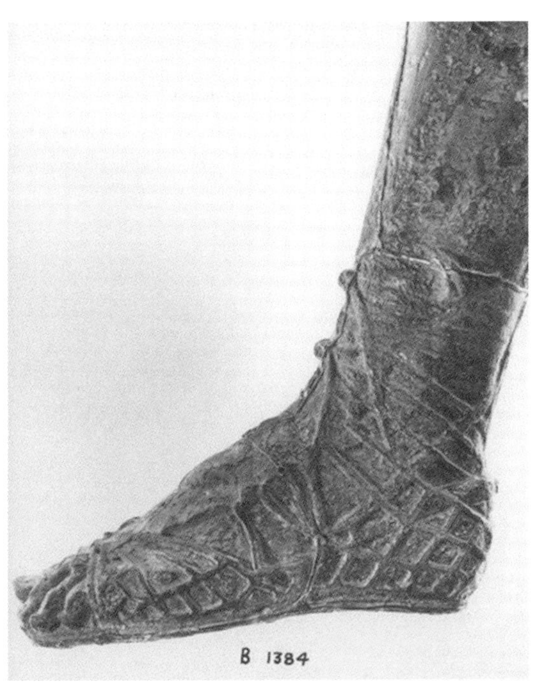

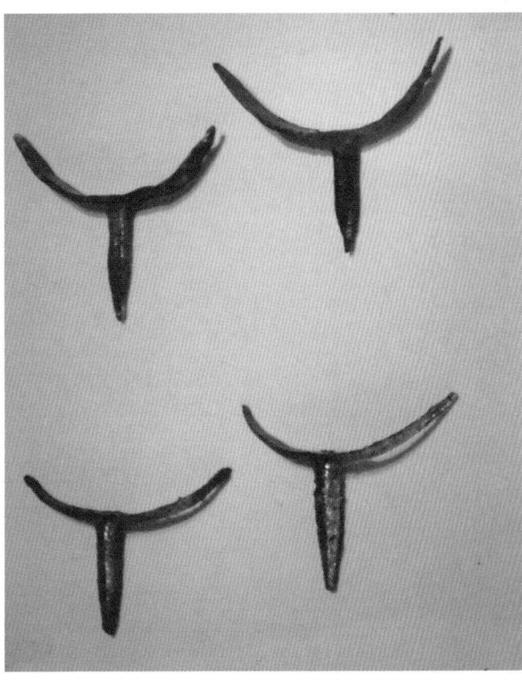

28 (*above left*). Left leg shod with a *krepis* from the same gilded bronze statue as in Plate 27. Note the complex lacing, netting arrangement at toe and heel, and evidence of a spur.

29 (*above right*). Sets of bronze spurs of varying sizes, to be attached to the heel of a rider's boot using leather straps.

30 (*below left*). Hetairoi members wearing the Afghan *pakol* or *chitrali* (serving as *kausiai*) with *chlamydes* fastened at the shoulder or breast.

31 (*below right*). Detail of a fresco from Boscoreale, c. 60–30 BC, depicting a figure in *kausia* with Macedonian *sarissa* and shield.

32. Detail of a 'Jason and Pelias' fresco from the House of Jason, 1–25 AD. The man leading a bull to sacrifice on the left and the figure of 'Jason' on the right could preserve clothing styles and colours worn by Macedonian cavalrymen and infantry officers.

33a (*above left*) and b (*above right*). Two views of a painted vase from Amphipolis depicting warriors on foot fighting with Amazons.

34. Detail showing 'Ajax' from a 'Sack of Troy' fresco (third quarter of the first century AD), possibly copying a Hellenistic work of the early-third century BC.

35. 'Alexander' medallion of the so-called 'decadrachm' series. Obverse: Alexander rears up with *xyston* at Indian bull elephant. Reverse: *Nike* crowns Alexander who wears full Macedonian battle panoply while grasping spear and thunderbolt.

36a (*above left*) and b (*above right*). Fragments of a mosaic from Palermo, c. 100 BC, showing a horseman from the left side of the mosaic, identified as Alexander, and a remnant of another horseman with Persian foot archer from the right side.

37. Panoramic view of the New Halos site in southeast Thessaly. The town may have been founded upon a military camp of Demetrius Poliorcetes.

Figure 41. Bronze objects identified as a surgeon's loop, probe and spatula. Excavated on the site of Olynthus, c. 350 BC.

soldiers can be adumbrated from stray literary references, and the overall picture is not unimpressive.[211]

The medical section was stationed well behind the combat line during engagements, probably within the royal enclosure in the centre of camp.[212] Following the teachings of Empedocles and Aristotle, it seems that Alexander was aware of airborne infection caused by putrefying corpses so he would have ensured his men were withdrawn from battlefields as soon as possible.[213] Among the infantry, casualties were helped by servants or fellow soldiers or carried on their shields if seriously impaired.[214]

To raise morale, Alexander regularly visited the wounded after battle, even if wounded himself, and sometimes examined their injuries.[215] He acquired drugs locally and supplied his troops with free medicines.[216] At one instance, a three-ton consignment from Greece arrived in camp during preparation of the Indus flotilla. The fact that they came by ship from such a distance is a vivid testimony to the value Alexander attached to keeping the troops in good condition. Macedonian kings might also order their personal physicians to treat officers with long-term ill-health.[217] Occasionally aid was offered to enemy leaders captured after battle, probably as a gesture of good will.[218]

The long-term sick were required to enter their names on a roll and were then transferred away from operations, while officers thought to be feigning illness were given a medical check-up.[219] On occasion, or perhaps as usual procedure, a field hospital was left behind at a locality thought safe from enemy attack in order not to hamper the army with wounded in movements leading up to a major clash, as during the manoeuvres before Issus and Gaugamela.[220] From Arrian it has been further posited by Holt that men were left to recuperate at leading cities and 'staging points'. When fit to travel again, they joined up with the main army or performed non-combat duties.[221] Those genuinely considered no longer capable of service were settled in new colonies,[222] their numbers, surely including amputees and invalided veterans returning to Macedonia, are described being well provided for with a gratuity of a talent each.[223]

When transporting casualties, Arrian relates how wagons were deployed as ambulances, although this may only be an incidental use of the usual supply wagons when crossing desert country.[224] Aeneas the Tactician certainly approved the usefulness of wagons in conveying injured men.[225] Transport animals would, however, be more commonly applied, probably in the form of wood framed 'armchair' packsaddles set sideways on the back of a horse or mule, with a horizontal foot-rest. Mules are shown carrying human effigies side-saddle in fifth-century BC Attic red-figure ware and are described in literature transporting invalids in the early-fourth century BC. The custom even continued in rural parts of modern Greece.[226] The Macedonians used a similar or identical type, adapted to carry the sick.[227]

The highest ranking of course had access to better transport. Alexander is portrayed by Curtius being conveyed in a military litter,[228] and also by Arrian, while recovering from a serious wound inflicted by the Malli.[229] Plutarch describes how, during his campaign against Antigonus, an unwell Eumenes was taken in a litter outside the ranks where the relative quiet would help him sleep.[230] The term *phoreion* in Arrian and Plutarch is loosely defined as a 'litter', 'palanquin' or 'sedan chair'. It is later mentioned for one occupant, with curtains on each side and a team of foot bearers, who in the Macedonian army were soldiers temporarily employed for the task. In Alexander's case, we find the infantry and cavalry vying with one another for the honour of carrying him.[231]

The Punishment Unit

Although soldiers could be well looked after, discipline was a fundamental requirement for a successful campaign. Just as loyalty was rewarded, opposition to Alexander's policies could mean isolation within a special unit organised for the purpose. We will complete this chapter by examining this unit in as great a detail as the literary sources allow.

After the plot and execution of Philotas (330 BC), Alexander purged from the ranks any man whose letters home were discovered by intercepting agents to have comments expressing discontent.[232] A new body called the *ataktoi* was then assembled comprising those who had displayed undue distress at the murder of Philotas' father Parmenion, as well as those who had generally criticised the King's policies. The term *ataktoi* has been variously translated as the 'unruly', 'disorderlies', 'unassigned' or 'disciplinary unit'.[233] It was, in short, the Macedonian counterpart of a twentieth-century penal battalion, led by Leonidas, an officer with an otherwise unknown pedigree (although he may have served later under Antigonus Monophthalmus and Ptolemy).[234]

Curtius informs us that the malcontents were required to encamp separately from the rest of the army:

> And he [Alexander] ordered a cohort of those who had chanced in their letters to have complained of the irksome military service to encamp apart

> from the rest by way of disgrace, saying that he would use their bravery in war, but would remove loose talking from credulous ears.[235]

Justin mentions how the *ataktoi* were to be used 'with the intention either to destroy them, or to distribute them in colonies in the most distant parts of the earth'.[236] The first part attests to their deployment for particularly dangerous missions, as a kind of 'forlorn-hope' or 'suicide squad' in the brunt of an attack. Their numbers and nature of equipment are not given but may be provisionally reconstructed from what little is known. Both Justin[237] and Curtius[238] have them as one cohort strong – the Latin interpretation of an original Greek source term. A cohort as envisioned by the Roman readership of these writers meant a body of about 500 men, so the *ataktoi* were of comparable strength.

Given its diverse origins, the unit would have drawn men not only from infantry but cavalry as well. On one occasion, a royal page is described having his horse confiscated as punishment – presumably horses freed up for service to other troopers would have been welcome.[239] We can therefore see these dissidents as a hotchpotch band operating on foot, with threadbare clothing, minimal armour and only light weapons: swords and spears or javelins. Given the Macedonian delight in owning expensive panoplies, they may have been issued with scratch items to emphasise their shameful, degraded status.[240]

In an army which set professional pride and *esprit de corps* at a premium, the *ataktoi* were, however, capable of fighting with desperate courage to reclaim the respect and confidence of their peers and commander.[241] In short, the stigma of 'letting the side down' transformed them into a crack force. As Curtius tells us, 'nothing was more enthusiastic for war than those men; their valour was enhanced both from a desire of wiping out disgrace, and because brave deeds could not be concealed among a few'.[242] The same ferocious display was demonstrated by Soviet penal companies who often spearheaded armies in the shock offensives of the Great Patriotic War.

Although evidence has not survived, the *ataktoi* were probably disbanded well before Alexander's death, perhaps during or soon after the Indian campaign, the reconciled soldiers being reassigned to their original units. After all, Alexander could ill afford to be seen humiliating Macedonians after the Opis mutiny or waste their skills and experience when he required all available veteran manpower in Asia for his ongoing plans of conquest.[243]

Chapter 11

Marching

For all their fighting abilities, the gory mayhem of battle nonetheless constituted but one part of a veteran's existence. Another major facet of his life naturally consisted of marching, punctuated by routines and recreational habits, the food consumed, and the basic necessities required for daily living. It is appropriate that we now address these often ignored aspects as fully as the evidence warrants.

Baggage Train

From the reign of Alexander's father onward, Macedonian soldiers became familiar with the drudgery of long, grinding hours wandering from one location to another, and all that it entailed.

Some authorities accept that Philip II devised the first systematic logistics arm in Greek warfare, setting the benchmark for future armies in the West.[1] Given the huge logistical challenges offered by the Persian Empire, it is not surprising that Alexander's army incorporated a well-organised baggage train under the jurisdiction of a transport officer (*skoidos*).[2] It is thought that one of Parmenion's functions in Asia was to operate in this capacity, and as a trusted veteran general he would have had twenty or more years exposure as a logistician by the time Alexander was King.[3]

High levels of logistical skill were first effectively achieved during Philip's Balkan campaigns. After Parmenion's death his post was probably filled by either Craterus or Erigyius. As stated by Engels, such officers were to care for the baggage train's security, animals, march order and 'the correct packing and balance of the saddle bags'.[4] They might in addition have been responsible for the repair of wagons and checking that all the men were adequately provisioned before the march commenced.[5] When Alexander's army was on the move in the East, food supplies held centrally in the baggage train were impressed with the royal seal (to discourage famished troops from breaking into it).[6]

The sources clearly imply time after time just how skilfully managed provisions must have been under Alexander in Asia, even though the internal technicalities are shrouded in mystery. The levels of expertise are well identified by Fuller, who rightly concludes that without a highly efficient supply column Alexander would have been unable rapidly to traverse thousands of kilometres of often sparsely populated land – the mountains of the Hindu Kush and desert of Makran.[7] His successful negotiation of the Sinai from Syria can be favourably compared with that of Napoleon's march from the direction of Egypt in 1799. Unlike the Macedonians, the desperate French were reduced to eating their own mules and camels to stay alive.[8]

Management efficiency stands out in the manoeuvres leading up to the battle of the Hydaspes. Here supply and transport officers were able to collect and prepare enough river craft and have them successfully launched with speed well before dawn as part of Alexander's battle plan and under adverse strategic circumstances. This was an achievement far superior to either William the Conqueror's invasion of England (1066)[9] or Henry V's disembarkation of troops at Harfleur (1415).[10] In terms of supply considerations, Alexander was able to succeed more roundly in contrast to the failure of the Roman generals Crassus and Mark Antony and the Emperor Julian the Apostate, along with the later Ottoman sultans in their frontier wars with Safavid Persia.

In the Macedonian army wagons were employed when travelling on roads. Pack animals, on the other hand, served in terrain with no tracks and were provided with their own transport guard of possibly trained drivers, most of whom went unarmed.[11] The animals usually consisted of mules or horses[12] and each handler led a pack line of up to five beasts.[13] Philip II was one of the first commanders in the Western world to recognise the advantage of employing horses in this role.[14]

Donkeys and asses were also used, depending on what was available, while camels were increasingly introduced from the occupation of Egypt (332 BC).[15] The introduction of camels signifies a deliberate logistical preparation for the hard campaigning which, as Alexander knew through intelligence, awaited his troops in the more arid terrain beyond Mesopotamia. Camels are ideal for desert country in that they require only minimal arrangements for upkeep, can drink brackish water and consume virtually any kind of vegetation to hand – even skin and bone in times of severe hardship. Aristotle's surviving zoological studies of camels suggest that the Macedonians were well aware of their usefulness as draught animals.[16] The use of camels for transport is demonstrated by an incident when spies deliberately travelled on them to report to Eumenes and Peucestas in 317 BC.[17] Antigonus Monophthalmus followed Alexander's example when he resorted to using these animals in his attempted invasion of Egypt (306 BC), this time with the help of Arab communities around the Dead Sea and Sinai regions.[18] Their value has been recognised throughout military history and they were still being deployed as beasts of burden by Soviet troops as late as 1943.

Engels observes that the number of pack animals with non-food supplies in Alexander's army in Asia was smaller than that in corresponding Roman armies. This was due to the fact that many non-food items were carried by servants rather than beasts.[19] The deliberate provisioning of a smaller baggage column had the advantage of creating less of a problem acquiring sufficient animals and fodder among populations engaged in subsistence agriculture.[20] However, with a reduction in the numbers used, Macedonian generals still recognised the importance of pack animals within a well-run supply system and, wherever equitable and for the sake of efficiency, set their welfare at a high level of priority when on the march. This is elucidated from an anecdote about Philip II, of admittedly dubious provenance, which describes how when he was about to pitch camp at a suitable location he bemoaned that it had to be abandoned as there was no suitable grass for the mules.[21]

During his eastern campaigns Alexander collected transport animals from the nearest communities and continued the Achaemenid precept of requiring subject-peoples and satrapies to send an annual fixed levy of baggage animals to a central round-up point.[22] After the gruelling march through the Makran, the King is noted having received beasts of burden and racing camels from the satraps of Aria, Drangiana, Parthia and Hyrcania. Alexander proceeded to distribute them proportionally to all officers, infantry and cavalry sub-units, and among non-Macedonian troops.[23] He is further described collecting large numbers of pack animals to replace those that had died, including 2,000 camels, flocks of sheep and herds of cattle, which were shared out equally among units and sub-units.[24] These routines reflect Iphicrates' methods, in which supplies received from cities were distributed evenly to each unit and file: some to cavalry, others to hoplites and light infantry.[25]

Pack animals were attached to Macedonian infantry at *dekas*, or squad, level and horses, mules and donkeys were capable of hauling a 60–100kg load.[26] Utilising a wood frame packsaddle or twin panniers, the burden was made up of large water pots or skins.[27] Other articles were tents, mills, straps, plunder, fodder, firewood and perhaps freshly cut palisade stakes if the environment to be marched through was known to be without trees. Some packed items were medical provisions and probably the full complement of a squad's all important construction tools – axes or adzes, saws or sickles, mattocks, picks and shovels.[28]

Marching with Weapons

When marching, *pezhetairoi* would slope their *sarissai* over the right shoulder, grasping it with both hands, with left arm across the body to hold the shaft stable. This posture unsurprisingly required an appreciable degree of strength and stamina if done for a long time. Polybius recognises the fatigue caused by the weight of the *sarissa* on the march.[29] Premature weariness is deduced from Curtius, who mentions a Macedonian soldier marching in freezing temperatures and finding it difficult to carry his weapons.[30] Alexander recognised that prolonged foot-slogging of this sort could damage morale. Arrian and the *Itinerarium Alexandri* both describe him sharing the toils and tribulations of his men, sometimes marching alongside them burdened with fighting gear.[31]

Using Sekunda's tool of 'inherent military probability', we can extrapolate how *sarissai* were carried on the march by looking at comparable historical cases, the pike-armed soldiers of the European Renaissance springing to mind as the obvious choice.[32] Literature of this period, such as Gervase Markham's *The Souldiers Accidence. Or an Introduction into Military Discipline* (1625), recommended that, for reasons of comfort, the helmet be left off and fastened to a hook on the breastplate when marching over great distances. Another tip is found in James Acheson's manual *The Military Garden, or Instructions for All Young Souldiers*

(1629), which advises that each man 'always shoulder his picke [pike], either just or sinking . . . and coming thorow [through] any port or gate, he is to port his pick [pike]'.

Reenactment marches with authentic Swiss and German pikes prove what a nuisance they could be. The angle of slope meant that the weapon might prick men marching a few ranks behind, with the shaft bouncing and vibrating uncomfortably on the shoulder.[33] Its cumbersome length meant that the pike would get tangled in vegetation overhead. This corresponds to a passage in Livy in which low hanging trees are recounted as an impediment to infantry armed with *sarissai*.[34] It is fairly clear that special training and expertise gained through experience must have been required to handle the *sarissa* competently while marching.

We can assume that *sarissai* were trailed on the ground or stacked on baggage wagons, as or when enemy contact was not expected. If this was practised, the wagons were presumably located close to the men to make them quickly accessible. Although it has been suggested that both Philip and Alexander followed a policy of keeping wagons to a minimum, they must have been employed in significant enough numbers when travelling on Persian roads for pack animals to become ancillary.[35] Otherwise, there is little way of apprehending how such weapons with the rest of the baggage were effectively transported across wide lands and all types of terrain. Defining the method used for carrying such arms during long drawn out campaigns is a conundrum still open to debate and conjecture.[36]

As far as Companion cavalry is concerned, Markle concluded through practical experiments with a 4.5m facsimile that the *xyston* would be sloped back over the right shoulder for a march.[37] This was refuted by Manti, who submitted that the motion of a walking or trotting horse would cause the shaft to vibrate so that within the space of 5km it would discomfort or even injure the shoulder and act as a hazard to nearby men and animals.[38] Other tests on horseback, most recently by Corrigan, demonstrate unequivocally that the weapon could be comfortably leaned against the shoulder utilising the proper method with a 50 or 60 per cent grip.[39]

During a march the kit-laden trooper could follow Xenophon's stricture that cavalrymen periodically mount and dismount to ease discomfort and take care of the horses' backs.[40] This was also useful for relieving pressure on hooves in an age devoid of metal horse shoes.[41] Remaining unmounted would be commonplace during long marches. Alexander is described going on foot – which can be interpreted to mean that he was often dismounted when marching.[42] Pliny records a painting of Antigonus Monophthalmus in which the general marched on foot alongside his horse.[43]

Given that the Companions dismounted for lengthy marches, it seems more than likely that on Alexander's eastward venture a groom was delegated to carry his master's lance on foot. The weapon would then be transferred to the cavalryman once he was mounted and ready for combat.

Personal Kit

Macedonian troops took along a substantial amount of kit with them when marching, as evident from literary sources (both as direct statement and inference) for Alexander's operations. Earlier texts describing Greek hoplite warfare occasionally offer additional insightful evidence. Even so, experimental archaeologists with reproductions of equipment may be required to help resolve what a phalangite, already encumbered with a *sarissa*, was able to shoulder cross country.

Most items seem to have been provided on a general issue basis.[44] Foot soldiers could have followed the widespread tendency of taking along a kit bag with bedroll, possibly converted into a hammock;[45] a square wickerwork ration box attached on a leather strap across the body under the shield or close to the chest; and a cooking pot tied to the end of the bedding, itself bound to the shield.[46]

There is written and artistic confirmation that this assortment was carried by soldiers and civilians elsewhere in the Greek world during the fourth and third centuries BC.[47] That the Macedonians did the same is made apparent from part of a terracotta bed decoration found in Pydna for this period featuring a young man carrying a bedroll over the shoulder using a strap.[48] A Roman fresco dated to the second or first centuries BC, and possibly copying a Hellenistic original, may offer further clues (Fig. 42). In this case, the man illustrated has been identified as the Cynic philosopher Crates of Thebes (365–285 BC), a contemporary of Alexander and the Successors. His luggage incorporates a food box and large bedroll slung at the shoulder.

Figure 42. Detail of a Roman fresco possibly portraying Crates of Thebes. From the Villa Farnesina.

To avoid tarnishing and constant cleaning, shields were secured by the men in their waterproof leather (or fabric) covers and slung at the back by the neck baldric or an arm band. Menander's plays sometimes relate the use of a shield case.[49] Troops had personal wallets containing pay, a few choice selections of booty[50] and some form of bronze or earthenware drinking vessel. Pillaging Macedonians might account for the paucity of evidence for metal containers found in the Achaemenid treasury building at Persepolis.[51] Lee observes that the vessels and implements carried by soldiers were never as perfect as those depicted by artists 'in vase paintings and sculpture'. He rightly goes on to say that in a world where most items were handmade there would have been wide variation 'even amongst objects of the same nominal type'.[52]

Other items carried would be: a sponge, strigil, oil jar and shaving tackle;[53] writing materials for sending letters home;[54] gaming dice;[55] spare clothing; and perhaps everyday essentials like a needle and thread (for, among other things, stitching up flotation devices at river crossings).[56] Strabo alleges that the men made use of cotton for making pillows in India, so this may have become a part of personal baggage too.[57]

It is rational to suppose that, beyond basic needs, soldiers in each *dekas* were apportioned specific items, such as medical supplies, for the use of the squad as a whole rather than all soldiers carrying completely identical belongings. Among stored items was included a measure of oil for shining metallic components. Well-polished and shining gear was part and parcel of the Macedonian heroic mentality but must have been increasingly difficult during laborious wars in the East, especially when confronted with dust-storms and monsoons.[58]

The *Cyropaedia* recommends that each soldier include with his kit a *xuele*, meaning a rasp or whittling-knife. Xenophon goes on to state that he who 'whets his spear whets his courage, ... for a man must be overcome with shame to be whetting his spear yet feel himself a coward'.[59] Knife blades (or blade fragments) have been discovered in Macedonian gravesites from the fourth century BC onwards, such as the burial mound at Chaeronea, the cemetery and Great Tumulus at Vergina, graves at Stavroupolis and tombs at Pydna. A well-worn whetstone with a hole for a retaining cord was found intact at the Chaeronea site.[60] The fact is that sources describe *sarissai* with razor sharp heads, demonstrating gruesome stabbing power – splitting shield and corselet together, piercing lungs and impaling thighs, and puncturing victims from groin to buttock.[61] Such murderous efficiency proves that they were scrupulously maintained.[62]

Engels further posits that Macedonian troops carried iron tent pegs about their persons and, as with later Roman legionaries, building tools.[63] Superficially this sounds persuasive, given the frequent construction tasks demanded of Macedonian troops in wartime. Responsibility might be spread across the squad, with each member burdened with different implements. Yet there is no literary evidence to confirm such items as part of a soldier's *personal* kit. Moreover, given all the other paraphernalia taken along, it would surely mean a crushing load and Engels appears to be the only authority to raise it. We can for now lay aside the suggestion until direct evidence is uncovered.

While out on special missions, baggage was kept to a deliberate minimum to aid mobility.[64] Apollodorus of Gela's *Sham Ajax* (c. 300 BC) baldly enumerates such kit in a general context as sword, spear and blanket.[65] For long distance marches in arid terrain men were sure to take along a waterskin or canteen containing a fresh supply, preferably from a mountain stream.[66]

The Madman of Diphilus of Sinope (late-fourth to early-third century BC) has a few lines, exaggerated for comic effect, which portrays a soldier ludicrously overburdened with a variety of provisions, many assigned to the servants and pack animals:

> And then keg, brazier, blankets, sleeping-mat / Bag, wallet, frying-pan! / No one with all that / Would take you for a soldier; when you come / They'll

> think a Round [the area of a market where household goods were sold] has got up to / make for home.[67]

In Menander's play *The Flatterer* a soldier is portrayed with 'his own pack, lunch bag, helmet, javelins, sheepskin, as heavy a load as the poor old donkey carries'.[68] This second list of objects is less than the one in the preceding extract and points to the sort of portage a Macedonian on more short-term operations would consider.

Food Provisions

The kit of each Macedonian soldier obviously included food rations. Troops were sometimes ordered to bring along pre-cooked food with their packs (avoiding the burden of cooking utensils). These instructions were most often associated with mobile operations, the rations being required to last fixed periods of two, three, four or ten days. The ten-day requirement is the most frequently attested, intended for operations expected to cover an unspecified period of more than four days.[69] We should keep in mind that modern military nutritionists consider pre-boiled food ideal for marching soldiers because it can be quickly prepared, more digestible and free of harmful micro-organisms.

Victuals for common rankers were moderate and plain.[70] Grain formed the staple and it has been suggested that some pre-cooked foodstuffs were consumed in the form of hard tack biscuit, not unlike those of Roman legionaries or the large weevil-infested specimens of British Napoleonic sailors.[71] It is thought that Greek mercenaries (400 BC) consumed *maza* (roasted barley cakes), *chondra* (barley groats) and *ptisane* (gruel).[72] The Macedonians almost certainly ate comparable concoctions.[73] Consumables which required no cooking at all are referred to in the *Suda* as fruit, vegetables and 'unfired' wine. Excavation of wells at the south and west stoas of the Pella Agora reveals that the inhabitants enjoyed olives, grapes, seeds and nuts (walnuts, hazelnuts, chestnuts and almonds). At least some of these may reflect popular culinary preferences, embracing those of military men.[74] To supplement their diet, troops incorporated rice and dried fruit such as mulberries, figs and dates, which could be conveniently eaten while marching, the high sugar content being considered suitable for active soldiers.[75]

For animal protein, some took the opportunity to purchase salted fish and Alexander is recorded distributing dried fish among his men.[76] On leaving the Makran Desert, the army received livestock such as sheep and cattle from nearby satraps. They were rounded up from royal hunting parks or scoured from surrounding lands.[77] Supplies of meat were obtainable from the King's sacrificial observances, with beasts slaughtered, then cured, smoked and dried for storage.[78] Xenophon specifies salted meats as essential to a soldier's diet: 'We must pack up and take along only such as are sharp, pungent, salty; for these not only stimulate the appetite but also afford the most lasting nourishment.'[79]

As with Xenophon's Ten Thousand in the Middle East, we should not dismiss Macedonians gaining local knowledge from their native-born women, servants and captives when gathering unfamiliar fruits, vegetables and edible plants. Many veterans would have become wise to what was best taken for a demanding march, further adapting themselves to the food habits of the indigenous peoples they lived amongst or were influenced by.[80] Alexander's expedition botanists were the first to introduce Mediterranean cultures to a number of delicacies like the citron, peach and pistachio and it is credible that these had already been tested with relish by his hungry troops.[81] In Hellenistic times Philon of Byzantium recommended his own 'hunger-and-thirst checking pill' for soldiers (similar in concept to the K-ration of the United States army in the Second World War). Philon's olive-sized pills were composed of sesame, honey, oil, and squill, providing protein and carbohydrate and serving as a tonic and appetite suppressant.[82]

The diet of wealthy Macedonians proved just as rich and varied as that of later Roman aristocrats. Some literary sources describe their formal banquets as a gourmet extravaganza. This is most vividly seen from a letter written by the Macedonian Hippolochus describing the menu of a marriage feast for his fellow countryman, Caranus (c. 300 BC). The menu lists bread, goat, chicken, duck, goose, pigeon, turtle-dove, partridge and an abundance of other fowl along with stuffed pork, eggs, fish, oysters and scallops in plenty.[83] Being a distinct social event, it would be misguided to endorse this as evidence of military diet, but it reflects that the Macedonians had few food taboos and were not averse to eating large quantities of meat, at least on special occasions.[84]

As far as shellfish is concerned, Arrian describes how the soldiers accompanying Nearchus on his voyage back from India in 325 BC at one time built a stone wall around their camp and hunted for mussels, oysters and razorfish at Alexander's Harbour.[85] French archaeologist Monique Kervran identifies this location as Tharro in modern Pakistan, the site to this day showing traces of two rubble defensive walls and large middens of oyster shells.[86] Another extract survives describing how Antigonus Gonatas witnessed the poet Antagoras of Rhodes stirring a casserole of conger eels in camp. This has been used as an example of the food provisions that were available, in particular for leading officers.[87]

Servants and Grooms

Servants were employed as baggage support to Macedonian infantry as it marched across Asia. In this they were particularly useful as, unlike other forms of transport, men on foot were capable of traversing rough or snow covered ground without reliance on trails or roads.[88]

Given everything which has been discussed so far, the weight of kit carried by Macedonian infantry must have been considerable. The idea that soldiers during the Successor campaigns continued to haul substantial burdens can be appreciated from an anonymous fragment of New Greek Comedy where a camp servant comments on his master's heavy baggage load: 'You don't see that a soldier who won't lack, / Must carry everything upon his back'. To this his soldier master retorts: 'How can you talk like this? You're only a serving man / Who carries round a miserable shield / His master's sole protection on the field'.[89]

To understand the haul taken along by each man, scholars have cited pack weights for soldiers in the American War of Independence, Napoleonic Wars, Russo-Japanese War, and the First and Second World Wars (1776–1945), these ranging from 20kg to 40kg. In comparison, Engels estimates the entire burden carried by each *pezhetairos* consisting of about 23kg for arms, armour, utensils and bedding and 13kg of food rations; the maximum weight that can be carried for protracted periods without causing injury amounting to about 36kg.[90] Anything in excess was shared by servants and families.

Similarly to soldiers, the servants carried their own bedding and rations. While on campaign, they probably slept inside tents along with the troops they accompanied.[91] Each servant was responsible for a hand mill and bore tent pegs and guy-ropes, used for bridge building and as pitons when rock climbing. Other items might include a kettle and earthenware fire pot with some dry kindling for making fires and cooking meals.[92] Some of these were suspended from a curved shoulder yoke. Xenophon describes carrying poles in his *Anabasis*,[93] and travelling servants appear to use some form of yoke elsewhere in classical Greek art.[94] Such implements are distantly related to the forked pole used later for provisioning by Roman legionaries.[95]

Pediment A of the Alexander Sarcophagus portrays an unarmed youth, possibly a servant, rescuing a fallen soldier (Plate 1a).[96] He wears a tunic faintly tinted pink or purple with a dark vertical stripe reminding us of a possible groom on the Agios Athanasios Tomb painting discussed earlier in Chapter 6. The tunic would be girded up for the march by doubling over the belt, drawing it up higher and thus keeping the legs free and the hem out of the way of dust and mud. When in full kit, servants wore a cloak, socks and boots. A stout staff would have lent practical help as a trekking pole, supporting the back and lower limbs when carrying heavy weights. Short, knotted versions are seen with slaves on Greek painted pottery from the fifth century BC[97] and walking sticks are mentioned, again in Xenophon's *Anabasis.*[98]

Within the *pezhetairoi*, servants came under the authority of the administrative aide (*hyperetes*) who was attached to each tactical unit of 256 men (*speira*).[99] Under Philip II, regulations required that a servant cater to the needs of each ten-man *dekas*.[100] The original Macedonian attendants in Alexander's army could have been promoted and absorbed into combat units, as casualties rose during the course of the eastern campaigns, but evidence is scant.[101] Assisting the troops in Asia, the servants themselves certainly sustained a considerable number of casualties by the end of the Indian expedition.[102] Replacements would be drawn from the large reservoir of captives taken on numerous campaigns. It is therefore evident that Philip II's strict rules pertaining to the numbers of servants and grooms became relaxed among well-off veterans after Alexander's death, and probably also much earlier.[103]

The social origins of Macedonian servants remain ambiguous through lack of evidence. It is reasonable to think that they represented a lower property qualification or were drawn from a class resident in Macedon but without the designation of *Makedones*, i.e. below the rights of full citizenship.[104] The background of captives who became servants following soldiers during the Successor Wars would be more diverse. The figure of Harpax ('robber' or 'snatcher') in the surviving comedy *Pseudolus* may offer a few intriguing insights into the social origins of these unfortunates. He bemoans his life as a servant since he was at one time a general in his homeland.[105]

A variety of servant duties can be gleaned from the sparse literary scraps that remain. Aside from their function as porters, they were to help in loading and then leading the baggage animals; pitching, striking and looking after tents; and gathering supplies of water, forage and firewood. They ground the grain ration into flour, prepared camp fires and food, tended the wounded, and helped soldiers with the cleaning and maintenance of their panoply using oil or wax. In one lively extract of Plautus' play *Miles Gloriosus* the character of Pyrgopolynices directs his 'parasite' Artotrogus to polish his shield as far as it will outshine the 'radiance of the sun in cloudless sky', bedazzling the eyes of oncoming foes.[106] It has even been suggested that during Alexander's eastern advance servants held some unspecified responsibility in the transportation and distribution of spare weapons.[107] Despite usually remaining unarmed, at times they could be pressed into battlefield service.[108]

It should be understood that grinding grain using mills (*cheiromulai*) was among the most important of a servant's tasks.[109] Whole grain was measured by *choinikes*,[110] carried over flour due to it having a lower risk of spoilage, weighing less and taking up the least space on the march.[111] Modern studies estimate that each soldier in Alexander's army would have required 3,600 calories per day, with 1.36kg of bread and 1.9 litres of water.[112] As each attendant catered to the energy needs of many men, long hours were spent preparing and cooking food.[113]

In contrast to infantry servants, grooms were a part of a range of ancillary services supporting the cavalry. They included horse breakers and trainers, and veterinary surgeons, who would have first and foremost taken care of the war horses, then pack animals and wagon oxen.[114]

Originally grooms may have been provided by horsemen from estates granted to them by the King, and Philip II required that one groom serve each Companion cavalryman.[115] However, it is obvious that this requirement became increasingly flexible following Alexander's conquest of Persia when cavalry were allowed two horses[116] or more. Amyntas, son of Andromenes, a probable officer of the Companions,[117] is described owning ten mounts, which strongly suggests that he had more than one groom to effectively handle them all.[118]

As discussed in Xenophon's *On Horsemanship*, the groom was responsible for ensuring that horses were watered and fed, their stables mucked out, and their halter and bridle harnessed and removed before and after an engagement. In icy weather grooms could even help protect vulnerable hooves by tying a form of leather boot over them for temporary relief.[119]

Although grooms in the Macedonian army were sometimes left behind in camp, they would be employed to keep surplus mounts at the rear, supplying them to the Companions. Alexander was himself quickly remounted with a fresh horse, whether in battle or during a pursuit. Curtius says how at Gaugamela he 'changed his horse – for he had tired out several',[120] while still later it is conveyed that 'the king's horse, which had been pierced by many shafts and was giving out, fell under him, rather dismounting him than throwing him off. And so, while he was changing his horse he pursued more slowly.'[121]

Lengthy cavalry pursuits are made mention of in sources documenting Philip II's Balkan wars.[122] Royal grooms following close at hand with spare horses may have therefore originated in these early campaigns. On a rapid march to Malus in Cilicia Demetrius Poliorcetes is described losing most of his horses to excessive hardship, with none of the sutlers or grooms capable of keeping up the pace.[123] Grooms were also responsible for exercising mounts and were assigned the task of riding them to this effect.[124]

For their tasks grooms used a *kentron*, or goad, with a variety of other instruments such as a *spathe* for cleaning the horse's coat, a *psektra* for combing and a *sorakis* for finally smoothing and shining the coat surface.[125] A curry comb, scraper or polisher is actually noted in a preserved papyrus from Ptolemaic Egypt.[126] During the investment of Nora (320 BC), grooms in the army of Eumenes were seen applying goads to stir cavalry horses to exercise, the beasts partly suspended using specially improvised contraptions secured to roofs with straps and pulleys. It reiterates the intensely inventive and professional approach followed by the Macedonians, arguably only ever eclipsed in the ancient world by the Romans.[127]

Hammond draws attention to the existence of different types of groom in the Macedonian army.[128] Sources refer to both the 'horse tender' (*hippokomos*) and 'mounter' (*anaboleus*), the latter head groom of a Companion with more than several horses, responsible for mounting and remounting the rider before or during an action.[129] Groom mounters were occasionally employed in combat roles. Arrian describes how at the Granicus Alexander called out to Aretis, a 'mounter' of the royal suite, for another lance. The latter was, however, 'hard pressed, though putting up a brave fight with the half of his broken weapon. Showing this to Alexander, he told him to call on someone else.'[130] From this it is clear that some of the king's grooms were obliged to fight and provide him during battle with a fresh lance if he had lost or splintered his own.

The fact that Aretis was so near at hand is a reminder of passages recounting Alexander being furnished with a fresh mount in battle. It may have been the case that grooms generally stood ready at the rear with spare mounts and weapons.[131] Grave *stelai* (late-fourth to third centuries BC) found in northern Greece, Bulgaria, Turkey and Egypt feature Macedonian horsemen, usually accompanied by one boy-groom apiece (in keeping with Philip II's code) carrying his master's helmet, shield, sword or lance.[132] An apparent relief likeness of a Macedonian trooper with groom from the environs of Beroea (dated to the fourth century BC) depicts the former in a helmet accompanied by the latter handling the bridle of his master's mount.[133] The *stele* of Nikanor, son of Herakleides, portrays a soldier on horseback with a groom on foot wrapped up in a thick blanket, braced against the cold (Fig. 17).[134]

A Test of Endurance

The extensive march-kit that soldiers laboured under (as outlined earlier) was a monument to Macedonian self-sufficiency and made Philip II's army unique for its time.[135] It may even be said that these changes introduced by Philip were something of a precursor to the reforms of Roman republican generals, such as Quintus Caecilius Metellus during the Jugurthine War after 109 BC[136] and the more famous Gaius Marius with his 'Marius' Mules' (from 107 BC).[137]

Alexander, aided by capable subordinates, retained and extended his father's organisational model wherever possible, despite being faced with vast tracts of barren terrain, the limitation of ancient transport and low levels of agricultural production.[138] His conquests were built on the outstanding hardihood and performance of men who campaigned with him in all places and environments, exhaustively marching in some of the most punishing geography in the world. This encompassed everything from open plain and desert to jungle and monsoon morass, forest, river valleys and rugged mountains.

All-season route marching gave the Macedonian army an impressive degree of mobility. Gabriel suggests that Philip could expect his troops to march 240km in ten days, embracing regions as far apart as the southern Balkans to Thebes, and the Illyrian coast to the river Nestus in Thrace.[139] Alexander's men are recorded covering some 400km of often precipitous mountain terrain in thirteen days between Lake Lychnitis and the walls of Thebes (335 BC).[140] A similar feat was achieved by Antigonus Monophthalmus in his lightning advance to Cretopolis in Pisidia (319 BC), in which he and a mobile force completed near to 500km in seven days flat.[141]

Dodge calculated that in the period from 336 to 323 BC the army covered over 35,000km, with an average of 2,580km per year.[142] As further contributed by Heckel, the later campaigns of the Silver Shields added at least another 8,000km onto this remarkable total.[143] It gives the Macedonian armies of Philip, Alexander and the Successors a strategic range only ever bettered in antiquity by Imperial Rome (1–300 AD).[144] Given these statistics, it is not surprising to find an abundance of circumstantial evidence which casts light on soldiers trained to the most exacting standards of endurance. This was starkly appreciated by historian and broadcaster Michael Wood, who attempted to retrace Alexander's route across Asia and witnessed all the discomforts of excessive heat (and cold), choking dust and steep slogging hill climbs in gaunt and inaccessible landscapes.

The exceptional case of Philip, brother of Lysimachus, illustrates the degree of hardship Macedonian soldiers were inured to. As a highly praised royal page or hypaspist, he is described jogging on foot alongside a mounted Alexander non-stop through Sogdiana to a distance of 500 stades (around 80–100km), even though he wore armour and carried his weapons. He then immediately engaged the enemy, before feeling faint and dropping dead from exhaustion. His feat reflects the potency of the competitive urge among some of Alexander's men, and the impulse to excel and impress both comrades and king.[145]

Even though packs were sometimes transferred to baggage animals during galling desert marches when men needed all their energy to cope with the hostile country around them, their load was still considerable.[146] In recounting how Alexander could on occasion order each man to carry four days' worth of water, in addition to his weapons and other supplies when crossing the desert, Hammond was moved to comment that Macedonian soldiers were physically tougher than any soldiers of the modern Western world. This could prove to be an accurate appraisal, as all the following information suggests.[147]

Current scholarship often tends to ignore the real danger of heat confronted by Macedonian soldiers in the Middle East. The average temperature in Mesopotamia and the Gulf region is usually in the low 30s degrees Celsius by April of any year. By June–July it soars into the 40s, while *in extremis* it may even exceed an enervating 50 degrees Celsius in the shade. Tamerlane's army experienced similar broiling conditions during the infamous forty-day siege of Baghdad, conducted over May–July 1401.

In the often strenuous and desultory warfare encountered in the eastern satrapies, Arrian notes one memorable pursuit which was 'sharp and distressing because of the great heat, so that all the army was consumed by thirst'.[148] During the hard campaigning endured on the Persian plateau (317 BC), Diodorus describes Antigonus as having 'lost a large number of men because of the extreme heat, for it was in fact the season when the Dog Star rises [late June]'.[149] Many troops probably resorted to wearing *petasos* hats as a means of warding off sunstroke when travelling in the merciless daylight hours, with helmets fastened at the shoulder. In another poignant excerpt from Diodorus the march 'was scorching hot because of the intensity of the sun's rays' so that 'many soldiers perished, and the army became discouraged'.[150]

The wearing of close fitting garments and heavy armour in such an oppressive setting gave rise to irritating skin rashes, boils and abrasions. It may be of passing interest to add that actors in the *Alexander* Hollywood movie epic soon found that they had to adapt themselves to the uncomfortable pinch of pre-cast breastplates while filming in the desert sands of Morocco. Other actors had to contend with the hazard of scorpion stings through their boots.[151]

From all the preceding sources we can justifiably conclude that in hot weather some Macedonian soldiers opted to discard elements of armour in order to achieve greater personal comfort. This is directly supported by modern European travelogues in Bukhara during the summer in which burning sand sunk into footwear, with gun barrels and all other metallic objects heated to such an extent that they could not be touched.[152] Interestingly, reenactors duelling under the sun of the San Joaquin Valley (California) with facsimile Greek or Roman equipage were reportedly exhausted after only about thirty minutes of mock combat.[153]

Among the surviving texts, the wearing of less or minimal armour in hot conditions is implied from what is known for Alexander himself. Although he is portrayed wearing a corselet, either while marching or fighting in the heat,[154] before Gaugamela he is specified by Curtius as having only rarely worn a corselet in battle.[155] Alexander is known for not allowing himself special privileges if it were to cause resentment or a diminution in morale, and if he discarded armour then many of the troops would do likewise. One of the celebrated episodes from his life was during the Makran desert march when he demonstratively refused to take water when the army was suffering from severe thirst.[156]

When considering the frequent demands and challenges overcome by Macedonian troops on campaign in similar climates, the results of an American Marine experiment conducted in the Twentynine Palms desert training area (California, 1984) offers further sobering data as to how hardy Alexander's old sweats must have been. Even with the scientific preparation and facilities of a modern army, and with limited periods of marching, 17 per cent of the total field force required some form of medical treatment over a fifteen-day period. 1,101 men suffered injuries that were serious enough to need special attention in battalion aid stations or further to the rear. Specific casualty statistics incorporate 228 men suffering from blisters, lacerations and abrasions, 152 with irritations to the nose or throat, 110 treated for heat exhaustion, 53 with disabling headaches, 31 with body cramps and nausea, 46 with nose bleeds due to dust and the same number with eye irritations, even though protective goggles had been distributed beforehand.[157]

Chapter 12

Camping

At the end of a demanding day's march, the army would set up camp. This was a vital requirement and we can propose that Macedonian camp methods were far more thorough and conscientious than previously thought.

Field Camp Methods

No effort has yet been made to investigate extensively the subject of Macedonian field camps, although some glimpses of the method applied can be construed from sifting scattered fragments in the usually reticent textual record and the gossamer traces of archaeology. Lack of modern analyses is puzzling since Alexander's encampment served as the all important base for most of his operational activities in Asia. It functioned as a virtual mobile capital, an active nerve centre where military and administrative decisions were implemented for the empire – implying *a priori* that it was carefully planned and laid out.[1]

As the full weight of technical utility was deployed in every other known facet of his military career, it seems unthinkable that a general with Alexander's aplomb and concern for organisation would have paid any less attention to marching camps. The King realised the immediate importance of a well-run base for the safety of the army when marching thousands of kilometres through unexplored lands. What is sorely needed is a careful review of all useful extracts found in the staple literary sources – namely, Diodorus, Curtius and Arrian.[2]

On the whole, Alexander's camp methods involved Greek or Macedonian techniques eventually combined with some eastern influence (see Diagram 2). As we shall see later, the King's own camp quarters was adopted from the Achaemenids. The overall spatial layout of the royal compound would have evolved throughout the period of Alexander's conquests (334–323 BC), as Persian features were introduced.[3]

The use of geometry for camp layout was approved of before Alexander's time. In his *Republic* Plato states that 'in dealing with encampments … an officer who had studied geometry would be a very different person from what he would be if he had not'.[4] Plato was an inheritor of Pythagorean discourse, and it is disclosed how during Philip II's hostage years in Thebes, both he and Epaminondas were students of the Pythagorean thinker Lysis of Tarentum.[5] This information need not be taken literally, but it remains an intriguing possibility that Philip could well have been influenced (directly or otherwise) by those who had been taught by Lysis, or had read books by him or his followers.[6] Even if Philip rejected Pythagorean celibacy, pacifism and vegetarianism, he was willing to integrate anything of use in building a model army.[7] Euphraeus of Oreus, an associate of

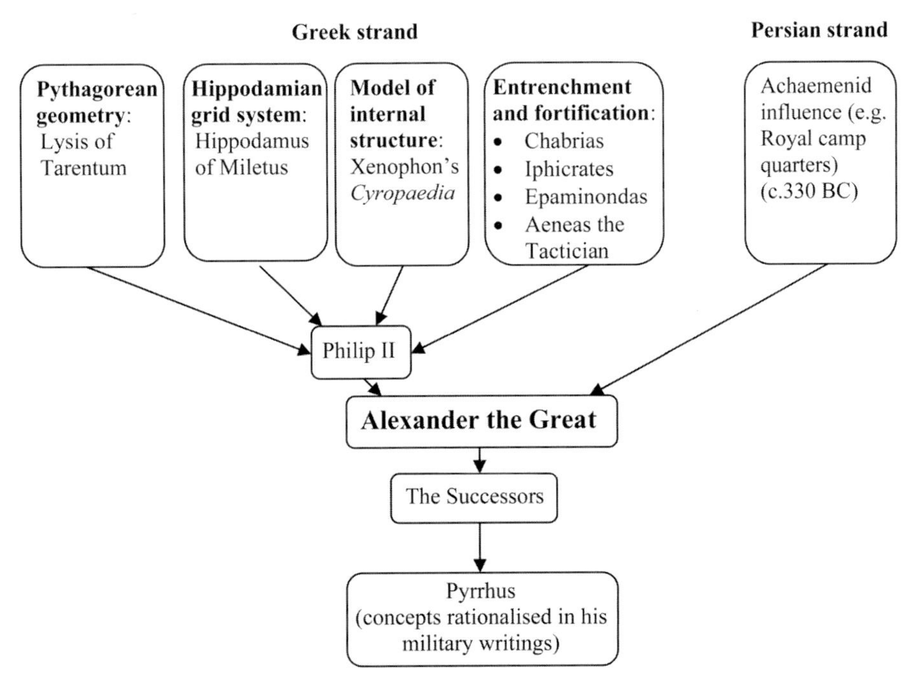

Diagram 2. A hypothetical reconstruction tracing the development of Macedonian field camp methods.

Plato, stayed at the court of Philip's older brother Perdiccas III and so dominated Perdiccas' choice of Companions that the philosopher allowed only those with some knowledge of geometry to sup with the King.[8]

Despite the last anecdote sounding like fable, we cannot rule out a filtered application of Pythagorean geometry in formation evolutions and the formulaic design layout of marching camps in the Macedonian army.[9] In the second century BC Polybius, for one, recommended that those who wished to be effective military leaders should study the basics of geometry, especially:

> For raising the scale of the divisions of a camp. For sometimes the problem is to change the entire form of the camp, and yet to keep the same proportion between all the parts included: other times to keep the same shape in the parts, and to increase or diminish the whole area on which the camp stands, adding or subtracting from all proportionally. On which point I have already spoken in more elaborate detail in my Notes on Military Tactics.[10]

The design of field camps became an established subject for intellectual interest and debate in the Greek world. Xenophon uses his *Cyropaedia* as a vehicle for endorsing a model of internal structure,[11] while his near contemporary, Aeneas the Tactician, composed a now lost work on planning encampments.[12]

A more precautionary move toward camp defences emerges among the Greeks in the early-fourth century BC, with an emphasis on entrenchment and fortification (when the imperative warranted it) by commanders such as Chabrias, Iphicrates and Epaminondas.[13] There is evidence for an opposing argument which, instead of constructing walls and ditches, advocated utilising natural terrain features. This layout was stressed particularly by Xenophon.[14]

Aeneas describes camps and walled cities being alike,[15] and the noun *charax* (defined as 'palisade') has been tendered as coming into general vogue, applied by Polybius and Diodorus to denote fortified camps, from the mid-fourth century BC.[16] It was further adopted by Philip II, who is known to have entrenched regularly with a palisade of stakes, and was one of the first in the Hellenic world to institutionalise the practice.[17] His routine of conscientiously fortifying camps was continued by Philip's veteran generals years after Alexander's death, Antigonus Monophthalmus being a prime example.[18] More than 100 years later ancient writers still described the Macedonians as excellent field engineers.[19]

It has been said that classical Greek camps generally had circular or quadrilateral alignments and depending on terrain Alexander's camp may have sometimes been square-shaped.[20] The internal layout of Macedonian encampments could have been based on an orthogonal town planning scheme originally formulated by Hippodamus of Miletus. If so, it would have incorporated straight, bisecting streets with main thoroughfares running north-south and east-west. This archetypal Greek and Hellenistic plan is witnessed in Pella and duplicated among cities, colonies and installations founded by Macedonian rulers in Greece and dotted across Asia Minor and the Middle East.

Alexandria Eschate offers an excellent case in point from the literary sources of a city (probably erected on a Hippodamian grid) plainly described being framed on the foundation of Alexander's army camp.[21] Winter recognises Antigonea, established by Antigonus Monophthalmus in Bithynia and re-founded by Lysimachus as Nicaea, as erected on the Hippodamian principle.[22] He and Reinders add New Halos (a town near the shore of Achaea Phthiotis, south-east Thessaly), based on a field camp of Demetrius Poliorcetes in 302 BC (Fig. 43 and Plate 37).[23] It is further contended that the monumental character, rectilinear design and fortifications of Dion in Macedonia (c. 300 BC) copied colonies built in the conquered Asian satrapies.[24] Some of many other examples of the Hippodamian scheme encompass Goritsa, Priene, Miletus, and Dura-Europos.

Through the excavations of French archaeologist Jean-Yves Empereur and his colleagues, Alexandria in Egypt offers the most stunning lesson for what was achievable in surveying and town planning, with a Macedonian king supervising a team of Greek engineers implementing the Hippodamian model.[25] This was a showcase for the sort of expertise Alexander and Ptolemy could expect to command.[26]

Before completing this analysis, we should now pause to reflect on the origins of the famous camping practices of the Roman imperial army, which have gone virtually unexplored by scholars. It is at least possible that the Romans made use of Greek, and particularly Pythagorean, mathematical methods for their own camping arrangements.[27]

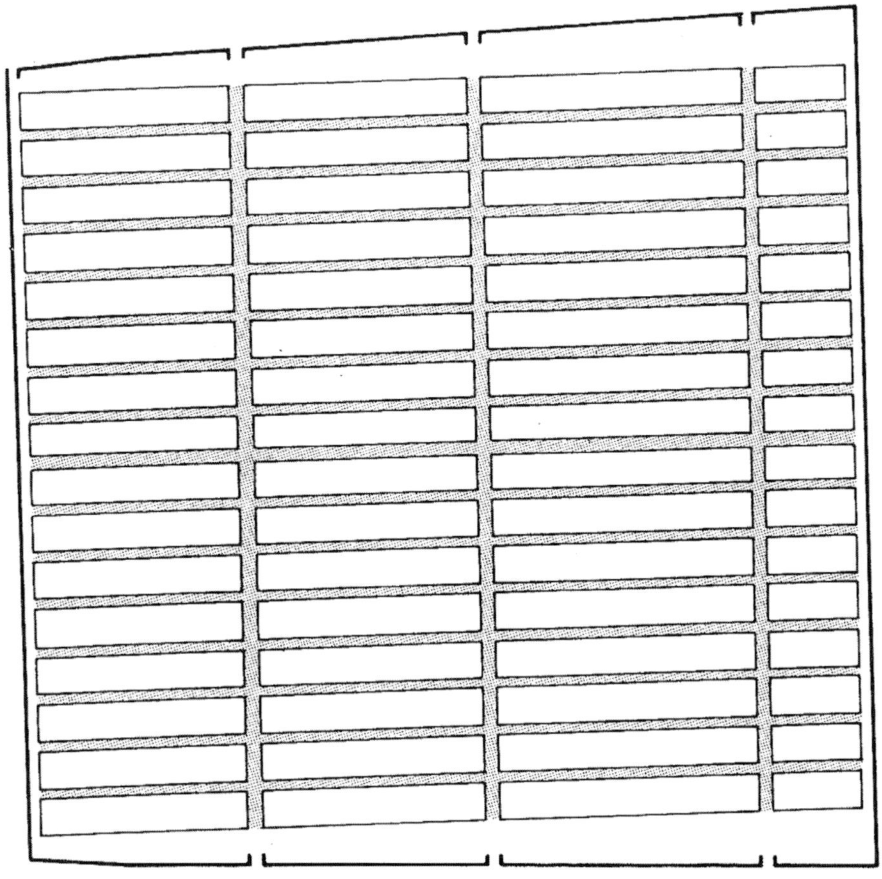

Figure 43. Residential quarters of New Halos, Thessaly.

As explained in Chapter 2, most Hellenistic military treatises have sadly been lost to posterity. Had they survived, we would be left in far less doubt as to what *some* origins of Roman practice were. A faint thread of literary tradition identifies Pyrrhus as the main inspiration for the Roman army's celebrated techniques. The Romans were certainly acquainted with Pyrrhus' lost military writings. According to Frontinus, the Epirote king was 'the first to inaugurate the custom of concentrating an entire army within the precincts of the same entrenchments. Later, the Romans, after defeating Pyrrhus, captured his camp [probably meaning at Maleventum, 275 BC] and noting its plan, gradually came to the arrangement which is in vogue today.'[28]

Frontinus is of course wrong in his assertion that Pyrrhus was the first general to enclose an entire camp within the same revetments – he was relying on faulty memory, had misread or misunderstood his source or consulted one already garbled. If Livy can be trusted, the Romans were familiar with the usefulness of fortified camps long before the Pyrrhic war.[29] Some extracts even portray

Pyrrhus or Philip V expressing their admiration for Roman camp layout.[30] However, it is clear from these extracts that Roman techniques were not necessarily considered superior and are simply described appearing unlike those of 'barbarians' to their Epirote and Macedonian royal eyewitnesses.

It is impossible overall to establish the accuracy of the above ancient literary passages. If Pyrrhus *was* ever a major influence on Roman camp organisation, then he developed his ideas based on career experience augmented with technical study.[31] Given his known passion for studying military matters virtually to the exclusion of all else, he would have been familiar with the lost manual on camps by Aeneas and Xenophon's camp layout extracts in the *Cyropaedia*.[32] There is an inkling that he had read up on Homeric *taktika* as well.[33] Fighting his wars within an overwhelmingly Macedonian orbit, Pyrrhus learnt his trade as an officer under Demetrius Poliorcetes, who was himself schooled in the doctrines of Philip and Alexander.[34] Almost certainly, Pyrrhus' camping routines were largely Macedonian in inspiration.[35]

Pitching Camp

The full set routine of camp construction from the conquest of Persia (c. 330 BC) onwards may be tolerably reconstructed, although procedures could greatly vary depending on available time and strategic factors including the whereabouts of enemy forces, or terrain environment.

Presumably, after intelligence patrols had done their work, armed scouts with surveyors were sent along first in a mounted party.[36] During the eastern campaigns Curtius mentions 'two of those who had gone ahead to choose a place for a camp' bringing water skins back to the main army for their sons.[37] This suggests that they had already selected a site close by a water source.

Aristotle (who was, after all, Alexander's tutor) advises that a commander deliberate over where he should pitch camp.[38] Greek campsites were ideally chosen for their dry, healthy and, if feasible, defensive location with good visibility on high ground and convenient access to nearby fresh running water, wood, food and fodder supplies.[39] Such measures may even have come as a response to a growing number of medical works emphasising the importance of sanitation.[40]

Arrian takes note of how Alexander regularly pitched camp about 4km from rivers when the army marched through the desert terrain of Makran. This ostensibly was undertaken in order to prevent the troops from gorging themselves on too much water at any given time and underscores the King's concern for the wellbeing of his men. Locating the camp in this way also helped avoid the fouling of water.[41] Alexander taking care of the physical wellbeing of his soldiers is reminiscent of Xenophon's belief that a general be acquainted with every aspect of his army: not only tactics, which constitute one part of incisive leadership, but also health, provisioning and discipline.[42]

Once a suitable site for a temporary field camp had been chosen, the perimeter was paced out, sometimes with a series of markers, and then subdivided appropriately. We read of a 'dummy' camp constructed by the loyal staff of Eumenes to deceive Antigonus in the manoeuvres before Paraitacene.[43] Similarly, in another rare instance, during his march to Maracanda in Sogdiana Alexander sent ahead a few men to measure out a camp, but they suffered a surprise attack and were killed.[44]

Carefully restored papyrus scraps recovered from Rifeh (Egypt) have been tentatively identified by Hammond as excerpts from Strattis of Olynthus's lost commentary on Alexander's 'day-to-day' royal journal.[45] The fragments reflect how during the King's Illyrian war (335 BC) a high-ranking officer named here as 'Corragus, son of Menoitas' was assigned the job of being responsible for the preparation of a camp in advance, with instructions detailing the camp watchword, whether a signalling post be set up and even what type of palisade stake be adopted for the outworks.[46] The signalling beacon functioned as a means of directing the army and any stragglers to the prearranged location – as Alexander did in a gruelling desert night-march to the Oxus (329 BC).[47]

The King's Quarters

With an appropriate site selected and measured, the first task was to dig in the King's tent (*skene*) at the centre of the royal headquarters complex (*basileia*).[48] This was assigned its own work party of special tent-pitchers under the jurisdiction of one Philoxenus in the post of *stromatophylax*, who was a 'guardian' of household items.[49] He was in charge of the royal tent and baggage, and was perhaps answerable to a superior officer.[50] The entire area would then be cordoned off from the main camp by a surrounding enclosure wall of four stades (over 700m).[51]

If military circumstances permitted, the royal tent could occasionally be transferred to a more prominent position, serving as an instrument of psychological propaganda in full view of enemy onlookers (as in the armies of other conquerors like Tamerlane).[52] This large and striking Achaemenid adopted structure may have been coloured white and purple, with cloth of gold.[53] Evidence shows that it served essentially as the medial point around which the rest of the camp was arranged.[54]

By 325 BC Alexander was protected by a protocol of Macedonian and Persian guard units, including a unit of armed bodyguards or royal hypaspists, dressed down without their full panoplies, who were usually required to stand before Alexander's tent.[55] Hypaspists were engaged in guarding his person and maintaining royal headquarters security as a kind of 'military police', possibly as a rolling rota of detachments. The literary sources impart that hypaspists were billeted conveniently close by, on the outer boundaries of the complex.[56] This is suggested in less studied realms of literature. For example, despite attribution to the tragedian Euripides, it is tenable that the play *Rhesus* was produced by someone at the court of either Philip II or Alexander (350–330 BC). It details instances of Macedonian military terminology and in its opening scene portrays sentries reporting to hypaspists stationed before the tent of the Trojan hero, Hector.[57]

Alexander's bedchamber was guarded by pages on a revolving rota,[58] while *somatophylakes* stood at the King's tent or chamber-door when he was ill.[59]

Companions seem to have had a general right of access to headquarters, though probably only linked to official engagements.[60] In contrast, people of lesser rank had to wait outside the tent before being escorted in by a bodyguard or the particular *somatophylax* on duty.[61]

Generals and higher ranking unit officers were required to report on a consistent basis in order to receive instructions or attend strategic and tactical planning meetings,[62] while some were required to visit twice daily.[63] Officers were called for a briefing in the early morning to receive the first orders.[64] When the army was led out on operations, a leading officer was temporarily appointed camp-commander or in charge of the camp-guard in the royal absence.[65] During the Successor Wars we also read of persons responsible for billeting troops in occupied towns.[66]

Entrenchments and Palisades

After the royal quarters had been established, Macedonian troops were assigned the task of digging camp entrenchments. When engaged in such fatigues during humid weather, arms were stacked, cloaks and armour shed, and the men clad in their tunics, often let down at the right shoulder for better ventilation and movement.[67]

The American Confederate general Robert E. Lee once said that 'there is nothing so military as labour'.[68] Almost 2,000 years before, the Roman commander Gnaeus Domitius Corbulo spoke of how victories were won with the pickaxe.[69] Greek armies, at least during the Hellenistic period, generally disliked digging ditches on campaign[70] but it is clear the Macedonians were quite different and well understood the concept of entrenching. This feature formed one of the chief strengths of the Macedonian soldier and stands out in contrast to the less practical approach of many other warrior peoples.[71] Indeed Polybius goes as far as to say that the Macedonians were 'industrious in digging trenches'.[72]

As first incisively realised by Engels,[73] the army carried palisade stakes on campaign, just as the New Kingdom Egyptians and Assyrians before, and the Romans after them.[74] Alexander is reported by Arrian fortifying a camp with ditch and palisade in the lead up to Gaugamela.[75] This was undertaken in a treeless region where it would be otherwise impossible to garner the necessary materials to make stakes.[76] The inference therefore is that stakes were prepared beforehand and brought along by the army, transported in stacks by the baggage train.[77]

Alexander and his officers were certainly conversant with the *Iliad*'s account of the Greek camp with its defensive wall, ditch and palisade.[78] The play *Rhesus* concedes the risks an enemy would have to face if attempting to attack such a fortified outpost at night, with unseen trenches and other obstacles.[79] Although camp fortifications were probably not erected by the Macedonians when in friendly territory or when danger was not high or forthcoming, they were common in Alexander's day.[80] The failure of the Illyrian king Cleitus to construct a ditch with palisade for the security of his men when facing the Macedonians at Pelium in 335 BC is condemned by Arrian (possibly repeating the opinion of

Ptolemy, his main source) as an imprudent oversight.[81] Fig. 44 presents an artist's impression as to what some Macedonian camp fortifications looked like: a ditch with upcast rampart crowned with a palisade of stakes.

As with the Romans, the Macedonians seemed to use different styles of palisade stake for entrenching in different situations. A papyrus fragment from Egypt describes Alexander directing the officer Corragus to enclose a camp with 'closed packed (ones)' (*pyknai charakes*).[82] It refers to an advance party ordered to prepare a fortified camp for the main army following up. These stakes may resemble ones recommended by Philon of Byzantium, who said that they be fixed upright and roped together for use when enclosing permanent stockades located outside towns. Here it was advised that there be 1,600 such stakes of medium size covering one stade (approximately 165–200m). When arranged in this way, the stakes would prove difficult to straddle and disrupt without first breaking the knotted ropes.[83] Alexander's specification for closely set stakes reflects that the Macedonians were familiar with erecting different camp fortifications on demand.

Various kinds of stake are mentioned elsewhere. For example, the Thracian Odrysae fenced a camp (399 BC) with stakes described as being at man height,[84] while the defences of Demetrius Poliorcetes at Rhodes were composed of large stakes, planted in dense array.[85] Polybius (followed also in Livy's narrative) describes the Macedonians (c. 200 BC) selecting tree branches with many sturdy

Figure 44. Typical Macedonian camp fortifications (reconstruction).

offshoots around the main stem for their stakes.[86] He has several criticisms for these: they were too large to be carried by soldiers encumbered with the *sarissa* and when secured in the ground could be easily uprooted by two or three of the enemy acting in concert. Their description does, however, resemble some of the obstacles used to form abatised bases in European warfare of the eighteenth and nineteenth centuries, effectively improvised by the Japanese as late as the jungle campaigns of the Second World War.

The well-ordered nature of Alexander's camp construction is evident from an episode at the river Hyphasis in north India, where, with the termination of campaigning, the King aimed to overawe nearby inhabitants with a momentous statement of propaganda. He thereby sorted his men into twelve work brigades and ordered them to build twelve stone altars, each one 23m high, consecrated to the gods of Olympus. Diodorus then recounts how Alexander traced 'the circuit of a camp thrice the size of the existing one', surrounded by a ditch 15m wide and 12m deep, the upcast packed hard into a large wall.[87]

Literary sources describe the troops of Antigonus Monophthalmus enclosing Eumenes' position at Nora with double walls, trenches and impressive palisades.[88] On a different occasion, Eumenes erected a winter camp with palisade and deep ditch while campaigning on the Iranian plateau.[89] At Rhodes Demetrius Poliorcetes had his troops build a long-term field encampment. Diodorus informs us that this was to be erected 'with a triple palisade', possibly the first recorded use of this type of defensive work in Greek warfare. It further involved demolishing nearby farmsteads, the stonework probably being recycled for the building of rubble walls and towers.[90] A corresponding event is recorded for 217 BC when, during Philip V's investment of Phthiotic Thebes, the King established his army in three fortified camps around the 3km circuit wall of the town. As explained by Polybius, 'the space between these camps he fortified by a trench and double palisade, and further secured them by towers of wood, at intervals of 100 feet, with an adequate guard'.[91]

A phase of strategic manoeuvres in Asia Minor over 302 BC clearly shows the familiarity Macedonian generals and soldiers had in erecting field fortifications. Lysimachus and his army marched over 60km to Dorylaeum, where an encampment with deep ditch and triple palisade was laid out,[92] the natural feature of a river granting further protection. Antigonus Monophthalmus surrounded this camp and sent for catapult artillery, while having his men counter the enemy fortifications with their own line of earthworks. Lysimachus then took the initiative, escaping with his army by filling some of the Antigonid trenches with debris during a stormy rain-racked night which offered poor visibility as convenient cover.[93] Frontinus captures the entrenchment skills of Lysimachus' troops in detail, identified by Lund as belonging to this campaign:

> [Lysimachus] dug a triple line of trenches and encircled these with a rampart. Then, running a single trench around all the tents, he thus fortified the entire camp. Having thus shut off the advance of the enemy, he filled in the ditches with earth and leaves, and made his way across them to higher ground.[94]

Soldiers' Tents

After constructing camp fortifications, soldiers' tents were set up.[95] It is possible that for pike armed infantry the routine was to leave stands of *sarissai* planted upright outside, as with their Renaissance counterparts.[96] Little is known of tent structures themselves, and the present study may well be one of few in-depth investigations into such a peripheral topic.[97]

It is uncertain exactly how many tent-mates made up a tent-party in the Macedonian infantry – a full-file, half- or quarter-file.[98] Several later Byzantine treatises considered it essential to provide a tent to each tactical file in order to nurture a strong team spirit. Given their highly organised methods of warfare which relied so much on superb unit morale, command and cohesion, it is credible that the Macedonians thought the same, but there is simply no evidence to confirm it.[99]

Sekunda suggests a sixteen-man tent divided into compartments for two men each.[100] In contrast, Heckel advances an ingenious proposal that two-man tents were the norm, based on the possible dimensions of the dummy camp erected by Eumenes.[101] This had a perimeter measured at 70 stades (around 12 or 13km), inside which were to be lit fires about 9m apart.[102] There are a couple of episodes in the literary sources that *may* help validate Heckel's thesis. Firstly, at the siege of Halicarnassus two Macedonian phalangites are portrayed drinking together while bivouacked, with no implication that there were any others in their tent.[103] Secondly, according to Diodorus, when building a camp of exaggerated dimensions at the Hyphasis, Alexander directed the infantry to construct shelters with two beds each.[104]

If two-man shelters were usual, they must have been fairly small and light.[105] This would help avoid the burden of transporting heavy tentage encountered by Xenophon's Ten Thousand – something that Philip, Alexander and their officers were aware of.[106] But in truth, there are only a few hints as to how tents were designed. They had door flaps;[107] were carried in waterproof leather or possibly goatskin covers, which acted as fly-sheets when they were erected; and were secured to the ground with guy-ropes and small iron pegs.[108] Portable tent poles made from saplings were another possible feature.[109] On one occasion Curtius makes the fleeting remark that during Dionysiac celebrations in 325 BC wagons were rigged out like tents, with curtains made from white cloth or some other costly material; however, nothing meaningful can be presumed from this and Curtius may anyway only have been thinking of gear familiar to his Roman readership.[110]

From the makeshift nature of Greek army shelters from before the late-fourth century BC in comparison to the fly-sheets, guy-ropes and iron pegs documented in sources for the Macedonians, we can speculate that those carried by Alexander's soldiers were an improved version, better adapted for long distance travel. As asserted by Anderson, before Alexander's campaigns Greek city-state armies had relied more on shelters which were probably in reality rough lodgings of timber and brushwood.[111] Despite Xenophon's mercenaries using leather tents, these articles were provided by their Persian employer[112] and down to the early-fourth century BC even a general as innovative as Iphicrates was satisfied with only single man lean-tos.[113] Thus, Alexander's apparently more conscientious provision was another facet of the King's concern for the comfort of his men and implies how assiduous his and his father's preparations had been.[114]

The idea that Alexander recognised good tent design is embodied with the famous 'tent to hold 100 couches', arguably acquired by Philip during his wars in Thrace, and described by Diodorus having been erected at the *Xandika* festival at Dion in 334 BC.[115] How remarkable such architecture was can be appreciated from contrasting it with the 'Hall of the Sixty Couches' used for banquets in Syracuse. Although praised as surpassing other Sicilian buildings for size and beauty, it had a smaller capacity than the Dion tent.[116]

Sadly, there are no direct artistic or archaeological traces for Macedonian army tents and little evidence of this sort exists for their appearance in Greece.[117] A few paltry fragments include scenes culled from a handful of painted Attic vases, generally of the fifth century BC, the lower register of a Hellenistic relief, and an extant fresco from Pompeii. The last is from the atrium of the House of the Tragic Poet, executed in the Fourth Pompeian style (about 62 – second century AD), portraying a scene from heroic literature, usually identified as Achilles releasing Briseis (Fig. 45). The figure of the seated 'Achilles' closely resembles artistic renditions of Alexander. Behind 'Achilles' stands a vigilant squad of guardsmen with bright bronze shields, and in the background is a pale brown pavilion open at one side, resembling a wall tent.[118]

As with mercenaries serving in Asia seventy years before Alexander's eastern expedition, after shelters were pitched in the evening the squad campfire served as the main anchor for interaction, with soldiers sitting around it on folding stools, on the odd pack-saddle or on the ground, with baggage animals hobbled near each tent. Macedonians are described usually caring for their bodies and cooking food around their campfires.[119] In Lee's study of Xenophon's Ten Thousand, the fire pit represented the hearth or focal point for communal life. It was warming in winter and helped ward off annoying insects in summer; and it was where groups of tent mates ate their meals, fraternized and bonded with one another.[120] Mary Renault's entertaining novel *Funeral Games* (on the plots of Alexander's Successors) recreates a believable portrait as to what the lifestyle was like: the soldiers accompanied by their multiethnic families, the capering children 'wary of a clout' from their fathers.[121]

Space Allocation of Units

It is unlikely that the grouping and integration of tents and living arrangements for infantry and cavalry were without proper form within Macedonian camps. As mentioned before, the tents of the hypaspists were definitely set in close proximity to Alexander's enclosure. There is an obscure reference to the highest ranking officers and administrators being assigned quarters with recourse to Alexander when disagreements over that location broke out.[122]

As with sources detailing the Macedonian royal compound, Xenophon's *Cyropaedia* recommends that the Persian king's headquarters should similarly be

Figure 45. 'Achilles and Briseis' fresco from Pompeii.

at the centre of the camp, with 'everything ... so organised that everyone knew his own place in camp – both its size and location'. Near the king were tents for his immediate entourage:

> And next to them in a circle he had his horsemen and charioteers ... To the right and left from him and the cavalry was the place for the targeteers; before and behind him and the cavalry, the place for the bowmen. The hoplites and those armed with the large shields he arranged round all the rest like a wall.[123]

In maximising an army's tactical potential if faced with external attack, this model allowed cavalry to saddle up and mount unimpeded inside, with the hoplites forming a protective outer barrier supported by missile troops shooting over their heads.[124]

As the Macedonian army of the late-fourth century BC was one of the great combined arms forces in antiquity, such advice may not have been lost on the shrewd minds of Philip, Alexander and succeeding generals, given the obvious usefulness of Xenophon's scheme. As a *conjectural* scenario, units of Macedonian pike armed infantry, constituting the largest portion of the army, could have been adapted to perform the hoplite role in the Xenophontic camp infrastructure made without fortifications. In this way, a barbed thicket of *sarissai* facing outwards would function as a temporary and artificial 'breastwork', if required.

Another possibility is that a deliberate method of arranging units was practised for camps, with the most prestigious ones located closest to the King. It is hinted at in the case of the *ataktoi*, or punishment unit (see Chapter 10), which, according to Curtius, was ordered by Alexander to encamp separately to accentuate its disgrace.[125] This can be interpreted to mean that it was taken out of the camp array, 'demoted', as it were, from the army's internal camp distribution. Alternatively, this unit was simply being ordered to bivouac outside the main perimeter.

Two particular events suggest a ranking system of sorts, based on individual and group prestige. The first is Alexander's mass banquet at Opis where he arranged his followers around him in concentric circles consisting of Macedonians, then Persians, and thirdly, as Arrian puts it, 'any persons from the other peoples who took precedence for rank or any other high quality'.[126]

The second event is the grand feast prepared by Peucestas for the army of Eumenes at Persepolis (317 BC), described in detail by Diodorus, most likely taken from a reliable source such as Hieronymus of Cardia. A circular pattern can be discerned here, similar to the concentric form of a Macedonian shield blazon.[127] The outer circuit was measured by surveyors at 10 stades (almost 2km) and was set aside for mercenaries and allied contingents; the next ring in had a circumference of 8 stades (about 1.5km) composed of the Macedonian Silver Shields and 'those of the Companions who had fought under Alexander'; while the third inner circle of 4 stades (over 700m) incorporated 'the commanders of the lower ranks, the friends and generals who were unassigned and the cavalry'. The innermost ring measured 2 stades (approximately 350m) comprising 'each of the generals and hipparchs and also each of the Persians who were most highly honoured'. At the centre of all stood altars to Philip and Alexander.[128]

Although Diodorus' language for the latter dining arrangement has been subject to slight differences in interpretation, it may still preserve something of the unit allotment in Eumenes' field camp.[129] The reference to Alexander's altar in the middle, encircled by leading generals, reminds us of an anecdote from Polyaenus in which Eumenes erected a pavilion in his camp to duplicate Alexander's headquarters and then pitched his own tent beside it, followed by those of his officers.[130]

We can argue from such accounts that if soldiers and units of a Macedonian army were carefully allocated space for state occasions and banquets, then the overall internal structure of marching camps – which was of far greater utilitarian value – was similarly hierarchical and well-organised. Having said this, what is eminently rational to modern thinking is not necessarily the same for people living in the past, but even so the assumption is cogent.

With the foregoing in mind, it should be remembered that when Eumenes constructed a camp in 317 BC he assigned an area to each of his followers with campfires to be placed at regular 20 cubit (just over 9m) intervals across the site,[131] implying a systemised configuration of units and tents.[132] Polybius refers to Macedonian troops in Alexandria (c. 200 BC) billeted in tents just beyond the surrounding wall of the royal compound, these being pitched close together towards one part of the city.[133] As far as rapid mobilisation was concerned, a regulated layout meant that soldiers knew exactly where to run to and muster once trumpets were blared and orders barked. It helped minimise panic and maximise order and efficiency. Equally, routine layouts had the benefit of giving troops the secure sense of a familiar home when surrounded by an alien environment. This was of especial relevance to Alexander and his almost perpetually wandering and sometimes homesick or mutinous men.[134]

Setting ourselves squarely behind the evidence, we can now make an informed speculation as to what the spatial orientation of units in Alexander's field camps was like. It is known that the disposition of units on the battlefield was based on elite standing and combined arms tactical factors. A hypothetical 'rough draft' could have Alexander at the centre with the secretariat located in a separate tent, near to or within the royal headquarters area.[135] Accommodation for pages was placed within, while leading officers, bodyguards and hypaspists might be closest to and surrounding the compound. Beyond them was space reserved for the Companions and other cavalry units, with the outer perimeter possibly occupied by the phalangite units. Those with the least battle honours and the disgraced pitched their tents the furthest out.

It is more difficult ascertaining the light armed and missile troops within this overall scheme but, following Xenophon's proposals (which Alexander would know), their tents may have taken up space between the cavalry and Macedonian foot. What is certain is that camps included streets dividing units from one another,[136] with sectors for soldiers and contingents to stack arms (*ta hopla*) near their tents,[137] for latrines, and for cavalry horses and pack animals to graze.[138]

The Use of Unit Standards

As the army arrived in camp, some role in pinpointing the space allocation of units was made using *semeioi*, or battle standards.

Suggestions of the use of standards for unit space orientation can be found in the literary corpus. On the eve of Macedonia's rise to predominance, Xenophon's account of his model army includes a *semeion* with the baggage marching on ahead of each body (*taxis*). When encamped, officers' tents were surmounted with a mast from which flew a banner for unit identification.[139] Seleucus is said to have trained his men to dine and sleep in full panoply in order of battle (that is, bivouac overnight in tactical formation).[140] It is hard to accept that this was achieved without standards determining where each group of men should be. Livy recounts how troops in the army of Philip V planted standards before grounding arms and taking a hasty meal. He also describes soldiers under Perseus withdrawing to camp with their standard-bearers leading the way.[141]

Although the Persians are known to have used standards before Alexander incorporated them, the chronological time frame as to when they were introduced into the Macedonian army remains obscure.[142] Modern authorities often make the mistake of doubting the existence of *semeioi* in Macedonian or Hellenistic armies.[143] Nevertheless, Alexander's army was the first to use them on a regular basis in the Greek world.

Arrian's passing reference to Hephaestion's cavalry with a *semeion* (see below) suggests that the adoption of standards had taken place over 331–329 BC. We can speculate that they were initially introduced as a means of promoting greater battle cohesion and manoeuvrability in mounted formations in action across the open plains of Asia, and in preparation for difficult steppe and mountain warfare that, as the King knew from intelligence reports, awaited his army even further east. Alexander is recorded having at least experimented with more efficient visual signal aides for field camps around this time.[144]

Curtius refers to the use of standards in Alexander's army before 331 BC, but this has been correctly dismissed as an embellishment from later Roman practice.[145] Some translations of an entry in surviving Babylonian astronomical diaries describe how on the morning of Gaugamela the 'king of the world' raised his 'standard', after which there is an unfortunate break in the text.[146] It is therefore highly doubtful whether this can be used as proof for the use of unit standards in the Macedonian army by the time this battle was fought.[147] Thus, the first authentic, or fairly indisputable, reference can be cited from Arrian (probably following Ptolemy's narrative), who remarks how after Hephaestion's death in 324 BC 'Alexander never appointed any one in place of Hephaestion as Chiliarch over the Companion cavalry, so that the name might never be lost to the unit; the chiliarchy was still called Hephaestion's and the standard [*semeion*] went before it which had been made by his order'.[148] This unique and invaluable passage implies that the unit device was usually selected by the unit leader, and that it was changed if that officer was transferred or killed.

As it stands, it is tempting to tie in the introduction of the Macedonian cavalry *semeion* with the period when new camp signalling methods and the first phase of Companion unit re-organisation (with each *ile*, or squadron, subdivided into two *lochoi*) took place: more precisely, in the later months of 331 BC, or in the winter of 331/330 BC, on the road between Babylon and Susa.

An equally valid option is that standards were first used with the change to *hipparchiai* in early 329 BC.[149] Events detailing Alexander's warships on the Indus in late 326 BC shed a dim, though pertinent, light on the concept of unit standards having taken hold by this time. Pliny remarks on how the extravagant dyeing of linen sails was first commenced in Alexander's fleets, 'his generals and captains having held a sort of competition even in the various colours of the ensigns of their ships'.[150] Similarly, the exceptional golden wreath displayed atop Alexander's catafalque is confirmation that the Macedonians fully understood the use of emblems for distance recognition purposes by 322 BC.[151]

At what time standards were extended to the infantry remains unknown. Given the lack of anything to the contrary, an argument from silence could endorse

Alexander introducing standards for foot and horse units, of which only a single stray example for the cavalry happens to survive.

Until evidence is forthcoming, however, it may be more reasonable to say that the period for using standards among the infantry was the forty-year span of the militarily active but all in all poorly documented Successor Wars. Dionysius of Halicarnassus gives an enigmatic reference to how just before Asculum, fought between Pyrrhus and the Romans in 279 BC, battle signals were hoisted and the troops chanted war cries, but this is of dubious reliability.[152] It may be that by the time of the *Diadochoi* it was thought rational to extend standards to infantry, given the fact that near identical armies were often engaged with each other. Thus, if each unit were to carry its own unique emblem, it would be more easily identifiable on the field. This also meant a concomitant improvement in the tactical interplay of units and a more effective system of battlefield communication.[153]

We can assume that the man who carried the unit standard was handpicked for reliable conduct, strength and bravery. In his Hellenistic-based treatise Arrian mentions that each 256-man tactical unit should have its standard-bearer, probably selected by its lead officer.[154]

The argument that standard-bearers in Alexander's army were specially chosen complements other literary evidence for a system of promotion based on merit outlined earlier in Chapter 10. Over a century later, at the battle of Cynoscephalae (197 BC), Livy describes how coordination was imperilled due to the Macedonian rank-and-file's difficulty in locating their unit standards in the dense fog. Polybius also mentions the bad weather but not the detail on standards, raising the possibility that Livy's inclusion of them derived from his own superimposed assumptions based on Roman military routines.[155] Nonetheless, the highly responsible post of standard-bearer would have necessitated someone who could be relied on. This brings to mind the honour of junior officers who were chosen as colour-bearers in the American Civil War, selected for courage in order to keep the colour aloft and visible during an engagement.

According to Arrian, Hephaestion's *semeion* was kept on the occasion of his death as a gesture to honour his memory, but – given Alexander's intense grief at losing his closest friend – it is possible that it also had some religious overtone. Perhaps it was thought that by keeping Hephaestion's *semeion* his heroic shade would continue to accompany his former unit as a charm, although this is an admittedly hypothetical proposition.[156] In fact, there is no direct evidence that the Macedonians granted any symbolic or emotional significance to their standards. They are only ever dryly described in theoretical manuals and seem to have been recognised for their immediate practical signalling value rather than a physical manifestation of the pride or 'soul' of the units they headed.

Security Measures and Signalling Routines in Camp

A number of routine timing and signalling procedures were followed in camp. Night-time was divided into either three or four watches, beginning at sunset and ending at sunrise, and there was an order for lamps to be lit when the light faded.[157] The Greeks of the fourth century BC are known to have been able to time with the aid of a *clepsydra*, or water clock, and – whether this was used by the Macedonians or not – each watch was at least marked by a trumpet call.[158]

The Macedonians, similar to Iphicrates, were careful enough to establish *prophylake*, or sentry outposts, which were routinely inspected on the approaches to their camps through a system of passwords and countersigns.[159] Alvarez-Rico has proven that by the fourth century BC some Greek encampments attained sophisticated levels of external security with tiers of guards, advance guards, look outs, and contrasting systems of day and night-time defence, integrating watch and patrol duties.[160]

Like Iphicrates,[161] or, indeed, leading warriors in the *Iliad*,[162] Alexander is at times recorded inspecting the checkpoints in person,[163] while enemy armies encamped without regular sentries were criticised for not providing adequate precautions.[164] *Rhesus* graphically illustrates the hazard arising when a camp was left devoid of proper defence measures with weapons stacked in little or careless order.[165] During important councils of war, Alexander posted mounted guards at the main camp gate, with nearby roads patrolled by cavalry vedettes to curtail messages being smuggled out to the enemy. It is conceivable that officers in charge of external security were able to detect and intercept letters coming into camp from enemies, sealed with unrecognised devices.[166]

Breaking Camp

When the King passed the order to break camp, the royal trumpeter would sound the call from the central position of the royal headquarters' tent.[167] One of the most celebrated of these figures was the colourful Herodorus of Megara, who accompanied Demetrius Poliorcetes on at least one occasion. Herodorus was famed for his enormous lung capacity and appetite, won many prizes for the excellence of his trumpet playing and blew two trumpets simultaneously.[168]

Trumpet blasts were transmitted to unit colleagues stationed throughout the encampment, with Sekunda envisioning their lower-ranking equivalents attaching a unique prefix to their signals, not unlike armies in more recent times.[169] Evidence in fact exists for trumpets in the classical world being played with distinct modulations and note-patterns.[170] It is alleged that when Herodorus blew on both his trumpets, the resultant noise was so deafening that no one could approach him. Epistades, another champion trumpeter, could be heard up to 50 stades (roughly 10km) away when he blew on just *one* instrument.[171] But these cases are exceptional or exaggerated, as trumpets were far from infallible.[172]

Following Gaugamela, Alexander adopted a signalling beacon when the army broke camp. As Curtius presents it:

> He used to give the signal with a trumpet, the sound of which was often not readily enough heard amid the noise made by the bustling soldiers; therefore he set up a pole on top of the general's [royal] tent, which could be clearly seen from all sides, and from this a lofty signal visible to all alike was watched for, fire by night, smoke in the daytime.[173]

This reminds us of a crystal globe atop a pole used above the tented accommodation of Darius when breaking camp in the lead up to Issus. The Persian device appears as a form of clever optical light technology with the globe magnifying the emblem inside, focusing rays of light from any objects nearby.[174]

Gabriel has made the exotic suggestion that Alexander's beacon was instead inspired by something the King saw in Egypt, although there is no independent evidence to corroborate this. A relief from the Luxor Temple depicts two apparent smoke signal poles used in the military camp of Ramesses II before the battle of Kadesh (c. 1275 BC).[175] The Macedonian device offers an intriguing resemblance to the 11m-tall *akash diya* ('sky lamp') documented for Mughal army bivouacs in the sixteenth and seventeenth centuries. It was similarly used for signalling and as a directional aide for soldiers in finding their way back to their own tents from another part of the encampment.[176]

Curtius proceeds to paint a picturesque account of the army hastening to be on the move:

> The soldiers scattered like madmen and prepared their baggage for the journey – one might have thought a signal had been given for a general packing-up of the camp – and the bustle of men looking for their tent-mates or loading wagons came to the king's ears.[177]

A comparable scene of commotion is conjured up in *Rhesus*, with soldiers ordered to take up spears, the royal squadron of cavalry summoned to bridle their horses, and the mustering of light troops and archers.[178] Lee proposes that the period given to breaking camp between waking and falling into ranks for some Greek armies took up the better part of an hour.[179] We hear of horn (*keratos*) signals for troops led by the mercenary Clearchus (400 BC), with first blast to pack up, second for piling baggage on pack animals and third to commence the march. This resembles Roman methods during the Jewish War five centuries later (c. 70 AD).[180]

Killing Time in Camp

The experiences of war for a Macedonian soldier would inevitably oscillate between camp boredom and the adrenaline of battlefield stress.[181] Troops attempted to relieve the former in a number of ways.

To the shocked sensibilities of citizens from Greek city-states, the Macedonians inhabited a coarse world where a propensity for violence was common. Officers and Companions were salaciously portrayed by Demosthenes with the temperament of drunken bandits and flatterers who enjoyed obscene dances and unmentionable attention-seeking displays.[182] In an even more vitriolic monologue, Theopompus discloses that those around Philip were a bawdy bunch of ruffians who idled their time drinking and gambling, goaded in their debauched antics by the King himself. Many preferred to be clean shaven when a beard was the social norm. Some scandalised the Greeks by having excess hair removed from their legs, although this was probably done in order to ride their horses comfortably.[183]

Drinking bouts were especially popular. The braggadocio of Macedonian soldiers in this regard is proffered by Arrian, who describes two drunken Macedonian *pezhetairoi* at the siege of Halicarnassus arguing in their tent over who was the better fighter and deciding to prove it by arming themselves and setting out to attack the town's walls. The two roused the ire of the town defenders, killing several who approached at close quarters while hurling javelins at others.[184] The incident serves to emphasise yet again the outstanding *esprit de corps* of Alexander's troops in wartime.

It is clear that the Macedonians had a notorious reputation among Greeks for quaffing excessive quantities of neat wine, with possibly 15–16 per cent alcohol content.[185] A consumer of such a beverage was labelled *philopotes*, i.e. a 'boozer', and soldiers used communal carousing as a means of bonding with one another. These drinking bouts represented well-earned physical and psychological recuperation from the constant discipline demanded of them on operations and in battle.

A predilection for hard liquor pervades the literary sources. As far as Philip II's habits are concerned, Theopompus fosters an image of him as a madcap who was 'inclined to rush headlong into danger, partly by nature and partly because of drink; for he was a deep drinker and was often drunk when he sallied forth into battle'.[186] Demosthenes even sneeringly condemned him as a sponge.[187] Yet Philip's conduct was far from unusual as drinking contests were frequent. Ephippus criticises Macedonians as already drunk before the first course was served at feasts.[188] The officer Proteas is singled out by Athenaeus for having a sturdy physique despite the fact that he spent most of his free time drinking. He is said to have beaten Alexander in a competition where he pledged the King's health, consuming the contents of a cup containing many litres of wine, twice over.[189]

These colourful episodes are reminiscent of a surviving extract from Menander's *The Flatterer*. Here the soldier Bias gloats: 'in Cappadocia / three times I drained a beaker made of gold, / Brimful. It held five pints.' To this his 'parasite' Strouthias responds: 'You've drunk more than / King Alexander!', to which Bias swears: 'On my oath, it was / No less!'[190] It will therefore come as no surprise that of two New Comedy theatrical masks recorded in later lexicons, representative of military figures, one is described with a florid complexion, which points to a character of high temper and the telltale flush associated with alcoholism.[191]

An even more eloquent example of the familiarity officers had with alcohol involves Hephaestion, who on one fateful occasion devoured for breakfast an entire boiled fowl and imbibed two litres of chilled wine while recovering from an illness.[192] This irresponsible consumption helped trigger a relapse, ending in his premature death soon after. Chares of Mytilene preserves the most excessive spree of all in which Promachus was proclaimed 'champion', after downing a voluminous draught of unmixed wine. The predictable outcome was that he died three or four days later from this overindulgence. But astonishingly he was only one of a swath of casualties, as thirty-five died of an immediate chill and six others later when attempting to recover in their tents.[193] It becomes increasingly obvious that the ability to absorb huge quantities of liquor was seen, alongside prowess in war and the chase, as the truest measure of a man in Macedonia.[194]

Their love for drinking contests stresses the fact that Alexander and his men kindled an extreme competitive streak which reached into almost every corner of their lives. A preoccupation with power struggles, thuggery and skulduggery had been a staple of the headstrong Macedonian royal and baronial lines for centuries. The ultimate expression of this was Alexander's supposed dying wish that only the best succeed him, with the cryptic prediction of many fiercely contended funeral 'games' after his death anticipating the great rivalries of the Successor Wars.[195] The King had plenty of opportunity to see these conflicts coming, as his ambitious officers sometimes fell short of the Shakespearean Henry V's 'band of brothers' while in each other's company. In one memorable incident heated rivalry sparked off a duel among officers under royal command. Plutarch recounts how after the Hydaspes resentment grew between Craterus and Hephaestion for Alexander's approval until at last they bandied fighting words, then drew swords, with friends from both sides joining in the tussle.[196]

Given their magnetic attraction to risk taking, it comes as little surprise that dice-play and gambling was another common, if not addictive, pleasure enjoyed by Macedonian soldiers, particularly appealing when they had won so much spoil during the lucrative Asian venture. Alexander is known to have occasionally disapproved of this tendency as detrimental to discipline and criticised some officers who apparently regarded games as more than a mere diversion.[197]

References to dice games and gambling are found among the ancient literary sources. When Philip II released all his Athenian prisoners without ransom after Chaeronea, many took advantage of his generosity by asking for the return of their clothing and bedding and, what is more, dared complain of their treatment at Macedonian hands. The King retorted by laughing and exclaiming to his subordinates, 'Does it not seem to you that the Athenians think they have been beaten by us in a game of knucklebones?'[198] In another story given by Plutarch, Antigonus Gonatas (when probably in the congenial company of his generals) likened Pyrrhus' military career to a dice player who makes many lucky throws but without the ability to exploit them well.[199] Large numbers of 'natural' knucklebones, bone dice, different coloured glass counters and even a tile engraved with a board game have been retrieved from Macedonian sites, dating from the fourth century BC and Hellenistic era.[200] Some of these artefacts suggest games of strategy, one similar to backgammon.

A more placid leisure activity was to pen a letter to loved ones back home in Macedonia. Arrian describes many soldiers at the Hyphasis longing to see their parents, wives and children.[201] As Heckel comments, the turnaround period for letters written in Asia would be considerable.[202] Letters written from even the Iranian satrapies were first censured by Alexander's agents before being delivered in packets to Europe through a relay of dispatch riders, on horses or racing camels, using the excellent Achaemenid post station network.[203]

Atkinson expresses surprise at these letter-writing activities because they presuppose a high literacy level among Macedonian troops.[204] This is not so remarkable when it is recalled that Alexander demanded that boy recruits in Asia be taught how to read and write Greek as part of their educational curriculum. Many Macedonian commoners would have surely achieved literacy before this time.[205] Examples of ivory styli have in fact been discovered at the Pella Agora over 2006/2007 and iron specimens retrieved recently at Koukkos – one of which was personally handled by the author. Moreover, a cylindrical bronze fold-up case for writing implements was found in a male grave in Stavroupolis. This comprises an elegant three compartment holder for an ink-pot in one half-cylinder, with space for styli in the other.[206]

The content of private missives was probably little different from those found at Karanis (Egypt) and Vindolanda (close to Hadrian's Wall) written by Roman soldiers.[207] The theme of a Macedonian soldier sending letters to family members later crept into the genre of prose fiction. In his work *The Incredible Things Beyond Thule*, Antonius Diogenes (second century AD) introduces text which he purports to have extracted from one such letter written by a Macedonian officer called Balagros to his wife Phila, in which Balagros transcribes for her some ancient cypress tablets taken from a grave vault during Alexander's seizure of Tyre.[208]

PART V

INGENUITY

Intense training and near constant campaigning under Philip II and his son sharpened the skills of Macedonian veterans on and off the battlefield. What most ancient armies viewed as insurmountable obstacles, were for Alexander and his men competitive challenges to be overcome in their obsessive pursuit of military *arete*. We will therefore cast our net wide here to include analysis of the Macedonian army's superb out-of-battle expertise in road construction, field engineering and crossing rivers.

Apart from the many achievements of technical specialists, we shall also not fail to consider some lesser known combat units in Asia. These serve as a good example of the flexible and imaginative combined arms ethic which generated the Macedonian army's operational success.

Chapter 13
Technical Expertise

It is clear from the literary sources that Alexander's expedition to the East included a significant corps of scientists who were hired from all over the Hellenic world and even further afield. Unlike the savants of the Institut de France who accompanied Napoleon to Egypt (1798–1801), these specialists were often integral to the army's performance.[1] As far as currently known, Alexander's invasion force was the first in history to employ scientific specialists to quite such a major extent, systematically harnessing and then synthesising the full potential of Greek ingenuity in the fourth century BC to help achieve its objectives.[2] In some ways it resembles the Mongol use of Chinese technical know-how in the thirteenth century.[3] These capabilities, when allied to the dexterity of soldiers' training and Alexander's frenetic energy, would have gone a long way in extending operational reach.

Iphicrates once likened a well-led and tactically balanced army to the human anatomy where all constituent parts harmonise, complement and lend strength to one another, with the general as the head; the light-armed the hands; phalanx as the trunk; and cavalry the feet.[4] Taking this useful analogy a step further, the comprehensive array of non-combatants that accompanied the Macedonian campaign in Asia was equivalent to the sinews and blood-pumping circulatory system. In other words, this largely disregarded, sometimes invisible technical 'support crew' helped coordinate the overall enterprise. They represented the integral mechanisms that kept Alexander's terrifying juggernaut of war on track for continual conquest.[5]

From an examination of evidence culled from a pool of literary extracts it is apparent that Alexander and his Successors expected their engineers and other specialists to be capable of supervising a wide span of tasks in terrain which could vary from marshland to desert drift and from open plains to mountain ranges. Non-combatant technicians could divert watercourses, sink wells, fill ravines, repair canals and build altars, pyres, burial tumuli and harbours.[6] To this burgeoning list can be added (among other things) the construction of siege mines and causeways, and the design of siege engines, and artillery. Hammond rightly assesses that no Greek army (relying on slaves) and no contemporary armies (with reliance on machinery) have achieved such peaks of skill.[7]

Road Construction

The making of straight roads, perhaps adopting the Achaemenid Persian model, had been undertaken in Macedonia since at least the reign of Archelaus, and Philip II succeeded in improving what he inherited, constructing more roads as the kingdom expanded.[8] An effective network of roads helped bind the Macedonian state together and keep it secure.

From Philip II onward, Macedonian rulers girded their fortress-kingdom within a deliberate strategy of defence-in-depth, taking full advantage of natural terrain features and protective mountain ranges.[9] The fact that the Romans found Macedonia such a frustratingly difficult nut to crack during their wars with Philip V and Perseus was due in part to this effective defence strategy.

The overall scheme consisted of colonies of transplanted soldier-citizens, doubling as 'border-guarding sites', forming a barrier along the frontier on the first or outer tier, followed by forts covering vulnerable passes and access routes by land and along the coastline on the second. In the north, there was also the occasional proactive policy of laying waste zones of borderland as a means of discouraging profitable enemy incursions.[10] The third, or innermost, tier saw cities and even towns of lesser importance (Pella, Aegae, Dion, Philippi, Amphipolis, Torone and others) defended by massive 2–8km long circuit walls, with towers and bastions boasting the latest architectural advances,[11] and garrisons offering further support and protection.[12] These emplacements comprised either mercenaries or royal guard units, depending on the military, political, economic or cultural significance of each site. By c. 200 BC Philip V was deploying as many as 15,000 troops for garrison duty in all his territories.[13]

Macedonian kings to the fall of the Antigonid dynasty were responsible for running road arteries ranging from as far west as Albania and Dyrrhachium on the Adriatic coast to the Hebrus river on the modern border with European Turkey in the east. In narrating Macedonia's wars with Rome, Polybius, Livy and Plutarch all imply or sometimes directly reveal that those who controlled corridors leading into the kingdom would inevitably penetrate it.[14] Accordingly, the plan was, wherever advisable, to fortify and hold watch and ward over these localities, keep enemies out, and carry the war to them as offensive operations.

The arterial infrastructure encouraged unity and cooperation between regions and well suited the launching of defensive strategies and aggressive campaigns abroad, helping expedite troop transit to vulnerable pressure points, the inland borders and coasts. Such infrastructure is suggested in ancient literature and is occasionally found linking inland urban centres and forts, mines and ports, running along hills, through valleys and forests.

Macedonia's geographic position made this infrastructure vital if it were to survive as an ongoing concern. To the south, the kingdom marched with the quarrelsome city-states of the Greek peninsula. It also occupied the middle ground between the Mediterranean world and the interior of Europe, surrounded in a great arc on its western, northern and eastern approaches by warlike and restless Balkan peoples: namely, Illyrians, Dardanii, Paeones, Celts and Thracians, who were attracted by land, climate, hunting and pasturage. Given these peoples' preoccupation with raiding for plunder in the form of moveable wealth, slaves and livestock, a compact and efficient all-weather road system was essential for speedy mobilisation.[15] As Livy informs us of the Macedonians, 'their barbarian neighbours make them still more ferocious by sometimes familiarising them with war'.[16] The Macedonian kingdom thus shares with the Roman Republic the honour of mounting a bold response to the ever-present threat posed by northern tribes.[17] Their success as a peripheral bulwark facing untamed nations was made all too apparent to Greek neighbours and Roman successors along the unstable frontier in the 100 years after the kingdom's collapse.[18]

A distance marker from northwest of Philippi affirms a road to the east during the fourth or third century BC.[19] Distance stones were also retrieved from the vicinity of Pella and Eordaea, the latter inscribed '100 stades from Boceria'.[20] These verify a *basilikos odos* (royal highway) on the route of the later Roman *Via Egnatia*. Their military character is confirmed by Yugoslav excavations at modern Bitola at the northern entry to the 'Iron Gates' (Demir Kapu) of the Axius. There was a square tower with long circuit wall on a steep hill here, probably built by Archelaus as a means of blocking invasion routes from the north into Amphaxitis. Cut through the narrow defile were remains of a road made for vehicles with 10cm wide wheel ruts spaced at regular 1.25m intervals. A 1.20m wide flagged track was also found in central Albania (running through uplands from Lake Ohrid via Orake to the Shkumbin valley) which took the form of a small limestone paving made apparently for cavalry patrols and pack animals, and this route was no doubt supported by side-roads to forts located in Illyria. It has been theorised that itineraries on road distances were preserved at Pella, later consulted by Polybius and used in his extant histories.[21]

Ober insightfully illustrates the emotion and psychology of troops driving roads through mountainous borderlands in the Greek world. Citizens working on farmland, surrounded by family members and neighbours, were transformed into soldiers cooperating in detachments, the grunt-work of manual tasks familiar in their daily lives away from military service.[22]

Road construction continued as an important feature for the Macedonian army during Alexander's Asian venture. In his *Cyropaedia*, Xenophon recommends an engineering department for making roads on campaign. He even summarises the modus by which veteran troops who had reached the end of their service were to be re-employed as pioneers. A superintendent of the engineer corps was to ensure that former spearmen were to be equipped with axes, bowmen with mattocks and slingers with shovels. With these tools they were to move in squads ahead of the baggage wagons, clearing and smoothing out roadways.[23]

Sources for Alexander's army reveal that duties prescribed by Xenophon were assigned to the lighter units. An evident body of Thracian pioneers, probably drawn from the Thracian infantry, is shown by Arrian cutting steps and widening the otherwise impassable gorge at Mount Climax in Cilicia: 'the Thracians had made a road, the approach being otherwise difficult and long'.[24] Other pioneers were deployed to clear rubble and other debris, making narrow tracks into roads, when the army wound its sometimes precarious way over the rugged Hindu Kush (326 BC). This was an event sharply brought into focus by Michael Wood, who followed Alexander's movements up the Panjshir Valley in Afghanistan, with the route blocked for a night by a landslide.[25] In 322 BC, back in the heart of his empire, Alexander's magnificent catafalque was escorted westward, accompanied by teams of road menders.[26] Later, the Macedonian army's road- and bridge-building abilities continued to impress Roman commentators down to as late as Philip V's reign.[27]

The best illustration for road plans from Alexander's time can be seen in his aim to launch a major series of campaigns in the West in order to annex both Carthage and the Iberian Peninsula. In achieving these objectives, the King was expecting to secure lines of supply and communication via a road extending from Egypt to the Straits of Gibraltar, with the construction of ports and shipyards at suitable points along the route.[28] Had Alexander lived to implement such vast ambitions, an all-weather military road projecting roughly 3,500km to the Atlantic Ocean would have been a formidable demonstration of the resources, and building and engineering expertise, which he was able to marshal and put into effect with complete confidence. Its sheer scale would have served as propaganda, overawing opposition in the western Mediterranean theatre. In this we observe military engineering forming part of the foundation for an effective imperial strategy.

Extensive plans for the making of roads and supply depots also appear as intriguing, though adumbral, glimpses in the *Alexander Romance*, preserving a few surprisingly accurate details which are omitted from the more acceptable sources.[29] On one occasion, orders were issued for each satrap to open a road, with destination markers placed at regular intervals and organised arsenals where all weapons and shields were to be stored.[30] The *Romance* extract describes a network of prepared armouries established in the satrapies, probably as a continuation of original Achaemenid sites. The plausibility of Alexander or his Successors as the first to refurbish the Persian system of post roads with new distance stones is suggested by archaeological evidence and the excavation of a bilingual 'mile-marker' in Greek and Aramaic, dated to the time of the Seleucid Empire in the early-third century BC.[31] The implication from more orthodox texts is that Achaemenid roads continued to be maintained under Alexander's administration.[32]

Surveyors and Runners

Roads in Macedonia and Asia outlined above were effectively laid out by scientific specialists called *bematistai* or similar. They were established in Macedonia before Alexander, introduced by his father or even earlier.[33] Alexander's army was accompanied by a number of them but their exact function is under debate.

The *bematistai* can be literally defined as 'pacers' or 'those who measure by pacing' and they have been cautiously interpreted as men of varied skills, usually land surveyors, though sometimes cartographers, topographers, geographers or even ethnographers.[34] It is posited that Eratosthenes (c. 250 BC) planned his famous work on world geography by consulting the reports of Alexander's *bematistai* in their original form.[35] Though far from proven, some academics suggest that they were the direct forerunners of surveyors in the Roman army.[36]

For surveying, *bematistai* used a stylus and wax tablets for tabulating measurements and an abacus for arithmetical calculations. It is possible that a simple form of mechanical odometer was employed for counting paces, based on the account given by Heron of Alexandria.[37] Given the minute accuracy of distances made in Asia, preserved in sources such as Pliny and Strabo with a mere 0.4–5 per cent deviation from modern estimates, a device of some kind might have been utilised by technical staff in the Macedonian army.[38] However, it is more conceivable that the *bematistai* worked out distances by simply pacing them out on foot, following the army's march step – thus the literal origin of their name.[39] Alexander's *bematistes* Diognetus is known to have published a study on distance stages and other aspects of data. Other works covering Alexander's eastern expedition include those of Amyntas and Biton.[40] In addition, Archelaus from Cappadocia is recorded specifically as a *choragraphos* (chorographer) employed by Alexander and primarily involved in distance measuring and map making.[41]

Bematistai were accompanied in their tasks by colleagues called *hemerodromoi*, variously translated as 'distance-runners', 'all-day runners' or even 'ultra-long distance' runners.[42] Evidence for them derives from both literary and epigraphic material and they are defined within a general context by the *Suda* as couriers who were 'young men, a bit older than the ephebes, near the age of their first down, that are equipped only with bows and arrows, slings and javelins, the weapons most useful for their marching'.[43] Some intriguing hypotheses have been developed for these specialists through the diligent efforts of Tzifopoulos, from whose analysis much of the following has been gratefully drawn.[44]

Source extracts show that *hemerodromoi* were employed by some Greek city-states and armies predating Macedonia under Philip and Alexander. The most often cited case is that of the famous Athenian messenger Pheidippides who ran to Sparta to request aid before the battle of Marathon (490 BC).[45] *Hemerodromoi* are mentioned again in 480/479 BC, when the Argives employed them to warn the Persians that the Spartan army had taken the field.[46] In 362 BC Cretan *hemerodromoi* were sent from Tegea to Sparta to warn Agesilaus of the approach of Epaminondas and his troops.[47]

In the East, Plutarch mentions Persian royal messengers,[48] and a pseudo-Aristotelian treatise refers to the Great King having communiqués delivered to him on a daily basis through an empire-wide network of outposts manned by *hemerodromoi*, along with scouts, message-bearers and overseers of beacon towers.[49] Diodorus adds the use of look out posts.[50] A vague, though tantalising, excerpt from Lucian explains how during Macedonian rule in Asia 'postmen had to run to every quarter of the realm carrying Alexander's orders', which implies that foot runners were actively employed by Alexander for military and administrative purposes.[51]

For the employment of *hemerodromoi* in Macedonian service, the main body of evidence rests with one Philonides and his career under Alexander. This primarily takes the form of an inscribed base from near the southwest corner of the Altis at Olympia where two copies were recovered. According to the translated text of both, 'King Alexander's *hemerodromos* and *bematistes* of Asia, Philonides the son of Zoitos from the Cretan city of Chersonesos, dedicated to Olympian Zeus (his statue)'.[52] Of five probable or certain *bematistai* known to have served with Alexander, only Philonides is further qualified as a *hemerodromos*.[53]

The above fragments hint that within a military context *hemerodromoi* were often linked to persons of Cretan background. Xenophon emphasises the skills and versatility of Cretan archers serving with the Ten Thousand.[54] Given the subject matter under consideration, it is revealing that when organising sacrifices and athletic games as a celebration of their survival, more than sixty Cretans in the mercenary force were recorded having entered the long-distance race.[55] It is also intriguing to recall that in his history of Alexander Curtius specifies a Cretan soldier delivering a letter.[56]

Foot races played an important role in the social and educational fabric of Cretan cities and were obviously essential for military preparation, given the island's rugged landscape. Inscriptions from Crete often refer to the terms *dromos* and *dromeus*. The former has been interpreted as 'gymnasium' (or 'track') and the latter as 'runner' (meaning an adult Cretan citizen). Sifting entries from literary and epigraphic texts, evidence seems to point to the *dromos* as more an educational 'institution' with military overtones. The *dromeus* was therefore a male who completed his training and was eligible for army service but with limited rights, from 20 to 30 years old. This definition corresponds to the *Suda*'s entry for *hemerodromoi*.[57]

Leading Cretans, such as Nearchus, were intimately associated with Alexander's expedition and delegated responsibility over journeys of exploration.[58] It is plausible then that Cretan archers within Alexander's army also provided crack *hemerodromoi* when required, falling out from their 20- to 30-year-old age group. Philonides and Nearchus may have been entrusted with some form of authority over not only messengers but also experts in topography and communications, i.e. the *bematistai*.[59]

Although considered apocryphal, an anecdote from Alexander's precocious childhood pictures him posing foreign diplomats questions of military significance and even asking them which roads were the shortest in the Persian Empire travelling from the coast inland.[60] Data compiled by experts such as surveyors or runners – the direction and length of travel, the quality of roads including alternate routes, the location of wells and water sources, and the names of places passed – appear to have been digested and reproduced as itineraries for logistical use. The importance of such intelligence was not lost on later writers. Referring to Greek works, Strabo asserts the importance of geographic and topographic knowledge to successful generalship, as does Vegetius (c. 400 AD) using more specific military detail.[61]

Building Cities and Fortifications

Macedonians were obligated to undertake public duties or services (*leitourgeiai*) within their kingdom. Through Philip's pursuit of a vigorous policy of 'modernisation', this meant being drafted to help build cities and frontier posts. Under the supervision of architects, they were even able to cut and shape stone blocks for permanent civic buildings, city walls and towers. Soldier-citizens mobilised to labour for the collective good would have felt pride and incentive to help defend the society that they were made to feel such an important part of.

Out on campaign, Alexander made frequent use of his troops in construction projects – not unlike the three Roman legions (II Augusta, VI Victrix and XX Valeria Victrix), ordered to erect Hadrian's Wall. A capacity for hard work in all forms of climate and terrain resulted in a broad range of building feats.[62]

Noteworthy examples of the remarkable talent involved include the construction of the aforementioned colony of Alexandria Eschate on the Jaxartes river. As Curtius says:

> Alexander ... surrounded with a wall all the space he had occupied with his camp; the wall of the city measured sixty stadia [about 12km] ... The work was completed with such speed, that seventeen days after the fortifications were raised the buildings of the city also were finished. There had been great rivalry of the soldiers with one another that each band – for the work was divided – might be the first to show the completion of his task.[63]

This passage reveals that large-scale building projects were highly organised and coordinated affairs, with the soldiers both familiar with and adept at construction. Antiochus III's refurbishment of Lysimachia seems to have been undertaken in a similar vein.[64]

There are many other episodes which clearly substantiate the Macedonian flair for erecting permanent structures. These are sometimes supported by physical evidence on the ground. Ongoing archaeological projects in Uzbekistan may eventually bring to light campaigning and fort-building activities.[65] Polybius reveals that military engineering exploits continued under the *Diadochoi* and Antigonid kings and, sure enough, this is considered by some academics to be scored into the landscape of Greece to this day. Candidates encompass the Kastro Ditnata at the base of Olympus, and the Antiforitis Wall in Boeotia.[66]

Archaeologically, the impressive Antiforitis Wall (estimated at 11km in length) has been recognised as possibly built by Ptolemaeus, a general of Antigonus Monophthalmus, to function as a base of operations against Cassander (313/309 to 308 BC). Its various attendant features, and in particular the fort at Kastro (with rooms for administration and stores), have received high praise from Bakhuizen for their arrangement and construction, integrating an observation post, barracks, bastions, sally and signal ports.[67]

A number of fortification remains have been further traced and mapped across modern Attica, some of which may date to the Chremonidean War (c. 264 BC) when Macedonia under Antigonus Gonatas engaged Sparta and her allies supported by Ptolemy II.[68] Ptolemaic bases and outposts have been suggested at Koroni, Voliagmeni, Patroklou Charax and other sites.

The 'Dema', one of the most frequently studied ancient Attic fieldworks, has been identified as an Antigonid structure, although this is now a largely outmoded interpretation. A recent, perhaps more persuasive, theory asserts that it was built by the Athenians themselves (425–375 BC).[69] It may be useful nonetheless to describe its appearance as a 4.3km long rubble wall with better preserved sections standing 1.5 to 2m high. The construction was carefully prepared for an angled defence, comprising fifty-three separate lengths of wall with sally ports.[70] An admiring McCredie has concluded that the Dema's apparent 'superior planning and execution', when compared to tentative Ptolemaic and Spartan defence works from elsewhere, express Macedonian superiority in fortification.[71]

A later demonstration of the engineering expertise so far described, almost uniquely combining literary and archaeological evidence for the one event, can be witnessed with Philip V's campaign in Acarnania (219 BC). Polybius reports how the King first ordered his troops to dismantle the wall at Paeonium, with rafts constructed from house timber, the debris then floated down the Achelous river to Oeniadae.[72] Observing that this locality occupied an ideal position for embarkation to the Peloponnese,[73] he then had his troops fortify the citadel separately and 'surrounding the harbour and dockyards with a wall he intended to connect them with the citadel, using the building material he had brought down from Paeonium for the work'.[74] The site at Oeniadae still preserves evidence for defences, including the activities of Philip's army. The remains comprise foundations for walls and five towers on the citadel, including a combination naval storage building with five slip ways at the north end of the town and some harbour fortifications.[75]

Philip V's thirst for military glory and confidence in his army's abilities in overcoming obstacles such as these is captured in a flattering (or sarcastic) epigram of Alcaeus of Messene, 'Make higher the walls of Olympus, Zeus. Philip can scale everything. Close the bronze gates of the blessed ones. Earth and sea lie subdued beneath Philip's sceptre. All that remains is the road to Olympus.'[76] The Acarnanian campaign sheds light on the resourceful and disciplined spirit required – the audacious ease with which a Macedonian army could be galvanised into an almost ant-like workforce, with unperturbed soldiers readily submitting themselves to toiling labour in order to achieve a strategic goal.

Confidence in Siege Work

Much ink has been expended by academics describing Alexander's sieges, and his and his generals' use of the latest technological advances in siege craft.[77] Greatly expanding and refining techniques inherited from Dionysius I of Syracuse, the kingdom of Macedonia soon became the epicentre for the development of siege science, so much so that a widely accepted working hypothesis considers the Romans as only derivative.[78] Cities under siege became a depressingly familiar sight by the time of the Successor Wars.[79] So imposing in fact was Macedonian dominance in this activity that a statistical analysis of the period from around 322 BC and especially 307–303 BC has shown that from a total of twenty-one such operations conducted by Macedonian armies, twenty were successfully concluded by the besiegers, with the city taken.[80] This was all but a part of the intense professionalism ushered in by Philip II, his son and their officers.[81]

The number of siege technicians who are believed to have been patronised by Macedonian generals during the mid- to late-fourth century BC is impressive. These can be enumerated as Polyidus of Thessaly, Diades of Pella (who published an *Engineering Compendium* of his inventions),[82] Aristobulus of Cassandreia, Epimachus of Athens, Isidorus of Abydus, Callias of Aradus, Hegetor of Byzantium and possibly Dionysius of Alexandria. The plethora of names only serves to remind us how Macedonia became a magnet for diverse military specialisms during and after Philip's reign. To this roll call can be added Posidonius, Charias, Philippus son of Balacrus, and unnamed Phoenician and Cypriot engineers employed by Alexander at Tyre.[83]

Given such specialist back-up, we can only imagine the trepidation felt by opponents who had the misfortune to be besieged by Macedonian armies. In this regard we read of Philip II confiscating 180 Athenian ships, the timbers of which were reused to construct assault engines on the spot during his attack on Byzantium (340 BC). The conversion process was almost certainly undertaken by units of Macedonian soldiers, and the Byzantines who watched them from the safety of their walls must have been not a little anxious about what their purposeful enemies were up to.[84]

An animated picture of the vigour of Macedonian troops during siege operations emerges from other specific events. During construction of a causeway at the siege of Tyre native citizens approached in boats to mock and insult Alexander's men as 'those famous warriors carrying loads on their backs like pack-animals', but Curtius was swift to clarify that 'their jeers served to fuel the soldiers' enthusiasm'.[85] Arrian notes that 'the Macedonians were very eager for the work, like Alexander; he was present directing each step himself, inspired the men with his words and encouraged their exertions by gifts to those who did work of exceptional merit'.[86] Likewise, over a century later at the siege of Palus on Cephallenia (off the Gulf of Corinth) in 218 BC, Philip V exhorted his troops to bring up assault engines against the town walls. The Macedonians responded and 'worked with such goodwill that about 200 feet of the wall was soon undermined'.[87]

The fact that Alexander had complete confidence in his siege specialists and men is well evinced with his ability to breach the almost impregnable Sogdian Rock (327 BC) in the face of defenders who mocked that they could only be conquered by winged soldiers.[88] For its seizure the King relied on the skills of 300 young selectees drawn from squads within the Agrianes or Macedonian infantry units from the upland regions. These men were adepts in clambering up cliffs with experience of rock climbing in previous sieges[89] or were from shepherd localities accustomed to driving flocks over mountainous and impassable terrain.[90]

Kalita argues that, rather than mere volunteers, the soldiers employed in climbing the Sogdian Rock formed a distinct group permanently embedded in the ranks and called out when required. This seems to complement possible detachments of slingers; rowing gangs for waterborne assaults; road-builders from among the Thracians; foot-runners from Cretans in the archers; and perhaps even designated medics drawn from the infantry as a whole.[91] Whether Kalita is right or not, such an arrangement fits in well with Philip and Alexander's known principles of flexible training and on-the-job adaptability.

The steps Alexander's troops cut into the side of Mount Ossa in Thessaly (to allow the army across in 336 BC) could still be seen by travellers in the second century AD and to this day there is real or alleged physical evidence for the labours of Macedonian soldiers, traced from Albania to the periphery of India.[92] It is even suggested among some modern scientists that the immediate coastline of Tyre was permanently altered by Alexander's engineers.[93] At Susa, an Iranian archaeological team led by Mir Abedin Kaboli in 1982 discovered 160 artillery balls and 400 arrowheads, indicative of a Macedonian combined arms assault. The startling find of an incendiary ball (identified as a burnt composite of mineral covered with flammable coniferous tree resin) was found in the ditch of a fort in Charsadda (Pakistan), associated with a siege by Alexander.[94] A stone catapult ball unearthed in 2002 along the Shurob river (Afghanistan) is thought to be another relic from one of his offensive operations.[95]

Crossing Water Obstacles

As has already been shown, on campaign in Asia Alexander was able to consult experts with scientific knowledge ranging over a variety of fields. One aspect of consultation concerned the rapid crossing of water obstacles, with individual flotation devices, larger rafts and pontoon bridges utilised. It is possible that Strattis of Olynthus accompanied the Macedonian army on its eastern expedition and, according to the *Suda*, even complied a work usually translated as *On Rivers, Springs and Lakes.*[96]

Although advice on getting over rivers was heeded, disasters did occur. In his offensive against Ptolemy in Egypt (321 BC), Perdiccas attempted to cross the Nile near Memphis by using a variation of Alexander's method in traversing the Tigris ten years previously.[97] He therefore posted elephants up river to break the current and horsemen down river to catch any helpless men, as his troops waded in. In spite of initial success, the movement of men, horses and elephants caused the sandy bottom to be disturbed and to disintegrate, inevitably increasing water depth at the crossing point. Over 2,000 men were consequently drowned or swept away by the fast flowing water; some were snatched by the ravening jaws of crocodiles and hippopotami.[98]

Such lessons were not forgotten by later technicians. Vegetius warned that 'when crossing rivers careless armies often get into serious difficulties. For if the current is too strong or the river-bed too wide it is likely to drown baggage-animals, grooms and even the weaker warriors.'[99] Diodorus gives a detailed example of a successful river crossing during the operations of Eumenes against Antigonus Monophthalmus at the Coprates river in Persia:

> This river, running from a certain mountainous region, enters the Pasitigris, which was at a distance of about 80 stades [14–15km] from Eumenes' camp. It is about four plethra in width [around 120m], but since it is swift in current, it required boats or a bridge. Seizing a few punts, he [Eumenes] sent

> some of the infantry across in them, ordering them to dig a moat and build a palisade in from it, and to receive the rest of the army.[100]

The soldiers who were required to erect fieldworks reached the opposite bank with weapons and construction tools. The main force crossed by pontoon bridge and managed to surprise an almost 10,000-strong Antigonid force. The Antigonid army in this case included 6,000 of those troops who were 'in the habit of crossing [water obstacles] in scattered groups in search of forage',[101] perhaps using skin floats filled with air or straw, depending on circumstance.[102]

Given Philip and Alexander's doctrine of speed and mobility, tent covers were multipurpose items, and their most common application was as individual floats or collections of floats at water crossings.[103] Though not directly proven, it is possible that tent equipage was useful in other circumstances as well, such as being applied as leather screens for siege-towers at Tyre.[104] Arrian describes tent covers for fording rivers, stuffed with matter or dry (and therefore light) chaff, tied at the top, then stitched as tightly and precisely as possible to make them waterproof.[105]

Strabo's account of the training of the *kardakes* (an Achaemenid military youth corps) describes them learning to negotiate watercourses effectively while keeping their armour and clothing dry.[106] This implies the use of inflatable skins, which certainly had an ancient pedigree in the region (and in fact as far west as Albania). Assyrian soldiers are shown in bas-reliefs from Nimrud from the reign of Ashurnasirpal II in the ninth century BC lying naked upon inflated animal bladders while crossing moats, and Xenophon describes troops fording the Euphrates on skins filled with air.[107] The Macedonians could have easily adopted such techniques, and to this day inflatable goatskins are commonly applied at rural crossings in the Middle East and parts of South Asia, some supporting up to 180kg in weight.[108] Fig. 46 illustrates the use of one such animal skin as a flotation device.

Rafts made from numbers of stuffed or inflated skins were made use of at the Jaxartes (329 BC), according to Curtius' narrative.[109] The soldiers are described here preparing 12,000 rafts in three days flat.[110] Facing a barrage of Scythian arrows shot from the far bank were light infantry (archers and slingers) on separate floats, followed by rafts for infantry and cavalry, with most horses held by the reins and swimming astern.[111] Some of these transports were large enough to have steersmen and crews to handle them. In the foreparts of the vessels were contingents of infantry kneeling with their shields, probably held up for better defence. Behind them were catapults manned by professional artillerists (perhaps the same specialists as used for a plausibly attested naval siege unit), covered by troops at their flanks. More soldiers were stationed behind forming a tortoise-like formation, thus shielding the unprotected oarsmen.[112] The whole scene offers a formidable demonstration of Macedonian organisational expertise, discipline and combined arms awareness.

The construction method of some floats sketched under Curtius and Arrian appears to be preserved through the filter of the *Itinerarium Alexandri* in remark-

Figure 46. Inflatable animal skin used for crossing the Tigris river, Iraq.

ably minute detail, though incompletely.[113] Bloedow was rightly moved to note the excellence of the Macedonian transports which ensured the well-executed transfer of nervous animals across a substantial river.[114] Modern parallels have been sought by academics for the larger rafts or collections of floats and these encompass the Ethiopian *jandi*, and Turkish, Kurdish or Iraqi *kelek*.[115] The latter can be observed in Fig. 47, which shows a large *kelek* made from hundreds of

Figure 47. A large *kelek* on the Tigris river, Iraq.

timber beams, lashed together and buoyed up by inflated goatskins, being unloaded on the Tigris river at Baghdad in the 1930s.

Apart from rafts, the capacity to construct usable bridges was a skill the Macedonian army practised as it roamed across the geography of Asia. Sappers successfully spanned the Euphrates with two bridges in 331 BC.[116] An even better example is the pontoon bridge constructed over the Indus river (326 BC).[117] The Indus waterway today is anywhere from 300m to 1,250m wide (sometimes over 2km wide during the dry season)[118] but it presented no insurmountable obstacles to the army with its technical abilities.

The bridge spanning the Indus was built by an advance group under Hephaestion (who often led these operations, described by Heckel as Alexander's 'utility-man')[119] or under joint command of Hephaestion and Perdiccas.[120] Specialists were employed playing a role akin to that of Napoleon's indispensable *pontonniers* 2,000 years later. Accounts specifically describe the Indus bridge consisting of boats or other vessels moored together to form the basis. These were easily assembled prefabricated items, constructed with bolts: smaller ones cut into two, larger into three pieces, then conveniently transported on wagons when not in use.[121] The semi-mythic *Itinerarium Alexandri* obscurely remarks on how the requisite materials for making a bridge were brought from both near at hand and a long way off, and with them the army effected a river crossing.[122] This merits note because a pontoon bridge built by British Royal Engineers over the Indus while campaigning in 1838 met with a similar problem. Here adequate lumber had to be floated over 300km down river from Firozpur, and this was supplemented by inferior stuff in the form of split date palms. As no rope was found, 500 cables of woven grass acted as a substitute, brought from over 160km away.[123]

By the second century AD the original bridge building methods employed by Alexander's engineers had been forgotten or subsumed under later advances. This is borne out by Arrian, who claims not to have been able to make out with certainty the exact techniques used at the Indus.[124] He therefore delivers an excursus on the methods applied by Roman engineers, either as simple means of filling in his narrative or as a rough template to Macedonian practices.[125]

Arrian's description of Roman military bridging does seem, in some particulars, to approximate to pontoon structures seen in Asia. A form of floating bridge was used by the Assyrians as early as the ninth century BC, and pontoons utilised under the Achaemenids.[126] Witnesses noted a pontoon bridge over the Tigris, north of Baghdad, in 1916, and some are built in Iraq, rural parts of Iran and Uzbekistan to this day. They were often employed in Mughal and British India (for example, see Fig. 48 for a traditional structure of a bridge of boats spanning the Indus at Kushalgarh, now in Pakistan), and the same type is seen on the subcontinent in modern times, e.g. a bridge near Attock (Pakistan) in 1967 and one in use at Mithankot.

These bridges *might* resemble those erected by Alexander's technicians, although for now we really have no way of knowing their exact configuration.[127] If alike, then the Indus structure was made by first setting down planks forming

Figure 48. Bridge of boats on the Indus in 1883.

a gangway across a chain of aligned vessels upon which was placed a surface padding of earth blended with straw. The cover was necessary to protect the wooden surface from weakening under the incessant pounding and shifting of animal hooves.

On encountering the Oxus, Alexander and his engineering team initially contemplated a bridge, but were unable to proceed as pilings could not be secured to the river's sandy bottom and neither was timber immediately available.[128] It has been hypothesised that Alexander's specialists were able to erect different bridge designs. One such is the structure planned (but never actually made) for crossing the Tigris during the return march of the Ten Thousand. It was to consist of 2,000 inflated animal skins joined together, anchored to rocks in the river with pack animal harness. Brushwood and earth were to be finally piled on top of the skins to prevent the soldiers from slipping off.[129] Something similar, made up of locally hewed date palm trunks and inflated skins, was successfully built by Nadir Shah when crossing the Tigris in 1733.[130]

Well-preserved clusters of close-set pointed timber stakes driven vertically into the sandy soil of a river bank, used as a means of strengthening the substructure, have been found at the site of Amphipolis, dated to the fifth or fourth centuries BC.[131] As with later medieval Arab practice, Macedonian bridging techniques may sometimes have involved chains for securing wooden anchorages at each bank, but there are only a few traces which could support this argument. Pliny mentions a remarkably preserved iron chain which still survived in the first century AD at Zeugma and was said to have been used by Alexander when crossing the Euphrates.[132] A village was demolished to provide materials for erecting a more substantial bridge over the Araxes river, the inference being a timber edifice erected over dressed stone piles.[133]

The almost entirely forgotten bridge building experts were themselves undoubtedly a mix of Macedonians and Greeks, using techniques from across the Hellenic world, the Balkans and Thrace. A surviving Babylonian document describes the existence of a corps of 'bridge-builders' in Achaemenid times who were allotted land near Nippur.[134] Middle Eastern specialists would have been employed in the later years of Alexander's reign, although explicit evidence is lacking.

Chapter 14

Little-Known Combat Units in Asia

Before leaving this part of the book, we must not fail to consider the fact that great depth in technical ingenuity ran parallel with tactical versatility, or the use of combined arms, in which different troop types were integrated for maximum tactical benefit. Advocated by Iphicrates, it served as a primary hinge for the battlefield success of the new Macedonian army.[1]

Having witnessed its last full-blown occurrence in the land forces of the Neo-Assyrian Empire, the combined arms ethic was rediscovered in the fourth century BC and put to best effect by Philip and Alexander. As Demosthenes warned his fellow Athenians: 'nothing has been more revolutionised and improved than the art of war . . . you hear Philip marching unchecked, not because he leads a phalanx of heavy infantry, but because he is accompanied by skirmishers, cavalry, archers, mercenaries, and similar troops'.[2]

According to Arrian, Alexander reiterated the importance of this approach in a speech to his men before Issus, in which he praised the variety and excellence of his troops in cavalry, archers and slingers in comparison with those of Greek armies before.[3] Writing in the second century AD, and with the benefit of hind-sight, Dio Chrysostom explained more candidly that the young conqueror required 'his Macedonian phalanx, his Thessalian cavalry, Thracians, Paeonians and many others if he were to go where he wished and get what he desired'.[4] The principle espoused embraced everything from small operations to full scale battles with an emphasis on flexibility – a lethal coherence of cavalry with infantry, light with heavy, shock with missile support.[5]

Of the full range of units used by Alexander in the East, the following is an overview of some lesser studied combatants which helped enhance this combined arms method – namely, the *dimachae* (mounted infantry) and *katapeltaphetai* (artillerists).

Mounted Infantry

From around or after the conquest of Persia (c. 331 BC), Alexander brought together a little known unit called *dimachae*. These men could be called out of the ranks at any time as an ad-hoc body to render a specific purpose when their skills were required.[6]

Curtius records how in order to promote a more rapid and effective 80km cross-country pursuit of Darius and Bessus, Alexander arranged that 6,000 cavalry were to be supported by 300 *dimachae* who 'rode horses'. However, 'when circumstances and the terrain required it they fought as infantry'.[7] They are mentioned in larger numbers by Arrian in later operations. This suggests that

Alexander was more than satisfied with their performance, having them expanded to at least 800 by 327 BC.[8]

Hesychius refers to *dimachae* as *hamippoi* (usually defined as infantry skirmishers mixed in with and supporting cavalry).[9] Pollux more fully defines them as:

> Another type of horseman, *dimachai*, Alexander's invention, having lighter equipment than hoplite infantry and heavier equipment than cavalry; they were trained to fight in both ways: from the ground and from horseback; when they were in terrain suitable for riding, they would ride; but if they came to terrain unsuitable for riding, they would not be entirely unable to fight nor would they suffer the Lydian misfortune; but a servant, following for this very reason, carefully took the horse; and the one who had dismounted from the horse was straightway a hoplite.[10]

The tactical role of the *dimachae* was therefore as a mobile force used to run down enemy troops who had entered or were ensconced within rough terrain, unsuitable for cavalry alone. They were meant to combine the speed of horsemen in getting to the opposition with the effectiveness of infantrymen when about to engage.[11]

Heckel postulates *dimachae* drawing significant numbers from the hypaspists, trained in horsemanship.[12] This is persuasive given that guardsmen were often selected for tough or difficult missions. Corrigan asserts that Alexander was limited in his choice of *dimachae*, as he would need to consider such factors as whether the strength of pursuing infantrymen was not already exhausted, and if they could ride a horse well and proceed at speed.[13] They were probably armed with a spear or javelins and sword, equipped in a corselet with boots, shield slung from the back for ease of movement.[14]

A puzzling fact has been raised that even though the *dimachae* knew how to ride, the frequency of them on horseback would seem inadequate (as they were essentially foot soldiers) for their leg muscles to be in good enough condition to cover an 80km mounted pursuit, as Alexander initially required of them.[15] The importance of leg conditioning for troops was certainly recognised by other generals, like Eumenes, during later campaigns.[16] The rapidity of some mounted pursuits carried out by Macedonians never dropped below 50km per day and one 70km stretch took only twelve hours. As Hyland rightly posits, the recorded speed is not that significant, but with factors of inadequate food, water, rest, tack and a lack of horse shoes, it was a remarkable achievement.[17]

Artillerists

Apart from *dimachae*, it should not be forgotten that Alexander was one of the first generals in history to utilise artillery on the field, fundamentally as defensive and offensive covering fire for his troops during amphibious assaults. Episodes include their enfilading enemy forces at the Eordaicus and Pelium, and at the Jaxartes (as noted in Chapter 13).[18]

Each of the lightweight prefabricated torsion powered Macedonian catapults (*katapeltai Makedonikoi*)[19] in the artillery train would have been built from scratch by a Greek master mechanic (*mechanopoios*), with skill in carpentry and metal-work, and specialist knowledge of technical dimensions.[20] The machines were further manned by teams of *katapeltaphetai*, or professional artillerymen. The term is found in the surviving Florentine manuscript of Diodorus, concerning the capture of eleven technicians serving under Demetrius at Rhodes.[21] Due to its extreme rarity, it was considered corrupt and amended by scholars, although the original reading may have to be restored.[22]

Experts such as these were hired from across the Hellenic world. From epigraphic texts, one of them in Athens is noted as a Mysian from Asia Minor. All indications are that catapultists in Alexander's army were few in number, perhaps a hundred or so, each well acquainted with his machine.[23] This is because duties involved not only the catapult's sighting and shooting, but general maintenance and repair.[24]

Catapult specialists saw employment throughout Hellenistic times, and another inscription, this time from the first century BC, even has *katapeltaphetai* manning Rhodian ships.[25] This use of catapultists in sea warfare could have been pioneered by the Macedonians during the Successor Wars, possibly first introduced with the navy of Demetrius Poliorcetes at the battle of Salamis in 306 BC.[26]

Epilogue

Macedonian Militarism and the Impact of Pydna

One of the chief hallmarks of the Hellenistic Age is its characterisation as an era of 'ubiquitous war'.[1] The *Makedones* were primarily in demand as soldiers at the throbbing heart of this restless and often unstable civilisation. In a century of rule even under the Antigonid kings there is hardly any time that the Macedonians were not involved in armed conflict, punctuated now and then by anaemic distillations of peace.[2] A policy of active aggression and localised bids for conquest was an electric current which ran through the kingdom's leadership up to and including the reigns of Philip V and Perseus.[3]

Polybius writes of Ptolemaeus, the son of a high ranking Ptolemaic minister, who on his return to Egypt from the court of Philip V thought the pampered city-dwellers of Alexandria inferior to the Macedonians he had mixed with.[4] As conveyed in a digression by Cornelius Nepos, 'in those days the Macedonian soldiers had a reputation that the Romans now enjoy, since those have always been regarded as of the greatest valour also rule the whole world'.[5] Alexander's men were seen as a standard of excellence by subsequent generations who thought they had a claim to fighting prowess. This is reflected in an incident from 299 BC when in his conquest of Corcyra (Corfu) the Sicilian tyrant Agathocles of Syracuse confronted the troops of Cassander. The Sicilians made a dogged go of it as they were determined not only to prove themselves victorious over Italians and Carthaginians but also, as Diodorus generalises, 'to show themselves in the Greek arena, as more than a match for the Macedonians, whose spears had subjugated both Asia and Europe'.[6]

In the third century BC the fear the Macedonians engendered as renowned fighters was comparable to that shown for the Spartans centuries before. This is confirmed by incidents read across the surviving literature. During a campaign prosecuted by Philip V in the Peloponnese (218 BC), advance parties of Eleans and Macedonians converged on the same strategic pass. The Elean general had heard report that the approaching enemy were Macedonian, and keeping the intelligence secret, promptly left his men to their fate. The abandoned soldiers misidentified the oncoming foe as Achaean but, realising their mistake, threw away their shields and fled pell-mell, some falling down high precipices in panic, with others killed or captured by the Macedonians following up.[7]

In another incident, Polybius recounts how contingents of Macedonian soldiers sought to breach the town of Messene (214 BC). When the inhabitants discovered from their speech and equipment who the attackers really were, fear initially prevailed because of their enemy's 'training in warfare and the good fortune which they saw they enjoyed in all their ventures'. The Messenians, however, soon plucked up courage, rallied and then repelled the intruders who were confessed to be 'brave and experienced troops'.[8]

Polybius concedes that in certain particulars the Macedonians of his time were man for man more effective as soldiers than any contemporaries. In a generally overlooked though telling passage regarding Macedonian troops during the Social War (220–217 BC), he qualifies them as 'most intrepid' soldiers 'in regular battles on land' – alternatively translated as 'gallant', 'valiant' or 'courageous' – and compares them to Hesiod's sons of Achaeus 'joying in war as if it were a feast'.[9] This admiration eloquently evokes the popular opinion of the day and re-emerges elsewhere when Polybius sketches some unique characterisations of Greek peoples including Aetolians and Thessalians. Of Cretans Polybius comments that they are 'cowardly and downhearted in the massed face-to-face charge of an open battle'. He adds: 'It is just the reverse with the ... Macedonians.'[10]

In contrast, in his counterfactual digression on whether Rome would have vanquished Alexander, Livy sets out to praise Roman military superiority over the Macedonian army in numbers, tactics and powers of endurance.[11] The overall passage is misleading as it shows that Livy was either poorly informed on Macedonian strengths or that he knew of them but refused to allow their true recognition.[12] On the other hand, the Polybian statements given above are remarkably high praise from a generally astute and dependable historian. They help form an image of the Macedonian as a first-rate soldier, excellently trained and technically proficient.[13]

It is worth appending that Polybius was in a relatively comfortable position to be well informed. During service as a cavalry officer he saw the Macedonians in action and read contemporary military handbooks, many of which we now know nothing of. He pursued an interest in conscientiously studying the warfare of his times while producing at least one lost, though influential, work on tactics. It is noteworthy that such views flow from the pen of a man who had voyaged widely in the Mediterranean and had seen contrasting cultures and conflicts from the Atlantic to Asia Minor and Egypt.[14] Polybius is also acknowledged as a great advocate of 'serious' politico-military history with a definite educational purpose based exclusively on facts and divorced from distortion arising through emotional colouring or ethnic bias.[15] Given that he wrote for wealthy patrons, had intimate contacts with the great families of the Scipiones and Aemilii and had pro-Roman sympathies, his estimation of the Macedonians is striking.

We can be excused for hypothesising that when the Romans first emerged on the world platform as a tangible power (c. 270 BC), many Macedonians, being the haughty, purblind heirs of Alexander, saw them of limited interest or consequence. The kingdom of Macedon was embroiled within the fractious blood-letting politics of the Balkans and Greek peninsula and Roman outsiders were viewed by the Macedonians as a brave, obscure and non-Hellenic people from a foreign land.[16]

However, as a typical Macedonian soldier-king, Philip V sniffed out opportunities in the unfolding drama of the Hannibalic War and after Rome's defeat at Lake Trasimene in 217 BC was tempted to seek domination in the West.[17] After the crushing victory of Cannae he took the fateful step of securing a treaty of mutual benefit with Hannibal (215 BC) by which Macedon would invade Italy in support of Carthage while the latter recognised the kingdom's expansionist interests over Illyria.[18] When direct confrontation with the puissant Romans became imminent, the Macedonians still seemed confident of achieving success over the upstart military power arising from across the Adriatic Sea.[19]

Confidence stemmed from the one indomitable trait of the Macedonian character which stands out more than any other, frequently seen in extant texts: namely, the intense *elan* that they manifested in their waging of wars. The basis of this psychology is well defined by Billows, who explains how from the mid-fourth century BC the army's central role in the state, the excellence of its organisation and the spirited discipline of its officers and men led to ongoing conquests.[20] Macedonian military pride is clearly seen in their rivalry with Greek mercenaries fighting for the Persians at the Granicus in which the Macedonians, according to Arrian, 'sought to ... preserve the reputation of the phalanx, whose sheer invincibility had been on everyone's lips'.[21]

In tandem with a Homeric warrior ethos, conquest fuelled a staunch propaganda of military self-glorification – embodied in the emblems of spear (hence 'spear won' land) and crescent-patterned shield (a metaphor for 'national' glory and identity).[22] The shield as ethnic emblem is in fact frequently found wherever the Macedonians settled as colonists in Asia. It lasted for centuries and was used as a means of binding their communities together while at the same time setting them apart from neighbouring peoples.[23] The descendants of Alexander's soldiers loudly boasted of themselves and their ancestors and this was, as Billows relates, pushed in every credible medium: from coinage and tombstones to public art, inscriptions and literature.[24] We see a glimmer of the same attitude as late as the second century AD when in the preamble to his *Strategemata* Polyaenus prides himself on his Macedonian lineage and brags that if he were only in his prime he would be an eager steadfast soldier just like his ancestors.[25]

Certain other symbols stand as proof for this form of arrogant chauvinism and proto 'ethno-nationalism'. As shown earlier in this book, eagles, wreaths and thunderbolts all appear as motifs on various features of armour and may even be part of an evolved imperial iconography, promoting the Macedonians as the warrior caste *par excellence* of a cosmopolitan world stretching from Greece to northern India. It is exemplified with a surviving carnelian gemstone (300–250 BC or first century BC) cut with a heroically nude Alexander, left arm resting upon a shield, surrounded with the attributes of Zeus: oak wreath, eagle and thunderbolt (Fig. 49).[26] Similar emblems were later used by the Romans as allegorical of their own uncontested power. The subjugated populations were thus left in no doubt as to who their masters were.

Before the advent of Rome, the notion of the Macedonians being supplanted by an intrusive rival appeared unthinkable to the self-deceived. But it was a

Figure 49. Red carnelian depicting a probable Alexander. Signed 'Neisos', who may be the artist or former owner.

diminished people that the Romans had the good fortune to face, living off the exhilarating memory of Philip and Alexander, under whom the kingdom had been an engine for forging a vast empire. If Rome had faced a master exponent of tactics and strategy in the form of the Macedonian great captains – Alexander or the much vaunted *Diadochoi* – the outcome could have been very different.[27] Then again, the study of history is replete with such hypotheses, so it is a fruitless exercise to ponder over them.

Nevertheless, ancient literature unanimously attests to the importance of the contest between Macedonia and Rome, portraying it as a collision of the two main European powers of the age.[28] Justin makes much of how Roman wars with Macedon were of greater renown than those waged against Carthage, because the Macedonians 'surpassed the Carthaginians in honour' and the Macedonian army had a 'reputation of being invincible'.[29] Just as Carthage was the Mediterranean basin's greatest naval power, so, as stated by Dionysius of Halicarnassus, Macedonia was reputedly the strongest nation on land.[30] Although Dionysius exaggerates, his statement at least helps to confirm the kingdom's high repute.

A mutual admiration is traceable between the Macedonians and Romans during the Second Macedonian War and by the time of the Cynoscephalae campaign in 197 BC.[31] In a pre-battle pep talk, the Proconsul Titus Quinctius Flamininus is said by Plutarch to have described his enemies as 'the bravest of antagonists'.[32] A year previously, Flamininus had himself witnessed how his Romans had failed to breach Atrax in Thessaly, defended with tenacity by a detachment of *sarissa*-armed Macedonian troops. What vexed the Roman general most was that 'he was allowing a comparison to be made between the tactics and weapons of the contending armies'.[33] A lofty regard for the Macedonians re-appears when before the battle of Thermopylae (191 BC) Livy puts into the mouth of the consul Manlius Acilius a speech to Roman legionaries comparing the Seleucid army of Antiochus III with that of Philip V. Here Philip's army is described 'composed of a somewhat better grade of soldiers' and the Macedonians are categorised as one of several 'most warlike nations', along with their neighbours the Illyrians and Thracians.[34]

The supreme confidence of Macedonian *esprit de corps* can be deduced from Plutarch's account[35] in which the massed ranks of phalangites filled the hills at Pydna in 168 BC with resounding cheers as they debouched from camp.[36] This recalls Onasander two centuries later advising that a general should send his troops into battle at a run, yelling and clashing their weapons to dampen enemy resolve.[37] It is almost certain that the pike armed units seen that fateful day could trace distinguished bloodlines to the redoubtables who accompanied Alexander. The clamorous cacophony confounded Roman onlookers and left a deep impression on their commander, Lucius Aemilius Paullus, who would recall for many years that the inexorable advance of the Macedonian phalanx, with serried 'groves' of *sarissai*, was the most stirring sight he could remember.[38]

Such an awestruck reaction was anything but new. Diodorus recounts how Alexander set out to 'overawe' the Thebans by appearing with his army 'in full battle array'.[39] A similar example of the lethal fusion of high morale and efficient training is given by Arrian, who describes the Illyrian Taulantii's reaction to veteran *taxeis* in 335 BC:

> The enemy, long bewildered both at the smartness and the discipline of the drill, did not await the approach of Alexander's troops, but abandoned the first hills. Alexander ordered the Macedonians to raise their battle-cry and clang their pikes upon their shields, and the Taulantians, even more terrified at the noise, hastily withdrew.[40]

Pydna, nonetheless, heralded the death knell of this ingrained soldiering tradition, as the kingdom's fate was effectively sealed within only an hour from the start of battle.[41]

It must be said that the encounter at Pydna turned out to be an almost wholly infantry affair, awkward and unimaginative. Indeed, in a larger context, Macedon's conflicts with Rome, rather than epic struggles, often had the air of a deflated anticlimax about them. Following Roman penetration of the opposing line, 3,000 of the crack Antigonid infantry, ironically not unlike the Theban Sacred Band of 300 men at Chaeronea, recognised that victory was now beyond hope and so fought and died to a man – preserving a reputation for 'valour' to the end. Maybe they felt that the Romans left them with little choice.[42]

Perseus fled the field in dismal ignominy, after having initially insisted that he would overcome or go down fighting.[43] The shattered and scattered remnants of infantry openly upbraided the cavalry with bitter recriminations for not being properly committed and, amazingly, in their rage even wrestled some of these compatriots from their mounts.[44] Perseus' lacklustre performance must have had a shattering psychological effect on his soldiers as the Macedonians expected their kings to lead them to victory in war.[45] The troops had not been allowed to perform as they had under Alexander, that is, as a tactically fluid force of combined arms composed from multiple specialist units in mutually supportive roles.

At the close of his study on the Macedonian phalanx, Polybius exclaims with an undisguised hint of exasperation that 'so many Greeks on those occasions when the Macedonians suffered defeat regarded such an event as almost incredible and many will still be at a loss to understand why and how the phalanx proved inferior by comparison with the Roman method of arming their troops'.[46] The tactical system devised by Macedonian commanders had for over 150 years pervaded the eastern Mediterranean and Middle East. However, complacent disdain gradually set in during the Hellenistic Age and, as a consequence, the once celebrated potency of the Macedonian system suffered, becoming an ossified and lubberly caricature of the army led by Alexander. Rigid over-reliance was placed on the pike armed infantry component to the detriment of cavalry and other support. To borrow Count Raimondo Montecuccoli's metaphor (when he was writing of pike formations in seventeenth-century Europe), the Macedonian phalanx became the 'castle of the battlefield'.[47]

Following on from Antigonid and Seleucid defeats inflicted by the Romans at Cynoscephalae and Magnesia, Pydna proved to be the *terminus ad quem* of pike phalanx dominance.[48] It marked one of the most important clashes in ancient history: a major shift, whereby the legion definitely eclipsed the phalanx as the foremost tactical formation of the classical world.[49] The demoralising effect this defeat had on public and private debate among the Hellenistic governing classes and 'armchair' military intelligentsia of those days would have been enormous. They vainly indulged in disputing the relative merits and weaknesses of the two styles of warfare, attempting to explain how and why the phalanx broke at the hands of the legion, with the ever growing shadow of Rome looming ominously over them.[50] Hannibal, for one, despised purely academic theorists as fools.[51]

Before defeat meted out by Rome the Macedonian army was generally thought to be qualitatively the strongest military force found anywhere in 'civilised' Europe and the Middle East. Following their victory, the Romans thought that their universal hegemony was from now on, in Polybius' view, no longer in dispute.[52] The shamed and baffled Macedonians had to concur reluctantly. It must have been a bitter pill to swallow for such a combative people, used to the fear and admiration afforded them by others.[53] The calamity the Macedonians now experienced, with their ensuing abrupt exit from the world stage, was a timely reminder of the predictions of Demetrius of Phalerum, who in his lost treatise on Fortune written soon after Alexander's death addressed his readers thus:

> I ask you, do you think that fifty years ago either the Persians and the Persian king or the Macedonians and the king of Macedon, if some god had foretold the future to them, would ever have believed that at the time we live, the very name of the Persians would have perished utterly – the Persians who were masters of almost the whole world – and that the Macedonians, whose name was formerly almost unknown, would be the lords of it all? But nevertheless this Fortune ... by endowing the Macedonians with the whole wealth of Persia, she has but lent them these blessings until she decides to deal differently with them.[54]

The presumptuous descendants of Philip and Alexander's 'conquest state'[55] were destined to learn the harsh lessons of *hubris*, dispatched by western 'barbarians', in the long run more bellicose than they were. Thus, the vigorous power created under the auspices of Philip II ended its history as it was begun: entirely dependent on the strength of the army and the keen points of its spears.

At a single stroke half the royal army was exterminated at Pydna and the ability to continue the war crippled. The blow to such a small population was irreversible, spelling doom before the many-headed Hydra of Latin manpower.[56] Still, a broader survey of events shows that the obdurate Macedonians would only slowly yield to the Roman heel: the full process of subjugation lasted about seventy-three years, with four wars and at least two insurrections (215–142 BC).[57]

In the aftermath of Pydna, the Macedonians found the proverbial rug pulled from under their collective feet, overcome by a fast developing Roman hyperpower. These were the new and painful realities. The Roman imperial army may have eventually offered a viable outlet for some Macedonians to continue martial pursuits, similar to ethnic Sikhs or Highland Scots serving in regiments of the British Empire.[58] Even so, Macedonia's preeminent position was usurped forever. With the army-kingdom brought low with defeat, the Roman Senate now moved to strip it of all independent action. Its national monarchy was abolished (after five centuries of existence), officer class deported to Italy, infrastructure demilitarised and dismantled, and the country divided into four autonomous tribute-paying republics. There was to be no common currency or trade and even marriage was forbidden between the strictly separate sectors. All this retarded unity and economic recovery, negating any serious resurgence. Outraged Macedonians complained that their carved up land resembled a dismembered beast.[59] By 148 or 146 BC the kingdom had been officially incorporated as a province of the Roman Empire.

Following Arnold Toynbee's proposal of history occasionally repeating itself 'up to date in a significant sense', Echols claims to see a parallel between the Allied handling of Germany (and end of Prussia) in the aftermath of the Second World War and, on an obviously much reduced scale, the dissolution of Macedon as a unified concern.[60] Whether an acceptable parallel or not, Rome's actions certainly speak volumes for its determination not to be impeded by 'peer-competitor' threats within a newly expanded sphere of influence.[61] The sheer thoroughness with which Rome went about eradicating Philip II's legacy is testament to how organised and dangerous it had been or potentially could be. By the end of the first century AD a rather underwhelmed Dio Chrysostom could report that the once royal capital of Pella, the birthplace of Alexander the Great, appeared to have vanished entirely with only a pathetic scattering of pottery shards to remind curious passersby that it was ever there.[62] Macedonia's fate stood as a dire warning to any other militarily confident people who dared to oppose Roman diktat.

From the discovery of inscriptions it is poignant that following the downfall of the kingdom, Macedonian prisoners fetched the highest prices at the booming slave-marts of Delos and Naupactus. This proves that their contemporaries valued them highly as captive labour. After all, soldiers and slaves are similar in that they are both required to be obedient, industrious and self-disciplined. The standard price for a slave of any nationality at the time was 3 or 4 minas. Macedonians take their place at the top with an average of 5¾ minas for men; Romans and other Italian males captured by Hannibal during the Second Punic War went for under 5¼ minas per head.[63]

Although defeated and conquered, the enviable Macedonian military reputation still lingered on as late as the first century AD. Josephus, for example, refers to them as a people of strong character and hints at the long-standing imprint they left behind.[64] He shows how at the siege of Jerusalem in 70 AD Antiochus IV of Commagene assailed the Jewish defended walls with a bodyguard of 'Macedonians', despite his derogatory aside that 'most of them lack any claim to belong to that race'. Antiochus led these men in a bold though vainglorious assault on the enemy position which saw many wounded and all captured. Afterwards the despondent survivors mused that 'even genuine Macedonians, if they are to conquer, must have Alexander's fortune'.[65] This sentence betrays feelings as to how a group of soldiers bearing the appellation 'Macedonian' were expected to behave whatever their true origins: that is, they should live up to such a fortuitous title by fighting bravely.[66]

When we review the Roman and Macedonian legacies, the Macedonian army can be arguably considered the second best of antiquity, though others may prefer that the terrifyingly efficient Assyrians occupy this space. As George Orwell put it, it is an inevitable truism that winners write history, while the vanquished or 'also-rans' are relegated to a back seat or are even forgotten. Yet, in the minds of Field Marshal Viscount Montgomery and other modern generals the army under Alexander formed the most tactically cohesive force the ancient world ever saw, and its successes must be regarded as hard to equal, let alone surpass, even with the more famous Roman army following in its wake.[67]

Appendix 1

Some Military Figures in Macedonian Funerary Art

Great Tumulus, Vergina: Grave *Stelai*

Hundreds of marble funerary monument fragments were recovered by archaeologist Manolis Andronikos from the Great Tumulus at Vergina.[1] These debris items were originally from the cemetery that surrounded the tumulus site, used to fill in the mound probably sometime after the pillaging of royal tombs at Aegae by Pyrrhus' Celtic mercenaries in 274/273 BC.[2]

According to Professor Chrysoula Saatsoglou-Paliadeli, the debris includes over fifty individual grave *stelai* spanning the fourth–early third centuries BC.[3] Almost all of them have patronyms giving the names of the deceased. Little known to many researchers is that a few of the *stelai* appear to show Macedonian soldiers or military features. As these rare finds complement the overall evidence given in this book, the author felt it necessary to provide a review of four of the more relevant *stelai* for the sake of completeness. Most information was compiled from Saatsoglou-Paliadeli's definitive Greek language work on the subject, of which only one copy is available in British academic libraries.

The first *stele* has the subject standing to the right in *kausia* with clearly-delineated hat band, tunic and *chlamys*, accompanied by a small child. The man's right hand extends in apparent farewell to a seated female figure to the left.[4]

The second *stele* consists of only one large fragment. It depicts a man standing or sitting in a heroic pose grasping the visible shaft of a spear in his right hand, with his right arm or shoulder and the right side of his upper chest all bare. To the immediate left of this figure is a non-crested Phrygian helmet, the top part – with its distinctive high medial cockscomb – clearly visible.[5]

The third *stele* has only the upper section surviving. The scene appears to incorporate a figure standing to the left in *kausia*, tunic and *chlamys*. The fragment cuts off, omitting the lower legs. Faded objects can be discerned on a shelf at the top including the hilt of a *xiphos* and a possible helmet on the far right side.[6]

The last *stele* consists of a single large fragment. On the left is a seated man of whom only the tunic and legs can be made out. He holds a spear or javelin and his footwear consists of shin or calf high socks and *krepides*, the left of which still clearly preserves details of strap netting. A figure stands before him on the right wearing an above knee length tunic, a possible *kausia* and *chlamys* with socks and *krepides*.[7]

Many of these *stelai* were painted. Preserved pigments have been described overall as red, blue, purple, yellow, green and black.[8] The author had the opportunity to examine personally colour pigments of two additional *stelai* from the same tumulus site, which are included in Appendix 2 under catalogue nos. 5 and 6.

Bella Tomb II, Vergina: Painting

The three tombs covered by the Bella Tumulus at Vergina were originally excavated by Andronikos (1981/1982). The second tomb (dated 300–275 BC) comprises one chamber measuring 3m × 3.5m with a height of almost 5m. Although it had been looted, this was to some extent compensated by an intact painting located above the marble doorway which had been treated with a thin coat of white plaster helping to preserve its colour pigments. The composition consists of an individual in full battle panoply flanked on each side by a deity (Plate 5).[9]

The youthful, clean shaven man at the centre represents the heroised tomb occupant (Fig. 50). He stands in triumphant pose gazing into the distance, with his body turned to the viewer, weight supported by the right leg and upright spear. The stance is remindful of *doriktetos chora* – the concept and propaganda of 'spear won land'.[10]

The central figure may be interpreted as an officer of Guard or Companion cavalry. He is depicted in a pale blue tunic over which is a white corselet with light pink-purple or violet shoulder yoke edged in red; a vague ochre-gold coloured Gorgon's head ornament is located on the upper chest. Laces are discerned tied to the front of the shoulder pieces with an apparent ochre (or gold) tinted object, barely seen at the end of the visible left guard. A smaller gold object with a red strand is located lower down on the left side. The hips and midriff are obscured by a scarlet *himation*, and the corselet can be seen to integrate two layers of white *pteryges*, the lower row (or both) with a single thin line of red trim. Also apparent are a pair of high, dark brown boots (*embades*) enclosing the foot and calf. Specific details can still be made out, such as scalloped leather turnovers and long dangling laces.[11]

Figure 50. Soldier in heroic pose from Bella Tomb II.

The upright spear shaft in the soldier's right hand has a butt on the reverse end portrayed as metal (bronze?) picked out using a darker hue. As suggested to the author,[12] the object appears virtually identical in symmetry and design to a surviving Macedonian (cavalry) butt preserved at the Shefton Museum.[13]

To the right of the soldier, there is a tall, slender and richly dressed female figure, shown extending her right arm and offering a yellow-gold pigmented wreath to the soldier. The woman could be an allegorical personification of Macedonia or symbolic of the deceased's *arete*. The wreath may even suggest the award of a posthumous military award.[14]

On the opposite side, to the left of the soldier, is a young man seated in profile upon a trophy of bronze shields, with another shield presented full face (Fig. 51). The figure wears a faded purple *himation*, with left hand resting upon a sheathed sword with white hilt and brown scabbard. The man could be Ares or a deified Alexander. The painted shield almost certainly belongs to the tomb occupant, the conspicuous manner of its appearance and the striking emblem signifying membership of a prestigious unit.[15] The shield face is proudly displayed, clearly meant to be seen by the onlooker. It is painted as bronze with a central orange-red disc within which is an ambiguous device done in yellow-gold: an eagle in the act of stretching its wings, gazing to the right with a thunderbolt grasped in its talons. Its identification closely resembles a miniature shield ornament found in Alexandria (dated to the third century BC), the eagle with thunderbolt motif becoming an established Ptolemaic dynastic emblem.[16]

Macedonian Combat Scenes in Hellenistic Art

Macedonian soldiers in combat scenes were popular subject matter for Hellenistic artists. Even before Alexander, this genre was beginning to develop at Philip II's court, as seen from a decoration on a gold and ivory couch found in the antechamber of Philip's supposed tomb at Vergina. The painstakingly reconstructed

Figure 51. Figure of a young man left of the central soldier from Bella Tomb II.

fragments (now on display at the site museum) revealed a number of battling warriors. Gilded reliefs from another couch were found at Stavroupolis. These are exhibited at the Archaeological Museum in Thessaloniki and feature several 'hoplite' figures duelling with Amazons.

Other than the more famous combat scenes (e.g. Alexander Mosaic and Sarcophagus), Pliny lists a series of battle paintings famous in his own time for displaying Alexander's officers. In one extract the artist Apelles of Cos (fl. 340–300 BC) is described having completed:

> A *Cleitus*[17] with his horse hastening into battle, and an armour-bearer holding out a helmet to a man who had asked for it. How many times he painted Alexander and Philip would be superfluous to enumerate ... He also made a *Neoptolemus* fighting against Persians from horseback ... and an *Antigonus* [Monophthalmus] armed with a breastplate and marching along with his horse. Those who are connoisseurs of the art prefer before all his other works a portrait of the same king seated on a horse.[18]

Another painter from the late-fourth century BC called Protogenes of Rhodes likewise produced a military portrait entitled *King Antigonus.*[19]

In the mid-third century BC, according to an extant Athenian inscription, one Herakleitos of Athmonon, a citizen in the service of Antigonus Gonatas and acting general of the Macedonian garrison at Piraeus, honoured Antigonus by commissioning either a written historical narrative or a series of paintings for a stadium at Athens. If we can assume the painted form, it portrayed the King defeating barbarians who were probably Celtic invaders (278–277 BC).[20] As conveyed by Rice, 'Presumably the paintings were historical in character and depicted his specific exploits; there would not be much point to the honour otherwise'.[21]

Centuries later travellers in different parts of the world were still able to view military monuments with combat scenes from the period of Alexander and the Successors. In his account of the life of the neo-Pythagorean sage Apollonius of Tyana (first century AD), Philostratus describes a particularly impressive and striking series of battle scenes featuring Macedonians viewed on a temple in Taxila in which 'tablets of bronze were let into each wall, depicting the feats of Porus and Alexander: elephants, horses, soldiers, helmets, and spears were represented in orichalque [orchicalcum], and silver and gold, and black bronze; lances, javelins, and swords, all in iron'.[22] Similarly, in the second century AD Pausanias saw a memorial to King Pyrrhus at Argos integrating a panorama 'carved in relief' which displayed 'the elephants and his other instruments of warfare'.[23]

Tomb of Judgement, Lefkadia: Battle Frieze

The so-called 'Tomb of Judgement' was found by chance in 1954 during the widening of a country road linking the town of Mieza with Pella. The site was subsequently investigated by archaeologist Photios Petsas. The striking facade measures 8.60 × 8.68m, and the overall structure dated from 325–300 BC or 300–275 BC.

In the upper storey of the facade, located below a row of eight Doric half-columns and above an entablature with scenes displaying a combat of centaurs, can be viewed a stucco relief battle frieze with a cavalry and infantry combat between Macedonians and Persians (Plate 6). For a photographic detail of the frieze, see Plate 7. This combat may represent an actual incident from the tomb occupant's life as a soldier who took part in Alexander's eastern wars.

The miniature relief figures were cut and sculpted using lime mortar attached to the background wall with nails. In their original eye-catching condition they numbered as many as seventeen individuals. Some of them appear to be in groups. The far right figure on the frieze is identified as a possible Macedonian cavalryman on foot in a tunic with traces of red, a yellow-brown cloak and boots retaining some red pigment. He is lunging at a Persian on the left, dressed in a red tunic and pale green trousers, defending himself with a round shield which has a red interior. This Persian soldier appears the best preserved of all the figures. The next group of combatants can be described as a Macedonian on a horse, with cloak and some red pigment on the saddlecloth, and a Persian lying wounded underneath him. Next to them is seen a possible Macedonian cavalryman on foot in a draped cloak of a grey-brown pigment. Yet another group is revealed as a Macedonian on a rearing mount, hurling a javelin at a Persian on his knees, with left arm raised. The Persian is wearing a tunic which has a patch of pale blue.

The central area of the frieze has unfortunately been damaged, but to the left of this largely erased section we can still view a Persian in red trousers on foot with his left arm out to one side, presumably fighting a Macedonian cavalryman whose figure is now gone, with only one leg of his horse visible. To the Persian's left is observed a group of fighting soldiers consisting of a Macedonian on a rearing horse and a Persian wearing a cloak and trousers, on a wounded mount. The next group of combatants is too fragmentary to make out individual figures and only a single Persian on foot (the far left figure on the frieze) wearing trousers and holding a shield with a flecked red design on its interior can still be recognised.

The emphasis on cavalrymen among the Macedonian figures is significant, implying that the tomb occupant was a member of a cavalry unit, or usually fought mounted.

It is interesting to note how the scenes from the 'Tomb of Judgement' frieze are in some ways akin to those executed on the Alexander Sarcophagus. Painted sculptures of Macedonians doing battle with Persians therefore became a fashionable Greek artistic theme in funerary art, especially during the late-fourth to early-third centuries BC. Diodorus recounts how as early as 324 BC Hephaestion's funeral pyre had a display of 'Macedonian and Persian arms, testifying to the prowess of the one people and to the defeats of the other'.[24] According to Diodorus, Alexander's catafalque similarly depicted 'groups of armed attendants, one of Macedonians, a second of Persians of the bodyguard, and armed soldiers in front of them'. This extract goes on to list 'elephants arrayed for war who followed the bodyguard' and 'troops of cavalry as if in formation for battle'.[25]

Tomb of Judgement, Lefkadia: Portrait of Soldier

At the ground level of the facade, between the lower Doric columns, there are four painted panels which represent the 'judgement of the dead' in Hades. The far left panel incorporates a portrait of the tomb's soldier occupant wearing a corselet and holding a spear (Plate 8). For a description of the soldier's clothing, see Appendix 2, catalogue no. 22. Some intriguing deductions have been made for this man, including the fact that he is not shown wearing a helmet or carrying a shield, suggesting that he did not die in battle. It has even been mooted as a possible likeness (and hence final resting place) of Alexander's officer and royal shield-bearer Peucestas of Mieza.

The figures in the remaining three panels (left to right) are identified as Hermes Psychopompos (Escorter of Souls) and two Underworld judges; namely, Aeacus and Rhadamanthys. According to Plato, Aeacus judged the souls from Europe and Rhadamanthys those from Asia.[26] Their depiction here may be an allusion to the deceased having lived out his life in both Europe and Asia, again intimating an officer in Alexander's army.

The military background of the tomb occupant is further stressed with the discovery of two circular relief shields (diameter: 0.78m), coloured predominantly white and symbolising trophies of victory found on each side of the door in the dividing wall leading to the burial chamber.[27]

Appendix 2

Evidence for Cavalry Clothing Colours

Both literary and artistic evidence suggest a distinct 'imperial' phase of Macedonian cavalry clothing apparent in the latter years of Alexander's reign (330–323 BC), which combined an expensive purple and similar tinted, often long-sleeved Persian, tunic worn with a saffron-coloured cloak bordered purple. Despite Alexander's sartorial ambitions for his troops, many found the new fashions distasteful. The soldiers complained that they could not face their families and neighbours back home, as Curtius puts it, 'dressed like captives' (meaning the Persians they had conquered).[1] Elements of the cavalry in particular found Iranian garments anathema, even though they did not dare refuse any of the articles impressed upon them.[2] Craterus was a spokesman for the army's grievances in these matters as he opposed the introduction of Persian influence and supported Macedonian customs.[3]

In a philosophical discourse 'On the Fortune or Virtue of Alexander', Plutarch goes on to pour scorn on the narrow minded inability of some to adapt to a different mode of garb:

> Conversely it were the mark of an unwise and vainglorious mind to admire greatly a cloak of uniform colour and to be displeased by a tunic with a purple border, or again to disdain those things and to be struck with admiration for these, holding stubbornly, in the manner of an unreasoning child, to the raiment in which the custom of his country, like a nurse, had attired him.[4]

In spite of being known for his traditional values, Craterus eventually wore purple garments on his return to Macedonia (as described in Chapter 8). In the eyes of contemporaries, his inspiring appearance contrasted favourably to the modest and diminutive Antipater, whose unprepossessing appearance harkened back to an earlier and poorer age.[5] Antipater's sober and modest attire may hint that plain colours were more evident among Philip II's officers and perhaps army.[6] According to Plutarch, Theophrastus described Philip as a king more free of personal luxury than others of his time.[7] Antipater remained in Europe and was one of Philip's oldest generals, whereas Craterus belonged to Alexander's promising younger generation and accompanied him throughout Asia.

The disparate threads of evidence presented in the catalogue below reveal certain patterns and consistencies for cloak colours in the years after Alexander embracing a wide spectrum from pale yellow through shades of brown to red. We also find Macedonian cavalrymen wearing tunics in mainly shades of red or bright blue and occasionally white.

These findings independently concur with those of Post, who with more of an emphasis on Ptolemaic evidence outlined the popularity of red, then white, blue and yellow among forty-six examples of tunic worn by Hellenistic infantry and cavalry soldiers from the third to first centuries BC. Most common cloak colours from twenty-eight surveyed were recorded by him as red, brown, yellow and pink.[8] There are in fact *stelai* from Alexandria possibly dated to the third century BC (held at the Louvre) which feature soldiers with tunics depicting white or yellow traces, while their cloaks appear with hints of yellow, brown or ochre-red.[9]

It is tempting to consider that in the armies of Alexander and the Successors cavalry colours were indicators of rank or squadron. But this may be stretching credulity more than it can bear, unnecessarily burdening the evidence with too great a degree of modern assumption. The fact that five different tunic colours are worn by six figures identified from the Agios Athanasios tomb as possible cavalryman colleagues (cat. nos. 16–21) suggests that colours were sometimes an individual's choice, though within certain culturally prescribed parameters.

In compiling a catalogue of extant evidence for Macedonian cavalry clothing colours in the armies of Alexander and the Successors, the selection has been entirely the author's own, based on a balance of personal interpretation as to what may be defined as horsemen, allied to observations made by those academics who have studied the artefacts firsthand. A surprising abundance of archaeological material survives depicting Macedonian and Hellenistic soldiers, but unfortunately these valuable fragments have been dispersed in public and private collections around the world and remain largely neglected or inaccessible. Griffith recognised the problem as early as 1935.[10] The same opinion was upheld by Snodgrass thirty years later when he expressed the hope that all pieces be systematically collated and made available to an interested readership.[11] The first significant and valuable steps in achieving this goal were made by Post in 2010.[12]

Before any attempted examination it is helpful to first consider the following cautionary points:

- Colour terminology used in the literary extracts (cat. nos. 2–4) requires some explanation as different nuances may be attached to interpreting original Greek and Latin expressions. For instance, according to Liddell and Scott's *Greek-English Lexicon*, the term *porphyrous* can be defined as 'a distinct colour, dark red, purple or crimson', whereas the Latin term *purpureus* is occasionally translated as 'red', 'rose-red' or 'violet'.[13]
- Existent artistic material outlined here may represent not only Companions, but a variety of mounted units, including off-duty cavalrymen and horsemen in civilian or hunting attire unrelated to military activities (examples include cat. nos. 5, 7, 9, 12–15, 23, 24, 30).
- Artistic fragments cover a wide geographic span from Greece to Egypt. Thus, colours popular in the Ptolemaic Egyptian army (cat. nos. 11, 26–29, 34–41) are not necessarily commensurate with those worn in the armies of Alexander and the Successors, although, given that the Ptolemies were one of the chief

inheritors of a debased Macedonian military tradition, it is entirely possible that they were.

- Some pigments were only used for shading rather than as a means of showing the primary clothing colour.[14] Moreover, colours could be selected for use by the artist solely as a matter of his own or his patron's personal taste, preference or imagination, rather than on an accurate basis of clothing dyes worn in real life. Paint colours were restricted by what the artist happened to have at hand or was able to afford, or by a particular artistic creed or aesthetic which required only a specific range of pigments. This last factor is best understood from the Alexander Mosaic (cat. no. 10), the artist of the original painting upon which the mosaic was based limiting himself to a four-colour palette comprising red, black, white, and yellow.[15] In a mosaic from Palermo (cat. no. 13) the artist followed a similar combination of colours. Even though this raises a question about historical accuracy, such artefacts still provide potentially valuable data for Macedonian dress.
- Faded or damaged colours increase the risk of misinterpretation or over-analysis. Hence, the same figure on any given artefact can be described in a variety of divergent colours or at least subtle colour variations, as differences in light and photography may have an impact on the individual conclusions drawn by research scholars. When confronted with this problem, it is useful to select the first account as the most reliable (as surviving colour pigments would be freshest at time of discovery). These accounts are especially relevant if based on the researcher's close *personal* scrutiny of the artefact in question. Modern scientific investigation utilising advanced forms of technology such as optical microscopy (OM), electron microprobe (EMP) or X-ray diffraction methods can potentially give us an even more accurate appraisal.[16]

Catalogue of Evidence

Cat. No.	*Source Type*	*Provenance*	*Date/Period*	*General Description*	*Tunic*	*Cloak*
1	literary passage	Curtius, 6.6.7	c. 330 BC	on Alexander's attempts to re-clothe Macedonian army units	'He compelled ... the cavalry ... to wear the Persian dress'.	
2	literary passage	Justin, 12.3.9	c. 330 BC	on Alexander's attempts to be accepted by his Asian subjects	'In order to avoid excessive animosity ... he also instructed his friends to wear long gold and purple robes'.	
3	literary passage	Diodorus, 17.77.5	c. 329 BC	on Alexander's distribution of cloaks to the Companions	'He distributed to his Companions cloaks with purple borders'.	
4	literary passage	Athenaeus, 12.539f–540a	324 BC?	on Alexander ordering purple dye from Chios and Ionia	'For he wanted to dress all his friends [*hetairoi*] in garments dyed with sea purple'.	
5	grave stele	Great Tumulus, Vergina	350–275 BC	Bareheaded soldier or huntsman standing with a javelin, wears tunic, *chlamys* and *krepides*[17]	blue tinted with a vertical red stripe down one side (author's personal observation)	yellowish-brown with a broad crimson or purple border
6	grave stele	Great Tumulus, Vergina	350–275 BC	dismounted cavalryman leading a horse, wears a crested helmet, a possible corselet, tunic, *chlamys* and *krepides*[18]	russet or red pigment (author's personal observation)	brownish
7	tomb-painting	Royal Tomb II, Vergina	336–335 BC? or 316 BC?	dismounted cavalryman? hurling a javelin (second from the right), naked but for *kausia*, *chlamys* and *krepides*[19]	N/A	brown with scattered traces of a faint reddish pigment on the lower end
8	sculpture	Alexander Sarcophagus, Istanbul	330–310 BC	mounted cavalryman in Boeotian helmet, arm raised to strike a Persian on foot	medium purple	gold-yellow with medium purple border

Cat. No.	*Source Type*	*Provenance*	*Date/Period*	*General Description*	*Tunic*	*Cloak*
9	sculpture	Alexander Sarcophagus, Istanbul	330–310 BC	mounted cavalryman hunting with Persians[20]	medium purple	gold-yellow with medium purple border
10	mosaic	Alexander Mosaic, House of the Faun, Pompeii	c. 100 BC, possibly based on an original painting from 330–300 BC?	two mounted Macedonian cavalrymen to the left of Alexander (fragments)[21]	red or red-brown tunic sleeve (far left figure)	indeterminate colour for cloak edge of figure to the right; possibly some shade of pinkish, red or red-brown
11	grave stele	Shatbi Cemetery, Alexandria	325–300 BC	mounted cavalryman on a rearing horse, with boy-servant running behind, holding the tail[22]	described as white or greenish – early records even refer to or depict traces of sky blue	although sometimes described as yellow-green, appears yellow in modern colour reproductions; crimson-purple border
12	wall painting	House of Jason, Pompeii	1–25 AD, possibly based on an original from 325–300 BC	young man leading a bull to sacrifice, wears a green wreath, long sleeved tunic with *chlamys* and high boots[23]	pale blue	saffron-yellow coloured with a broad light-mid purple border
13	mosaic	Piazza della Vittoria, Palermo	125–75 BC, possibly based on an original painting 325–275 BC?	horseman (from the upper right hand register), wears a tunic, *chlamys* and *krepides* with a spear and a possible javelin in his rein hand[24]	pale pinkish	light yellow
14	tomb-painting	Kazanluk tomb (western frieze), Bulgaria	325–275 BC	two mounted figures on the far left of the frieze tentatively identified as Macedonian; both armed with spears and wearing crested helmets with tunics, *chlamydes* and boots	red or possibly rose	sky blue

Cat. No.	*Source Type*	*Provenance*	*Date/Period*	*General Description*	*Tunic*	*Cloak*
15	tomb-painting	Kazanluk tomb (eastern frieze), Bulgaria	325–275 BC	three putative mounted Macedonians (from left to right on the frieze); one bareheaded, one in possible *kausia* and another in helmet; all three wear tunic, *chlamys* and Thracian pointed shoes[25]	one red and two sky blue (from left to right)	two sky blue and one red (from left to right)
16	tomb-painting	Agios Athanasios, near Thessaloniki	325–275 BC	dismounted cavalryman (to the right of a banquet scene) with *kausia*, corselet, *krepides* and spear	pink-red	orange-yellow bordered purple
17	tomb-painting	Agios Athanasios, near Thessaloniki	325–275 BC	dismounted cavalryman (to the right of a banquet scene) with *kausia*, corselet, *krepides* and spear	bright blue	crimson bordered with purple
18	tomb-painting	Agios Athanasios, near Thessaloniki	325–275 BC	dismounted cavalryman (to the right of a banquet scene) with *kausia*, corselet, *krepides* and spear	white	orange-yellow bordered purple
19	tomb-painting	Agios Athanasios, near Thessaloniki	325–275 BC	dismounted cavalryman (to the right of a banquet scene), unarmoured with *kausia*, *krepides* and spear	bright blue	orange-yellow bordered purple
20	tomb-painting	Doorway, Agios Athanasios, near Thessaloniki	325–275 BC	dismounted cavalryman? (to the left of the doorway) armed with long spear, wearing *kausia*, tunic, *chlamys* and boots	light blue or grey-mauve with a single, narrow vertical white stripe down the side	light pink-red with very broad white border

Cat. No.	*Source Type*	*Provenance*	*Date/Period*	*General Description*	*Tunic*	*Cloak*
21	tomb-painting	Doorway, Agios Athanasios, near Thessaloniki	325–275 BC	dismounted cavalryman? (to the right of the doorway) armed with long spear, wearing *kausia*, tunic, *chlamys* and boots[26]	brown-coloured with a wide vertical white stripe	dark brown with a very broad white border
22	tomb-painting	Tomb of Judgement, Lefkadia	325–275 BC	dismounted cavalryman? bareheaded, in corselet with boots and spear	mid-red	light brown with purple-red border
23	stucco relief battle frieze	Tomb of Judgement, Lefkadia	325–275 BC	dismounted cavalryman? (portrayed between a horseman and a wounded Persian)	N/A	dark charcoal or grey-brown pigment (perhaps only the fabric of the stucco material)
24	stucco relief battle frieze	Tomb of Judgement, Lefkadia	325–275 BC	dismounted cavalryman? (on extreme right of frieze)[27]	traces of red	yellowish-brown
25	tomb-painting	Bella Tomb II, Vergina	325–275 BC	dismounted cavalryman? in corselet with boots and spear, bareheaded[28]	pale blue	scarlet-red
26	grave stele	Alexandria	320–275 BC	dismounted cavalryman? in long sleeved tunic, *kausia*, *chlamys* and *krepides*[29]	includes elements of yellow	described as yellow, although colour reproductions show it as obviously red with a very broad border rendered pale blue
27	tomb-painting	Mustafa Pasha Tomb I, Alexandria	300–275 BC	Mounted man wearing a helmet and cuirass pouring libations at an altar	red-violet	yellow
28	tomb-painting	Mustafa Pasha Tomb I, Alexandria	300–275 BC	Mounted man wearing a *kausia* and cuirass pouring libations at an altar	pink	N/A

Cat. No.	*Source Type*	*Provenance*	*Date/Period*	*General Description*	*Tunic*	*Cloak*
29	tomb-painting	Mustafa Pasha Tomb I, Alexandria	300–275 BC	Mounted man wearing a *petasos* hat and cuirass pouring libations at an altar[30]	yellowish	appears to be dark coloured
30	tomb-painting	Kinch Tomb, near Naoussa	325–275 BC? re-evaluated c. 250 BC	mounted cavalryman in helmet with long spear charging from the left, attacking Persian[31]	pale blue-grey upper chest section, pale yellow sleeves, pink-red mid chest, pale blue-grey under tunic with dark blue-grey band	pinkish-red
31	marble bed decoration	Soteriades Tomb, Dion	320–250 BC	Macedonian on far right wearing a tunic with his arm raised[32]	pale blue, possibly worn on top of a light yellow under-tunic	colours described as faded
32	literary passage	Plutarch, Demetrius 44.6	288 BC	on the escape of Demetrius Poliorcetes from the Macedonian camp when campaigning against Pyrrhus		disguised himself in a 'dark', 'black' or 'shabby' soldier's (presumably cavalry) cloak
33	literary passage	Dionysius, 19.12.6	280 BC	on Pyrrhus exchanging his royal cloak at Heraclea for that of Megacles, one of his 'most faithful ... Companions'		dull brown cloak
34	loculus slab	Soldiers' Tomb, Ibrahimiya Cemetery, Alexandria	third century BC	dismounted cavalryman? in cuirass, spear in right hand, large round shield on ground to the left, bareheaded[33]	white	dark crimson or blood-red

Cat. No.	*Source Type*	*Provenance*	*Date/Period*	*General Description*	*Tunic*	*Cloak*
35	loculus slab	Soldiers' Tomb, Ibrahimiya Cemetery, Alexandria	third century BC	dismounted cavalryman? with spear in right hand and round shield resting at feet, bareheaded[34]	light ochre	dark yellow-ochre
36	grave stele	Hadra Cemetery, Alexandria	third century BC	mounted cavalryman on rearing horse; servant approaching from left, taking helmet; *kausia* (or bareheaded?), cuirass and spear[35]	dark red or brownish red	dull yellowish
37	grave stele	Hadra Cemetery, Alexandria	third century BC	dismounted cavalryman? with two boy-servants; cuirass, with staff and banner, shield behind feet, bareheaded[36]	white	light brown
38	grave stele	Alexandria	third century BC	dismounted cavalryman? in cuirass with *pteryges*, bareheaded, handed plumed helmet by boy-servant[37]	red? otherwise not visible	described sometimes as greenish; a colour reproduction appears to depict greyish-brown (perhaps with a very broad white border)
39	grave stele	Shatbi Cemetery, Alexandria	third century BC	dismounted cavalryman? in cuirass and boots, hurling javelin in the right hand, bareheaded[38]	blue with brown traces in the lower part	red
40	grave stele	Cemetery of al-Manara, Alexandria	third century BC	mounted horseman, portrayed in profile in belt and hat (or 'conical helmet'), with boy-servant[39]	N/A	perhaps violet

Cat. No.	*Source Type*	*Provenance*	*Date/Period*	*General Description*	*Tunic*	*Cloak*
41	grave stele	Soldiers' Tomb, Ibrahimiya Cemetery, Alexandria	third century BC	dismounted cavalryman, in tunic and *pilos* hat attending rearing horse with servant[40]	yellowish-white	N/A
42	grave stele	Pagasai-Demetrias, Volos	third century BC?	dismounted cavalryman? with close-cropped hair, bareheaded, wearing a corselet with *pteryges* and *krepides*[41]	barely recognisable – perhaps originally pinkish, red or purple? A white sash is tied over the tunic	an obscure pale brown with possible faint traces of a red border at the end
43	tomb-painting	Amphipolis	third century BC?	dismounted cavalryman? wearing a corselet with shoulder pieces and at least one layer of *pteryges*, shin high boots, bareheaded	no colour pigments clearly discerned from photographs, though occasional dark blue patches where the tunic should be (author's personal observation)	N/A

Notes

Preface

1. Hammond, 1991a, 205.
2. Sekunda, 2001c, 40. For a sampling of theses see Nutt, 1993; Lopez, 1997; Boardman, 1999; English, 2003.
3. Bose, 2004; Alonso-Nunez, 2003, 175; Hanson, 1999a.
4. Hanson, 1999a, 405.
5. Baker, 2005, 373.
6. Griffith, 1935, 322; n. 2; Snodgrass, 1967, 129.
7. Nash, 1987, 216.

Part I: Origins and Perspectives

Chapter 1: The Macedonian Army's Place in History

1. Lonsdale, 2004, 15–17; 40; 52; Lonsdale, 2007, 38–44.
2. Hanson, 1995, 43; Chaniotis, 2005, 78.
3. Tarn, 1930, 43–44.
4. Scholarly analogies drawn between Macedonia and Prussia, including their military leaders, are not infrequent and date back to the works of Gillies (1789) and Yonge (1858), continuing unbroken to the present – see Billows, 1995.
5. White, 1996, 63–64.
6. Edson, 1970, 44; Fuller, 1954, 84; Billows, 1995, 16; 18; Borza, 1968, 9.
7. Ellis, 1976, 28; Borza, 1990, 282.
8. Ellis, 1977, 103–104; Cawkwell, 1978, 15; 27; Carlton, 2001, 56; Hatzopoulos, 1996, 267. Macedonia is definable as both militaristic and a 'garrison-state' – see Lasswell, 1941, 455–468.
9. Warfare has often played a fundamental role in the forging of group consciousness throughout history, seen for example with the Mongols, Marathas and Zulus.
10. Ellis, 1977, 103–104; Worthington, 2004, 19; 29; Austin, 1995, 205–206; Buckler, 1996; Bloedow, 2003; Brosius, 2003, 227–238.
11. Heinl, 1966, 257.
12. Gabriel, 2010, 7.
13. Demosthenes, *On the Chersonese*, 11.
14. Hammond and Walbank, 1988, 87; Hammond, 1989c, 193; White, 1996, 63–64.
15. Appian, *History*, Preface 10.
16. Gabriel, 2005a, 297; Justin, 9.8.4–5; 9.8.21.
17. Justin, 13.1.12: 'Never before, indeed, did Macedonia, or any other country, abound with such a multitude of distinguished men … Who then can wonder, that the world was conquered by such officers?'
18. Hammond, 1989b, 63.
19. Thomas, 2007, 45; Hammond, 1991a, 205.
20. Turchin & Adams, 2006, 223.
21. Lonsdale, 2004, 198.
22. Hanson, 1999b, 178; Hanson, 2001, 74–90.
23. The best recent study on Iranian influence is Olbrycht, 2004, 77–204.
24. Arrian, 7.1.4.
25. Livy, 9.17.1–19.17; 9.19.1–8; Morello, 2002; Ligeti, 2008.

26. Arrian, 7.1.4.
27. Lamb, 1946, 381.
28. Arrian, 7.26.3; Curtius, 10.5.5; Diodorus, 17.117.4; 18.1.4; Justin, 12.15.8.
29. Demetrius, *On Style*, 284; Plutarch, *Moralia*, 181f; 336f; attributed to Leosthenes.
30. Bennett & Roberts, 2009, xiii.
31. Gaebel, 2002, 199; Hornblower, 1981, 223.
32. Plutarch, *Pyrrhus*, 12.2.
33. Grainger, 2007, 189–193.
34. Billows, 1995, 146–182. For a review of questions relating to migration, see Hamilton, 1999b.
35. Investigated by Marrinan, 1998. See also Blasius, 2001.
36. Ottoni, 2011.

Chapter 2: Transmission of Military Knowledge

1. For the army as the chief recipient of state finances, see Diodorus, 16.8.6; Griffith, 1956; Griffith, 1965, 127–129; n. 4.
2. Noguera Borel, 2007.
3. Diodorus, 16.3.2–3; Polybius, 18.28.6–7; Homer, *Iliad*, 13.130–33; Wheeler, 1988, 180; Blumberg, 2009; Aelian, *Tactics*, 1.
4. Aristophanes, *Frogs*, 1034–1036.
5. Gabriel, 2000, 96; Gabriel, 2005, 287; Borza, 1999; Gaebel, 2002, 7; 149; 157–158.
6. Anson, 2010b, 58; Worthington, 2008, 19–20; Hammond, 1994, 26.
7. Polybius, 11.8.1–2.
8. Gaebel, 2002, 150.
9. Aelian, *Varia Historia*, 4.19; Aulus Gellius, 9.3.1–3.
10. Heckel, 1992, 13–23; 38–49; Heckel, 2006c, 35–38; 190–192; Plutarch, *Moralia*, 177c; 179b; Plutarch, *Phocion*, 29.2; Curtius, 4.13.4; 7.2.33; Athenaeus, 10.435; *Suda* s.v. *Krateros*; s.v. *Antipatros*; Kanatsoulis, 1958/59; Baynham, 1994.
11. Diogenes Laertius, 9.48; Delebecque, 1957, 242–245; Burliga, 2008, 94.
12. Whitehead, 2001, 13–16; Tejada, 2004, 142; Burliga, 2008.
13. Justin, 6.9.7; 7.5.2–3; Plutarch, *Pelopidas*, 26.5; Diodorus, 16.2.2–4; Gabriel, 2010, 25; Hammond, 1997c; Pritchett, 1974, 90–92.
14. Athenaeus, 1.3a.
15. Polybius, 3.6.10–14.
16. Arrian, 2.7.1
17. Christesen, 2006, 63; Gaebel, 2002, 150.
18. Cicero, *Letters to his Brother Quintus*, 1.1.23; Cicero, *Tusculan Disputations*, 2.26; Cicero, *Letters to Friends*, 9.25.1; Suetonius, *Julius Caesar*, 87; Pease, 1933/34; Farber, 1979, 497–574.
19. Renault, 1975, 69–70.
20. McGroarty, 2006, 105–124.
21. Plutarch, *Alexander*, 4.4; 8.1; Brown, 1967.
22. Polybius, 12.22.2–3: Justin, 9.8.18, for father and son's mutual interest in literature.
23. *Suda* s.v. *Marsyas*; Heckel, 1980.
24. Nefedkin, 2000a.
25. Plutarch, *Alexander*, 8.1; 26.2; Plutarch, *Moralia*, 327f–328a.
26. Dio Chrysostom, 2.3–6; Pearson, 1960, 10; 46.
27. Athenaeus, 4.167a–c; Demosthenes, *Second Olynthiac*, 18–20. For contrast, see Polybius, 8.9.6–13. For Artabazus, Sisines, Menapis and also Mentor and Memnon of Rhodes at Philip's court, see Engels, 1980, 328–329 and n. 10.
28. Xenophon, *Memorabilia*, 3.1.1–11.
29. Xenophon, *Anabasis*, 2.1.7; Plutarch, *Agesilaus*, 26.2–3.
30. Pritchett, 1971, 152: Diodorus, 16.3.2, for Homer's *Iliad* as the original source.
31. Jacoby, 1923, 115 F 81.
32. Webster, 1953, 43; from the play *Philip*, 441/7–10.

33. Tarn, 1930, 44. Macedonian grave epigrams often proclaim military *arete* in general – see Saatsoglou-Paliadeli, 1984, no. 6, pl. 15, 19.
34. Athenaeus, 10.421b–c; Xenophon, *Hellenica*, 3.4.17.
35. Justin, 9.8.4–5.
36. Modern perceptions are germane here; namely, Gaebel, 2002, 7; 157–158; 149.
37. Gaebel, 2002, 308.
38. Marsden, 1977, 212; 223: Diodorus, 14.42.1, spoke of Dionysius of Syracuse's improvements in military technology as having been achieved because the best craftsmen 'had been gathered from everywhere into one place.' This statement is equally, if not more, applicable to Philip II of Macedon.
39. Plutarch, *Pelopidas*, 26.4.
40. Polybius, 8.9.6–13; Justin, 13.1.12.
41. Demosthenes, *Second Olynthiac*, 17–18.
42. Demosthenes, *First Olynthiac*, 4; Hamilton, 1995; Winter, 1995, 84–95; Schulz, 1999, 281–310.
43. Hanson, 1999, 162–163.
44. Garlan, 1974, 210; Burliga, 2008, 94.
45. Another writer was one Iphicrates, not to be confused with the famous Athenian mercenary general of the same name. See also Hanson, 1999, 162–163; Whitehead, 2008, 139–155.
46. Rostovtzeff, 1941b, 1083.
47. Bury, *et al.*, 1923, 35; Polastron, 2007, 25.
48. For example, see Carney, 2003, 47–63.
49. Pearson, 1960, 243.
50. Aelian, *Tactics*, 1.2; Spaulding, 1933, 659.
51. Meissner, 1992, 355–356; 476.
52. Polybius, 18.29.1–5; 18.29.7; 18.30.2; 18.30.4; Devine, 1995, 40–44; Lendon, 1999, 282–285; Stadter, 1978, 117–128.
53. Garoufalias, 1979, 465–468; n. 6.
54. Aelius Donatus, 783; Polyaenus, 6.6.3; Frontinus, 2.6.10; 2.3.21.
55. Plutarch, *Flamininus*, 21.3–4; Appian, *Syrian Wars*, 2.10–11; Plutarch, *Pyrrhus*, 8.2.
56. Appian, *Syrian Wars*, 2.10.
57. Plutarch, *Pyrrhus*, 8.2.
58. Lucian, *Hippias*, 1.
59. Cicero, *Letters to Friends*, 9.25.1.
60. Livy, 35.14.6–12.
61. Devine, 1994; Whitby, 2000, 15–17.
62. Plutarch, *Pyrrhus*, 11.4.
63. For Pyrrhus' all consuming obsession with the study of warfare, see Champion 2009, 20.
64. Wilcken, 1906, 141; Wheeler, 1983b; Jones, 2001, 28–32.
65. Cornelius Nepos, *Hannibal*, 13.2–3; Brizzi, 1991, 207–210.
66. Pausanias, 4.35.4; Hernandez, 2010, 119–132; Ammianus Marcellinus, 24.1.3.
67. Cicero, *Letters to Friends*, 9.25.1.
68. Martel, 2007, 24–26.
69. Plutarch, *Aemilius Paullus*, 28.6.
70. Isidore, 6.5.1.
71. Plutarch, *Aemilius Paullus*, 22.4.
72. Polybius, 31.23.3.
73. Sallust, *Jugurthine War*, 85.12; Cicero, *Lucullus*, 2; Cicero, *For Cornelius Balbus*, 47; Cicero, *Letters to Friends*, 9.25.1; Plutarch, *Brutus*, 4.4; Plutarch, *Mark Antony*, 45.6; Campbell, 1987, 13–29.
74. Spaulding, 1933, 660; Nap, 1927; Pliny, Preface 30.
75. Sallust, *Jugurthine War*, 85.12.
76. Feugere, 2002, 11–13; Lonsdale, 2007, 80.
77. Plutarch, *Julius Caesar*, 11.5; Suetonius, *Julius Caesar*, 7.1; Suetonius, *Gaius*, 52; Suetonius, *Nero*, 19.2; Cassius Dio, 37.52.2; 78.7.12; Herodian, 4.9.4–5; *Historia Augusta*, *Severus Alexander*, 50.5; Appian, *Mithridatic Wars*, 17.117. See also Spencer, 2002.

78. Polybius, 6.25.11; Diodorus, 23.2.1–2; Arrian, *Tactics*, 33.2; 32; Sallust, *Catiline War*, 51.37–38.
79. Cicero, *Lucullus*, 2; Cicero, *For Cornelius Balbus*, 47; Cicero, *Letters to Friends*, 9.25.1; Plutarch, *Brutus*, 4.4; *Mark Antony*, 45.6; Lendon, 1999, 281; Campbell, 1987.
80. As argued in Lendon, 2005, 266–268; 278–280; 285–286; 378, n. 8; 379, n. 20; 380, n. 28.
81. Zuckerman, 1990, 217–219.
82. Neill, 1998; Kleinschmidt, 1999, 605–606.
83. De Madariaga, 2005, 88.

Part II: Preparation

Chapter 3: Training Soldiers

1. For the most comprehensive analysis, see Hammond, 1990, 275–284. See also Psoma, 2006, 291.
2. Diodorus, 16.1.16.
3. Dio Chrysostom, 49.4; Aelian, *Varia Historia*, 4.19.
4. *Suda* s.v. *Marsyas*; Heckel, 1980; Carney, 2003.
5. Aristotle, *Politics*, 1337a–1339a.
6. Jaeger, 1944, 251–258; 208.
7. Aelian, *Tactics*, 1.7.
8. Aristotle, *Politics*, 1338b.
9. Diogenes Laertius, 5.27: for publishing books of letters to Philip and Alexander.
10. Nefedkin, 2000a; Chroust, 1972; Chroust, 1973, 155–176.
11. Hammond, 1990, 275; Arrian, 7.6.1; Plutarch, *Moralia*, 328e; Fraser, 1996, 1–46; 171–201; 239–243; 99.
12. Plato, *Laws*, 7.804c.
13. Bar-Kochva, 1979, 28; Strabo, 16.2.10.
14. Hatzopoulos, 2001, 101; 133–140; Psoma, 2006, 288; Rzepka, 2008; Juhel, 2011.
15. Forbes, 1945, 37–38; Pritchett, 1974, 213–219; Rostovtzeff, 1941c, 1588, n. 23; Cohen, 1978, 36; Chankowski, 2004, 58–61.
16. Girtzy, 2001, 272; Psoma, 2006, 289–293.
17. Lazaridis, 1997, 52–59 and fig. 29–33. For Pella, see Chrysostomou, 1996; Launey, 1950, 869–874.
18. Plato, *Critias*, 117c–d; Whitley, 2005, 70; Whitley, *et al.*, 2007, 55; Polybius, 5.36.4–6; 15.25.3; 15.29.1–2; Delia, 1996, 45 and n. 29.
19. Akamatis, *et al.*, 2004, 137; Thomas, 2007, 186.
20. Livy, 40.6.1–7; 44.32.6; Vickers, 1972, 161, fig. 4; 166–168. For Cyinda in Cilicia, see Simpson, 1957, 503–504; Bing, 1973.
21. Whitley, 2005, 87; Whitley, *et al.*, 2007, 55.
22. Arrian, 7.6.1.
23. Hatzopoulos, 1996, 372–396.
24. Diodorus, 17.108.2.
25. Arrian, 7.12.2; Diodorus, 17.108.2–5; 17.110.3.
26. Plato, *Laws*, 7.813e.
27. Hammond, 1989c, 162–163; Dahm, 2008; *Alexander Romance* (Greek version), 1.25.
28. Lucian, *How to Write History*, 37; Plato, *Laches*, 182b–183a; Launey, 1950, 815–835.
29. Diogenes Laertius, 9.48.
30. Wheeler, 1981; Wheeler, 1983b; Jesse, 2007, 58; Dahm, 2008.
31. Arrian, 7.9.2, a possible rhetorical construct.
32. Hammond, 1989c, 153.
33. Pliny, 5.62; *Acts* 16; 9.
34. Pélékidis, 1962, 207; Chaniotis, 2005, 46–51; 97–99; Psoma, 2006, 292; Livy, 44.10.6.
35. Austin, 1981, 201–202; 117; 211–212; 120; Hatzopoulos, 2001, 137–138, n. 5; Launey, 1950, 815–835; Pritchett, 1974, 208–231; Strabo, 16.2.10.
36. Polyaenus, 4.2.1.
37. Curtius, 3.2.15.

38. Chaniotis, 2005, 50.
39. Chaniotis, 2005, 50; Hatzopoulos, 1994, 55–61; Crowther, 1991. For Xenophon on *eutaxia*, see *Anabasis*, 5.18.13–14; 3.1.38; 3.2.29–31; Crowther, 1985.
40. Plutarch, *Moralia*, 342e; Diodorus, 16.3.3; 17.2.3; Curtius, 7.6.26. Hellenistic inscriptions survive containing prices of arms and armour rewarded as prizes in military contests; Lendon, 2005, 126–131; 356–357; n. 25–31; 126–127.
41. Curtius, 13.13.–16; Heckel, 1992, 296; 297.
42. Kosmidou and Malamidou, 2004/05, 142–143, fig. 27.
43. Lonsdale, 2004, 40–41.
44. Hammond, 1990, 278. However, see Psoma, 2006, 292: for age categories of youth in Hellenistic Macedonia being *paides* (14–18), *ephebes* (18–20) and *neoi* (20–30). Young men entered the army at 20.
45. Hatzopoulos, 2001, 109–118; Diodorus, 17.110.3; Plutarch, *Alexander*, 47.6; *Moralia*, 328d.
46. Polybius, 5.63.14–64.7.
47. Diodorus, 16.3.1–2.
48. Polyaenus, 15.2.7; 4.2.10.
49. Plutarch, *Moralia*, 178c; Livy, 45.30.7.
50. Plutarch, *Alexander*, 40.2; Johnstone, 1994.
51. Plutarch, *Alexander*, 4.11.
52. Lindsay Adams, 2007b, 129–130 and n. 21; Lindsay Adams, 2008, 63 and n. 29. For calming effects, see Tritle, 2007, 177–178; Launey, 1950, 813–874.
53. Polyaenus, 4.2.6; Plutarch, *Moralia*, 602d; Lindsay Adams, 2008, 67; Kertesz, 2007, 328–329.
54. Plutarch, *Alexander*, 40.1.
55. Strabo, 15.1.67.
56. Butler, 2008, 81; 95.
57. Peatfield, 2007, 32.
58. Athenaeus, 8.348e–f.
59. Psoma, 2006, 288.
60. Polyaenus, 4.3.32.
61. Plutarch, *Moralia*, 180a; Plutarch, *Alexander*, 22.4–5.
62. Curtius, 9.7.16–17.
63. Plutarch, *Moralia*, 192c–d; Cornelius Nepos, *Epaminondas*, 5.4; Plutarch, *Alcibiades*, 11.3; Athenaeus, 12.534b; Diodorus, 17.11.4.
64. Cornelius Nepos, *Epaminondas*, 2.4–5; Brown, 1977; Slowikowski, 1989; Hammond, 1997c.
65. Plutarch, *Pelopidas*, 15.1–3.
66. Plutarch, *Pelopidas*, 32.7.
67. Dio Chrysostom, 49.5; Diodorus, 15.67.4; Justin, 7.5.2–3.
68. Plutarch, *Philopoemen*, 3.2; Plato, *Republic*, 3.403d–404b; Williams, 2004.
69. Polyaenus, 4.2.7; Frontinus, 2.1.9; Demosthenes, *Third Philippic*, 52; Vidal-Naquet, 1968.
70. Plutarch, *Alexander*, 4.11; Poliakoff, 1987, 64–67 and 100.
71. Plutarch, *Alexander*, 30.2–5. See also Xenophon, *Cyropaedia*, 2.3.17–8; Onasander, 10.4.
72. Livy, 40.6.6–7.
73. Carter, 2006.
74. Kosmidou & Malamidou, 2004/05, 142, fig. 27.
75. Diodorus, 16.3.2.
76. *Alexander Romance* (Greek version), 1.13. Historical figures closer to us in time, such as Shivaji Bhosle and Tsar Peter the Great, are known to have done the same. The *Romance* was itself originally written in Greek as early as the third century AD using a conflation of genuine documents and fictitious components. Latin, Armenian and Syriac versions also exist dating from the fourth to sixth centuries.
77. Diodorus, 17.2.3.
78. Plutarch, *Philopoemen*, 7.4–5. Cf. Diodorus, 16.3.1–2 already discussed above; Polybius, 5.63.14–64.7.
79. Plato, *Laws*, 8.833d–e; Jacoby, 1923, 73 F 1; Hammond, 1989c, 24–26.

80. Polyaenus, 3.9.34.
81. Ellis, 1968, 7.
82. Aeschines, 2.26–29; Cornelius Nepos, *Iphicrates*, 3.2.
83. Arrian, 2.15.4; Curtius, 3.13.15.
84. Diodorus, 15.44.1–4; Cornelius Nepos, *Iphicrates*, 1.1–3; Arrian, 5.3.6; 5.8.3; 5.20.2.
85. Athenaeus, 1.14a–15a.
86. Miller, 1991, 116–119.
87. Janssen, 2007, taf. 27.101; Plutarch, *Alexander*, 39.5; 73.7; Plutarch, *Moralia*, 182b; Athenaeus, 1.14a; Theophrastus, *Characters*, 5.9–10, 'Obsequiousness'.
88. Plato, *Laws* 7.813d–e.
89. Diodorus, 17.100.4–7; Curtius, 9.7.16–26.
90. Diodorus, 17.11.3–4; Curtius, 3.11.4–5.
91. Diodorus, 17.11.3–4; Curtius, 3.11.4–5; 6.7.4; 7.1.9; 7.9.7; 9.7.19; 9.7.22; 10.7.18; Arrian, 1.21.1–3; 3.26.3; 5.17.3; Diodorus, 17.25.5; 17.88.3; 17.100.6–7; 18.7.9; Polyaenus, 2.38.2; Cassius Dio, 78.7.12. Javelin heads were also recovered from the Macedonian burial mound at Chaeronea and in the Vergina Cemetery. Side B of the surviving gymnasiarchal law *stele* from Beroea stipulates that those below 22 years old were to train in javelin throwing and archery every day; Markle, 1977; Markle, 1980a; Markle, 1982.
92. Arrian, 1.2.4; 4.2.3; 4.30.1; Curtius, 3.9.9; 4.3.15; Diodorus, 17.42.7; Paunov & Dimitrov, 2000.
93. Robinson, 1941, 418–439; Lee, 2001.
94. Dezso, 2006, 108; Cheesman, 1914, 132; Griffiths, 1989.
95. Griffiths, 1992, 2; 4. See also Plato, *Laws*, 834a, for competitions in throwing stones by hand or sling.
96. E.g. Plutarch, *Alexander*, 55.7; Arrian, 4.14.3.
97. Polyaenus, 4.6.4; Polybius, 31.39.13.
98. Psoma, 2006, 292.
99. Ueda-Sarson, 2000, 33.
100. Polybius, 10.24.7; Lendon, 2005, 150.
101. Arrian, *Tactics*, 5.23.
102. White, 1996, 71.
103. Bayliss, 2011, 12, 216, n. 3.
104. Manti, 1992, 31, n. 3: one Attic *poda* equals 30cm.
105. Wheeler, 1983a, 233; n. 51. See also Polybius, 4.20.12–21.1; Reed, 1998, 4–5; 22–30.
106. Xenophon, *Anabasis*, 6.1.7; Hesychius, s.v. *telesias*, *karpaia*; Athenaeus, 14.629d; Faure, 1982, 88.
107. Cicero, *On the Orator*, 2.75.
108. Plutarch, *Philopoemen*, 4.4.6; Pausanias, 8.49.3. Like Pausanias, the *Suda* also describes Philopoemen studying military manuals in order to emulate the generalship of Epaminondas.
109. Diodorus, 16.3.2; Plutarch, *Alexander*, 71.1.
110. Wheeler, 1983b; Roesch, 1982.
111. For a mere selection of the many passages outlining the high quality of infantry drill on Alexander's campaigns, see Arrian, 1.1.9; 1.2.4; 1.4.2; 1.6.1–4;1.6.6; 1.6.10; 3.13.6; Diodorus, 17.57.6.
112. E.g. Aelian, *Tactics*, 24–41; Arrian, *Tactics*, 20–32.
113. Diodorus, 17.65.4; Arrian, 3.9.8; Curtius, 3.2.13–16; Carney, 1996, 23.
114. Curtius, 3.2.13–14; Arrian, 3.9.8; Lloyd, 1996, 172; Onasander, 10.2.
115. Diodorus, 17.65.4.
116. Polybius, 4.64.3–11; Echols, 1953, 215–216. See also Ferrill, 1985, 180–181.
117. Plutarch, *Alexander*, 4.11.
118. Plutarch, *Alexander*, 23.2–3; 57.3; Plutarch, *Moralia*, 338d; Curtius, 8.3.16. For training with the bow in Antigonid Macedon, see Psoma, 2006, 292.
119. Xenophon, *Cyropaedia*, 1.2.10; Wood, 1964, 51–55; Renault, 1975, 69–70; Athenaeus, 1.18; Briant, 1986; Carney, 2002.
120. Polyaenus, 15.2.7.
121. Polyaenus, 4.2.10; *Itinerarium Alexandri*, 5.13; 12.27.

122. Frontinus, 4.1.6: the impact of this reform can be appreciated from the fact that even fifteen-day rations represented almost double the Greek city-state hoplite norm; Engels, 1978, 20, n. 29–30.
123. Faure, 1982, 178; Neumann, 1971; Engels, 1978, 153–156; Bose, 2004, 12–14.
124. Plutarch, *Alexander*, 40.3.
125. Aelian, *Varia Historia*, 14.48.
126. Frontinus, 3.12.2; Curtius, 4.13.21.
127. Xenophon, *Hellenica*, 6.1.5–6; Plato, *Republic*, 3.404a.
128. Hatzopoulos, 2001, 36; 46, n. 5; 117; Appendice epigraphique no. 2 II, L.52; Psoma, 2006, 292; Gauthier & Hatzopoulos, 1993, 77–78; 162; Busolt & Swoboda, 1926, 1121; Livy, 44.11.7; Chaniotis, 2005, 46.
129. Arrian, 2.23.2–3.
130. Murray, 2008.
131. Polybius, 5.2.4–5; 5.5.11–14; 5.109.4; 1.21.1–3.
132. Griffith, 2001, 54.
133. Arrian, 7.9.2; Dio Chrysostom, 33.26; 2.8–9.
134. Livy, 45.30.7.
135. Howell, 1952, 136.
136. Dio Chrysostom, 2.9.
137. Diodorus, 17.95.1–2; Arrian, 4.2.1; 4.21.3; 4.29.7.
138. Polyaenus, 3.9.35; Davies, 1968.
139. Plutarch, *Philopoemen*, 4.1–2; Pausanias, 8.49.3.
140. Onasander, 9; 10.1.
141. Hammond, 1997b, 185; Hammond, 1981a, 261.
142. Diodorus, 17.108.1; Plutarch, *Alexander*, 71.1. It is possible that the recruits were partly formed from the basis of the Achaemenid youth corps, the *kardakes*. See also Arrian, 2.8.6; Curtius, 3.9.2–3; Cornelius Nepos, *Datames*, 8.2; Briant, 2002, 1036–1038, point 3; Hammond, 1990, 278–279.
143. Curtius, 5.8.5.
144. Hammond, 1998a; Thomas, 1974; Bertrand, 1974.
145. Hammond, 1990, 287.
146. Hammond, 1990, 279; Hammond, 1991a, 147: perhaps as many as 200,000 by 323 BC.
147. Hammond, 1990, 279; 289; Hammond, 1999a, 371–372; Hammond, 2000, 149; Hammond, 1993a.
148. Legras, 1999, 133–154; 195–236; Aristotle, *Athenian Constitution*, 42.3–4; Kent, 1941, 349–50.
149. Hammond, 1990, 278; Berve, 1926, 186–191; Hatzopoulos, 1994, 89; Hatzopoulos, 2001, 119–122; Psoma, 2006, 299.
150. Plutarch, *Alexander*, 40.3; Onasander, 28; Diodorus, 16.3.1–2; Appian, *Macedonian Wars*, 11.1; Plutarch, *Aemilius Paullus*, 8.4–5; Livy, 42.52.11.
151. Diodorus, 17.108.1–2; Arrian, 7.6.1; 7.8.3; Devine, 1997.
152. Sekunda, 2011b, 19, for one interesting reconstruction.
153. Koch, 1912, 98–99, abb. 128.
154. von Rohden, 1880.
155. Merker, 1965; Fol & Marazov, 1977, 45; Head, 1982, 112 and 113: 40a; Sekunda, 2011b, 18–19; 22: for one of the coin versions showing a figure in a *kausia*, possibly indicating a member of Alexander's new 1,000-strong Persian hypaspists or Silver Shields; Arrian, 7.11.3; Diodorus, 17.110.1. The Paeonian coin images have been most recently addressed under Wright, 2012.
156. Kinch, 1920, 283–288; Couissin, 1932, planche I–II; Devine, 1997.
157. Hackim, 1954: fig. 263.
158. Plutarch, *Moralia*, 274d–e; Plutarch, *Philopoemen*, 3.2.
159. Plutarch, *Alexander*, 71.3.
160. Cf. disparaging, misinformed comments given by Livy, 9.19.5; 9.19.10–11.
161. Hammond, 2000, 148; Arrian, 7.10.7; Curtius, 7.1.29; 10.7.9; Justin, 14.4.3–8; 24.5; Plutarch, *Eumenes*,12.2; Cornelius Nepos, *Eumenes*, 10.2; Polybius, 15.25.11.
162. Arrian, 3.6.6; Carney, 2001; Baldwin, 1967, 42–43.

163. Arrian, 7.10.7; Justin, 14.4.1–14.
164. For the form of execution, see Arrian, 3.26.3; Curtius, 6.11.10; 6.11.38.
165. Curtius, 6.11.20.
166. Eitrem, 1947, 37–38.
167. Polybius, 15.25.11; Curtius, 10.9.11–15; Livy, 40.6.1–7; 43.21.5; Plutarch, *Moralia*, 290d; Herodotus, 7.39; *Book of Genesis*, 15: 9–10; 17–18; Nilsson, 1906, 404–406; Hellmann, 1931, 202–203; Pritchett, 1979, 196–202; Faraone, 1993.
168. E.g. Diodorus, 18.16.1
169. Noguera Borel, 2006, 228–229; Rzepka, 2008, 55–56; 53; Juhel, 2011. For earlier work, see Berve, 1926, 186–191; Tarn, 1948, 142; Hammond & Griffith, 1979, 427, n. 1; Hammond, 1981a, 27; Anson, 1985; Billows, 1995, 18.
170. Beloch, 1922, 296.
171. Ellis, 1976, 34; Ross, 1990.
172. Gabriel, 2010, 83.
173. Thomas, 2007, 49.
174. Livy, 39.24.3; Polybius, 25.3.10; Walbank, 1940, 224 and n. 5.
175. Hamilton, 1999b.
176. Arrian, 3.16.11.
177. Curtius, 7.1.37; 7.5.27; Diodorus, 18.12.2; Plutarch, *Demetrius*, 44.5–6; Plutarch, *Pyrrhus*, 11.4.
178. Lindsay Adams, 1985; Bosworth, 1986a, 1–12; Bosworth, 1986b, 115–122; Bosworth, 2002, 64–97; Hammond, 1989b, 56–68; Billows, 1995, 184–206; Rubio, 2011, 41.
179. Hatzopoulos, 2001, 119.
180. Justin, 11.6.4.
181. Diodorus, 17.27.1; Curtius, 8.1.36–37.
182. Curtius, 4.6.30; 7.1.37; Diodorus, 17.49.1; 19.52.5; Justin, 12.11.4. But see Arrian, 7.12.4. See also Griffith, 1965, 132–133.
183. Arrian, 1.24.1–2; 1.29.3–4.
184. Frontinus, 4.1.10.
185. Hatzopoulos, 2001, 109–118; 153–156; 157–160. For a pertinent analogy, see Baldwin, 1967.
186. See Errington, 2002, 22; 23: 'This gradually improving knowledge of the efficiency of the state apparatus gives us a good idea... how the Macedonian army was capable of offering the Romans such a significant challenge'.
187. For recruitment drafts in Alexander's army, see Arrian, 1.24.2; Curtius, 4.6.30; 7.1.37; Diodorus, 17.49.1; 19.52.5; Justin, 12.11.4; Griffith, 1965, 132–133.
188. Diodorus, 19.17.7; Sekunda, 2011b, 20.
189. Diodorus, 19,17.6.
190. Diodorus, 19.57.5.
191. See Hammond, 1980, 256: citing Diodorus, 19.20.2–4; 19.55.1; 19.58.2.
192. See Livy, 42.51.1.
193. Livy, 37.7.12–13; 44.43.1; Plutarch, *Aemilius Paullus*, 8.4–5.
194. Appian, *Macedonian Wars*, 11.1.
195. Plutarch, *Alexander*, 16.2; Demosthenes, *On the Chersonese*, 11; Polybius, 4.66.7.
196. E.g. Diodorus, 17.16.3–4; Dio Chrysostom, 2.2; Livy, 40.6.1–7; 43.21.5; Baege, 1913, 10–12.
197. Livy, 33.3.1–4.
198. Hammond & Griffith, 1979, 392; Corvisier, 1999, 221–222; Landucci, 2003.
199. Livy, 40.6.2; Hammond, 1989a, 218, n. 8: 'The Loeb edition reads *arma insignia* translated 'the arms and the standards'.
200. Hammond, 1989a, 218; Hammond, 1991a, 32.
201. Livy, 42.52.4.
202. Livy, 33.3.4–5; 42.52.1–53.3; Hesychius, s.v. *Xanthika*; Polybius, 23.10.7; *Suda* s.v. *diadromai* and s.v. *enagizon*.
203. Demosthenes, *On the Crown*, 235. For epigraphic evidence on the sophisticated nature of recruitment in at least Antigonid Macedon see Hatzopoulos, 2001, 109–122.
204. Florus, 1.28.12.1–3.

Chapter 4: Supplying Arms, Armour and Cloth

1. Willems, 1986, 32.
2. Montgomery, 1985; Hammond, 1995a.
3. Tsigarida & Ignatiadou, 2000, 31–47.
4. Livy, 45.18.3–5; 45.29.10–11.
5. Davies, 1932; Hammond, 1972, 12–18; Hammond, 1979, 140; Borza, 1982; Davies, 2010.
6. Theophrastus, *Characters*, 2.3: 'Fraudulence', Theophrastus, *Enquiry into Plants*, 1.9.2; 4.55; 5.2.1; 7.1–3; Johnson, 1927; Meiggs, 1982, 128–129; Borza, 1987c.
7. Greenwalt, 1999, 174.
8. Strabo, 15.1.30.
9. Strabo, 11.14.9; Hammond, 1996a, 134; Heckel, 2006c, 166.
10. Borza, 1987c, 40, n. 36; Diodorus, 16.8.6; 16.53.3; Shepherd, 1993, 103–108; Forbes, 1963, 142.
11. Borza, 1987c; Strauss, 1984.
12. Thucydides, 2.100.2; Diodorus, 12.68.5.
13. Diodorus, 16.3.1–2; Polyaenus, 4.2.10; Griffith, 1956/7, 9.
14. Errington, 1993, 25–26.
15. Rostovteff, 1941b, 1220–1221; Garlan, 1975, 142–145; McKechnie, 1989, 80–85; McKechnie, 1994; Whitehead, 1991; Trundle, 2004, 124–131; Bertosa, 2003.
16. Treister, 1996, 205; Treister, 2001, 381; Pflug, 1988b; Herodotus, 7. 119.
17. Diodorus, 12.68.5.
18. For more on this system, see MacMullen, 1960.
19. Pounder, 1983, 245 and n. 40; Thompson, 1939; Mattusch, 1977, 357–358.
20. E.g. Lysias, 12.19.
21. Demosthenes, *Against Aphobus*, 1.9; Demosthenes, *For Phormio*, 4.
22. *Itinerarium Alexandri*, 7.18: taken from Hans-Josef Hausmann's conjecture for the corrupt *bicoris*. The *Itinerarium Alexandri* is of anonymous authorship, produced c. 340 AD; Lane-Fox, 1997.
23. Borza, 1982, 9.
24. Faklaris, 1985, 1–16; 283–284.
25. Akamatis, *et al.*, 2004, 138; Livy, 45.33.5–6.
26. Stojcev, 2001.
27. Greenwalt, 1999, 174.
28. Diodorus, 19.52.2–3.
29. Diodorus, 22.5.2.
30. Livy, 44.32.6; Vickers, 1972, 161, fig. 4; 166–168.
31. E.g. Chaniotis, 2004a.
32. Dio Chrysostom, 77.13.
33. Archibald, 1985, 179; Ognenova, 1970.
34. Demosthenes, *Third Philippic*, 9.26; Jacoby, 1923, 115 F 27; Millett, 2010, 490.
35. Ellis, 1969.
36. Polybius, 5.11.3–4; Chaniotis, 2005, 170.
37. *Alexander Romance* (Greek version), 1.25. See also Diodorus, 18.12.4; Lindsay Adams, 1985: for Antipater's preparations at Lamia in 323 BC; Xenophon, *Hellenica*, 3.4.17: for a more atmospheric account of artificers at Ephesus, 395 BC.
38. See general statements in Andronikos, 1988b, 146; Hammond, 1994, 180; Yalouris, *et al.*, 1980, 27.
39. Waurick, 1988, 176–178 and fig. 64; Treister, 1996, 300; Schafer, 1997, 83–89.
40. Hatzopoulos, 2008.
41. Sokolovska, 2011, 46, fig. 18.
42. Clement, 1.16.76; Hammond, 1993, 41, n. 15.
43. Delemen, 2006, 266–267; 268, fig. 19.
44. McKechnie, 1989, 142–177.
45. Sekunda, 2001c, 29–30. See also Theophrastus, *Enquiry into Plants*, 5.8.1.
46. Plutarch, *Alexander*, 8.3.
47. E.g. Plutarch, *Timoleon*, 21.3; 24.4; 27.6; Diodorus, 17.76.6; Heckel, 2006c, 107.

48. Diodorus, 14.43.2–3.
49. Diodorus, 14.41.1–46.1.
50. Diodorus, 16.9.2; 14.10.2; Caven, 1990, 88–97; 243–247; Trundle, 2004, 127; Treister, 1996, 225–229.
51. Livy, 44.10.3–4; 44.42.6; 44.46.7–8; Diodorus, 30.11.
52. Whitley, 2005, 87; Whitley, *et al.*, 2007, 55.
53. *Alexander Romance* (Greek version), 1.25.
54. Livy, 42.12.9.
55. Livy, 31.23.7.
56. Appian, *Syrian Wars*, 1.1; Diodorus, 20.29.1; Pausanias, 1.9.8.
57. Appian, *Syrian Wars*, 4.21.
58. Appian, *Syrian Wars*, 6.28. We gain a more vivid insight into the character of siege apparatus stored at Lysimachia from a detailed extract by Philon of Byzantium recommending that all articles be tested during peacetime to ensure they be in working order.
59. Pounder, 1983, 232; 234; 238 and n. 11–12; Cimok, 2001, 15; Bernard, 1967, 71–95; Bernard, 1980, 51–63 and pl. 21–22.
60. von Szalay & Boehringer, 1937. See taf. 8–25; 27; 32; 37–42; Kohl, 2004, 190–192.
61. Robinson, 1941, 382ff; Marsden, 1977, 213–215.
62. Sekunda, 1984a, 29. The interpretation of the pikes on the mosaic as Macedonian is followed by the majority of scholars, e.g. Pernice, Schone, Rumpf, Holscher, Robertson, Andreae, Schefold and Calcani. For *contra*, see Moreno, 2001, 21–22.
63. Sekunda, 1984a, 28; Sekunda, 2001b, 34–35, fig. 7; Shefton, 1978, 13: inv. no. 111.
64. Andronikos, 1988b, 111, fig. 67; Briant, 1991.
65. Wooton, 2002, 269 and fig. 12.
66. Pandermalis, 2004, 55, fig. 5.
67. Hammond, 1996b, 366; Adam-Veleni, 1998, 19, pl. 1 and 20, fig. 2.
68. Dakaris, 1993, 19, fig. 3.
69. Juhel & Temelkoski, 2007b; Grozdanova, 2007.
70. Oikonomides, 1988, 81, fig. 9.
71. Hammond, 2000, 149.
72. Gagsteiger, 1993, 54–63; 89, entry 23 and taf. 21a–b; 22.
73. Kroll, 1977b, 141–146; Bertosa, 2003; Couvenhes, 2007.
74. Plutarch, *Demetrius*, 43.3–4; Plutarch, *Pyrrhus*, 10.5.
75. Plutarch, *Demetrius*, 20.1; Justin, 16.2.1.
76. Plutarch, *Aemilius Paullus*, 8.4–5.
77. Livy, 42.12.7–10; 42.53.2–3; Justin, 33.1; Plutarch, *Aemilius Paullus*, 13.3; Errington, 1993, 25–26.
78. Garlan, 1975, 141.
79. Welles, 1938, 251–254; Le Bohec, 1993, 305–311; pl. VIII; Ma, 2002, 177 and n. 11.
80. Florus, 1.28.12.4–5.
81. Livy, 42.52.3; Appian, *Macedonian Wars*, 11.1.
82. Ma, 2008, 77.
83. Polybius, 5.8.9.
84. Lee, 2007, 127–132.
85. Athenaeus, 203a; Appian, Preface 10; Rice, 1983, 123–126.
86. Polybius, 4.69.4–6.
87. Wallinga, 1991; Briant, 2002, 422–471; Schmidt, 1957, 91–111.
88. Dezso, 2006, 97, 108–109, 117.
89. Curtius, 3.3.6.
90. Diodorus, 17.39.3–4; 17.53.1–3; 17.64.2; Curtius, 4.9.3–5; *Codex Sabbaiticum*, 29.12 in Robinson, 1953, 267–268.
91. For a recent re-evaluation of dates, see Primo, 2009.
92. Julius Africanus, 1.1.45–50.
93. Litvinskij & Picikjan, 1999, 104.

94. Thompson, 1988, 57, n.152; 72; Post, 2010, 19.
95. Xenophon, *Cyropaedia*, 8.3.6; Curtius, 7.1.19; Athenaeus, 5.194e–f; 197f; 203a; 12.539d–e; Polyaenus, 4.3.24; Aelian, *Varia Historia*, 9.3; Polybius, 30.25.8–10.
96. Diodorus, 17.74.5; Plutarch, *Aemilius Paullus*, 18.3.
97. Curtius, 9.3.10.
98. E.g. Diodorus, 17.77.5; 17.69.8; Curtius, 5.5.24; Justin, 11.14.11–12.
99. Athenaeus, 12.540a; Plutarch, *Alexander*, 56.2; Plutarch, *Moralia*, 11a; Curtius, 5.2.18–20; 9.3.10; 8.1–2; *Itinerarium Alexandri*, 44.101; 45.102.
100. Brinkmann, 2003.
101. Winter, 1912, taf. 1, 5 and 13; Head, 1999a; Tsibidou-Avloniti, 2002b, and tav. vi.2; vii.2.
102. Athenaeus, 11.484d; 12. 538c; Curtius, 6.6.7; 10.2.23; Justin, 12.7.5; Diodorus, 17.77.5; Plutarch, *Eumenes*, 14.4; Aelian, *Varia Historia*, 9.3; Polyaenus, 4.3.24; Jacoby, 1923, 81 F 4.
103. Grote, 1913. For more modern discussion, see Faure, 1982, 251–252.
104. Bennett, 1997, 49–50; 92–96; 107.
105. van Wees, 1988; 1996; Marinatos, 1970; Cohen, 1995.
106. Curtius, 9.8.4–7; Arrian, *Indica*, 24.1–9; Plutarch, *Eumenes*, 14.4; Diodorus, 17.100.3–5; 20.83.2–3; Plutarch, *Aemilius Paullus*, 18.7–8; 32.5–7; Theocritus, 93–94; Alciphron, 4.16.1; Onasander, 1.20.
107. Plutarch, *Alexander*, 16.19.
108. Curtius, 3.10.9–10.
109. Diodorus, 17.35.4.
110. Curtius, 3.13.7; 13.10–11.
111. Diodorus, 17.64.3.
112. Curtius, 5.1.10.
113. Bloedow, 2003; Brosius, 2003.
114. Plutarch, *Alexander*, 36.2–4; Alciphron, 3.10.4; Stephanus, s.v. *Hermione*.
115. Plutarch, *Alexander*, 36.2–3.
116. Lee, 2007, 118–119.
117. Olmstead, 1948, 70; Briant, 2002, 21.
118. Diodorus, 2.53.1–2; Vitruvius, 7.13.1–3; Strootman, 2007, 376–379. Many thousands of molluscs are needed to extract a tiny quantity of dye. Obviously, if the numbers of shellfish were drastically denuded, substitute dyes would have been applied. See also Blum, 1998, 191–267; Stutz, 1990; Robinson, 1971; Cooksey, 1995.
119. For infantry tunic colours in various reds or purples, blue or white, see the evidence in Winter, 1912, taf. 2, 5–6, 13–14; Bresciani, 1980, 140–141, fig. 164–165; Hanfmann, 1983, 247 and tav. XLV.1; Brecoulaki, 2006, pl. 94; Plutarch, *Philopoemen*, 6.2; 11.1–2; Plutarch, *Aemilius Paullus*, 18.3; Josephus, *Jewish Antiquities*, 12.7.312; I. *Maccabees*, 4.23–24.
120. Sumner, 2009, 114–118.
121. Pliny, 19.5.22.
122. Curtius, 5.6.3–5.
123. Diodorus, 17.70.2.
124. Diodorus, 17.69.3; 17.69.8; Curtius, 5.5.5; 5.5.24; Justin, 11.14.11–12.
125. Diodorus, 17.74.5; Justin, 12.1.1.
126. Curtius, 6.6.14–17; Plutarch, *Alexander*, 57.1–2; Polyaenus, 4.3.10.
127. Curtius, 8.12.16.
128. Curtius, 9.3.10.
129. Diodorus, 17.94.2.
130. Kingsley, 1991, 72.
131. Curtius, 9.8.1–2.
132. Diodorus, 17.94.2; Kingsley, 1991; Kingsley, 1984, 67 n.20 and 68; Fredricksmeyer, 1994, 137–138.
133. Arrian, *Indica*, 16.17; Strabo, 15.1.71.
134. Plutarch, *Alexander*, 67.3; Curtius, 9.10.25.
135. Curtius, 9.10.26.

136. Curtius, 10.2.23–24.
137. Athenaeus, 12.539f –540a; Plutarch, *Moralia*, 11a.
138. Athenaeus, 12.540a; Homer, *Iliad*, 5.83.
139. See also Plutarch, *Moralia*, 340d; Faure, 1982, 251–252; Prestianni Giallombardo, 1991, 276–280.
140. Athenaeus, 11.484d.
141. Appian, *Mithridatic Wars*, 17.117; Preface 10; Dio Chrysostom, 4.4.
142. Dio Chrysostom, 33. 27. For more on the easy accessibility Macedonians under Alexander had to purple and similar dyes in Asia, see Athenaeus, 12.539c; Pliny, 19.22; Plutarch, *Alexander*, 67.1; Curtius, 9.7.15.
143. Sekunda, 1992b, 8–10; 30–61, pl. A, E–L.
144. Winter, 1912, taf. 2, 5, 10, 12–14 and 17.
145. Brecoulaki, 2006, pl. 94; Tsibidou-Avloniti, 2002b, and tav. vi.2; vii.2.
146. Hatzopoulos & Loukopoulos, 1991, 67, fig. 46; Calcani, 1989, 39; fig. 16; 58–59.
147. Saatsoglou-Paliadeli, 1984, 152–159: for text; pl. 43: no. 20.
148. Rouveret, 2004, 45–46, fig. 3.
149. Brown, 1957, 26, cat. no. 21, pl. xi.
150. Brown, 1957, 52–53, cat. no. 34, pl. xxiv.1.
151. Sekunda, 2003, and pl. x–xi.
152. Wooton, 2002, 269, fig. 10.
153. Pliny, 21.31–34.
154. Robert, 1963, 181–184.
155. Curtius, 6.6.8–10; Plutarch, *Eumenes*, 6.2; Athenaeus, 4.155c–d.
156. Virgil, *Aeneid*, 11.768–80. The dye is known to have provided the main basis of colour for the dress of Roman Vestal Virgins. See also Virgil, *Eclogues*, 4.43–44.
157. Virgil, *Aeneid*, 9.598–620.
158. Arrian, *Indica*, 5.8–9; Diodorus, 17.2.38; 17.3.63; 17.4.3; Strabo, 11.5.5; Athenaeus, 5.198d; *Suda* s.v. *krokotos*; Hartmann, 1965; Christesen & Murray, 2010, 431–432.
159. Jameson, 1993, 50 and n. 10.
160. Curtius, 8.8.15.
161. Arrian, 5.2.1; 5.26.5; Plutarch, *Moralia*, 326b–c, for Alexander's aim of surpassing Dionysus.
162. Gaukowsky, 1981, 3–19; Jost, 1999; Gilley, 2004, 11–13.
163. *Alexander Romance* (Greek version), 2.11; 21.
164. Arrian, 3.6.8; Bosworth, 1974; Kasher, 2011, 135–136.
165. Curtius, 4.9.20.
166. Kuhrt, 2007, 679, n. 5; 748–749.
167. Xenophon, *Anabasis*, 3.4.17.
168. Engels, 1980, 328.
169. Plutarch, *Alexander*, 32.8; Diodorus, 17.35.1; Arrian, 2.11.10; Curtius, 3.13.2–17. Nor were commanders such as Alexander averse to wearing armour sent to them as gifts, made by famous master craftsmen: see Plutarch, *Alexander*, 32.8–11; Plutarch, *Demetrius*, 21.3–4; *Alexander Romance* (Greek version), 3.23.
170. Borza, 1987a, 112.
171. Justin, 11.13.5.
172. Jacoby, 1923, 15 F 225b; Polybius, 8.11.5–13; Athenaeus, 6.260d–261a; Plutarch, *Alexander*, 15.2; Justin, 11.13.11; *Suda* s.v. *Leonnatos*; Plautus, *Curculio*, 633–634; 634–638.
173. Schmidt, 1957, 97–101 and pl. 75–77; Cahill, 1985, 384–385; Hammond, 1992, 359; Diodorus, 17.70.1–2; Curtius, 5.6.2.
174. Plutarch, *Alexander*, 62.7; Diodorus, 17.95.2.
175. Diodorus, 17.79.4–5; Curtius, 6.7.23.
176. Markle, 1980b.
177. Curtius, 8.5.4: if this is an authentic event rather than an anachronism for the later supply of equipment in India, then the mention of silvered shields probably relates to the hypaspists only.
178. Justin, 12.7.5.

179. Badian, 1965, 181, n. 1.
180. Curtius, 9.3.10–11.
181. Diodorus, 17.94.2.
182. Diodorus, 17.95.4; Curtius, 9.3.21.
183. Curtius, 9.3.21–22.
184. The veracity of written evidence is best summarised by Lush, 2007, 20. See also Faure, 1982, 246–250; Dihle, 1964.
185. Chugg, 2009, 106: 11.40.
186. McCrindle, 1960, 231, n. 1.
187. Curtius, 9.4.18.
188. Diodorus, 17.100.3–5.
189. Euripides, *Rhesus*, 380–385.
190. Curtius, 9.8.4–7.
191. Arrian, *Indica*, 24.1–9.
192. See also Diodorus, 20.83.2–3, for the gleam of armour witnessed at the siege of Rhodes, 305 BC.
193. Homer, *Iliad*, 5.293–295; 12.451–471; 18.213–214; 19.444–445; Arrian, 1.14.4; Diodorus, 17.98.6; Plutarch, *Moralia*, 343e; Plutarch, *Alexander*, 63.4–5; Curtius, 4.4.10; Polybius, 11.9.1; Strabo, 15.1.33; *Suda* s.v. *Leonnatos*; Diodorus, 19.8.1.4.
194. Diodorus, 17.105.6; Chugg, 2009, 134: 12.49, prefers 'surpassed all in the art and practice of warfare'.
195. Curtius, 9.10.23; 26.
196. Arrian, 7.6.1; Diodorus, 17.108.2; Hammond, 1990, 279.
197. Arrian, 7.12.2; Diodorus, 17.110.3, describes the number of children as being about 10,000; Justin, 12.4.5–8.
198. Atheneaus, 12.538b; Jacoby, 1923, 126 F 5; Tarn, 1948, 354, n. 2.
199. Heisserer, 1980, 194 and n. 73; Herman, 1987, 85–86; Heckel, 2006c, 127; Spawforth, 2007.
200. Hammond, 1990, 279.
201. Arrian, 7.28.2.
202. *Suda* s.v. *Alexander*.
203. Curtius, 6.6.7; 6.6.9–10; Plutarch, *Eumenes*, 6.2.
204. Diodorus, 17.80.3–4; Strabo, 15.3.6; 3.9; Justin, 13.1.9; Engels, 1978, 35; 79; Faure, 1982, 240–252; De Callatay, 1989; Lindsay Adams, 1996; Holt, 1999a; Holt, 2010, 359: perhaps equivalent in Alexander's day for enough money to support one million soldiers for ten years.
205. Bellinger, 1963, 30–38; 71; 73–74; Milns, 1987, 254; Aperghis, 2004, 205.
206. Németh, 1995.
207. Juhel, 2009, 353, n. 53.
208. Juhel, 2009, 353, n. 53–55.
209. Chrysostomou, *et al.*, 2001; Chrysostomou, *et al.*, 2002.
210. Vokotopoulou, 1996a, 134; Tsigarida & Ignatiadou, 2000, 23.
211. Morgan, 2008, 68.
212. Despini, 1980, 204–209, fig. 5–7, 9.
213. Themelis & Touratsoglou, 1997, 194–195 and A30, A71–72, A90.
214. Catling, 1989, 69.
215. Choremis, 1980, 10–13; 18–20: fig. 4–8; Rakatsanis & Otto, 1980, 55–57 and fig. 3.
216. Andronikos, 1988b, 138–139, fig. 95–96; 140–144; Andronikos, 1981.
217. Deleman, 2006, 263 and n. 124. See also the Thracian prince in Euripides, *Rhesus*, 305; 370–371; 380–385.
218. Whitley, 2005, 71.
219. Bronze silver-gilt greaves, Athenaeus, 5.202f; Winter, 1912, taf. 6; von Graeve, 1970, taf. 68.1; Andronikos, 1988b, 146; 216–217, fig. 185. Dedication of panoply, Liampi, 1998, 3; Themelis, 2003, 165.
220. Plutarch, *Eumenes*, 14.4.
221. Polyaenus, 4.12.14.
222. *Suda* s.v. *Leonnatos*; Heckel, 1992, 91–106; Heckel, 2006c, 147–150.

223. Diodorus, 20.51.1.
224. Athenaeus, 5.195f.
225. Athenaeus, 5.202e.
226. Asclepiodotus, 2.8.
227. I. *Maccabees*, 6.1–2.
228. Pollux, 1.175; Sekunda, 1994a, 14–15; Sekunda, 2001b, 91–92.
229. *Historia Augusta*, *Severus Alexander*, 50.5; Pollux, 1.175.
230. *Alexander Romance* (Greek version), 2.26.
231. Originally written in Greek, possibly as early as the third century AD, then translated to Latin and epitomised into Greek recensions of the *Alexander Romance*; Berg, 1973; Gunderson, 1970, 355–359.
232. *Alexander's Letter*, 142, 9; Heckel, 2006c, 273.
233. Bar-Kochva, 1989, 326 and n. 61–62; Burnett, *et al.*, 1982. See also Born and Junkelmann, 1997, 189–193: for the Roman use of brass and tin in lieu of gold and silver.
234. Xenophon, *Anabasis*, 3.2.7; Aelian, *Varia Historia*, 3.24.
235. Onasander, 29.2.
236. Billows, 1995, 24–55.
237. Lush, 2007, 28.
238. Curtius, 8.8.15–17.
239. Hammond, 1989a, 217.
240. Homer, *Iliad*, 19.13–27.
241. Plutarch, *Eumenes*, 13.4–5.
242. Plutarch, *Philopoemen*, 9.6–8.
243. Gilliver, 2007, 15–16.
244. Athenaeus, 5.195f; 5.202e; Diodorus, 31.8.11; Polybius, 30.25.5; I. *Maccabees*, 6.39; Pausanias, 5.10.5; Onasander, 1.20; Aulus Gellius, 5.5.2; Pollux, 1.175; *Historia Augusta*, *Severus Alexander*, 50.5; Plutarch, *Philopoemen*, 9.5–8; *Aemilius Paullus*, 18.7–8.
245. Plutarch, *Sulla*, 16.2–3; Peachey, 2005, 24. Mithridates often had himself represented as a second Alexander – see Mayor, 2010, 65–67; Oikonomides, 1958: in Greek with English summary.
246. Suetonius, *Julius Caesar*, 67.2; Green, 1978. See also the advice at Onasander, 1.20; 28.

Chapter 5: Cavalry Horses

1. Thomas, 2007, 28.
2. Evelyn-White, 1970, 157; Corrigan, 2004, 22–23.
3. Euripides, *Bacchae*, 575.
4. See also Diodorus, 16.34.5.
5. Grattius, 522–526.
6. For an excellent overview, see Corrigan, 2004, 13–21.
7. Theophrastus, *Causes of Plants*, 5.14.5.
8. Corrigan, 2004, 18–19.
9. Corrigan, 2004, 19–20 and n. 23–25.
10. Herodotus, 8.137–138.
11. Thucydides, 2.100.2.; *Alexander Romance* (Greek version), 1.13; Kroll, 1977a, 87–88; 109, fig. 2, pl. 33; 111, fig. 2, pl. 34; Braun, 1970, 258–259, Z18; Raymond, 1953, 78, n. 5; 126, n. 108 and no. 110–111; Kremydi-Sicilianou, 1999, 651–652; Hammond, 1989c, 55.
12. Plutarch, *Alexander*, 3.8; 4.9; Plutarch, *Solon*, 9.16; Justin, 12.6.6; Kertesz, 2007.
13. Lindsay Adams, 2008, 62–63.
14. Jacoby, 1923, 15 F 225b; Polybius, 8.11.5–13; Athenaeus, 6. 260d–261a; Plutarch, *Alexander*, 15.2. For mention of Companion squadrons, or *ilai*, with contingents raised from Amphipolis and Apollonia in Mygdonia as well as Anthemus and around Olynthus, see Arrian, 1.2.5; 1.12.7; 1.14.1; 1.14.6; 1.15.1; 2.9.3; 3.11.8; Hammond & Griffith, 1979, 352–353; 367–371; Spence, 1993, 27–30.
15. *Alexander Romance* (Greek version), 1.13.
16. Akamatis, *et al.*, 2004, 137.

17. Plato, *Critias*, 117b–c.
18. Livy, 44.28.5; Hatzopoulos, 2001, 42–43; Lévêque, 1968, 269; Xenophon, *On Horsemanship*, 4.1–5; Anderson, 1961, 89.
19. Plutarch, *Aratus*, 6.2.
20. Griffith, 2006, 197.
21. Polybius, 18.11.4–7.
22. Diodorus, 19.67.2; Plutarch, *Philopoemen*, 9.6; Briscoe, 1978; Bakhuizen, 1970, 55.
23. Plutarch, *Aratus*, 24.1.
24. Plutarch, *Aratus*, 18.1–2.
25. Polybius, 5.37.7–12; Rostovtzeff, 1932, 730.
26. Strabo, 16.2.10; Clutton-Brock, 1992, 111–112; Bikerman, 1938, 92.
27. Polybius, 10.27.1–2.
28. Diodorus, 4.15.3; Semple, 1922, 22–24; 27.
29. Corrigan, 2004, 64–66; Griffith, 2006, 196.
30. Calcani, 1989, 92–119 and fig. 49–66; Corrigan, 2004, 58–81; 172–231; Anderson, 1961, pl. 26a; Anderson, 1930, 3–7; 8–10; Hyland, 1968, 15.
31. Herodotus, 8.115.
32. Draganov, 2000; Hammond, 1972, 75–115. See also Grose, 1926, pl. 120: 1–11; 12–17; 18–21; pl. 121: 4–8; pl.122: 1–8.
33. Xenophon, *Hellenica*, 5.2.38; 5.2.40; 5.3.1; Corrigan, 2004, 23–57.
34. See also Gardiner-Garden, 1989, 32–36.
35. Justin, 9.2.14–16.
36. Justin, 9.3.1–2; Lucian, *Octogenerians*, 10; Corrigan, 2004, 58–60; Gaebel, 2002, 24; 157.
37. Hammond & Griffith, 1979, 582, n. 6.
38. Pliny, 8.56.166; Pausanias, 1.21.6; Corrigan, 2004, 185–185.
39. This was certainly practised much later: Livy, 44.28.5.
40. Pliny, 8.56.163–164.
41. Oppian, 1.170.
42. Arrian, *Cynegetica*, 1.4.23.2; Strabo, 7.4.8; Rice, 1957, 70–71; Rolle, 1989, 101–102; 108.
43. Polybius, 3.6.1–14.
44. Arrian, 4.25.4–5; *Itinerarium Alexandri*, 46.105; Howe, 2008, 226; Bosworth, 1995, 166.
45. Ridgeway, 1905, 305–306; Bokonyi, 1968, 39.
46. *Itinerarium Alexandri*, 35.79.
47. Curtius, 9.3.11.
48. Arrian, 3.15.6; 3.20.1; Justin, 12.1.2.
49. Arrian, 4.5.5; 6.24.5; Diodorus, 17.94.2; Hyland, 2003, 66–70; 156–159; 161–162.
50. Curtius, 9.10.11–12; Arrian, 6.25.1.
51. Arrian, 3.3.6; 3.20.1–2; 3.21.6; 6.25.4–5; Diodorus, 17.94.2; Curtius, 8.2.34; 9.10.12; *Itinerarium Alexandri*, 27.64.
52. For more on the system inherited by Alexander in Asia, see Hyland, 2003, 120–121; Herodotus, 3.90; Strabo, 11.13.8; 14.9; Xenophon, *Anabasis*, 4.5.24; 4.5.34.
53. Arrian, 1.26.2; 1.27.4.
54. Arrian, 3.17.6; *Itinerarium Alexandri*, 28.66.
55. Herodotus, 1.192.
56. Pliny, 6.25.29; Cook, 1983, 54; Azzaroli, 1985, 89; Edwards, 1987, 65–66; Sidnell, 2006, 86–88.
57. Diodorus, 17.110.6; Arrian, 7.13.1; Strabo, 11.14.9; 11.13.7; Aelian, *On Animals*, 17.17.
58. Plutarch, *Eumenes*, 8.3; Curtius, 4.9.3–4. For other satrapies as centres for probable state or private herds, see Curtius, 10.1.24; Diodorus, 19.20.3–4.
59. See also Briant, 2002, 463–465.
60. Plutarch, *Eumenes*, 8.3; Gaebel, 2002, 209.
61. Turner, 1955, 127–128.
62. Hunt, *et al.*, 1938, 32–33.
63. Plutarch, *Eumenes*, 4.2–3; Briant, 1982, 30–50.
64. Diodorus, 19.91.5; Corrigan, 2004, 441–453.

65. Hammond, 1998b, 425.
66. Diodorus, 17.49.3; Curtius, 4.7.9.
67. Strabo, 17.3.21; Oppian, 1.290–295; Pindar, 4.17; Athenaeus, 3.100f, quoting from *The Unhappy Lovers* of the fourth-century BC poet, Antiphanes.
68. Arrian, 3.7.3; Hammond, 1996a, 133–135, n. 19.
69. Strabo, 11.14.9.
70. Curtius, 5.1.21; mention of 'horses' is also given in a contemporary Babylonian astronomical diary for Alexander's entry into Babylon, although surrounding breaks prevent identification of these as gifts – see van der Spek, 2003.
71. Curtius, 9.8.1; Arrian, 6.14.3. For more on chariots and Alexander, see Plutarch, *Alexander*, 23.2; Arrian, 3.30.4; *Indica*, 34.6; Curtius, 5.1.33; 4.6.29; Suetonius, *Gaius*, 19.2; Dionysius, 59.17.3.
72. Curtius, 8.12.16; Diodorus, 19.20.4; Athenaeus, 5.194e–f; Polybius, 30.25.6–9.
73. Arrian, 3.24.1–3; 5.19.4–6; Curtius, 6.5.11–21; Diodorus, 17.76.3–8; Plutarch, *Alexander*, 44.2–3; 45.3; Plutarch, *Moralia*, 341b.
74. Curtius, 7.9.16.
75. Arrian, 3.30.6. The replacements perhaps resembled the modern Turcoman breed – see Sidnell, 2006, 87–88.
76. Arrian, 7.15.1–3; Diodorus, 17.111.4–6; Plutarch, *Alexander*, 72.3; Strabo, 11.13.6; 16.1.11; Hammond, 1978, 138.
77. Arrian, 3.19.5–6.
78. Curtius, 7.1.15; Heckel, 2006c, 39: for the questionable historicity of Antiphanes.
79. Curtius, 7.1.34–35.
80. For some possible surviving fragments, see Athenaeus, 11.782; 784; 13.608a.
81. Braun, 1970, 129–132; 198–269; pl. 83–92.
82. Kroll, 1977a, 84–85; Camp, 1998, 37–38: fig. 51–52; Bugh, 1982.
83. Braun, 1970, 256–264: for a full list of 57 brand marks.
84. Aristotle, *Athenian Constitution*, 49.2.
85. Aristotle, *Athenian Constitution*, 49.1; Xenophon, *Cavalry Commander*, 1.13; 3.1; 9; Hatzopoulos, 1994, 55–61; Manakidou, 1996. Fourth/third century BC inscriptions from Macedonia describe *hippon dromos*, 'horse-races', possibly connected to military training or funerary rites.
86. Kroll, 1977a, 88–89.
87. Kroll, 1977a, 97–99.
88. Thucydides, 2.100.2; Sekunda, 2010, 467–468.
89. Rostovtzeff, 1922, 167–168.
90. Hatzopoulos, 2001, 47–49.
91. Hyland, 2003, 31.
92. Dio Chrysostom, *Orationes*, 3.130.
93. Hyland, 1990, 171; Hyland, 2003, 29–31; Lugli, 1972, 112; Grattius, 532; Andronikos, 1988b, 104–105, fig. 60–63; 107, fig. 64; 109, fig. 66; Polybius, 30.25.6; Athenaeus, 5.194d–e; Herodotus, 7.40; 3.106; Aristotle, *History of Animals*, 632a30: Aeschylus, 317–318; Oppian, 1.310–314; Philostratus, *Imagines*, 2.5; Ammianus Marcellinus, 23.6.30; Dio Chrysostom, 26.45.
94. Curtius, 7.7.35–36; Arrian, 3.30.6; Engels, 1978, 102; Naval Intelligence Division, 1945, 449.
95. Curtius, 7.4.30; Hyland, 2003, 22; Sidnell, 2006, 86.
96. Curtius, 9.10.17; 9.10.22.
97. Diodorus, 17.110.6; Arrian, 17.13.1; Strabo, 11.14.9; 11.13.7.
98. Sidnell, 2006, 88.
99. Justin, 12.4.8.
100. Plutarch, *Pyrrhus*, 11.2.
101. Arrian, 7.3.2–4; Plutarch, *Alexander*, 59.3.
102. Pekridou, 1986, Abb.1; taf. 5; Picard, 1964; Heckel, 1992, 171–175; 2006c, 8–9.
103. Palagia, 2003, 149; Tsibidou-Avloniti, 2002a, pl. 23A and frontispiece.
104. *Suda* s.v. *Leonnatos*; Arrian, *Successors*, F12.
105. Plutarch, *Eumenes*, 8.3.
106. Diodorus, 19.20.2–4.

Part III: Dress and Panoplies

1. For one of the most recent see Anson, 2010b.
2. E.g. Andronikos, 1970/71; Connolly, 2000; Sekunda, 2001c; Corrigan, 2004, 453–520.

Chapter 6: Cavalry

1. Saatsoglou-Paliadeli, 1993, 141–142.
2. Saatsoglou-Paliadeli, 1993, 123–126; 127; 136; Post, 2010, 15–16.
3. Kingsley, 1981a; 1991; Fredricksmeyer, 1994; Fredricksmeyer, 1986; Prestianni Giallombardo, 1993.
4. Saatsoglou-Paliadeli, 1993, 137–139; 140.
5. Rotroff, 2003, 217–221 and fig. 7–10. The most thorough investigation to date is Janssen, 2007, 250–251 and taf. 22–29, 32, 34–37 and 39.
6. Sokolovska, 1986, pl. 57.7.
7. Brown, 1957, 24–25, cat. no. 16, pl. x; 52–53, cat. no. 34, pl. xxiv.1.
8. Janssen, 2007, taf. 30–33; 40.
9. Saatsoglou-Paliadeli, 1993, 140.
10. Polyaenus, 5.44.5; Diodorus, 17.7.8: does not include the details imparted by Polyaenus; Heckel, 1992, 355–357; Heckel, 2006c, 74–75.
11. Pollux, 10.162.
12. Plutarch, *Pyrrhus*, 11.6; Plutarch, *Demetrius*, 44.6.
13. Kingsley, 1981a, 42; Fredricksmeyer, 1986, 222–223. See also Dionysius, 19.2.6.
14. Gow & Page, 1965, 37: Antipater of Thessalonike, *Garland of Philip*, 41; *Suda* s.v. *kausia*.
15. Saatsoglou-Paliadeli, 1993, 124–125; 147.
16. Plautus, *Miles Gloriosus*, 1178–1181.
17. Andronikos, 1988b, 111–113, fig. 68–69.
28. From photographs of the Kazanluk tomb, courtesy of Christopher Webber.
19. Uhlenbrock, 1990, 116.
20. British Museum Terracotta 3551.
21. British Museum Terracotta 3233.
22. Brown, 1957, 24–25, cat. no. 16, pl. x; Rouveret, 2004, 45–46, fig. 3.
23. Kingsley, 1981b; Kingsley, 1981a, 39 and n. 4.
24. Fredricksmeyer, 1986, 224; Post, 2010, 15; Theocritus, 14.66; Saatsoglou-Paliadeli, 1993, 143–145; Tarbell, 1906; Robinson, 1939; Plutarch, *Alexander*, 26.5; Strabo, 17.8; Pliny, 5.11.
25. Kingsley, 1991, 66; no. 15 and n. 52; Bruno, 1977, fig. 6: Hermes from the second panel of the Tomb of Judgement; Petsas, 1965.
26. Palagia, 1998, and fig. 1–2, 4–5, 12, 14; Dohrn, 1968, pl. 26, 29 and fig. 8–9; Lattimore, 1975.
27. Sumner, 2009, 96.
28. Kingsley, 1981a, 45, n. 54; Durham, 2001, 108.
29. Post, 2010, 16; Theocritus, 15.6; Delia, 1996.
30. Post, 2010, 16.
31. Morrow, 1985, 74; 88; 149; 182, n. 62;186, n. 46.
32. Moreno, 1987, 89; fig. 40–41.
33. Plutarch, *Moralia*, 187a; Aristotle, *Rhetoric*, 1.7.
34. Diodorus, 15.44.1–4; Pollux, 7.89; Athenaeus, 11.471b; Morrow, 1985, 179, n. 35. For fuller discussion, see Webber, 2011, 128–129.
35. Houser, 1980; Houser, 1987, 251–281, fig. 16.1–16.6; Leslie Shear, 1973, and pl. 36. For *contra*, see Brogan, 2003, 198–203.
36. Morrow, 1985, 107–108; Leslie Shear, 1973, 165–166; Houser, 1980, 231 and n. 6.
37. Calcani, 1989, 80 and n. 316; 126; fig.72–73.
38. For instances of Roman soldiers recorded in texts wearing what is described as *crepidae*, see Sumner, 2009, 203.
39. Lee, 2007, 120.
40. Pollux, 7.85.
41. Curtius, 4.8.8: he died from exhaustion on reaching the bank; Blyth, 1977, 137.

42. Blackman, 1997, 79; Touloupa, 1973.
43. Faklaris, 1986, 50: fig. 30. For comparison with Roman patterns, see Sumner, 2009, 196: fig. 138.
44. For tread patterns on the soles of boots – see Kesner, 1995, 118 and 119: fig. 5.
45. Morrow, 1985, 46–48; 62–63; 72–75; 107–114; 145–146.
46. Lee, 2007, 120; 132.
47. Morrow, 1985, 123.
48. For a modern analogy, see Davis, 1983, 163.
49. Morrow, 1985, 121.
50. Theophrastus, *Characters*, 21.8; 'Petty ambition'.
51. Blackman, 2000, 82.
52. Plutarch, *Alexander*, 40.1; Pliny, 33.14.50; Athenaeus, 12.539c; Arrian, *Indica*, 18.8; Heckel, 2006c, 128.
53. Pasinli, 2000, 33; Winter, 1912, taf. 10. For Demetrius in gold embroidered boots, see Plutarch, *Demetrius*, 41.3–5; Athenaeus, 12.536a. See also Xenophon, *Horsemanship*, 12.10.
54. Andronikos, 1988a; Andronikos, 1989.
55. Polybius, 16.22.4–5.
56. Justin, 38.10.
57. Kingsley, 1984, 68.
58. Calcani, 1989, 79; fig. 39a–b; 81; 40a–b; 83; 41–42; 85; 44; 87; 46; Wallace, 1940.
59. Heuzy & Daumet, 1876, pl. 26; pl. 31.4; Kinch, 1920; Saatsoglou-Paliadeli, 1984, 120–122; pl. 27–28: no. 11a–b.
60. Kosmidou & Malamidou, 2004/05, 141 and 143: fig. 26, L537.
61. Holt, 2003, 70: pl. 3–5; 119: fig. 5; Stewart, 1993, fig. 51.
62. Crawford, 1979, and pl. 67.
63. Vokotopoulou, 1982, 516–520; Vasilev, 1984, 15–17; Archibald, 1998, 254; Alexinsky, 2005, available in Russian with a summary in English at www.xlegio.ru/ancient-armies/armament/early-hellenistic-helmets-with-tiara-shaped-crown/, last accessed June 2012.
64. Waurick, 1988, 176–177 and fig. 64; Choremis, 1980, 13–14; fig. 7–8; 18–20; Schröder, 1912, 327: Form 5 and Beilage 12, 6; Delemen, 2006, 263.
65. Hatzopoulos & Juhel, 2009, 430 and 433: fig. 10.
66. Saatsoglou-Paliadeli, 1984, 55–64: for text; pl. 12–13: no. 4a–b.
67. Besios, 2010, 275; Vasic, 2010, 39; 55: pl. IV.2; Lazar, 2009, 18; Adam-Veleni, 2004, 54, fig. 3–4; Pflug, 1988b, 141; Ninou, 1978, 106 and pl. 60: no. 458.
68. Moreno, 2001, 52; pl. VI; Andronikos, 1988b, 143: fig. 101.
69. Ceka, 2001, 113: *konos* type with cheek pieces in excellent state of preservation. Dated to around 314–312 BC, although this seems a little too precise.
70. Plutarch, *Alexander*, 32.8; Lee, 2007, 130.
71. Ma, 2008, 77; Lee, 2007, 129–130.
72. Pollux, 1.149. However, the situation of a single helmet form seeing widespread, even ubiquitous, service was not entirely alien – as witnessed with abundant archaeological evidence for the use of the Illyrian helmet in Macedonian armies of the 5th century BC: see Elpida Kosmidou's video, 'The Early Macedonian Army: Military Technology and Army Structure, with Particular Reference to the Army of Alexander I', available at http://simpozion2011.bibliotecametropolitana.ro/video-detail-en.aspx?cid=68&vid=600, last accessed June 2012.
73. Moreno, 2001, 51: pl. V, 60: pl. X.
74. Winter, 1912, taf. 2 and 4; Dintsi, 1986, taf. 3.3; Richter, 1954, no. 140, pl. 102.
75. Rizzo, 1925/26, and fig.7; Poulsen, 1993, 165.
76. Sekunda, 2011a, 18.
77. Vermeule, 1984, pl. 18.
78. Wilkinson, 1978, 9, inv. 1977. 256.
79. For the established typology, see Dintsis, 1986b, Beilage 1. For *contra* to the accepted theory of the Boeotian's identity, see Jarva, 1991.
80. Aelian, *Varia Historia*, 3.24; Frazer, 1922.
81. Xenophon, *Horsemanship*, 12.3.

82. Allard Pierson Museum, height: 24.8cm; Louvre, height: 30cm.
83. Hamiaux, 1998, 214–215: cat. 236; Pflug, 1989; Moorman, 2000, 187; 192; pl.90c–d.
84. Frazer, 1922, 106.
85. Moreno, 2001, 46–47: pl.III.
86. See the Early Riding Group's experiments and deductions, available at http://earlyriding-group.org/research_petasosandboeotianhelmet.html, last accessed June 2012.
87. Bull, 1991, 25; Snodgrass, 1967, 94–95; 125; Rumpf, 1943; Schuchhardt, 1978, 76–77; Feugere, 1994, 29; Waurick, 1988, 151–180; Dintsis, 1986a, 1–21; Frazer, 1922.
88. Oikonomides, 1983, 70; Jarva, 1995, 141 and n.992; Everson, 2004, 187; 192.
89. Jarva, 1995, 20–32; Zimmerman, 1979; Richardson, 1996. See also Hagemann, 1919, 36–43; von Merhart, 1954.
90. Jarva, 1995, 33–51; Everson, 2004, 140–159; 193–195. For Macedonian armour see also Petsas, 1966, 118–122; Calcani, 1989, 49–91; Kunzl, 1997; Cadario, 2004, 19–107: tavola III–XI.
91. Pandermalis, 1997, 49; Pandermalis, 1999, 208.
92. This assumption influenced by Xenophon, *Horsemanship*, 12.1–6.
93. Felten, 1989, 424: fig.7.
94. Chrysostomou, 1998; Miller, 1993b, 52; Makaronas & Miller, 1974.
95. Carter, 1970, especially pl.29 and fig.4.
96. Nielsen, 1984, 12–13.
97. Calcani, 1993; Bergemann, 1990, 72–78; taf. 27–32.
98. Moreno, 1995, 52–53, fig.4.18.2.
99. Calcani, 1989, 80 and n.316; 126: fig.72–73.
100. Andronikos, 1988b, 138–139, fig.95–96; 140–144; Vokotopoulou, 1996a, 156–157; Everson, 2004, 187–189; 192–193. There are vague reports claiming the excavation of a possible iron corselet dating to the late-fourth century BC from near Marvinci, although this remains unconfirmed.
101. Head, 1982, 103; Sekunda, 1992a; Vermuele, 1974, 5–8: B.1A; Borza, 1987a, 111–112; Everson, 2004, 192–193.
102. Calcani, 1989, 44: fig.11.
103. For the weight of armour see Diodorus, 18.31.2–3; 19.3.2; Plutarch, *Demetrius*, 21.4; Plutarch, *Philopoemen*, 6.3–4; Snodgrass, 1967, 123; Hanson, 1989, 78–79; Jarva, 1995, 132–139; Everson, 2004, 187–189. For the defensive qualities of Alexander's and his officers' corselets see Arrian, 4.23.3; 4.24.4; 6.10.1; Curtius, 4.6.18; 8.4.19; 9.5.9; Diodorus, 17.20.4; Plutarch, *Alexander*, 16.4–5; 63.3; Plutarch, *Moralia*, 327b; 344d and f. The effectiveness of the linen corselet will be addressed by G. Aldrete and S. Bartell in a work provisionally titled *Reconstructing and testing ancient laminated linen body armor*. See also Markoulaki, 2010.
104. Curtius, 4.13.25.
105. Curtius, 4.6.18; 7.5.16; 7.8.3; 9.5.9. See also Arrian, 5.1.4.
106. Winter, 1912, taf. 1 and 2; Kinch, 1920.
107. Bergemann, 1990, 72–78; taf. 27a, d, and e; Kopcke & Moore, 1979, pl.xxxviii.2–3; Newell, 1927, pl.vii.15; viii.2–3; x.8–10; xvi.13; xvii.18.
108. For Dionysius I of Syracuse in iron armour – see Diodorus, 14.2.2. See also Plutarch, *Timoleon*, 28.1: for the Carthaginians.
109. Archibald, 1985, 176–177; 179; Webber, 2011, 41; Sprague de Camp, 1977, 115–116.
110. Justin, 9.2.10–16; Badian, 1983.
111. Tsibidou-Avloniti, 2002b and pl.VI–VII; Brecoulaki, 2006, pl.94.
112. An early period purple-coloured corselet is seen on the Ariston of the Philaidae pottery fragment preserved at the British Museum; Hanson, 1991, 141.
113. Curtius, 9.3.22.
114. Tornkvist, 1969; Jarva, 1995, 43; 135.
115. Bardunias, 2010, 51–52; Bardunias, 2011: citing *argolis* in Aelian.
116. Curtius, 8.5.4; 9.3.21; Diodorus, 17.95.4; Justin, 12.7.5.
117. For some examples, see Themelis & Touratsoglou, 1997, 47a: A19; 50–51: A71–72; pl.52: A19; pl.59: A97; Ninou, 1978, 35–36: no.22–25; 180–183; 231 and pl.28; Greek Ministry of Culture,

1988, 278–279, fig. 226–228; Despini, 1980, 204–209, fig. 5–7, 9; Despini, 1988, 141–143; Treister, 2001, 383 and 425: fig. 54.1–2.

118. Ognenova-Marinova, 2000; Ognenova, 1970.
119. Xenophon, *Horsemanship*, 1.11.
120. Gaebel, 2002, 27; Corrigan, 2004, 216–217; Hyland, 2003, 35.
121. Xenophon, *Horsemanship*, 12.8–9; Xenophon, *Cavalry Commander*, 1.18; Corrigan, 2004, 216–217.
122. Xenophon, *Cyropaedia*, 8.8.19.
123. Hyland, 2003, 52.
124. Vandorpe, 1997, 987–988; Lewis, 1986, 88–103; Lane-Fox, 2004b, 134.
125. Strabo, 15.1.20.
126. Vandorpe, 1997, 987–988; Corrigan, 2004, 215.
127. Winter, 1912, taf. 1, 11.
128. Ridgway, 1990, 350.
129. Hatzopoulos & Loukopoulos, 1991, 70, colour pl. 48; Brown, 1957, 26, cat. no. 22, pl. xiv.
130. Pandermalis, 1999, 13; Miller, 1993a, 116, fig. 2.
131. Petsas, 1966, 159–170; Rhomiopoulou, 1997, 24–29, fig. 19–23.
132. British Museum Terracotta 3544.
133. Wooton, 2002, 269 and fig. 10. By contrast, the undercloth has a rear dagged edge and is dark green enhanced with a narrow pale yellow border. For a similarly dagged saddlecloth in a Thracian tomb painting from Alexander's time, see Kitov, 2009, 63: fig. 87.
134. Pliny, 19.22; Plutarch, *Alexander*, 67.3; Plutarch, *Eumenes*, 14.4; Florus, 1.23.16; Athenaeus, 12.539c.
135. Vandorpe, 1997, 987.
136. Apuleius, 10.18.
137. Sekunda, 1984a, 17; 37.
138. Doherty, 2004, 40. See also Xenophon, *Cyropaedia*, 8.3.6: perhaps implying uniformity in cavalry furniture.
139. Winter, 1912, taf. 2; Pasinli, 2000, 23; Pfrommer, 2001, 61 and Abb.69.
140. Diodorus, 17.77.5.
141. Justin, 12.7.5.
142. Curtius, 8.5.4; Arrian, *Successors*, F12; *Suda* s.v. *Leonnatos*.
143. *Alexander's Letter*, 143, 11.
144. Curtius, 3.13.10–11; Schmidt, 1957, 97 and 100; pl. 78–79.
145. Moreno, 2001; 18–19; 45: pl. II; 46–47: pl. III; 50: pl. V; Andreae, 1977, 44 and Abb.3; Cohen, 1997, 9; fig. 4; Sidnell, 2006, 84–85.
146. Themelis & Touratsoglou, 1997, 45–48; 84–85; 209; 194 and 48, A121[delta]: for remains of a harness found in Tomb A; Faklaris, 1986, 30: fig. 22; 32: fig. 23; 33: fig. 24; 38: fig. 26; Hyland, 2003, 49–60; Anderson, 1961, 40–49; 64–88; Hrouda, 1969, and pl. 40–41; Treister, 2001, 121–124; Corrigan, 2004, 213–231; Calcani, 1989, 119–127.
147. Athenaeus, 5.194e–f; Polybius, 30.25.6–9; Aulus Gellius, 5.5.3.
148. E.g. as seen in Moreno, 2001, 34; 44–45: pl. II; 46–47: pl. III; 51: pl. V.
149. For all horsemen, see Winter, 1912, taf. 1–2, 4, 10–11; Schefold, 1968, pl. 43.
150. Hatzopoulos & Loukopoulos, 1991, 70, colour pl. 48; Rostovtzeff, 1941a, 150, pl. 19; Brown, 1957, 26, cat. no. 22, pl.xiv.
151. Moreno, 2001, 54–55: pl. VIII.
152. Moreno, 2001, 44–5: pl. II.
153. Xenophon, *Horsemanship*, 5.6–8; Plutarch, *Alexander*, 72.1; Schefold, 1968, taf.28; Napoli & Pontrandolfo, 2002, tav.xxiii. 2.
154. Winter, 1912, taf. 9: to the right; Pasinli, 2000, 33; Coleman Carter, 1975, 68: 185 and pl. 30C.
155. Pottier, 1905; Bieber, 1964, 35–36, pl. II.19; XII.20–21; Calcani, 1989, 32: fig. 6; 106: 61; 45: 11; 102: 56; 104: 57–58; 109: 62; 117: 65; 123: 68; 126: 73.
156. Faridis, 2001, 12.

157. Plutarch, *Alexander*, 32.7; Plutarch, *Eumenes*, 3.5; Arrian, 4.13.1; Curtius, 5.1.42; 8.2.35; 8.6.4–5; 10.5.8; 10.8.3; Diodorus, 17.76.5; 17.65.1; Aelian, *Varia Historia*, 14.49.
158. Diodorus, 18.45.3; 19.28.3; 19.29.5. See also Plutarch, *Eumenes*, 3.5; Anson, 1988, 132, n. 11.
159. Hammond, 1990, 266; 284; Curtius, 5.1.42.
160. E.g. Arrian, 4.24.3, see also Curtius, 8.2.35–36.
161. Heckel, 2003, 205–206; Heckel, 1992, 237–244; 250–252; 289–295; Scholl, 1987, 110–111; Hatzopoulos, 1994, 99–102; Carney, 2008a.
162. Hammond, 1990, 270–271; Diodorus, 18.27.1.
163. Aelian, *Varia Historia*, 14.48; Arrian, 4.13.2; 4.16.6; Curtius, 10.8.4; Diodorus, 18.45.3; 19.28.3; 19.29.5.
164. Polybius, 16.22.4–5.
165. Plutarch, *Moralia*, 760a–b; Janssen, 2007, 230–235; 250–251.
166. Athenaeus, 5.197f; Walbank, 1967, 527.
167. Karunanithy, 2002; Miller, 1974; Fischer, 1994, 161–166; no. 197; 201; 207; 209; 211; 214; 221 and taf. 17–18.
168. For extensive discussion of the frieze see Briant, 1991; Prestianni Giallombardo, 1991, 280–282.
169. Petsas, 1965; Robertson, 1982.
170. Plutarch, *Moralia*, 760a–b.
171. Aristotle, *Politics*, 1297b; Leitao, 1995; Bonfante, 1989; Bonfante, 2000.
172. Palagia, 2003; Vermeule, 1984, pl. 18.
173. Brecoulaki, 2006, pl. 94; Tsibidou-Avloniti, 2002b, pl. VI–VII.
174. Saatsoglou-Paliadeli, 1984, pl. 12–13; no. 4a–b.
175. Kitov, 2001, 25, n. 10. For possible lower order figures in white stripes, see Kitov, 2009, 25: fig. 25; 32: fig. 37; 50: fig. 68. For an alternative theory on vertical stripes indicating men in paid service, see Sekunda, 2012b, 42.

Chapter 7: Infantry

1. Polyaenus, 4.2.10. These troops would have gradually been more fully furnished during the course of his reign following recovery and expansion of the Macedonian economy, embodied with the annexation and exploitation of rich gold and silver mines over 358–355 BC – see Diodorus, 16.8.6; Griffith, 1956; 1965, 127–129, n. 4; Rebuffat, 1983.
2. Cassius Dio, 78.7.1–2; Gow & Page, 1965, 37; Antipater of Thessalonica, *Garland of Philip*, 41; Karunanithy, 2001, 34–36. For hide helmets in Homeric epics, see Homer, *Iliad*, 10. 255–260; Lorimer, 1950, 245.
3. Archibald, 1998, 198–199; Venedikov & Gerassimov, 1973, 87; Kitov, 1996; 1999, 8; Ognenova-Marinova, 2000, 15–17; Delemen, 2006, 265.
4. Greenewalt & Heywood, 1992, 23, n. 13.
5. Curtius, 9.3.21–22. See also Curtius, 3.2.13; 3.2.15; 3.3.26–27; 5.4.31; 6.6.9; 10.2.23–24.
6. Hatzopoulos & Juhel, 2009, 427–428; Juhel, 2009; Sekunda, 2010, 458. For the typological development of the *pilos* into the *konos*, see Dintsis, 1986b, Beilage 3; Juhel, 2009, 347, fig.b.
7. 'Embroideries' or, metaphorically, 'Amulets'.
8. Julius Africanus, *Kestoi*, 1.1.45–50, in Hatzopoulos & Juhel, 2009, 426–427, n. 39; Vieillefond, 1932.
9. For the Spartans at Pylos in 425 BC wearing *piloi* offering poor protection against archery, see Thucydides, 4.34.3. To see examples of the Macedonian helmet with cheek pieces – see Liampi, 1998, taf. 23: M8e–M16a; taf. 24: M18b–M26a; M33a.
10. Schafer, 1997, 105–126; Dintsis, 1986a, 57–73; 75–76; 77–95.
11. Wheeler, 1997.
12. For Sparta's influence on ancient military science, see Cartledge, 1977.
13. Lilimpake-Akamate, 2011, 126–127; 232; 234.
14. Hatzopoulos & Juhel, 2009, 426 and 428: fig. 3.
15. *Suda* s.v. *Alexander*.
16. Juhel, 2009, 352.

17. Mathisen, 1979; Liampi, 1986, taf. 4–6; 1998, 100–109; taf. 23: M6a–M17b; taf. 24: M18a–M26a.
18. Arrian, *Tactics*, 3.5; Pollux, 1.149; Juhel, 2009, 345–346 and n. 16–17; Hatzopoulos & Juhel, 2009, 427, n. 40.
19. Polyaenus, 4.12.14.
20. E.g. Burstein, 1985, 88; .66; Austin, 1981, 137; .74; Asclepiodotus, 2.2; Dintsis, 1986c.
21. Homolle, 1882, 35–37 and 75–76; Feyel, 1935, 31, line 4 and 37, n. 2.
22. E.g. Hatzopoulos, 2001, pl. VI.
23. E.g. Callaghan, 1981, 117.
24. Kottaridi & Walker, 2011, 79: fig. 62; 238: cat. no. 61. A grave *stele* from Pella 425–400 BC, now in the Istanbul Archaeological Museum, also has a warrior in a *pilos* helmet – Sakellariou, 1988, 73, fig. 39; Besios, 2010, 235.
25. Lazaridis, 1997, 66–67 and 106: fig.74.
26. Touchais, 2000, 917 and fig. 169.
27. Lilimpake-Akamate, 2011, 98.
28. Winter, 1912, taf. 2, 5–8 and 14.
29. Tsibidou-Avloniti, 2002a; Brecoulaki, 2006, pl. 94.
30. Juhel, 2009, 354.
31. Praschniker, *et al.*, 1979, Abb.53; Head, 1982, 109: fig. 37.
32. Kosmidou & Malamidou, 2004/05, 138 and 141: fig. 21.
33. E.g. Brecoulaki, 2006, pl. 94: for examples from the Agios Athanasios tomb.
34. Winter, 1912, taf. 2, 5–7 and 14.
35. Deszo & Curtis, 1991, 125; Deszo, 2006.
36. Brown, 1957, 24–25, cat. no. 16, pl. x; 26, cat. no. 22, pl. xiv; 52–53, cat. no. 34, pl. xxiv.1; Fraser, 1964, pl. 10.1.
37. See a translation of this tablet by Bert van der Spek, available at www.livius.org/cg-cm/chronicles/bchp-ptolemy_iii/bchp_ptolemy_iii_01.html, last accessed June 2012.
38. Greenewalt & Heywood, 1992, 23, n. 16.
39. Winter, 1912, taf. 5–6 and 14.
40. Polyaenus, 4.2.10. The omission of corselets is curious, implying Philip did not initially have enough money to equip thousands of men with full body armour – see Diodorus, 16.3.1–2. Alternatively, armour could have been deliberately omitted as a means of providing greater mobility and shock impact to the new phalanx formation. See also Diodorus, 14.43.2–3; Jarva, 1995, 127–128.
41. For mention in the Amphipolis inscription c. 200 BC, see Burstein, 1985, 88: .66; Feyel, 1935; Hatzopoulos & Juhel, 2009, 426 and 428: fig. 3.
42. There are hints that the infantry were provided with corselets in the later years of Philip II's reign or the early part of Alexander's – see Demosthenes, *Third Philippic*, 9.49; Arrian, 1.20.10; 1.28.7; Diodorus, 17.44.2. See also Plutarch, *Philopoemen*, 11.1–2; Curtius, 4.3.26; 7.9.3. For Alexander's initial lack of financial resources, see Arrian, 7.9.6; Curtius, 10.2.24; Plutarch, *Alexander*, 15.2–3; Plutarch, *Moralia*, 327d–e; Justin, 11.5.5; Aeschines, 163; Hamilton, 1968, 36–37. For modesty of equipment at Alexander's accession and before annexing Persia, see Curtius, 3.2.13; 3.2.13; 3.2.15; 3.3.26–27; 5.4.31; 6.6.9; 10.2.23–24.
43. Polyaenus, 4.3.13.
44. Ginouves, 1993, 121 and fig. 107.
45. Praschniker, *et al.*, 1979, Abb. 53.
46. Winter, 1912, taf. 2, 5–6, 13–14.
47. Brecoulaki, 2006, pl. 94.
48. MacDonald, 1899, 348: pl. xxiv.8; Venedikov & Gerassimov, 1973, pl. 347; Liampi, 1998, 134: M83–M84, taf. 28: M83b, M85a, M87a; Merker, 1965, 45–46.
49. Sekunda, 1990, 21.
50. Curtius, 9.3.22.
51. Aeneas the Tactician, 29.4; Bardunias, 2010, 53.
52. Cassius Dio, 78.7.1–2.

53. Karunanithy, 2001. For Caracalla's Alexander obsession, see Cassius Dio, 78.7.4; 78.8.1; 78.9.1; 78.22.1; Herodian, 4.8.1–3; 4.8.9; *Historia Augusta*, *Antoninus Caracalla*, 2.2. For a different view, see Cowan, 2009, 31–35.
54. For late-fourth century BC design, see Bardunias, 2010, 50.
55. Everson, 2004, 193.
56. Lush, 2007, 23.
57. Praschniker, *et al.*, 1979, Abb. 53; on WI, III–IV; SII, IV, VI–VII; OV.
58. Catling, 1989, 69. The regions of Paeonia and Epirus were both famous for their cattle breeds, suggesting a ready abundance of hide, e.g. Herodotus, 7.126; Jacoby, 1923, 115 F284. See also Cicero, *Against Piso*, 36 [87]; Robinson, 1941, 446.
59. Sekunda, 2003, 33; Winter, 1912, taf. 2, 5 and 13.
60. Allemandi, 1998, 130–131, fig. 82; Caro, 1988.
61. Liddel and Scott define a *hemithorax* as a 'half [thorax]' or simply as the 'front plate' of the (metal) *thorax*; Connolly, 1988, 80; Snodgrass, 1967, 110; Griffith, 1956, 3–4.
62. Polyaenus, 4.3.13.
63. Polyaenus is the only ancient source to mention it. However, see Diodorus, 17.65.3–4: after Alexander's army left Babylon for Sittacene he appears to have introduced further improvements for soldiers. With no clue as to its origin and date, the Polyaenus extract has not surprisingly been rejected out of hand by historians such as M. Feyel and G.T. Griffith. See also Whitehead, 1991, 106.
64. Everson, 2004, 194–195.
65. Winter, 1912, taf. 6; Schefold, 1968, taf. 17; von Graeve, 1970, taf. 68.1; Pasinli, 2000, 48; 50. For *contra*, see Heckel, 2006b, 387.
66. Lumpkin, 1975, 194 and lower right face of pl. LXXVI; Heckel, 2006a, 15; 60; pl. C.5.
67. Hatzopoulos & Juhel, 2009, 426 and 428: fig. 3.
68. Pollux, 1.134.
69. Sprawski, 1999, 110–111; Westlake, 1969, 103–125.
70. Boardman, 1994, 183: fig. 202a.
71. Sprawski, 1999, 102–114.
72. Plutarch, *Moralia*, 596d: in 379–378 BC. Date of use implies development predating 380 BC.
73. A useful study comparing the military capacities of Philip II and Jason can be found in Etienne, 1999.
74. For Macedonia's close ties to Thessaly, see Graninger, 2010.
75. Polyaenus, 4.2.10; Asclepiodotus, 5.1; Aelian, *Tactics*, 12; Anderson, 1976; Liampi, 1998, 4–5; Hammond, 1996b.
76. E.g. Markle, 1994, 95, n. 30 and fig. 12.
77. Liampi, 1998, 53–56 and taf. 1.3, 2.1–2, 3 and 33 2–3; Pandermalis, 2002, 100–101.
78. Mitropoulou, 1993, 910; 912; 919; Maass, 1985, 318–321 and 320, n. 51.
79. O'Sullivan, 2008, 97 and n. 68–69.
80. Liampi, 1990, 160–161; Liampi, 1998, 13–15 and taf. 1.1.
81. Miller, 2001.
82. Liampi, 1998, 55–56: S4–S5; taf. 33.2–3; 99–100: M1–M4 and taf. 23: for posthumous tetradrachms of Philip II (336–315 BC) depicting small shields apparently showing five crescents surrounding a central medallion.
83. Markle, 1994, 90, fig. 10.
84. Chrysostomou, 1987.
85. Price, 1991b, no. 57: pl. 19; no. 804: pl. 40; Liampi, 1986, and taf. 4–6; Liampi, 1998, 252–253, n. 18.
86. Winter, 1912, taf. 2, 6–8; Winter, 1894.
87. Huguenot, 2008, 272–273. See also Wright, 1898, 147–148; Vermeule, 1965, 367–369, fig. 12–17; Yalouris, *et al.*, 1980, 151–152; no. 95–98.
88. Besios, 2010, 230.
89. Sarcophagus pediments, Winter, 1912, taf. 14.

90. Andreae, 1977, 56 and Abb. 10; 58 and Abb. 11; 64 and Abb. 14; 66 and Abb. 15; Moreno, 2001, 61; pl. 12.
91. Hartwig, 1898, and taf. XI; Pfrommer, 2001, 64–65 and Abb. 75.
92. Andronikos, 1988b, 199: fig. 160; Markle, 1999, 250.
93. Homer, *Iliad*, 11.41.
94. Paipetis & Kostopoulos, 2008, 200.
95. Andronikos, 1988b, 137.
96. Brecoulaki, 2006, pl. 94; Markle, 1999, 240–241.
97. Giuliani, 1977, 26: Abb. 1.
98. For shields used and held in reserve on rare occasions, principally for dismounted or extraordinary actions – see Arrian, 1.6.5–6; 2.27.2; 4.23.2; Plutarch, *Alexander*, 16.7; Diodorus, 17.20.3; 46.2; Curtius, 4.4.10–11; 8.1.20; Cassius Dio, 59.17.3; Lloyd, 1996, 192; Bosworth, 1980a, 72, n. 6. Cavalry shield use in battle first came into general vogue soon after Alexander, possibly by or around 300 BC – see Hatzopoulos & Juhel, 2009, 428–433, fig. 8 and 10; Nefedkin, 2000b.
99. Brecoulaki, 2006, pl. 90.1; Tsibidou-Avloniti, 2002b, 42.
100. Lendon, 2005, 124–129.
101. For some pattern analyses see Liampi, 1998, taf. 36–39.
102. Rossi, 1971, 114–118.
103. Sekunda, 1994b, 31 and 33.
104. Anson, 2010a, 88–89.
105. Juhel, 2010, 109–110.
106. Baldassarre, 1998; Pontrandolfo, 2002.
107. Vermuele, 1965, 368–369; Winkes, 1979.
108. De Franciscis, 1975a, and fig. 11–12; De Franciscis, 2000, 34 and 35.
109. Winter, 1912, taf. 5–6 and 8.
110. Andreae, 1977, 44 and Abb. 4; Moreno, 2001, 48, pl. IV.
111. Andreae, 1977, 44 and Abb. 4.
112. Cohen, 1997, 8–9; d. on fig. 4; Sekunda, 1992b, 53 and pl. I2.
113. Bresciani, 1980, 140–141, fig. 164–165.
114. Andronikos, 1988b, 145: fig. 103.
115. Polyaenus, 4.2.10; Diodorus, 16.3.1–2; Pausanias, 8.50.1.
116. Treister, 2001, 121; Heckel, 2006a, 16.
117. Curtius, 9.8.4–7; Arrian, *Indica*, 24.1–9; Plutarch, *Eumenes*, 14.4; Diodorus, 17.100.3–5; 20.83.2–3; Theocritus, 93–94.
118. Cf. Diodorus, 16.3.2–3: on Philip II's invention of the pike-armed battle formation with Homer *Iliad*, 13.130–133. It was this extract from Homer referred to in Polybius, 18.28.6–7 as an excellent description of the Macedonian phalanx. See also Curtius, 3.2.13.
119. Themelis & Touratsoglou, 1997, (Delta) 18; A85; pl. 7: A15; pl. 54: A15; pl. 95: B38; Andronikos, 1988b, 145–146, fig. 103; 216–217, fig. 185; Green, 1998, 163; Catling, 1989, 68–69; Blackman, 1997, 45.
120. Winter, 1912, taf. 2 and 6.
121. Everson, 2004, 195.
122. Lee, 2007, 115.
123. Winter, 1912, taf. 2; Andronikos, 1988b, 216–217, fig. 185; Athenaeus, 5.202f: for two outsize pairs of gold greaves recorded being paraded at Ptolemy II's accession in 283 BC.
124. Winter, 1912, taf. 6.
125. Andronikos, 1988b, 146; Andronikos, 1977, 66: fig. 26; Ninou, 1978, no. 88.
126. Kunze, 1991, 66; 76–77; Jarva, 1995, 96–97; 99–100; 136–137; 141–142 and n. 996.
127. E.g. Homer, *Iliad*, 3.330–331.
128. Kroll, 1977b.
129. Webber, 2011, 48.
130. Hanson, 1989, 76.
131. Lee, 2007, 115.

132. *Maurice's Strategikon*, 128; Sumner, 2009, 205.
133. Polybius, 11.9.4.
134. Lee, 2007, 119; Xenophon, *Anabasis*, 4.5.14; Xenophon, *Hellenica*, 2.1.1; Hesychius s.v. *karpatinion*.
135. Ma, 2008, 76, and pl. 6a.
136. Observed with figures on the Alexander Sarcophagus and Agios Athanasios tomb fresco.
137. Blyth, 1977, 137; Lee, 2007, 120.
138. For an appropriate Zulu analogy under King Shaka – see Morris, 1965, 52.
139. Xenophon, *Cavalry Commander*, 7.
140. See the article in *The Daily Telegraph*, 2 March 2004, by T. Leonard, 'Veni, Vidi, Veggie...' available at www.telegraph.co.uk/news/worldnews/europe/italy/1455814/Veni-vidi-veggie....html, last accessed June 2012.
141. Post, 2010, 16.
142. Winter, 1912, taf. 5–6, 13–14.
143. For nudity in artistic representations, see Bonfante, 1989; Bonfante, 2000.
144. Darnell & Manassa, 2007, 82 and n. 208–209.

Chapter 8: Officers

1. Plutarch, *Alexander*, 16.7; Arrian, 1.14.4; Plutarch, *Alexander*, 63.6. For other laudatory references to Alexander's gleaming panoply, see Curtius, 4.4.10; Arrian, 6.9.5; Plutarch, *Moralia*, 343e; Strabo, 15.1.33.
2. Diodorus, 18.61.1; Curtius, 10.6.4.
3. Kunzl, 1997, Abb. 26, 80–81; Tacitus, 2.63; Gissel, 2001. See also Suetonius, *Gaius*, 52.
4. Curtius, 7.1.18; Heckel, 2006c, 24–25; LIoyd, 1996, 171. Reminiscent of an incident recorded for a Roman army officer – see Tertullian, *On the Military Garland*, 1.3.
5. Diodorus, 18.31.1–2.
6. Plutarch, *Eumenes*, 7.4–5.
7. For the personalisation of shield devices used by Macedonian kings and officers, see the informative discussion in Borza & Palagia, 2007, 116.
8. Diodorus, 19.8.1.4; Plutarch, *Pyrrhus*, 16.7; Dionysius, 19.12.4–6; Homer, *Iliad*, 19.15.2; Andronikos, 1988b, 140.
9. Strootman, 2007, 364; Hammond, 1989a, 222–223.
10. Plutarch, *Demetrius*, 21.3–4.
11. For biographical information on Alcimus, see Billows, 1990, 366–367. For a vivid colour reconstruction of Demetrius and Alcimus in their armour at Rhodes, see Karunanithy, 2011, 43.
12. Jarva, 1995, 133–135; 139; Marsden, 1969, 96–97.
13. Homer, *Iliad*, 11.19–38; Jarva, 1995, 49–50; Hainsworth, 1993, 218–219.
14. Karageorghis, 1973, Tomb 79, 19, no. 129; 24, no. 188: as 129; Karageorghis, 1974, pl. CI, CII: no. 129; pl. G 1–2, pl. CIII–CV and CCLVII: no. 188.
15. Westholm, 1938, 169: fig. 5; 170: fig. 6; 171–173: fig. 7–8.
16. Plutarch, *Alexander*, 32.8–11.
17. Holt, 2003, 119; fig. 5 and 121.
18. Curtius, 8.13.21.
19. Plutarch, *Alexander*, 63.6.
20. Diodorus, 16.93.1; Hammond, 1989c, 68, n. 68.
21. Plutarch, *Demetrius*, 41.4–5; 44.6; Plutarch, *Pyrrhus*, 8.1; Athenaeus, 12.536.f; Diodorus, 20.92.3–4; Dionysius, 19.12. 4–6; Wace, 1950, 115; 116–117.
22. Suetonius, *Gaius*, 19.2; Cassius Dio, 59.17.3.
23. Wooton, 2002, 269; 273 and fig. 12.
24. *Alexander Romance* (Greek version), 3.23.
25. Eddy, 1961, 12. For ancient literary *testimonia* on Alexander's dress, see Stewart, 1993, 352–357: T32–48, comprising Diodorus, 17.77.5; Curtius, 6.6.4–5; Justin, 12.3.8; Athenaeus, 12.535f; Plutarch, *Alexander*, 45.2; Plutarch, *Moralia*, 329f–330a; Arrian, 4.7.4; 4.9.9; 7.6.2; 7.22.2–5. See also Plutarch, *Pyrrhus*, 8.1; Lucian, *Dialogues of the Dead*, 13; Herodian, 4.8.1–2.

26. Moreno, 2001, pl. VII–VIII.
27. Curtius, 5.2.18–20.
28. Curtius, 5.1.20. See also Diodorus, 20.93.4; Plutarch, *Demetrius*, 22.1.
29. Athenaeus, 12.537e.
30. Prestianni Giallombardo, 1991, 276–280; Kingsley, 1991, 73; Saatsoglou-Paliadeli, 1993, 140.
31. Curtius, 10.1.24.
32. Cicero, *Tusculan Disputations*, 1.43.102; Livy, 43.30.42; 44.26.8.
33. Aelian, *Varia Historia*, 9.3; Athenaeus, 12.539.f; Polyaenus, 4.3.24.
34. The eighth was Peucestas – see Arrian, 6.10.1; 6.28.4; 7.5.4–6; Heckel, 1992, 263–264; Heckel, 2006c, 203–205.
35. For the role of the *somatophylakes*, their origins, functions and potted career histories, see Berve, 1926, 25–30; Hammond, 1991b, 396–413; Heckel, 1978; Heckel, 1985; Heckel, 1986; Heckel, 1992, 237–288; Heckel, 2003, 206–208.
36. Arrian, 3.5.5; Heckel, 1992, 93.
37. Curtius, 3.12.7.
38. Heckel, 1992, 70–71 and n. 61; Heckel, 2006c, 133–137.
39. Diodorus, 17.37.5; Arrian, 2.12.6–7; Valerius Maximus, 4.7.ext.2; *Itinerarium Alexandri*, 15.37; Curtius, 3.12.15–18.
40. Another candidate is Parmenion, although the face appears simply too youthful to be that of Philip II's veteran general.
41. However, the cloak colours are different: Alexander's cloak is purple with a broad saffron-yellow band, while that of the 'Hephaestion' figure has these colours reversed – see Winter, 1912, taf. 1 and 2; von Graeve, 1970, 151–152.
42. Andronikos, 1988b, 108–109; 111, fig. 65–7, 70–71; Briant, 1991; Prestianni Giallombardo, 1991, 283–284; 284–286.
43. Saatsoglou-Paliadeli, 1993, 126. For Seleucid examples, see I. *Maccabees*, 6.15; 10.20.
44. Plutarch, *Eumenes*, 8. 6–7; Prestianni Giallombardo, 1991, 263–267; Saatsoglou-Paliadeli, 1993, 138.
45. Strabo, 15. 1.63–65.
46. Moreno, 2001, 22; 54: pl. VII; Saatsoglou-Paliadeli, 1993, 136. For identity, see Andreae, 2004.
47. Smith, 1994; Robertson, 1955, 61–62 and pl. XII.
48. Sekunda, 2003, 33.
49. Rouveret, 2004, 45–46, fig. 3; Rouveret, 1998b, 181–182, fig. 1a–b.
50. Athanassakeion Archaeological Museum, Volos (Greece) – see Arvanitopoulos, 1928.
51. Curtius, 4.1.21–23; Plutarch, *Moralia*, 340d.
52. Justin, 12.3.9.
53. Diodorus, 18.48.5; *Book of Esther*, 6: 9; Xenophon, *Cyropaedia*, 8.2.7; Xenophon, *Anabasis*, 1.2.27; Aelian, *Varia Historia*, 1.22; Lucian, *How to Write History*, 39; Widengren, 1956; Thompson, 1965.
54. For adoption of long-sleeved tunics on the Persian model, see Sekunda, 2003, 32; Hatzopoulos & Juhel, 2009, 428; 429. For influence of Achaemenid practices on the Macedonian court and army from as early as Philip II, see Griffith, 1976.
55. Curtius, 6.6.7.
56. Athenaeus, 4.155c–d; *Suda* s.v. *krokotos*; Olmstead, 1948, 70. At least one Macedonian foot figure on the Alexander Sarcophagus is naked in a cloak which is seemingly purple but for a broad saffron-yellow border – see Winter, 1912, taf. 12. The aforementioned 'Hephaestion' figure wears a saffron-yellow cloak with purple band.
57. Winter, 1912, taf. 12; Sekunda, 1984a, pl. B3; 37.
58. Athenaeus, 5.196f.
59. Polybius, 30.25.8–9; Athenaeus, 5.194.e–f; 6.245a; Herman, 1980/81; Speidel, 1997.
60. Virgil, *Aeneid*, 11.768–780.
61. Butler, 2008, 122–123.
62. Plutarch, *Eumenes*, 6.2.
63. Demetrius, *On Style*, 289.
64. Plutarch, *Eumenes*, 6.1; Heckel, 1992, 107–133; Heckel, 2006c, 95–99.

65. In a surviving extract from Arrian's lost *Successors*. See also Goralski, 1989; Plutarch, *Eumenes*, 7.5–6; Cornelius Nepos, *Eumenes*, 4.3–4; Diodorus, 18.30.5.
66. Plutarch, *Alexander*, 32.9.
67. Pritchett, 1979, 265; Homolle, 1890; 1891.
68. Andronikos, 1988b, 140: fig. 97; 144; Dintsis, 1986a, 40; Dintsis, 1986b, taf. 18.2; Hatzopoulos & Loukopoulos, 1981, 227, fig. 129; Spinazzola, 1953, 1004: fig. 1046.
69. Sokolovska, 2011, 46, fig. 18.
70. *Pompei*, 1990a, Ins. 7, 11, 690, fig. 120.
71. Choremis, 1980, 13: fig. 7; 14: fig. 8; 18–20.
72. Plutarch, *Alexander*, 16.7. For more allusions to his magnificent armour, see Arrian, 1.14.4; Curtius, 4.4.10; *Itinerarium Alexandri*, 12.28.
73. Plutarch, *Alexander*, 16.10; Plutarch, *Moralia*, 327a; 341b; Arrian, 1.15.7; Diodorus, 17.20.6; *Itinerarium Alexandri*, 9.21; Hammond, 1989a, 221.
74. Holt, 2003, 118–120 and fig. 5; Oikonomides, 1981, 81; Oikonomides, 1988, 83; Errington & Cribb, 1992, 54.
75. Bieber, 1949, 395: fig. 4; Plantzos, 1996, pl. 23b, 24; Kraft, 1965, and pl. 1–4.
76. Smith, 1980, 65, pl. 8.1–2; 157: cat. no. 7; Dintsis, 1986b, taf. 73.4.
77. Plutarch, *Pyrrhus*, 11.5: from *lophos*; Heuzey & Daumet, 1876, pl. 31; Pais, 1926; Winkes, 1992, 185, fig. 7–10.
78. Hadley, 1974; Houghton, 1980.
79. Morkholm, 1991, 73 and pl. viii. 142–143; Houghton & Stewart, 1999.
80. Sekunda, 1990, 19–20; De Carolis, 2001, 77; Leon, 1989, 22; 26–27, fig. 2a–b, 3c.
81. Livy, 27.33.2; Edson, 1934, 214. See also Crawford, 1974, 284–285: no. 259.
82. Homer, *Iliad*, 9.1–3; Pausanias, 10.23.5; Polyaenus, 1.2.
83. Homer, *Iliad*, 11.40–42.
84. Winter, 1912, taf. 1 and 17. For Alexander dressing up as Heracles, Hermes or Ammon, see Athenaeus, 12.537e; Anderson, 1928.
85. Bieber, 1964, pl. 17.30–32; 20.37–39b.
86. Hill, 1923.
87. Crawford, 1979, pl. 67.
88. Price, 1974, 29 and pl. XIII.74.
89. Ridgway, 2000, pl. 8.
90. Pflug, 1988a, 48: 21 and Abb. 47.
91. Vermeule, 1959/60, pl. II.7 and 33, n. 3; Hanfmann & Vermeule, 1957, 236, pl. 74.
92. Moreno, 2001, 24; 52: pl. VI.
93. Moreno, 2001, 45: pl. II; 54: pl. VII; 68: pl. XV; 60: pl. X. See also Miller, 1994, 110–111; Sekunda, 1984a, 16.
94. E.g. Asclepiodotus, 7.2. For Cleitus the Black as commander of the Royal Squadron near Alexander at the Granicus, see Arrian, 1.15.8; Diodorus, 17.20.7; Plutarch, *Alexander*, 16.11; Curtius, 8.1.20; Heckel, 1992, 34–37; Heckel, 2006c, 86–87. For commanders at the battle of Gaza observed at the head of their respective cavalry guard units, see Diodorus, 19.83.5.
95. Moreno, 2001, 18; 46: pl. III; Andreae, 1977, 44 and Abb. 3; Cohen, 1997, 9, fig. 4.
96. Winter, 1912, taf. 6, 8 and 13.
97. Juhel, 2009, 347, n. 28 and 29; Liampi, 1998, taf. 23: M6a; 7i–j; 8e–g; 9f–g; 12a–e; 13b; 14a; 15a; Sekunda, 2011b, 18.
98. Vegetius, II.13.
99. Schefold, 1968, pl. 21, 25 and 57; Winter, 1912, taf. 2; 5–8; 13.
100. Winter, 1912, taf. 14; Pasinli, 2000, 56; 58.
101. Vokotopoulou, 1982, 501: Abb.10–12; 502: Abb. 13.
102. Hoddinott, 1981, 107, pl. 101.
103. Plutarch, *Timoleon*, 31.3.
104. Head, 1999a, 37–38, Illustration 4.
105. Winter, 1912, taf. 6; Pasinli, 2000, 48–49.
106. Also encountered in Japanese, Italian and French forces.

107. Homer, *Iliad*, 18.598.
108. Polybius, 6.23.13; Gilliver, 2007, 9–10; 13–14.
109. Callaghan, 1981, 117; Markle, 1994, 95, n. 30 and fig. 12.
110. Asclepiodotus, 7.3; Pliny, 10.63; Cicero, *On the Nature of the Gods*, 2.125.
111. MacCary, 1972, 280.
112. Menander, *Perikeiromene*, 294.
113. Plutarch, *Aratus*, 31.3; Jacoby, 1923, 231 F 2.
114. Plutarch, *Aratus*, 32.1–2.
115. Curtius, 9.1.6; Diodorus, 17.89.3; *Itinerarium Alexandri*, 15.36; Plato, *Republic*, 5.468b.
116. Arrian, 7.5.4–6.
117. Arrian, 7.5.5–6; Arrian, *Indica*, 42.9–10.
118. For other references to crown award, see Arrian, 7.10.3; 7.23.5.
119. Oak leaves remain an emblem of bravery in today's armies and one is also reminded of the decorative gold wreaths inlaid upon the skip of peaked caps worn by modern generals. See also Athenaeus, 5.211b; I. *Maccabees*, 10.20; Lloyd, 1996, 183; Arrian, 1.16.4; Plutarch, *Alexander*, 16.15.
120. Roisman, 2003, 310.
121. Winter, 1912, taf. 4; Schefold, 1968, 48–49; Pasinli, 2000, 27; Sekunda, 1984a, pl. C2; 38.
122. Oikonomides, 1989.
123. Lyall, 1852, 283. For Alexander on a cameo wearing a helmet with wreath, see Laubscher, 1995, taf. I.1. See also Burstein, 1978.
124. Gardner, 1886, pl. I.3; Whitehead, 1943; Cunningham, 1971, back cover illustration.
125. Dintsis, 1986a, 163–165; Sekunda, 2001b, 122–124.
126. Six, 1891; Smith, 1988, cat. no. 5; Winkes, 1992, 177–179, fig. 1–3; 186–188.
127. Brown, 1995, 6.
128. Plutarch, *Pyrrhus*, 11.5.
129. Athenaeus, 5.194e–f; Dionysius, 19.12.4–6. See also Diodorus, 22.10.3.
130. Plutarch, *Pyrrhus*, 34.1.
131. Arrian, 7.5.4–6; Curtius, 9.1.6; Diodorus, 17.89.3.
132. Arrian, 7.10.3–4; Bosworth, 1988, 108, n. 64.
133. Austin, 1981, 15 and n. 8; 74; Hatzopoulos, 2001, 161–164, no. 3 A III, lines 2–4; Chaniotis, 2005, 94.
134. Chaniotis, 2004b, 378–386 and n. 48; Robert & Robert, 1983, 24–25; Sahin, 1999, 14, lines 48–49; 23, line 18.
135. E.g. Sandars, 1913, fig. 17k.
136. Russell, 1999, 185–186. For seal rings, see for example Baldus, 1987; Hammond, 1995b; Curtius, 3.6.7; 6.6.6; Plutarch, *Alexander*, 39.3; Plutarch, *Moralia*, 333a.
137. Sherwin-White & Kuhrt, 1993, 132; Mooren, 1998.
138. Curtius, 7.2.16; 7.2.25.
139. Plautus, *Pseudolus*, 55–58.
140. Plautus, *Curculio*, 424.
141. Curtius, 3.7.12; Sherwin-White & Kuhrt, 1993, 132.
142. Sekunda, 1984a, pl. C1; 16; 38. For a photographic close up of the 'bracelet' worn on the nearer or rein arm, see Schefold, 1968, pl. 51; see also the sleeves (bracelets or rolled cuffs) of Persians worn on the same monument at pl. 9 and 36.
143. *Book of Esther*, 6:9; Herodotus, 9.20; Xenophon, *Cyropaedia*, 8.2.7; Xenophon, *Anabasis*, 1.2.27; 8.28.9; Aelian, *Varia Historia*, 1.22; Lucian, *How to Write History*, 39.
144. Diodorus, 18.48.5; Polyaenus, 4.16; Polybius, 5.60.3; I. *Maccabees*, 11.57–58; Sherwin-White & Kuhrt, 1993, 132.
145. Chaniotis, 2005, 94; Empereur, 1981, 566–568, no. 7 and fig. 48.
146. Maxfield, 1981, 89–91.
147. Xenophon, *Horsemanship*, 12.2.
148. Lefebvre des Noëttes, 1931, fig. 241.
149. The chief works remain Faklaris, 1985; Archibald, 1985; and Krasteva, 2007.

150. Curtius, 7.4.36; Plutarch, *Eumenes*, 7.4–8; Plutarch, *Demetrius*, 40.3.
151. Archibald, 1985, 176 and 179.
152. Plutarch, *Alexander*, 32.5.
153. Hammond, 1993, 41: n. 15; Faklaris, 1985, pl. 11.
154. Hurwit, 1999, 60; 260; Kosmetatou, 2004.
155. Faklaris, 1985, pl. 11; 7. Bauer, Neuffer, Wuescher-Becchi, Jaeckel, Lumpkin and Dintsis all identify it as the famous Macedonian headgear. Faklaris presents a number of valid points in support of his argument.
156. Mansel, 1968, and Abb. 48 and 49.
157. Holt, 2003, 118.
158. Bieber, 1949, 395: fig. 4; De Grummond, 1974, pl. 87.1.
159. Faklaris, 1985, 283–284.
160. Hatzopoulos & Loukopoulos, 1981, 64: fig. 4; Faklaris, 1985, pl. 12.b; Andronikos, 1987, 10.
161. Andronikos, 1988b, 72: fig. 33; Andronikos, 1987a, 114, n. 33; Ninou, 1978, no. 98.
162. Andronikos, 1988b, 217.
163. Faklaris, 1985, pl. 4–5 and 7.
164. Archibald, 1985, 169–170 and fig. 4; Faklaris, 1985, pl. 12a; Despini, 1980, 205: fig. 5–6; Ognenova, 1961, 530.
165. Hammond, 1998b, 406.
166. Andronikos, 1988b, 188–189; Hammond, 1989a, 217, n. 5; Archibald, 1985, 165–166; 181.
167. Faklaris, 1985, 283–284; Head, 1999b; Archibald, 1985, 181; Andronikos, 1988b, 189.
168. Silius Italicus, *Punica*, 9.445; 9.462.
169. Buschor, 1958, pl. 54–58; Marx, 1993.
170. Moreno, 2001, 54–55: pl. VII; 57: pl. VIII; Carter, 1970, 132–134; pl. 29, fig. 4.
171. Vermeule, 1959/60; Vermeule, 1964; Vermeule, 1966, 52–53; Vermeule, 1974, 5–8: B.1A; B.2A; 1, 2, 164 and 229; 81; Cadario, 2004, 19–107.
172. Deacy & Villing, 2009; Andronikos, 1988b, 189–190.
173. Andronikos, 1988b, 37.
174. Swindler, 1929, 309 and fig. 487; Pfuhl, 1923, 278: Abb. 658. For fragments of a colossal statue possibly of the same king wearing a corselet with traces of red and gold paint pigment, see Hekler, 1919, 196–198, fig. 124–127; Winkes, 1992, 185, fig. 7–10.
175. Hekler, 1919, 202; fig. 129.
176. Andronikos, 1988b, 37; Calcani, 1989, 39: fig. 16; 58–59.
177. Royal tomb II, Andronikos, 1988b, 190; Kottaridi & Walker, 2011, 91: fig. 77; 240: cat. no. 91.
178. By analogy, see Ognenova-Marinova, 2000, 16, fig. 4.
179. Pekridou, 1986, 43–50, with reliance on Stupka, 1972.
180. Diodorus, 17.77.5; Plutarch, *Alexander*, 45.1–2; 47.3; Plutarch, *Moralia*, 330a; Curtius, 6.6.4–5. Jarva, 1995, 41. For a fuller analysis, see Reyes, 1996.
181. Plutarch, *Alexander*, 51.5; Stewart, 1993, 356 and T46.
182. Pekridou, 1986, 47.
183. Plutarch, *Alexander*, 16.15; Arrian, 1.16.4; Velleius Paterculus, 1.11.3–4; Calcani, 1993; Bergemann, 1990, 72–78; taf. 27–29; 31.
184. Moreno, 2001, 54–55: pl. VII.
185. For more discussion, see Nicgorski, 1995.
186. *Enciclopedia dell'Arte Antica*, 1961, 230–232, fig. 276; Cohen, 1997, 65, fig. 39.
187. Hartle, 1982, 160, fig. 3–4.
188. Stemmer, 1978; taf. 8, 4–5; 13; Kleiner, 1963, taf. III., Abb. 4a–b; 5; Pekridou, 1986, Abb. 1; Brown, 1957, 24–25, cat. no. 16, pl. x; Fraser, 1964 and pl. 10.1.
189. Plutarch, *Eumenes*, 17.2.
190. Pliny, 8.73.193.
191. Pekridou, 1986, 48.
192. Diodorus, 17.30.4.
193. Pekridou, 1986, 48.
194. Pekridou, 1986, 49–50.

195. Pekridou, 1986, 50.
196. Sekunda, 1984a, 17.
197. Hartwig, 1898, and taf. XI; Yalouris, *et al.*, 1980, 122–123: no. 45. Now held at the Museum of Fine Arts, Boston, no. 99.542.
198. Calcani, 1989, 44, fig. 11.
199. Holt, 2003, 125, fig. 6 and 126; Oikonomides, 1988, 83, fig. 18.
200. Price, 1974, pl. XI.56; Le Rider, 1977, 364; Prestianni Giallombardo & Tripodi, 1996, 353, fig. 8.
201. Kopcke & Moore, 1979, pl. xxxviii. 2–3; Newell, 1927, pl. vii.15; viii.2–3; x.8–10; xvi.13; xvii.18.
202. Couissin, 1932, planches I–II.
203. Ridgway, 1990, 350.
204. Schmidt, 1991, 17–18 and Abb. 43; Berger, 1990, 253–55; Moreno, 1994, 80–81, pl. 94–98; 273, Abb. 1; Beilage 27.1; .3; .5; 28.1; .4; Pfuhl & Möbius, 1979a, 339–342; Pfuhl & Möbius, 1979b, taf. 208.1429–1432; taf. 210.1439–1440.
205. Xenophon, *Cyropaedia*, 8.3.6.
206. Arrian, *Indica*, 15.4–5; Strabo, 5.1.44.
207. Curtius, 9.8.3; *Alexander Romance* (Greek version), 3.18.
208. Anderson, 1961, 80; Homer, *Iliad*, 3.16–17.
209. E.g. Siganidou & Lilimpaki-Akamati, 2003, 64–65, fig. 45.
210. Detienne, 1979, 37–38; Bosworth, 1996, 166–169; Calcani, 1989, 122–123; Gabelmann, 1996: for a detailed survey.
211. Liddell & Scott, 1897; Asclepiodotus, 2.9; Arrian, *Tactics*, 10.4; Aelian, *Tactics*, 9.4 (33); Russell, 1999, 149–150, n. 41. For ancient references to the closely associated *phoinikis*, see Diodorus, 13.77.4; 15.52.5; 17.115.2; 18.27.2; Plutarch, *Philopoemen*, 6.2; Polybius, 2.67.10–11.
212. Nylander, 1983, 29; 32; Rumpf, 1962, 235–236.
213. Dieulafoy, 1884, 19, fig. 22; Hill, 1922, 195–210, pl. xxix.6–7; Pope, 1939, pl. 126e–h.
214. Diodorus, 13.77.4; 17.115.2; Plutarch, *Philopoemen*, 6.2.
215. Markle, 1994, 95, n. 30 and fig. 12; Callaghan, 1981, 117.
216. For more artistic evidence for Hellenistic standards, see Swindler, 1929, 344–345: illus. 552; Brown, 1957, 24, cat. no. 15, pl. ix.1; Jaeckel, 1965, 101: fig. 21.
217. Dintsis, 1986b; taf. 31; Moreno, 1994, 423: fig. 541; 427–429: 'fregi d'armi'; Jaeckel, 1965, 101: fig. 21.
218. Bernard, 1969, 50; Sekunda, 1995, 28; Sekunda, 2001b, 148.
219. For mention of Hephaestion's standard, see Arrian, 7.14.10.
220. Diodorus, 17.115.2–4; McKechnie, 1995, 427–428.
221. Homer, *Iliad*, 12.200–207; Plutarch, *Timoleon*, 26.6.
222. Ridgway, 1990, 36–37; Pekridou, 1986, Abb. 8 and 12.
223. Justin, 7.1.
224. Head, 1879, 37: no. 2; 159: no. 2; 163: no. 1.
225. For some sample illustrations, see Price, 1974, pl. xi–xiv; MacDonald, 1899, pl. xx–xxiv; Grose, 1926, pl. 112–119; 120–137.
226. Appian, *Syrian Wars*, 9.56; Justin, 15.4.3; Grainger, 1990, 2; Baldus, 1978; Lund, 1992, 6–8; 160.
227. Lane-Fox, 1973, 31; 506.
228. Athenaeus, 5.198a; Peters & Thiersch, 1905, pl. vi.
229. Sinn, 1979, 138 and taf. 34.6, 35.4; MB101.
230. Swindler, 1929, 344–345: illus. 552.
231. Reproductions appear to show the colour as brown, yellow-brown, russet-brown or dark red – see Breccia, 1912, 12, no. 10, pl. 25 and 26; Pagenstecher, 1919, 53, 69, no. 51 and fig. 49; Bernard, 1995, 42.

Chapter 9: Swords

1. Vokotopoulou *et al.*, 1985, 57 (no. 66); 65 (no. 91–92); 70–71; 119 (no. 179); 131 (no. 212); 150–151 (no. 241–242); 168–169 (no. 269–271); 228–229 (no. 370); 250–251 (no. 409); 280–281 (no. 459, 461); 284 (no. 468); 303 (no. 507).

2. Soteriades, 1903, Beilage XLI: 9, 11–12; Pritchett, 1985, 136–139.
3. Touratsoglou, 1986, 624; Sekunda, 2000, 59 and pl. F13–17.
4. Sekunda, 2001a; Winter, 1912, taf. 2–3, 7, 12.
5. Praschniker, *et. al.*, 1979, Abb. 53: on WI, WIV, SIV, OI, OII.
6. Plutarch, *Alexander*, 32.10.
7. Fabricius, 1885; Lindsay Adams, 1983, 24, fig. 32a; Miller, 1993b, 55 and 57, n. 126; Besios, 2010, 197.
8. Holt, 2003, 120–121.
9. Grose, 1926, 67 and 132.21. See also Diodorus, 19.68.57; 19.77.6.
10. Houser, 1987, 259–261; Houser, 1980, 231; Leslie Shear, 1973, 166–167; Svoronos, 1903, 38, no. 14–15: 14; Pinkwart, 1965, 57, no. 242–243.
11. Touratsoglou, 1986, 624: for design reconstruction; Ninou, 1978, 41, no. 43, pl. 9; Karamitrou-Mentessidi, 1993, 49: fig. 22; 50: fig. 23; 67: fig. 36; 68: fig. 37.
12. Themelis & Touratsoglou, 1997: [Delta] 52; [Delta] 53; [Delta] 62.
13. Andronikos, 1988b, 142–143: fig. 99 and 100; 144–145.
14. Di Vita & Alfano, 1995, 230, fig. 22b; Adam-Veleni, 2004, 53.
15. For one, see Markle, 1982, 101, fig. 23.
16. Touratsoglou, 1986, 626.
17. Aelian, *Varia Historia*, 3.45; Head, 1999a, 37.
18. Blackman, 2002, 81; Adam-Veleni, 2004, 56; Whitley, *et. al.*, 2006, 89.
19. Litvinskij & Picikjan, 1999, 71: fig. 42.6; 82–83;100.
20. *Kopis* from the Greek verb *kopto*, 'to cut'.
21. Vandorpe, 1997, 988–989.
22. Snodgrass, 1967, 97; Blyth, 1977, 25; 138; Sandars, 1913, 237 and fig. 17h, j. See also Barnett, 1983.
23. Herodotus, 8.137.5: knife or sword.
24. Oakeshott, 1960, 49–50, fig. 24a–c; Gordon, 1958.
25. Xenophon, *Horsemanship*, 12. 2; 11.
26. Wilkins, 1992, 10: *On Wounds of the Head*, 11.
27. Curtius, 8.14.28–29. For late Antigonid infantry see De Sanctis, 1934, 520–521; Plutarch, *Aemilius Paullus*, 20.5. Makaronas & Miller, 1974; Miller, 1993b, 54–55 and pl. 9a–b, d; 12c–d; colour pl. III a–b.
28. Diodorus, 17.46.2.
29. Petsas, 1979, fig. 9 and 95–97.
30. Miller, 1993b, pl. 1f; Vollmoeller, 1901, 343.
31. Pekridou, 1986, 53, Abb. 3; taf 5.2 and 6.3.
32. Daux, 1961, 846–847, and fig. 5a; Lumpkin, 1975, 198.
33. Daux, 1961, 846–847, and fig. 5b; Dakaris, 1993, 46, fig. 29; Litvinskij & Picikyan, 1995, 127, fig. 11–12.
34. Vokotopoulou, *et al.*, 1985, 168, no. 269; Sokolovska, 1986, 37: fig. 4.
35. Adam-Veleni, 2004, 57, fig. 9: length 43cm.
36. Adam-Veleni, 2004, 57, fig. 8.
37. Choremis, 1980, 16: fig. 9–10. A scabbard for the sword was also discovered.
38. Litvinskij & Picikyan, 1999; Kruglikova, 1986, 26–27, fig. 24.4; Francfort, 1984, 12, n. 5; pl. 10 and III.
39. Andronikos, 1988b, 142–143: fig. 99 and 100; 144–145; Blyth, 1977, 25; 138 and n. 961; Lee, 2007, 115; Adam-Veleni, 2004, 57, fig. 8.
40. Swords of a distinctly Greek or Hellenistic appearance preserved in wall paintings of mythological scenes from both Pompeii and Herculaneum offer an as-yet largely untapped resource. Scabbards are in most instances painted cream-white.
41. Litvinskij & Picikyan, 1999, 52–77, fig. 7–50.
42. Xenophon, *Cyropaedia*, 2.3.10.
43. Anderson, 1970, 85–86; Cohen, 2003, 8; Cook, 1989.
44. Plutarch, *Pyrrhus*, 7.4–5; Plutarch, *Demetrius*, 41.3–4.

45. Sekunda, 2001a, 35.
46. Amberger, 1998, 170–172.
47. Plutarch, *Eumenes*, 7.5–6; Diodorus, 18.31.1–5.
48. Hanson 1991, 91; 125; 272–273; Atkinson, 2010, 11–14. For other combats see Arrian, 4.24.3–5; Curtius, 9.7.16–26; Diodorus, 17.100.1–101.5; 18.34.2–3; 19.83.5.
49. Curtius, 7.4.33–38; Arrian, 3.28.31; Diodorus, 17.83.4–6; Kebric, 1988, 301.
50. Plutarch, *Pyrrhus*, 8.1; Garoufalias, 1979, 464. Sobriquet.
51. Plutarch, *Pyrrhus*, 10.1.
52. For further passages outlining the terror evoked by Pyrrhus in battle, see Plutarch, *Pyrrhus*, 22.5–6; 24.2–4; 30.5–6. To this day Pyrrhus is commemorated in Albanian folklore with the title of *Birrus* (the 'Brave').
53. Livy, 31.34.4–5. For a general paraphrasis, see Florus, 1.23.7.9–10.
54. Webber, 2011, 102; Gaebel, 2002, 246; Briscoe, 1973, 140; Polybius, 6.25.5–11.
55. Arrian, 1.15.8; Diodorus, 17.20.6–7; Curtius, 8.1.20; *Itinerarium Alexandri*, 9.22.
56. Diodorus, 17.46.2.
57. Curtius, 4.6.15–16.
58. Curtius, 4.9.25.
59. Curtius, 8.14.29–30.
60. Arrian, 4.24.4.
61. Plutarch, *Pyrrhus*, 24.3–4; Adams, 2011.
62. Ma, 2008, 76 and pl. 4c.
63. Loades, 2010, 44.

Part IV: The Men

1. Lee, 2007, 104 and n. 160.

Chapter 10: Veterans and their Families

1. Lock, 1977; Anson, 1981; Anson, 1985; Anson, 1988.
2. Curtius, 8.5.4; Justin, 12.7.5; Diodorus, 17.57.1; Curtius, 4.13.26; Hesychius s.v. *argyraspides*; Hammond, 1983b, 147; 151–152; Hammond, 1996c, 48–49.
3. Németh, 1995.
4. See the interesting article with comments on silver shields by Alan Shaw, available at http://pubs.acs.org/cen/80th/silver.html, last accessed June 2012.
5. Robertson, 1955, 64; Hornblower, 1981, 191–192.
6. Polyaenus, 4.3.24; Athenaeus, 12.539c; Aelian, *Varia Historia*, 9.3. See also Hammond, 1996c, 45–49.
7. Arrian, 7.11.3–4.
8. Dorjahn, 1959; Culham, 1989; Mataranga, 1999.
9. Diodorus, 19.44.1.
10. Diodorus, 19.48.3; Plutarch, *Eumenes*, 19.2; Bosworth, 1995, 243.
11. Bosworth, 2002, 164.
12. Billows, 1990, 299–300; Polyaenus, 4.6.15; Diodorus, 19.44.4; Justin, 14.4.20.
13. Bosworth, 2002, 234–235; see also Roisman, 2012, 237–238, who rejects Bosworth in favour of a colony at Antigonea (Nicaea) in Bithynia.
14. Sekunda, 1989; Grainger, 1990, 95–113; 117; Champion, 2001, 42.
15. Present day authors regard them as the most effective infantry unit in the ancient Greek world (Gaebel) or even all ancient history (Lane-Fox).
16. Plutarch, *Eumenes*, 16.4; Sekunda, 2011b, 21.
17. Compare Plutarch, *Eumenes*, 16.3.–4; 18.1 with Frontinus, 4.2.4; Diodorus, 19.30.5–6; 19.41.1–2.
18. Curtius, 7.4.33–8; Arrian, 3.28.31; Diodorus, 17.83.4–6; Heckel, 2006c, 119.
19. Billows, 1990, 316; Plutarch, *Demetrius*, 29.4–5; Diodorus, 21.1.4; Frontinus, 4.1.10.
20. Justin, 17.1.9: of Seleucus and Lysimachus, 'But at this age they both had the fire of youth, and an insatiable desire of power'; Appian, *Syrian Wars*, 64; Jacoby, 1923, 434; Lund, 1992, 3: born between 361 and 351 BC.

21. Heckel, 1992, 309; Heckel, 2006c, 30–31; Heckel, 1982.
22. Diodorus, 17.9.3.
23. Incidentally, this complements Plutarch, *Eumenes*, 16.4: on the age of the Silver Shields in 317 BC.
24. Justin, 11.6.5–7.
25. Diodorus, 17.27.1–2.
26. Frontinus, 4.2.4; Hammond, 1984a, 51–52 and n. 6; Kebric, 1988. For a rejection of the sources, with a preference for the Silver Shields in their forties or fifties under Eumenes, see for example Billows, 1995, 18–19, n. 63; Anson, 1981, 119, n. 11.
27. Diodorus, 18.29.5; 18.30.4; 18.40.7; 18.53.3; Hammond, 1984a.
28. Hammond, 1984a, 52, n. 8. For a particularly excellent account of the physical hardships most likely sustained, see Gabriel & Metz, 1991, 81–109; Borza, 1979; Borza, 1987b; Mark, 2002.
29. Hammond & Griffith, 1979, 705: quoting Jacoby, 1923, 115 F 348; Diodorus, 16.4.5; *Suda* s.v. *agema*.
30. Frontinus, 4.1.3; Polybius, 5.79.4; 30.25.3; Livy, 42.51.5; Plutarch, *Aemilius Paullus*, 18.3; Josephus, *Jewish War*, 5.460; Appian, *Syrian Wars*, 6.
31. Polyaenus, 4.3.24.
32. Curtius, 7.10.4–9. As these men were the 'noblest born' they were possibly aristocratic recruits to the *Agema* of Companion cavalry. See also Arrian, 7.6.3: for selection of Persian *Euacae* cavalrymen based on recommendations such as outstanding reputation, beauty or excellent physique.
33. Ma, 2008, 76.
34. Hatzopoulos, 2001, 64; Kron, 2005, 72.
35. Musgrave, *et al.*, 1984, 78.
36. Blackman, 1997, 73.
37. Curtius, 10.2.10.
38. Curtius, 9.3.10.
39. *Itinerarium Alexandri*, 50.113.
40. Justin, 12.8.12–13; Milns, 1967, 511; Erskine, 1989; Hammond & Griffith, 1979, 414–418; 705–709; Polyaenus, 4.3.24.
41. Plutarch, *Eumenes*, 11.2.
42. Plutarch, *Moralia*, 339d; Diodorus, 17.45.6.
43. Plutarch, *Alexander*, 70.4–6; Plutarch, *Moralia*, 339c; Heckel, 2006c, 31.
44. Plutarch, *Demetrius*, 40.3.
45. Demosthenes, *On the Crown*, 18.67; Riginos, 1994.
46. Musgrave, *et al.*, 1984, and pl. Va–b; Prag, 1990, 244; fig. 3; 245; fig. 4.
47. Pliny, 35.90.
48. Suitably abstracted under Dio Chrysostom, 64.21.
49. For full source documentation, see Chugg, 2007, 205; Harding, 2006, 89; 241–242.
50. Ruffin, 1992, 469–472.
51. Strabo 7.7.8; Durham, 2001, 108.
52. Polyaenus, 4.3.2; Plutarch, *Moralia*, 180b; Plutarch, *Theseus*, 3.3; Athenaeus, 13.565a; Heckel, 2006a, 58. See also Boardman, 1973; Koehl, 1986; Troncoso, 2010, 21.
53. Plutarch, *Demetrius*, 6.1; Troncoso, 2010, 22–23.
54. Fermor, 1966, 12.
55. Polyaenus, 4.6.13; Justin, 14.3.3.
56. Quoted by Athenaeus, 11.484d. Cyinda was the location for the royal treasury – see Diodorus, 18.52.7; 18.58.1–2; 18.62.1–2; Plutarch, *Eumenes*, 13.1–4; Strabo, 14.5.10; Simpson, 1957; Bing, 1973. The ruins of Cyinda, with thick outer walls, watchtowers and a large storage hall, have been recently identified high atop the Karasis in the Taurus Mountains – see http://makedonia-alexandros.blogspot.com/2010/09/alexanders-treasure-at-kyinda-cyinda.html, last accessed in June 2012.
57. Theophrastus, *Characters*, 23.3.
58. Diodorus, 18.61.4.
59. Menander, *The Shield*, 31–33; 34–37; Plautus, *Miles Gloriosus*, 1060–1065.

60. For retention of booty, see Arrian, 5.26.8; Plutarch, *Alexander*, 20.11; 24.1–3; Diodorus, 17.35.1–4; 17.70.1–6; 17.94.3–4; 17.104.1; Curtius, 3.9.20; 3.13.10–11; 5.6.3–5; 8.1.34; Holt, 1999b, 30; n. 21; Lindsay Adams, 1996; Delrieux, 1999.
61. Plutarch, *Alexander*, 24.3; Justin, 11.5.9. For physical evidence of the rapaciousness of Macedonian plunderers at Persepolis, see Hammond, 1992, 358–359.
62. Curtius, 3.11.20; 13.10–11; 9.10.12.
63. Krasilnikoff, 1992, 33.
64. Arrian, 7.9.2; Curtius, 10.2.23–4.
65. Arrian, 7.5.3–4; Justin, 12.11.1–3; Diodorus, 17.109.2; Curtius, 10.2.10; Plutarch, *Moralia*, 339c; 343d.
66. Bosworth, 2000.
67. Gilliver, 2007, 11; 15; 16–17.
68. Arrian, 5.27.6; 5.27.9; Justin, 12.11.1.
69. Plutarch, *Alexander*, 71.8–9; Atkinson, 2010, 12–13.
70. Polyaenus, 3.9.13.
71. Pandermalis, 1997, 79.
72. Chanioits, 2005, 38.
73. Tod, 1948, 276–278; Heckel, 2006c, 31; 270.
74. Billows, 1995, 34; 213–214 and n. 63; Heckel, 2006c, 34. For more discussion see Roisman, 2012, 238–239.
75. Andrianou, 2006, 240 and 260. For more mention of the fabulous wealth of veterans, see Lindsay Adams, 1996, 34–35. See also Diodorus, 21.12.3–6.
76. Barr-Sharrar, 2008, 43–45: from the *krater* inscription, possibly 'Astioun, son of Anaxagoras, from Larisa'.
77. For Peucestas see Arrian, 6.30.2–3; 7.6.3; Diodorus, 19.14.5.
78. Curtius, 6.11.1. For suggestion of 'Bolon' as invented by Curtius, see Heckel, 2006c, 73.
79. Arrian, 2.12.1;1.26.3; 3.16.11.
80. Diodorus, 17.65.3; this excerpt may hint at reports being quite meticulous documents with critical accounts of the conduct of officers during operations. See also Atkinson, 1987.
81. Hatzopoulos, 1996, 458. For connection to Homeric rank ordering, see Lendon, 2005, 356, n. 25. For 'king's appointees', see for example Diodorus, 17.65.2–3; Arrian, 3.16.11.
82. Curtius, 5.2.5; Heckel, 1992, 304; Diodorus, 19.90.3.
83. Lendon, 2005, 127.
84. See for example Curtius, 5.2.5.
85. Lendon, 2005, 127–128; Blumberg, 2009, 22.
86. Onsander, 34.1–2; Hammond, 1981a, 164; Faure, 1982, 98–101; Roisman, 2003. The fluidity of promotion, transfer and re-assignment of officers in the Macedonian army is obvious from even a brief perusal of the definitive Heckel, 1992, 57–163.
87. Gunn & Gardiner, 1918, 49; 53; Ruiz, 2001, 207.
88. Diodorus, 16.3.3; 16.75.3–4; 16.53.2–3; 16.86.6; Ellis, 1976, 55. For possible precursors see Xenophon, *Cyropaedia*, 8.27; Xenophon, *Hellenica*, 6.16; Diodorus, 14.53.4; Polyaenus, 3.9.31; 14.7.13; Kendrick Pritchett, 1974, 276–290.
89. Arrian, 7.5.4–6; 4.18.7; Diodorus, 17.40.1; 17.46.6: 19.81.6; Curtius, 5.6.20; 7.11.12; Polyaenus, 4.6.15; *Itinerarium Alexandri*, 15.36; Onasander, 4.15; 34.1–2; Carney, 1996, 25; Roisman, 2003, 302–313.
90. *Alexander Romance* (Greek version), 2.21; *Alexander Romance* (Armenian version), 199; Diodorus, 18.48.5. For cups as wedding gifts, see Plutarch, *Alexander*, 70.3. For robes as Achaemenid presents, see Xenophon, *Cyropaedia*, 8.2.7; Xenophon, *Anabasis*, 1.2.27. See also Herodotus, 3.131; Lewis, 1987.
91. The subject has spawned voluminous discussion – see Price, 1991a; Miller, 1994; Lane-Fox, 1996.
92. Holt, 1998; 2003, 44; fig. 2 and pl. 2–13;141–143; 145–147; 167; Holt, 2005b; Holt, 2006; Holt, 2010, 360.
93. Holt, 2003, 149; 155.
94. Olbrycht, 2010, 361.

95. Arrian, 3.19.5–6; 7.12.1–2; Diodorus, 17.74.3; Plutarch, *Alexander*, 42.5; 71.8; Curtis, 1957, 74–75 and pl. X.3.
96. Although the earliest commemorative medals in the English-speaking world were a series struck under the auspices of Elizabeth I (1588), the custom of awarding gold coins as medals can be traced much earlier to eleventh-century Kievan Russia.
97. Ramsay, 1920, 107–112.
98. Diodorus, 19.16.1.
99. Diodorus, 19.16.2–5.
100. Waterfield, 2011, 72; Bosworth, 1996, 27; Heckel, 1979; Diodorus, 19.16.1–3.
101. Justin, 14.2.6.
102. Diodorus, 19.28.1.
103. Diodorus, 19.30.5–6; 19.41.2–3.
104. Diodorus, 19.43.1; Hammond, 1989b, 62. For other passages on Gabiene with the Silver Shields as 'athletes' of war, see Plutarch, *Eumenes*, 16.3–4; Justin, 14.3.5. See also Lendon, 2005, 142. For more references to expert soldiers as 'athletes', see Polybius, 2.20.9. For experience making soldiers formidable, see also Polybius, 3.35.8; 3.89.5; Diodorus, 25.9.1.
105. See the article by Heinz Hilbrecht, 'European Martial Arts – which people killed the knights', available at www.practical-martial-arts.co.uk/practical_martial_arts/hh_landsnecht.html, last accessed in June 2012.
106. Diodorus, 19.15.2–3.
107. Justin, 14.2.8–11.
108. Anson, 1981, 119; Roisman, 2012, 242; Diodorus, 19.22.2.
109. Tremendous morale was not associated with the Silver Shields alone but was a feature common to all of Alexander's veteran units. This is inferred from a fragment of Arrian's lost work on the Successor Wars, summarised by Photius – see Goralski, 1989, 95–96; Bosworth, 1978, 228; 237.
110. Cornelius Nepos, *Eumenes*, 8.2–3; Plutarch, *Eumenes*, 4.2: for Macedonian foot having become 'insolent and self-willed'; Roisman, 2012, 17–30: for the legitimacy of soldiers' concerns.
111. LeGrand, 1917, 94–97; 220–221; Hanson, 1965; Lape, 2004, 62–63; 171–201; MacCary, 1972.
112. Bieber & von Bothmer, 1956, pl. 67, fig. 6; Bieber, 1961, 175: fig. 235. For more artistic remains of Greek New Comedy soldiers, see Arnott, 1988, 11 and taf. I; Webster, 1995, 3, 1AT1; 4, 1AT2 a–b; 189, 3DM6.1; 193, 3DT 1a; 207, 3DT50; 411, 5NP10; 442, 5WT7.
113. Athenaeus, 13.557b; Polyaenus, 4.2.3.
114. Plutarch, *Alexander*, 72.3; Plutarch, *Eumenes*, 2.2.
115. Plutarch, *Alexander*, 48.4–5; Justin, 12.4.2–4; Arrian, 7.12.1–3; Diodorus, 17.110.3; Curtius, 3.11.21–23; 5.1.6; 6.2.2. See also Launey, 1950, 800–803.
116. Arrian, 6.26.5.
117. Justin, 14.3.6.
118. Plautus, *Truculentus*, 482–505; Plutarch, *Demetrius*, 9.6. Soldiers are frequently portrayed in the company of prostitutes over a number of plays, e.g. Terence, *Eunuchus*; Plautus, *Epidicus* and *Pseudolus*.
119. Roisman, 2012, 59–60.
120. Plautus, *Truculentus*, 535–536; 539–540; Plutarch, *Moralia*, 339b–c: for some of Alexander's officers, however brave in battle, being distracted by loot and women.
121. Droysen, 1885; Arrian, 1.29.4.
122. Justin, 14.3.7.
123. Diodorus, 19.41.1–3. For early examples of 'strategic rhetoric', see Hansen, 1993; Hansen, 2001; Hammond, 1999b, 250.
124. Plutarch, *Eumenes*, 16.4.
125. Curtius, 4.2.17.
126. Morrison, 2001. For more on psychology and stress in combat, see also Cilliers & Retief, 2000.
127. Vatin, 1970, 137.
128. Athenaeus, 13.557b; Plutarch, *Alexander*, 38.1–2; 70.3; Arrian, 5.27.6; 7.4.8; Curtius, 5.7.2; Justin, 12.4.8. For the free monthly grain dole, see Diodorus, 17.94.4. See also Plutarch, *Eumenes*, 18.1; Curtius, 10.2.27; Dio Chrysostom, 4.9. For polygamy – see Arrian, 5.27.6; Faure,

1982, 68–74. See also Roisman, 2012, 58: for apparently no evidence of couples registering marriages in the Greek world.
129. Plautus, *Truculentus*, 901–906.
130. Arrian, 1.16.5; 7.10.4.
131. Hatzopoulos, 1996, 403–404; Popazoglou, 1978, 154; Arena, 2003.
132. Arrian, 7.12.2; Diodorus, 17.110.3; Plutarch, *Alexander*, 71.9; Justin, 12.4.9; Pouilloux, 1954, 371–372: no. 141; Hammond, 1990, 277–278; Roisman, 2012, 58. See also Diodorus, 17.94.4; 2.84.3; Bengston, 1975, 481; Bagnall & Derow, 2004: no. 23; Austin, 1981, section 196 and n. 5; Virgilio, 1983; Chaniotis, 2005, 87–88; Stroud, 1971, 287–288.
133. Justin, 12.4.8; Roisman, 2012, 58.
134. Plautus, *Miles Gloriosus*, 1077–1078.
135. Arrian, 7.6.1; 7.12.1–2; Roisman, 2012, 59: daughters probably 'accorded the least of the king's attention.'
136. Justin, 12.4.8.
137. Arrian, 7.12.1–2; Diodorus, 18.4.4.
138. Haythornthwaite, 1988, 23.
139. Justin, 12.4.2–4; Green, 1991, 336.
140. Justin, 12.4.5–10.
141. Plato, *Republic*, 5.466e–467e.
142. For a recent brief discussion of this in a purely military context, see Lendering, 2010.
143. Diodorus, 18.30.5; 19.14.5; 19.27.6; 19.29.3; 19.40.3–4.
144. E.g. Billows, 1990, 292–305.
145. Justin, 12.4.5–7; Bosworth, 1996, 166–184; Bosworth, 1980b, 18; Hammond, 1990, 286–287; Badian, 1963, 201.
146. Livy, 39.24.3; Polybius, 25.3.10; Walbank, 1940, 224 and n. 5.
147. Psoma, 2006, before 285; Pomeroy, 2002, 12–19.
148. Anson, 2008, 22–25; Billows, 1995, 146–182.
149. Arrian, 2.12.1; 2.16.93; 7.12.1–2; 7.23.3; Diodorus, 17.64.6; Curtius, 5.1.45; *Itinerarium Alexandri*, 15.36.
150. Booty was apparently shared in equal proportion between the king and his men, although see Curtius, 4.14.6. See also Hatzopoulos, 2001, 161–164, no. 3 B I, lines 10–18; Juhel, 2002, pl. 37–38.
151. Arrian, 7.5.3; Diodorus, 17.109.2; Curtius, 10.2.9–11; Plutarch, *Alexander*, 70.3; Plutarch, *Moralia*, 339c; 343d; Justin, 12.11.1–2; Lloyd, 1996, 182–183; Berve, 1926, vol. II, 193–196; vol. II, 302–304; Milns, 1987; Faure, 1982, 74–77.
152. Roueche & Sherwin-White, 1985.
153. Carney, 1996, 26 and no. 40.
154. Arrian, 1.16.4–6; 2.12.1; 5.20.1; 5.25.6; Curtius, 3.12.12; 5.4.3–4; 7.9.21; 8.2.40; Diodorus, 17.14.1; 17.21.6; 17.40.1; 17.46.6; 17.64.3; 17.86.6; 17.89.3; 18.32.2; 18.36.1; 19.31.4–5; 19.34.7; 19.85.1–2; 19.85.4; Plutarch, *Eumenes*, 9.2; Plutarch, *Demetrius*, 17.1; Appian, *Syrian Wars*, 3.16; Livy, 36.8.4–6; Onasander, 36; Hammond, 1989b, 57.
155. Thompson, 1982; Rice, 1993; Chaniotis, 2005, 214–244.
156. Miller, 1993b, 51; Chugg, 2007, 206: for a papyrus extract from the second century AD which encapsulates the Macedonian identification with military imagery on tombs.
157. For sculptures from the council building in the Macedonian colony of Sagalassus in Pisidia, see Mitchell & Waelkens, 1987, 40–42 and fig. 2.
158. Watson, 1985, 133–137; Phang, 2001, 39.
159. Diodorus, 16.35.2; Arrian, 5.25.2; 7.8.2–3; Plutarch, *Alexander*, 71.8–9.
160. Curtius, 3.6.16–7; Hamilton, 1999a; Plutarch, *Eumenes*, 13.5–6; Arrian, 7.10.3–5.
161. Gabriel & Metz, 1992, 147; Gabriel, 2005a, 42; 43.
162. Ruffin, 1992, 467–468.
163. Greenwalt, 1986. For the ability of Pyrrhus to heal miraculously, see Plutarch, *Pyrrhus*, 3.4–5.
164. *Suda* s.v. *Hippocrates*; Heckel, 1981.
165. Xenophon, *Cyropaedia*, 1.6.15–16; Xenophon, *Constitution of the Lacedaemonians*, 13.7.

166. Cohn-Haft, 1956; Massar, 2001.
167. Xenophon, *Hellenica*, 6.1.6.
168. Greenwalt, 1986, 221.
169. Salazar, 2000, 68–74.
170. Plutarch, *Moralia*, 177f.
171. Diodorus, 17.31.6; Arrian, 2.4.8; Curtius, 3.6.1–17; 4.6.17–20.
172. Frontinus, 4.7.37.
173. Athenaeus, 7.289c–e; Aelian, *Varia Historia*, 12.51; Nutton, 2004, 46 and n. 62: *Anonymus Londinensis*, 14–20.
174. Pliny, 7.124; Prag, 1990, 239–243; Lascaratos, *et al.*, 2004; Miserachs Garcia & Castillo Campill, 2010.
175. Curtius, 9.5.25; Arrian, 6.11.1; Arrian, *Indica*, 18.7; Heckel, 2006c, 100.
176. Salazar, 2000, 233; Prag, 1990, 240–241: fig. 2a–c.
177. Arrian, 6.11.1–2; Plutarch, *Alexander*, 53.11–13; Curtius, 9.5.22–30; Phillips, 1973, 132–135.
178. *Suda* s.v. *Hippocrates*; s.v. *Drakon*; Heckel, 2006c, 116; 140; Chugg, 2006, 183.
179. Arrian, 2.7.1; 4.16.6; 5.29.3; 7.14.4; Curtius, 3.6.1; Justin, 12.13.6–8; Plutarch, *Alexander*, 41.3–4; 72.1–2; Gabriel & Metz, 1991, 134; Lee, 2007, 244. See also Plutarch, *Alexander*, 41.6–7; Pliny, 14.58.
180. Ruffin, 1992, 467–468.
181. Ruffin, 1992, 467.
182. Xenophon, *Cyropaedia*, 6.2.32.
183. Lee, 2007, 244.
184. Diodorus, 17.95.4; Curtius, 9.1.12.
185. Diodorus, 17.95.4; Ruffin, 1992, 468.
186. Lainas, *et al.*, 2005, 278.
187. Hobhouse, 1858, 142–143.
188. As indicated in the speech of Coenus at Arrian, 5.27.2–9; Borza, 1987b; Ruffin, 1992, 467 and 473–475.
189. Rigorous scientific enquiry undertaken on Alexander's far-flung campaigns led to several works of major historical significance including Aristotle's *History of Animals*. See also Pédech, 1980.
190. Strabo, 15.1.22.
191. Diodorus, 17.90.7; 17.103.8; Curtius, 9.1.12; Justin, 12.10.3.
192. Ruffin, 1992, 468; Aetius, 7.90.
193. Strabo, 15.2.10; Aelian, *Varia Historia*, 12.37; Curtius, 7.4.25; Arrian, 3.28.6–7. For Greek herbal knowledge in general, see Baumann, 1993.
194. Theophrastus, *Enquiry into Plants*, 4.4.5; Strabo, 15.1.21.
195. Curtius, 9.10.1–2.
196. Lee, 2007, 239.
197. Irby-Massie & Keyser, 2002, 311.
198. Curtius, 7.4.23.
199. Olufsen, 1911, 264.
200. Plutarch, *Alexander*, 41.7; Curtius, 9.8.21–22; Diodorus, 17.103.6–8; Strabo, 15.2.7; Justin, 12.10.2–3.
201. Diodorus, 17.103.4–6; 17.103.8; Curtius, 9.8.20; Justin, 12.10.3; Mayor, 2003, 86–91.
202. Plutarch, *Alexander*, 8.1; Adamson, 1973; Maxwell-Stuart, 1976; Lainas, *et al.*, 2005.
203. Mayor, 2010, 70.
204. Gabriel and Metz, 1991, 143.
205. Arrian, *Indica*, 15.11–12; Diodorus, 17.90.7; 17.103.4–6; Mayor, 2003, 86–91; Majno, 1975, 283; 285; Badian, 1998.
206. Thapliyal, 2002, 125.
207. It may be more realistic to say that this Greek expertise was brought to Afghanistan much later by Byzantine exiles in Persia.
208. *Suda* s.v. *Aristogenes*.
209. Garlan, 1975, 136.

210. Morgan, 2008, 65.
211. Berve, 1926, 196–198.
212. E.g. Arrian, *Indica*, 15.11–12; Plutarch, *Alexander*, 63.3–4; Plutarch, *Moralia*, 343d–344d; Justin, 12.9.10–11; Curtius, 7.6.4; 9.5.22; Pausanias, 8.49.6.
213. Lainas, *et al.*, 2005, 277.
214. Arrian, 6.11.1; Curtius, 6.1.5; 7.3.17; 7.6.9–10; 8.4.9; 9.5.15; 9.10.15; Lee, 2007, 246.
215. *Alexander Romance* (Greek version), 2.17; *Itinerarium Alexandri*, 15.36; Arrian, 1.16.5; 2.12.1.
216. Curtius, 9.1.12; Diodorus, 17.95.4; Salazar, 2000, 63.
217. Plutarch, *Pelopidas*, 1.1–2.
218. Curtius, 8.14.45.
219. Plutarch, *Moralia*, 181a; 339c–d; Plutarch, *Alexander*, 41.9.
220. Arrian, 2.5.6–2.6.2; 3.9.1; Diodorus, 17.32.4; Curtius, 3.7.2–15; 4.12.2–3. See also Arrian, 4.16.6.
221. Holt, 2000. See also Arrian, 2.7.1; 3.19.8; 4.16.6; 7.12.1–4.
222. E.g. Curtius, 7.3.23; Arrian, 4.4.1; 4.22.5; 4.24.7; 5.27.6; 5.21.3; *Itinerarium Alexandri*, 38.84.
223. Arrian, 7.12.1–2; Plutarch, *Alexander*, 71.8.
224. Arrian, 6.25.2–3; Gabriel & Metz, 1991, 134; Devine, 1979, 273.
225. Aeneas the Tactician, 16.15.
226. Griffith, 2006, 215, fig. 9.
227. Arrian, 6.25.2; Curtius, 9.10.15; Griffith, 2006, 202, n. 77 and 225; Anderson, 1961, 112; pl. 7; British Musem, 1920, 54, fig. 45a; Lysias, 24.12.
228. Curtius, 7.6.8–11
229. Arrian, 6.13.1–3.
230. Plutarch, *Eumenes*, 14.3; 15.1–2.
231. Plutarch, *Eumenes*, 14.4–5; Curtius, 7.6.8–11; Hall Sternberg, 1999, 192–193. See also Lucian, *Dialogues of the Dead*, 12.
232. Russell, 1999, 114–115; Engels, 1980, 336–337.
233. Diodorus, 17.80.4; Justin, 12.5.4–8; Polyaenus, 4.3.19; Xenophon, *Anabasis*, 5.4.21.
234. Curtius, 7.2.35; Lendon, 2005, 125; Billows, 1990, 397; Heckel, 2006c, 147.
235. Curtius, 7.2.37; Lendon, 2005, 125; Lloyd, 1996, 182.
236. Justin, 12.5.8.
237. Justin, 12.5.7–8
238. Curtius, 7.2.35.
239. Arrian, 4.13.2.
240. This may be hinted at in Curtius, 7.1.19, where a soldier on trial requests that his original unit dress be restored to him.
241. Atkinson, 2010, 11.
242. Curtius, 7.2.37–38.
243. For a general discussion on the *ataktoi* see Berve, 1926, 154.

Chapter 11: Marching

1. Gabriel, 2002, 188. The outstanding work on the subject of logistics is still Engels, 1978, although this must be read with caution. See also Holt, 1993; Mieghem, 1998; Bose, 2004, 197–213.
2. Hammond, 1983a, 31, n. 11; Kalléris, 1954, 262–264; Pollux, 10.16; Photius, s.v. *skoidos*; Arrian, 1.5.9; Asclepiodotus, 11.8; Lee, 2007, 132–135.
3. Arrian, 3.18.1; 3.19.7; Justin, 12.1.3; Gabriel, 2010, 247.
4. Engels, 1978, 35–36, n. 53–54.
5. Xenophon, *Cyropaedia*, 6.2.35.
6. Arrian, 6.23.4.
7. Fuller, 1958, 53. For some statistics on supply logistics, see Hanson, 1999b, 26; 172.
8. Engels, 1980, 327; 340.
9. Fuller, 1958, 188; 90.
10. Green, 1991, 394.

11. Arrian, 1.24.3; 3.14.5; 3.19.3–4; Hammond, 1983a, 30; Lee, 2007, 138.
12. Plutarch, *Alexander*, 37.2; 39.2; 42.4; Arrian, 6.23.5; 6.24.5; 6.25.1.
13. Lee, 2007, 138: citing muleteer studies.
14. Engels, 1978, 15–16; 24; Gabriel, 2005a, 291; Gabriel, 2010, 85–86.
15. Plutarch, *Alexander*, 37.2; Arrian, 6.27.6; Curtius, 4.7.12; 5.2.10; 5.6.9; 8.4.19; Lee, 2007, 136–137: for a full range of modern studies; Griffith, 2006, 229–233.
16. Aristotle, *History of Animals*, 498b8–49930; Engels, 1978, 129.
17. Diodorus, 19.37.6.
18. Diodorus, 20.73.3.
19. Engels, 1978, 17; 14–15, n. 10, 12.
20. Gabriel, 1990, 98–99.
21. Plutarch, *Moralia*, 178a; 790b.
22. Arrian, 3.17.6; Strabo, 11.13.8. See also Livy, 42.53.2–3; 44.26.5–6.
23. Arrian, 6.27.6; Bosworth, 2010, 95; Diodorus, 17.105.6–7; Curtius, 7.4.25; 9.10.12. For Antigonus distributing pack animals among troops who lacked them, see Diodorus, 19.20.1–4.
24. Curtius, 8.4.18–20; 9.10.22; Plutarch, *Alexander*, 56.3.
25. Polyaenus, 3.9.31.
26. Lee, 2007, 136; Griffith, 2006, 204 and n. 86.
27. Plutarch, *Alexander*, 42.7; Curtius, 4.7.12; 7.5.10; 7.5.14. See also Cornelius Nepos, *Eumenes*, 8.7.
28. Curtius, 3.13.16; 5.6.14; 6.5.17; 8.4.11; 8.14.28; 9.2.19; 9.5.19; 9.10.12; Plutarch, *Alexander*, 42.4; 42.7; Polyaenus, 4.2.12; 4.6.15; Diodorus, 17.71.2; 17.105.7; 18.90.7; Xenophon, *Cyropaedia*, 6.2.34; Engels, 1978, 17–18; Griffith, 2006, 214, fig. 8.
29. Polybius, 18.18.3–4.
30. Curtius, 8.4.15.
31. Arrian, 6.26.1; *Itinerarium Alexandri*, 12.27. See also Xenophon, *Cyropaedia*, 1.6.25.
32. Sekunda, 2001c, 40–41.
33. See for example Cooper, 2004, 28–29.
34. Livy, 31.39.10–12.
35. For Philip's *initial* banning of wagons, see Frontinus, 4.1.6. See also Plutarch, *Alexander*, 57.1–2; Curtius, 6.6.15–6; Polyaenus, 4.3.10; Xenophon, *Anabasis*, 4.1.12–3. For interpretation of evidence see Engels, 1978, 15, n. 13. However, to highlight this false premise see Devine, 1997, 273; Hammond, 1980a; Hammond, 1983a, 29–30.
36. E.g. Markle, 1978, 496–497; Manti, 1994, 88. For some useful comments on the perceived possibility of a 'coupling sleeve' for dismantling the weapon, see Dickinson, 2000, 53. The sleeve as a securing device is disputed.
37. Markle, 1977, 334–337 and fig. 1–4.
38. Manti, 1994, 82; 1983.
39. Corrigan, 2004, 498; 499: fig. 7.18.
40. Xenophon, *Cavalry Commander*, 4.1.
41. Diodorus, 17.94.2; Curtius, 8.2.34; Matthews, 2008, 60.
42. Arrian, 6.26.1; Curtius, 6.5.5; Frontinus, 4.3.10.
43. Pliny, 35.95.
44. E.g. Curtius, 4.9.20.
45. Diodorus, 17.90.7.
46. Aristophanes, *Acharnians*, 113b; Lee, 2007, 124–126.
47. Terracotta of a New Comedy mercenary of the third century BC in Ducrey, 1985, 141; fig. 98; Chamay, 1977, taf. 14.3.
48. Besios, 2010, 228.
49. Edmonds, 1961, 659. For other passing references, see Xenophon, *Anabasis*, 2.15–20; Diodorus, 20.11.2; Sparkes, 1975, 129–130 and pl. XIIIf.
50. Curtius, 4.9.19.
51. Plutarch, *Alexander*, 37.3; Arrian, 1.21.1; 6.6.2; Curtius, 7.5.10–11; Diodorus, 17.49.5; Cahill, 1985, 383.
52. Lee, 2007, 109–110.

53. Polyaenus, 4.3.2; Plutarch, *Moralia*, 180b; Plutarch, *Theseus*, 3.3; Strabo, 15.1.67; Athenaeus, 13.565a; Julius Africanus, 1.1.45–50. For shaving under Philip II, see Demosthenes, *Second Olynthiac*, 18; Lendon, 2005, 130; 357, n. 31.
54. Diodorus, 17.80.4; Curtius, 7.2.35–38; Polyaenus, 4.3.19; Justin, 12.5.4–8.
55. Plutarch, *Alexander*, 76.1; Plutarch, *Demetrius*, 52.1; Plutarch, *Moralia*, 177e–f; 181e.
56. Xenophon, *Cyropaedia*, 6.2.30; Arrian, 1.3.6; 3.29.4; 4.4.2; 5.9.3; 5.20.8; *Itinerarium Alexandri*, 34.77.
57. Strabo, 15.1.20.
58. Plutarch, *Alexander*, 40.3: for Alexander's recommendation that helmet and weapons remain serviceable.
59. Xenophon, *Cyropaedia*, 6.2.33; Onasander, 28.
60. Ma, 2008, 75; Markle, 1982, 101, fig. 24; Boardman, 1971; Anderson, 1974, 166.
61. Sharp heads, Arrian, 1.4.1.
62. Diodorus, 17.84.4; Lucian, *Dialogues of the Dead*, 22; Justin, 9.3.1–3; Plutarch, *Moralia*, 331b; Plutarch, *Pelopidas*, 18.5; Plutarch, *Alexander*, 51.9–11; Plutarch, *Flamininus*, 8.2; Plutarch, *Aemilius Paullus*, 19.1; 20.8.
63. Engels, 1978, 21, n. 31.
64. Diodorus, 19.93.2; 19.95.2.
65. Edmonds, 1961, 201.
66. Diodorus, 17.49.5; Curtius, 4.7.12; 7.5.10–11; 7.5.14. See also Xenophon, *Cyropaedia*, 6.2.26–27.
67. Maxwell Edmonds, 1961, 126–127. For a full definition and discussion of the noun *aposkeue*, or 'baggage', see Holleaux, 1942, 15–26; Bikerman, 1938, 91–92; Launey, 1950, 780–790.
68. Quoted in Campbell, 2009, 11.
69. Arrian, 3.21.3; Curtius, 5.4.17–18; Cornelius Nepos, *Eumenes*, 8.7; Polyaenus, 4.6.11;4.15; Diodorus, 17.49.5; 19.37.3; 96.4; 20.73.3; Pritchett, 1971, 30–52; Dalloy, 1992.
70. Polyaenus, 4.3.32; Plutarch, *Moralia*, 180a; Plutarch, *Alexander*, 22.4–5; Curtius, 3.2.16; 9.7.16–17.
71. Lee, 2007, 218–219, n. 82; Dalby, 1996, 164–65.
72. Lee, 2007, 216–217.
73. For barley meal, see Arrian, 3.2.1; Plutarch, *Alexander*, 26.8. For mention of bread, see Curtius, 4.2.14.
74. Akamatis, *et al.*, 2004, 138.
75. Arrian, 6.23.6; Diodorus, 19.13.6; Polyaenus, 4.6.11; Strabo, 15.2.5–7; Geller, 1990, 1–2. Philon of Byzantium also approved the use of sesame.
76. Arrian, 4.21.10; Curtius, 7.4.24: on the soldiers consuming fresh water fish with herbs. The Greek version of the *Alexander Romance* mentions fish and mushrooms.
77. Curtius, 5.5.24; 8.4.1.19; 18–20; 9.8.29; Diodorus, 19.21.3.
78. For example, Arrian, 7.24.4.
79. Xenophon, *Cyropaedia*, 6.2.3; Diodorus, 18.41.3.
80. Lee, 2007, 221; Garlan, 1975, 141.
81. Dalby, 2003, 4; 88; 252; 261; Singer, 1927, 2 and n. 4.
82. Codellas, 1948.
83. Athenaeus, 4.128–9.
84. Dalby, 1996, 155.
85. Arrian, *Indica*, 21.10–13.
86. Wood, 1997, 206–207.
87. Athenaeus, 8.340.e–f; Sekunda, 2002; Faure, 1982, 182–184; Lee, 2007, 208–231.
88. Hammond, 1983a, 28; 31, n. 11
89. Maxwell Edmonds, 1961, 319; 126–127; Polyaenus, 4.9.6; Webster, 1995, 85, 1TT9; 190, 3DP2.
90. Engels, 1978, 21, n. 31; Miesen, 2011, 27: 23.1kg. For criticism of Engels' calculations, see Mann, 1986, 39. See also Curtius, 4.9.19; 4.9.21; 6.2.16; 9.10.12; Frontinus, 4.1.6; Polyaenus, 4.2.10; Lee, 2007, 126.
91. Xenophon, *Anabasis*, 5.1.2; Pritchett, 1971, 51. See also Lee, 2007, 256–259.
92. Frontinus, 4.1.6; Polyaenus, 3.10.10; 4.8.4; Diodorus, 19.38.3.

93. Xenophon, *Anabasis*, 6.4.23.
94. Chamay, 1977, taf. 14.1–3; Hodges, 1917.
95. Frontinus, 4.1.7.
96. Winter, 1912, taf. 13; Schefold, 1968, taf. 17; von Graeve, 1970, taf. 68.1; farbtaf. II. I, Giebel c.
97. Bieber, 1961, 142, fig. 519.
98. Xenophon, *Anabasis*, 4.7. 26; Lee, 2007, 242.
99. For more on servants, see Gabrielli, 1995.
100. Frontinus, 4.1.6; Xenophon, *Anabasis*, 3.2.27–28.
101. Curtius, 9.3.10–11; Hammond, 1989c, 132.
102. Curtius, 8.4.13–14; 9.3.10–11; Arrian, 6.25.3.
103. Camp followers who were considered part of the *aposkeue* (or 'baggage') took no part in battles or in army assemblies – see Diodorus, 18.15.1; 19.42.3; 19.43.7–8; 19.80.4; Polyaenus, 4.6.13; Justin, 14.3.12; Plutarch, *Eumenes*, 9.3. For the *aposkeue* of Macedonian-Hellenistic armies, see Holleaux, 1942, 15–17; Parke, 1933, 206–209; Bikerman, 1938, 91–92; Launey, 1950, 780–790.
104. Ellis, 1976, 27; Anson, 2008, 19–20; Gabbert, 1988.
105. Plautus, *Pseudolus*, 1170.
106. Plautus, *Miles Gloriosus*, 1–8. Smiths may have used abrasive substances such as pumice to polish newly made shields: Burford, 1972, 192; fig. V. Several pumice stones have been excavated on the site of Athenian foundries; Mattusch, 1977, 353; n. 30–31; pl. 86.
107. Markle, 1978, 496–497. For *contra*, see Manti, 1994, 88. In support of Markle's thesis, see Diodorus, 19.80.4–5.
108. Arrian, 3.13.6; 3.14.5–6. See also Arrian, *Tactics*, 2.1.
109. Xenophon, *Cyropaedia*, 6.2.3. *Cheiromulai* may suggest the use of the saddle quern. This consisted of a large stone slab upon which grain was placed and rubbed using a small hand stone.
110. Diodorus, 19.49.2; Engels, 1978, 125. One Attic *choinix* was equivalent to about 1.1 litres.
111. Anderson, 1970, 49; Kendrick, 1971, 30–52; Foxhall & Forbes, 1982.
112. Engels, 1978, 123–126, Appendix 1, 144–145, Appendix 5: Tables 1–3.
113. Moritz, 1958, 104, n. 2; 109–110 and fig. 10; White, 1963; Storcke & Teague, 1952, 82; Gordon-Childe, 1943.
114. Strabo, 16.2.10.
115. Frontinus, 4.1.6; Sekunda, 2010, 468.
116. *amphippoi* – a term also met with in Arrian, *Tactics*, 2.3.
117. Diodorus, 17.45.7; Arrian, 1.8.2; 3.24.1; Heckel, 1975.
118. Curtius, 7.1.34.
119. Vigneron, 1968, pl. 10a and c; Lane-Fox, 1973, 73.
120. Curtius, 4.15.36; Pollux, 1.10.132.
121. Curtius, 8.4.34.
122. Hammond, 1978, 138.
123. Diodorus, 19.80.2.
124. Diodorus, 17.76.6.
125. Pollux, 1.185; Anderson, 1961, 89–97.
126. Vandorpe, 1997, 989–990.
127. Plutarch, *Eumenes*, 11.4; Diodorus, 18.42.3; Frontinus, 4.8.34; Cornelius Nepos, *Eumenes*, 5.4–6; Anderson, 1961, 94, n. 25; Podhajsky, 1948, 19; Gaebel, 2002, 210.
128. Hammond, 1990, 268; Heckel, 2006c, 43.
129. Curtius, 7.1.15; Arrian, 1.15.6; 3.13.6. On mounting horses, see Xenophon, *Cavalry Commander*, 1.5; 1.17; Xenophon, *Horsemanship*, 6.12; Anderson, 1961, 84; Hyland, 2003, 138. See also Pollux, 1.213; Diodorus, 17.76.6–7.
130. Arrian, 1.15.6; Heckel, 1992, 290; Heckel, 2006c, 43. If 'Aretis' and 'Aretes' are the same person, then perhaps the difference in office signifies a promotion: from head groom to heading the *Prodromoi*, or 'Mounted Scouts'. See also Bosworth, 1980a, 303–306.
131. Arrian, 3.13.6: for *hippokomoi* at Gaugamela engaged with Persian chariots, suggesting that they must have been armed; Hoffmann, 1906, 179; Berve, 1926, 170–171.

132. For painted *stelai* illustrating accompanying boys from the *Megali Toumba*, Vergina, see Prestianni Giallombardo, 1991, 270–272; 300, fig. 8; Saatsoglou-Paliadeli, 1984, 55–64; pl. 12–13: no. 4a–b; 65–70; pl. 14: no. 5a–b; 127–30; pl. 31–32: no. 13a–b.
133. Hatzopoulos, 2001, 50–51.
134. Hatzopoulos & Juhel, 2009, 432–433, fig. 10.
135. Polyaenus, 4.2.10; Frontinus, 4.1.6.
136. Sallust, *Jugurthine War*, 45.1–3; Frontinus, 4.1.2; Valerius Maximus, 2.7.2.
137. Plutarch, *Marius*, 13.1–2; Frontinus, 4.1.7; Gabriel, 2000, 96; Badian, 1979.
138. Engels, 1978, 22–24; 119–122; Holt, 1993; Mieghem, 1998; Bose, 2004, 197–213; Gabrielli, 1995; Jones, 2001, 45–62.
139. Gabriel, 2010, 87; 89.
140. Arrian, 1.7.4–7.
141. Diodorus, 18.44.2; Billows, 1990, 78–79 and n. 51.
142. Dodge, 1994, 680–683.
143. Heckel, 2006a, 20.
144. Gabriel, 2007, 107–108; Dodge, 1994, 439.
145. Curtius, 8.2.35–40; 7.6.8–11; Arrian, 2.27.6; 4.30.2; Diodorus, 17.82.6–7; Aelian, *Varia Historia*, 10.4. For other references to Alexander's speed of march, reflecting the tough enthusiasm and competitive nature of Macedonian troops, see Plutarch, *Alexander*, 26.7; Plutarch, *Moralia*, 342e; Curtius, 5.7.1; 7.4.1; 5.8.2; 5.13.5–6; Arrian, 1.7.5; 3.17.4; 3.25.7; 7.28.3; Hammond, 1981a, 33–34. See also Plutarch, *Eumenes*, 11.3–4.
146. Curtius, 9.10.12; *Alexander's Letter*, 143, 10.
147. Hammond, 1997b, 139; Gabriel & Metz, 1992, 34–35; Engels 1978, 101–102; Plutarch, *Moralia*, 341a–f: a eulogising monument to Alexander's (and his men's) toughness and powers of endurance.
148. Arrian, 4.4.8.
149. Diodorus, 19.18.1–2.
150. Diodorus, 19.19.2. For anecdotal references to the extreme conditions of the Babylonian region and Persia, see Plutarch, *Alexander*, 35.14; Theophrastus, *Enquiry into Plants*, 8.11.7; Strabo, 15.3.10; Naval Intelligence Division, 1945, 155–187; Olufsen, 1911, 253–263; Jarcho, 1967; Curtius, 7.3.13; Diodorus, 17.82.7.
151. Lane-Fox, 2004b, 117–118.
152. Olufsen, 1911, 259; 260.
153. Hanson, 1989, 56.
154. E.g. Arrian, 5.1.4; Curtius, 4.6.7; 7.5.16; 7.8.3.
155. Curtius, 4.13.25.
156. Arrian, 6.26.3. However, this incident may be apocryphal.
157. Gabriel & Metz, 1992, 34–35; Gabriel & Metz, 1991, 107; Kerstein & Hubbard, 1984; Jarcho, 1967; Vaughan, 1980, 306–307; Steinman, 1987.

Chapter 12: Camping

1. Weber, 2009, 84; 85.
2. For two recent outstanding studies see Alvarez-Rico, 2002; Lee, 2007, 176–194. See also Berve, 1926, 174–176; Faure, 1982, 180–182; Arrian, 4.4.1; Justin, 12.5.12; Polyaenus, 4.3.27; Diodorus, 18.70; 19.39; Curtius, 7.6.26; 9.10.5.
3. Spawforth, 2007, 97 and n. 38.
4. Plato, *Republic*, 7.526d.
5. Diodorus, 10.11.2; 16.2.2–3; Cornelius Nepos, *Epaminondas*, 2.2. See also Vidal-Naquet & Lévêque, 1986, 61–82. The period Lysis lived is uncertain, although he appears to have flourished in the late-fifth century BC.
6. Bekker-Nielsen, 2001, 122.
7. Worthington, 2008, 16.
8. Athenaeus, 12.508 d–e.
9. See the story in Stobaeus, 2.31.115; Plutarch, *Alexander*, 7.1–9.

10. Polybius, 9.20.4. See also Bekker-Nielsen, 2001, 127.
11. Xenophon, *Cyropaedia*, 5.2–14.
12. Aeneas the Tactician, 21.2.
13. Xenophon, *Constitution of the Lacedaemonians*, 12.1–7; Billows, 1990, 317 and n. 3; Strabo, 15.803; Diodorus, 15.32.2–5; Plutarch, *Moralia*, 187a; Polyaenus, 3.9.17; Xenophon, *Hellenica*, 5.4.38–54; 6.5.30; 7.5.8; Buchholz, 1881, 331–342.
14. Xenophon, *Constitution of the Lacedomanians*, 1.2; Xenophon, *Cyropaedia*, 8.5.8–14; Alvarez-Rico, 2002, 41–42.
15. Aeneas the Tactician, 22.1.
16. Pritchett, 135, n. 9; 136–138: Table 2.
17. Demosthenes, *Second Philippic*, 23; Demosthenes, *On the Crown*, 87; Aeschines, 140.
18. Billows, 1990, 317: citing Polyaenus, 4.6.19; Diodorus, 19.18.4; 19.41.6.
19. Polybius, 5.2.5.
20. Xenophon, *Constitution of the Lacedaemonians*, 12.1; Polybius, 6.42.3–4; Alvarez-Rico, 2002, 40; *Alexander's Letter*, 144, 15–16.
21. Curtius, 7.6.25–27; Arrian, 4.4.1.
22. Strabo, 12.4.7; Diodorus, 20.47.5.
23. Reinders, 1988, 181–182; 202; Winter, 1971, 46. More recently, Haagsma, 2010.
24. Pandermalis, 1999, 218–233.
25. For the careers of some engineers, see Heckel, 2006c, 99; 106.
26. Empereur, 1998.
27. Richardson, 2001, 183.
28. Frontinus, 4.1.14; Bekker-Nielsen, 2001, 129, n. 22: for Frontinus' inexplicable terminology.
29. Livy, 2.32.4; 2.45.5; 2.59.2–3. For *contra*, see Frontinus, 4.1.14.
30. Livy, 31.34.8; Plutarch, *Pyrrhus*, 16.4–5; Lévêque, 1957, 324.
31. Gilliver, 1999, 69.
32. Plutarch, *Pyrrhus*, 8.3.
33. Frontinus, 2.3.21.
34. Garoufalias, 1979, 21–22.
35. Champion, 2009, 26; Bekker-Nielsen, 2001, 126.
36. Engels, 1980, 330; 332 and n. 35; 333.
37. Curtius, 7.5.10.
38. Aristotle, *Eudemian Ethics*, 1227a.
39. E.g. Xenophon, *Anabasis*, 1.2.14; 1.2.21; Xenophon, *Cyropaedia*, 5.4.40; 6.1.23; Polybius, 6.42.1–4; Onasander, 8; Polyaenus, 2.30.3; Alvarez-Rico, 2002, 37–38. See also Dahm, 2007a, 45; Dahm, 2007b, 44.
40. Breitenbach, 1950, 73 and n. 10.
41. Arrian, 6.25.6; Ruffin, 1992, 469.
42. Xenophon, *Anabasis*, 1.6.14.
43. Plutarch, *Alexander*, 26.4–5; Arrian, 3.2.1; Diodorus, 19.38.3; Plutarch, *Eumenes*, 15.5–6; Cornelius Nepos, *Eumenes*, 9.3–5; Polyaenus, 4.8.4.
44. *Itinerarium Alexandri*, 35.80; Curtius, 3.8.19; Hanson, 2001, 80.
45. However, see Chugg, 2007, 236.
46. See Hammond, 1987a, for translation and discussion; Clarysse & Schepens, 1985.
47. Curtius, 7.5.13.
48. Curtius, 5.2.6–7; 7.7.9; 8.2.4; 8; 11; 9.5.22; Polyaenus, 4.8.2; Spawforth, 2007.
49. Plutarch, *Alexander*, 57.4.
50. Arrian, 4.16.6; 6.25.5. See also Heckel, 2006c, 195.
51. Athenaeus, 12.538b–539a; Xenophon, *Anabasis*, 1.6.4.
52. Curtius, 8.13.20; 9.6.1.
53. Perrin, 1990, 218; 219–220; Spawforth, 2007, 94–97; 112–120; Diodorus, 17.36.5; Curtius, 3.3.8; 3.12.3; 3.12.8; 3.12.10; 3.13.1–3; 5.2.7; Plutarch, *Alexander*, 20.6; 20.12–13; Athenaeus, 12.539d; Aelian, *Varia Historia*, 9.3.14–17; Florus, 1.23.9.

54. Polyaenus, 4.8.2; Briant, 2002, 188. For the later Antigonid army encampments, see Bustein, 1985, 88–89; Austin, 1981, 136–137.
55. Curtius, 3.12.3; 6.8.18; 8.13.20; Jacoby, 1923, 81 F 41; Polyaenus, 4.3.24; Athenaeus, 12.539e–f; Aelian, *Varia Historia*, 9.3. For Iranian elements in the court ceremony, see Wiesehöfer, 1994, 53–54; Andreotti, 1957, 120–166.
56. Plutarch, *Alexander*, 51.6–7; Arrian, 4.8.8; Curtius, 8.1.47. See also Curtius, 6.8.19–22; Arrian, 7.8.3; Hammond, 1991b, 398–399.
57. Liapis, 2009, 73–80; Euripides, *Rhesus*, 1–3.
58. Curtius, 8.6.10; 8.6.15–16; 8.6.19–20; Arrian, 4.13.1–4.
59. Curtius, 9.6.4.
60. Arrian, 5.28.2–3.
61. Curtius, 6.7.17; Arrian, 4.13.7.
62. Curtius, 4.10.4; Cornelius Nepos, *Eumenes*, 7.2–3.
63. Arrian, 3.26.2.
64. For the routines on the morning of Gaugamela when the King slept late – see Diodorus, 17.56.2; Plutarch, *Alexander*, 32.1; Curtius, 4.13.17–20.
65. Arrian, 1.8.1; Arrian, 3.14.5; 5.11.3; Curtius, 5.4.14; 5.4.29.
66. Frontinus, 4.1.10; Plutarch, *Demetrius*, 23.4; Plutarch, *Moralia*, 182b–c.
67. Arrian, 2.18.5; Polyaenus, 4.3.3; 4.9.1; Plutarch, *Cleomenes*, 37.2; Post, 2010, 15.
68. Fellman, 2003, 126.
69. Frontinus, 4.7.2.
70. Polybius, 6.42.2.
71. Rawlings, 1996.
72. Polybius, 5.2.5.
73. Engels, 1978, 17, n. 19
74. For an excellent introduction, see Gabriel, 2007, 43; 51; 53.
75. Arrian, 3.9.1.
76. Curtius, 4.9.10.
77. Engels, 1978, 17, n. 19; Livy, 33.5.6.
78. Homer, *Iliad*, 7.337–343. For other field works see 9.359–50; 10.180–189; 12.55; 15.1; 15.344; 18.177.
79. Euripides, *Rhesus*, 110–119.
80. Diodorus, 19.47.2. However, see Plutarch, *Moralia*, 187a; Polyaenus, 3.9.17.
81. Arrian, 1.6.9; Morgan, 1983, 53–54, n. 184. For carelessness in camp defence over 312 BC see also Diodorus, 19.95.3–6.
82. Hammond, 1987a, 334–335 and n. 9; Clarysse & Schepens, 1985.
83. Philon, I. 37–38; Garlan, 1973.
84. Xenophon, *Hellenica*, 3.2.2–5.
85. Lawrence, 1979, 162.
86. For the Polybian digression comparing Roman with Greek-Macedonian palisade stakes see Polybius, 18.18.3; Walbank, 1967, 574, 13. See also Livy, 33.5.5–12.
87. For the whole episode, see Diodorus, 17.95.1–2; Arrian, 5.29.1–2; Plutarch, *Alexander*, 62.7–8; Curtius, 9.3.19. See also Justin, 12.8.16–17.
88. Diodorus, 18.41.6.
89. Diodorus, 19.39.1.
90. Diodorus, 20.83.4; Sekunda, 2012a, 12.
91. Polybius, 5.99.9; Diodorus, 19.49.1–2.
92. Diodorus, 20.108.7.
93. Diodorus, 20.108.4–7; 20.109.1–4.
94. Frontinus, 1.5.11; Lund, 1992, 75.
95. *Skenai* is also defined as 'huts' or 'booths'.
96. Sekunda, 2001c, 30–31 and 32, fig. 6.
97. See also Alvarez-Rico, 2002, 50–53; Lee, 2007, 122–123.

98. Sekunda, 1984a, 25. The Roman army tent-group was the eight-man *contubernium* – see Junkelmann, 1986, 93–94.
99. McGeer, 1995, 182; Lloyd, 1996, 180.
100. Sekunda, 2010, 464.
101. Heckel, 2006a, 27.
102. Diodorus, 19.38.3; Polyaenus, 4.8.4.
103. Arrian, 1.21.1.
104. Diodorus, 17.95.1–2; Heckel, 2006a, 27.
105. Heckel, 2006a, 27.
106. Xenophon, *Anabasis*, 3.2.27.
107. Curtius, 7.8.2; Plutarch, *Moralia*, 182d; 457e; Seneca, *On Anger*, 3.22.2.
108. Arrian, 1.3.6; 3.29.4; Bloedow, 2002, 57.
109. Arrian, 4.19.1; Curtius, 7.9.13–14; Polyaenus, 4.3.29; *Itinerarium Alexandri*, 44.99.
110. Curtius, 9.10.25.
111. Anderson, 1970, 62; Drew, 1979, 205, n. 42.
112. Xenophon, *Anabasis*, 1.5.10; Lee, 2007, 122.
113. Polyaenus, 3.9.19; van Wees, 2004, 107–108.
114. Polybius, 3.6.12.
115. Diodorus, 17.16.4; Spawforth, 2007, 87; 90; 92–93; 96. See also Curtius, 6.2.16; 9.7.15; Athenaeus, 12.537d–540a.
116. Diodorus, 16.83.2.
117. Alvarez-Rico, 2002, 50.
118. Bergmann, 1994.
119. Diodorus, 19.38.3.
120. Lee, 2007, 183.
121. Renault, 1982, 255–257.
122. Plutarch, *Eumenes*, 2.2.
123. Xenophon, *Cyropaedia*, 5.2–14.
124. Alvarez-Rico, 2002, 41; 48–49; Lee, 2007, 179.
125. Curtius, 7.2.37.
126. Arrian, 7.11.8–9.
127. Oikonomides, 1988, 78–79; Pandermalis, 2002, 100–101.
128. Diodorus, 19.22.2.
129. Milns, 1981, 350.
130. Polyaenus, 4.8.2; Plutarch, *Eumenes*, 13.3–4.
131. Diodorus, 19.38.3; Polyaenus, 4.8.4.
132. Heckel, 2006a, 27.
133. Polybius, 15.29.1–2.
134. Justin, 12.4.2–3.
135. Plutarch, *Eumenes*, 2.3.
136. From analogy with Homer, *Iliad*, 10.66. See also Lee, 2007, 180 and n. 38–40.
137. Livy, 31.42.4; *Alexander Romance* (Greek version), 2.36; Alvarez-Rico, 2002, 33–36; Lee, 2007, 177; 180–181.
138. Plutarch, *Moralia*, 178a; 790b; Curtius, 3.8.19; Lee, 2007, 183 and n. 57.
139. Xenophon, *Cyropaedia*, 6.3.4; 6.5.13.
140. Polyaenus, 4.9.1.
141. Livy, 31.42.4; 42.59.11.
142. Xenophon, *Cyropaedia*, 6.3.4; 8.5.1.
143. Anderson, 1970, 82–83; Sekunda, 1995, 8; 27–28, section 7; fig. 11b, colour pl. 3.
144. Curtius, 5.2.6–7; Atkinson, 1994, 64; Russell, 1999, 149, n. 19.
145. E.g. Curtius, 3.6.3; 3.10.3; 4.4.14; 4.14.6. The term *ante signa* is a variant of the common Latin *antesignanus*, meaning 'before the standard'.
146. van der Spek, 2003; Lendering, 2004, 164–174.
147. Manning, 2009, 23.

148. Arrian, 7.14.10.
149. Perhaps during the first series of cavalry reforms in 331/330 BC – Curtius, 5.2. 6–7; Diodorus, 17.65.3–4; Brunt, 1963, 28; Daniel, 1992, 47; Aperghis, 1997, 139.
150. Pliny, 19.5.22; Arrian, *Indica*, 28.3–4; Price, 1974, 24–25; Grose, 1926, pl. 125.1–6; Burnett, *et al.*, 1998, pl. 5.1–24.
151. Diodorus, 18.27.2.
152. Dionysius, 20.80.2.
153. Standards in later Hellenistic armies are better documented: Livy, 33.10.8; 36.19.12; 37.46.3; 37.59.4. See also Livy, 31.24.11; 31.42.4; 32.6.7; 45.43.5. For the structure of supernumeraries in the Macedonian phalanx, see Wrightson, 2010.
154. Arrian, *Tactics*, 10.4.
155. Livy, 33.7.1–2. See also 31.24.11; Polybius, 18.20.7–8; Asclepiodotus, 12.10; Arrian, *Tactics*, 27; Aelian, *Tactics*, 35.1–5.
156. For the armour of at least deceased kings having sacred properties, see Hammond, 1989a.
157. Alvarez-Rico, 2002, 45: for a review of evidence for night watch shifts. For lamp lighting, see Diodorus, 19.31.1; 19.43.5.
158. Arrian, 5.24.2; Curtius, 3.8.22; 5.4.17; 6.8.17; 7.2.19–20; Aeneas the Tactician, 18.21; Pattenden, 1987, 64–74; Bosworth, 1995, 333. For *contra* see Atkinson, 1980, 202.
159. Polyaenus, 3.9.11; Xenophon, *Hellenica*, 6.2.29; Aeneas the Tactician, 24.16.
160. Alvarez-Rico, 2002, 43–45.
161. Frontinus, 3.12.2.
162. Homer, *Iliad*, 10.150–189.
163. Curtius, 4.13.21.
164. Arrian, 1.6.9; Diodorus, 19.68.7; 19.92.3; 19.95.3–5.
165. Euripides, *Rhesus*, 764–766.
166. Engels, 1980, 335–336 and n. 54.
167. Plutarch, *Alexander*, 51.1–4; Arrian, 4.8.7–8; Curtius, 8.1.28–29; 8.1.47; Arrian, *Tactics*, 10.4.
168. Athenaeus, 10.414f–415a; Pollux, 4.88–89.
169. Sekunda, 1984a, 12.
170. Plutarch, *Moralia*, 973c–e; Landels, 1999, 80; Lee, 2007, 91, n. 73.
171. Pollux, 4.85; 4.88; Arrian, *Indica*, 30.4–6; Diodorus, 17.106.7; Curtius, 10.1.11–12.
172. Asclepiodotus, 12.10; Arrian, *Tactics*, 27; Aelian, *Tactics*, 35.1–5.
173. Curtius, 5.2.6–7. Cf. Curtius, 3.3.8. See also Atkinson, 1994, 64; Russell, 1999, 149, n. 19. See also Polybius, 2.67.10–11.
174. Curtius, 3.3.8; Briant, 1988, 267 and n. 17; Temple, 2000; Xenophon, *Cyropaedia*, 5.13; Herodotus, 9.59.
175. Gabriel, 2003, 74–76, with illustration at fig. 3.2.
176. Ansari, 1963, 20–21; Phul, 1978, 219.
177. Curtius, 6.2.16–17.
178. Euripides, *Rhesus*, 24–33; Liapis, 2009, 78–79.
179. Lee, 2007, 141 and n. 8.
180. Xenophon, *Anabasis*, 2.2.4–5; Josephus, *Jewish War*, 3.5.4; Krentz, 1991, 110–120; Russell, 1999, 150; Anderson, 1965.
181. See for example Retief & Cilliers, 2005; Lloyd, 1996.
182. Demosthenes, *Second Olynthiac*, 18.
183. Polybius, 8.9.7–10; Athenaeus, 4.167a–c; Demosthenes, *Second Olynthiac*, 18–20. See also Stagakis, 1970; Heckel, 2003. During the filming of the Oliver Stone-directed *Alexander* epic, all horse riders were required to have their leg hair shaved.
184. Arrian, 1.21.1.
185. Davidson, 1997, 40.
186. Athenaeus, 10.435b; Justin, 9.8.15: for both Philip and Alexander being fond of drink.
187. Plutarch, *Demosthenes*, 16.4.
188. Athenaeus, 3.120e.
189. Athenaeus, 4.129a; 10.434a.

190. Also found in Athenaeus, 10.434b–c.
191. Pollux, 4.147; Simon, 1938, 7.
192. Plutarch, *Alexander*, 72.2.
193. Athenaeus, 10.437a–b; Plutarch, *Alexander*, 70.1–2; Aelian, *Varia Historia*, 2.4. For the seven-day mass revelry/Dionysiac rout of the army following the hardship of the Makran, see Plutarch, *Alexander*, 70.1; Arrian, 5.1.6; Diodorus, 17.106.1; Curtius, 8.10.11–18; 9.10.24–29.
194. Liappas, *et al.*, 2003; O'Brien, 1980a; O'Brien, 1980b; O'Brien, 1994, 6–8; Carney, 2008b.
195. Arrian, 7.26.3; Diodorus, 17.117.4; 18.1.4–5; Curtius, 10.5.5; Justin, 12.15.6–8.
196. Plutarch, *Alexander*, 47.11–12; Plutarch, *Moralia*, 337a.
197. Plutarch, *Moralia*, 181e.
198. Plutarch, *Moralia*, 177e–f.
199. Plutarch, *Pyrrhus*, 26.2.
200. Themelis & Touratsoglou, 1997, 193 and pl. 63: A54; Ignatiadou, 1999.
201. Arrian, 5.27.6
202. Heckel, 2006a, 28.
203. Diodorus, 17.80.3–4; Curtius, 7.2.35–38; Polyaenus, 4.3.19; 4.6.3; Borza, 1977, 299: reviewing a thesis by Rossi.
204. Atkinson, 1994, 259.
205. Plutarch, *Alexander*, 47.6; 71.1; Plutarch, *Moralia*, 328d–e.
206. Vokotopoulou, 1996a, 191, no. 7437.
207. Stephan & Verhoogt, 2005; Bowman, 1998.
208. Curtius, 7.2.36.

Part V: Ingenuity

Chapter 13: Technical Expertise

1. Green, 1991, 426; Pédech, 1980, 135–156; Faure, 1982, 65–66.
2. Hammond & Griffith, 1979, 448.
3. Well summarised by Gaebel, 2002, 308.
4. Polyaenus, 3.9.22.
5. Hanson, 1999b, 172.
6. Arrian, 1.20.8; 4.21.2–5; 5.29.1; 6.18.1–2; 6.20.1; 6.20.5; 6.21.3; 7.19.4; 7.21.2–7; Diodorus, 17.85.6; 17.95.1; Curtius, 7.10.14; 8.10.30–31; 8.11.8–9; 9.10.2; Strabo, 16.1.9; 16.1.11; Frontinus, 3.7.4.
7. Hammond, 1991a, 201.
8. Herodotus, 5.52–54; 7.115; 7.131; 8.98; Thucydides, 2.98.1; 2.100.2; 2.128.4.
9. For an excellent review, see Girtzy, 2001, 249.
10. Polybius, 28.8.2–4; Livy, 43.20.1.
11. Scranton, 1941, 131–32; 139–140; Cambitoglou, 2002, 32 and 39.
12. Hammond, 1981b; Hammond, 1997a; Lindsay Adams, 1997; Lindsay Adams, 2007a.
13. Hatzopoulos, 2001, 29–32; Griffith, 1935, 71–72 and n. 3.
14. E.g. Livy, 44.7.10.
15. Polybius, 9.35.2–3; 18.37.9–10; Livy, 33.12.10; 40.57.6; Appian, *Macedonian Wars*, 9.11; Justin, 11.1.16; Strabo, 7.5.6; Papazoglu, 1978, 210; Heskel, 1997.
16. Livy, 45.30.7.
17. Strabo, 7.5.6; Edson, 1970, 43.
18. Polybius, 9.35.2–3; 18.37.9–10; Appian, *Macedonian Wars*, 9.11.
19. Edson, 1951, 11.
20. Distance stones were divisable by 10 stades, hence the term *dekastadion* – see Thonemann, 2003, 95, n. 2.
21. Hammond, 1972, 56–57 and n. 3; 146–147; 156; Hammond, 1989c, 97; Hammond, 1991a, 48–49 and fig. 4; Hammond & Griffith, 1979, 140; Hammond & Walbank, 1988, 54; Girtzy, 2001, 256–259; Koukouli-Chrysanthaki, 1998.
22. Ober, 1999, 177–178 and n. 15–17.

23. Xenophon, *Cyropaedia*, 6.2.35–36.
24. Arrian, 1.26.1; Plutarch, *Alexander*, 17.4; Curtius, 6.6.26.
25. Arrian, 4.30.7; Wood, 1997, 139.
26. Diodorus, 18.28.1.
27. Appian, *Macedonian Wars*, 9.14.
28. Diodorus, 18.4.4.
29. *Alexander Romance* (Greek version), 2.21.
30. *Alexander Romance* (Greek version), 2.21; *Alexander Romance* (Armenian version), 199; Bose, 2004, 202–204; Engels, 1978, 119–122.
31. Stronach, 1978, 159–161; Graf, 1994.
32. Hatzopoulos, 1985; Forbes, 1955, 139.
33. Greenwalt, 1999, 168.
34. Hesychius, s.v. *bematizei* (from *bematisein)*; Kalléris, 1954, 130–131; no. 44–45. For *contra*, see Tzifopoulos, 1998, 144.
35. Fraser, 1996, 80.
36. Sherk, 1974, 535; Lewis, 2001, 22; Bekker-Nielsen, 2001, 124; Tzifopoulos, 1998, 137, n. 1.
37. Humphrey, *et al.*, 1998, 431–432: 10.34.
38. Pliny, 6.61–62; Strabo, 11.8.9; Engels, 1978, 68–69, Table 8; Robinson, 1953, 25–43.
39. Price, 1955, 1.
40. Athenaeus, 10.442b; 11.500d; 12.529e; 2.67a; Aelian, *On Animals*, 17.17; Strabo, 11.8.9; Pliny, 6.61–62; 7.11; 7.20.
41. Diogenes Laertius, 2.17; Strabo, 2.4.1; 5.2.7–8; 6.1.11; 6.2.11; 6.3.10.
42. Matthews, 1974, 161, n. 4.
43. Tzifopoulos, 1998, 146–147: *Suda* s.v. *hemerodromos*.
44. Tzifopoulos, 1998; Matthews, 1974.
45. Herodotus, 6.105–106; Plutarch, *Moralia*, 862a–b; Pliny, 7.84; Lucian, *Slip of the Tongue*, 3; Cornelius Nepos, *Miltiades*, 4.3; Pausanias, 1.28.4; 8.54.6.
46. Herodotus, 9.12.
47. Diodorus, 15.82; Xenophon, *Hellenica*, 7.5.10; Plutarch, *Agesilaus*, 34.4; Polybius, 9.8.6.
48. Plutarch, *Alexander*, 28.7; *Suda* s.v. *aggaros*.
49. Pseudo-Aristotle, 398a32; Livy, 31.24.4; Kuhrt, 1995, 692–693.
50. Diodorus, 19.17.6–7.
51. Lucian, *Professor of Public Speaking*, 5; Tzifopoulos, 1998, 147.
52. Tzifopoulos, 1998, 137; n. 1; 138–139; Bengtson, 1956; Blinski, 1959/60; Pausanias, 6.16.5; Pliny, 2.181; 7.84.
53. Russell, 1999, 99, n. 162; Matthews, 1974, 165. For more on communications in Alexander's army, see Tarn, 1948, 171–179; Borza, 1977; Rossi, 1973.
54. Xenophon, *Anabasis*, 1.2.9; 3.3.7; 3.3.15; 3.4.17; 4.2.28; 5.2.29–32.
55. Xenophon, *Anabasis*, 4.8.27.
56. Curtius, 3.7.12.
57. Tzifopoulos, 1998, 165.
58. Badian, 1975; Sofman & Tsibukidis, 1987; Heckel, 1992, 228–233; Heckel, 2006c, 171–173.
59. Tzifopoulos, 1998, 149; Arrian, 1.8.4; 2.9.3; 3.5.6; Agostinetti Simonetti, 1977/78; Spyridakis, 1981.
60. Plutarch, *Alexander*, 5.1; Plutarch, *Moralia*, 342b–c; Xenophon, *Cavalry Commander*, 4.6.
61. Strabo, 1.1.16–17; Vegetius, III.6; Tzifopoulos, 1998, 149.
62. Arrian, 4.4.1; 4.25.5; 4.28.5; 5.20.2; 6.15.4; 6.15.7; 6.17.1; 6.18.1; 6.20.1; 7.21.7.
63. Curtius, 7.6.25–27; Arrian, 4.4.1; Fraser, 1996, 151–161.
64. Livy, 33.39.10–14.
65. Romey, 2004.
66. Kastro Ditnata, Rizakis, 1986.
67. Bakhuizen, 1970, 105–130; 165–168; Bakhuizen, 1972.
68. Gabbert, 1982; Gabbert, 1983, 129–136.
69. Munn, 1993; Ober, 1985; Ober, 1989; Harding, 1988; Harding, 1990.

70. McCredie, 1966, 64.
71. McCredie, 1966, 114–115, n. 29; Jones, 1957.
72. Polybius, 4.65.4–5.
73. Polybius, 4.65.8.
74. Polybius, 4.65.11.
75. Powell, 1904; Sears, 1904; Scranton, 1941, 97.
76. Walbank, 2002, 128.
77. Tarn, 1930, 107; Marsden, 1971, 177–178, n. 106; Marsden, 1977, 212; 223; Winter, 1971, 318 and n. 100; Berve, 1926, 155–158. For a recent introductory treatment see English, 2010; Pimouguet-Pedarros, 2003.
78. Tarn, 1930, 107.
79. Best exemplified by Edmonds Maxwell, 1961, 75; Aristotle, *Politics*, 1331a.
80. McNicoll, 1997, 212; 47; table 7.
81. Hornblower, 1981, 223; Chaniotis, 2005, 78–101.
82. Murray, 2008, 35 and n. 22; Athenaeus Mechanicus, 10.10-11.
83. Arrian, 2.21.1.
84. Jacoby, 1923, 115 F 115; 328 F 162.
85. Curtius, 4.2.20.
86. Arrian, 2.18.4.
87. Polybius, 5.4.2–3.
88. Arrian, 4.18.4–19.6; Curtius, 7.11; Strabo, 11.11.4; Polyaenus, 4.3.29. See also Arrian, 2.26.3: as early as 332 BC Alexander insisted that his engineers take Gaza no matter what the difficulty, 'for the achievement would strike great terror into his enemies just because it was beyond calculation, while not to take it would be a blow to his prestige'; Arrian, 4.21.7: 'there was not a place in the world Alexander and his army could not take by force'.
89. Arrian, 4.19.1: Polyaenus, 4.3.29.
90. Curtius, 7.11.7.
91. Kalita, 1995, 29.
92. Polyaenus, 4.3.23: referred to as 'Alexander's Ladder'.
93. Marriner, 2008.
94. Ali, *et al.*, 2006. Alexander is said to have witnessed the effects of ignited naphtha in Babylonia, so he may have considered using flammable liquids in certain military operations – see Diodorus, 17.64.3; Plutarch, *Alexander*, 35.1–7; Arrian, 3.16.3; Curtius, 5.1.10–16; Strabo, 16.1.15. For Susa archaeological finds, see www.cais-soas.com/News/2006/April2006/09-04-susa.htm, last accessed October 2012.
95. Holt, 2005a, 83; Grentez, 2003, 31–32; Curtius, 8.2.22; 8.2.25.
96. *Suda* s.v. *Strattis*; Heckel, 2006c, 258: in agreement with Berve.
97. Arrian, 3.7.5; Diodorus, 17.55.1–6; Curtius, 4.9.15–21.
98. Diodorus, 18.34.6–35.6.
99. Vegetius, III.7.
100. Diodorus, 19.18.3–4.
101. Diodorus, 19.18.3–5.
102. Bosworth, 2002, 115–116 and n. 77; Echols, 1953.
103. E.g. Arrian, 3.29.2–4; Curtius, 7.5.17–18; 7.7.16; 7.8.6. See also the excellent technical and source-critical study in Bloedow, 2002.
104. Arrian, 2.18.6.
105. Arrian, 1.3.5–6; 3.29.3–4; 4.4.2–4; 5.9.3; 5.10.2; 5.12.4; 5.20.8; Polyaenus, 4.3.9; *Itinerarium Alexandri*, 34.77. Bloedow, 2002, 65, n. 42.
106. Strabo, 15.3.18–19.
107. Xenophon, *Anabasis*, 1.5.10; Lane-Fox, 2004a, 235–236. Similar techniques were used by a unit of *utricularii* ('bladder-bearers') employed in the Roman imperial army.
108. von Schwartz, 1906, 64 and taf. II. See also Hornell, 1946, 6–17; 20–34. For the process of manufacture, see Bloedow, 2002, 66, n. 45; 67, fig. 3; 68, fig. 4.

109. Curtius, 7.8.6. See also Curtius, 7.5.17–18; 7.7.16. Alternatively, for the use of punts by Seleucus and Eumenes after Alexander's death, see Diodorus, 19.12.5; 19.13.3; 19.18.4.
110. Curtius, 7.8.7; Bloedow, 2002, 61–62, n. 27.
111. Arrian, 4.4.5; Curtius, 7.8.6; 7.9.4. See also *Itinerarium Alexandri*, 34.77.
112. Bloedow, 2002, 63; Keyser, 1994; Murray, 2008; Curtius, 7.9.2–8. See also Curtius, 5.3.21; 5.3.23; Arrian, 1.1.10; Polyaenus, 4.3.11. These refer to 'shield linkage' in its broadest sense – perhaps merely a Latin equivalent to 'locked shields'; Heckel, 2005, 191–193; Dubs, 1941, 327–328, n. 18. See also Livy, 32.18.13.
113. *Itinerarium Alexandri*, 34.77. For a rigorous and illuminating critique on the later source passages, with alternate translations, see Bloedow, 2002, 64–66; 75.
114. Bloedow, 2002, 75. Other animals could also be ferried. For example, for waterborne elephant transportation under Cassander, see Diodorus, 19.54.3.
115. Rey, 1927, 114; Hornell, 1945, 73, n. 51; 74–79. For a variety of modern parallels showing river transportation, possibly resembling those adopted by Alexander, see photographic illustrations in Bloedow, 2002, 69–74 and 73–74, n. 58; 67: fig. 3; 68: fig. 4; 69: fig. 5; 70: fig. 6; 71: fig. 7–8. See also von Oppenheim, 1900, 193–195.
116. Arrian, 3.7.1–2; Curtius, 4.9.12.
117. Diodorus, 17.86.3.
118. Arrian, *Indica*, 3.10; Bosworth, 1995, 254.
119. Curtius, 8.12.4. For Hephaestion's specialisation in bridge building tasks, see Heckel, 1992, 77; Heckel, 2006c, 134.
120. Arrian, 4.30.9; 5.3.5.
121. Curtius, 8.10.2–3; Arrian, 5.8.4–5; 5.12.4; 7.19.3; Strabo, 16.1.11.
122. *Itinerarium Alexandri*, 48.109.
123. Macrory, 1966, 105.
124. Arrian, 5.7.1.
125. Arrian, 5.7.1–8.1; Bosworth, 1995, 254.
126. Xenophon, *Anabasis*, 1.2.5; 1.4.11–18; 2.4.13; 2.4.17; 2.4.24–25.
127. Green, 1991, 553, n. 56; Wood, 1997, 182; 184; Watanabe, 2003.
128. Arrian, 3.29.3–4; Curtius, 7.5.17.
129. Xenophon, *Anabasis*, 3.5.8–12; Bloedow, 2002, 58–60; includes n. 11–13; 59, n. 12. See also Thompson, 1952, 119.
130. Hornell, 2006, 129.
131. Lazaridis, 1997, 32–37 and fig. 17–18; Hammond, 1967, 140; 235–237 and fig. 3; Hammond, 1972, 25; 160–162.
132. Pliny, 34.150; Cohen, 2006, 192.
133. Curtius, 5.5.2–4; Arrian, 3.18.6; 3.18.10; Echols, 1953; Stein, 1942. For other references to bridges erected during Alexander's campaigns, see Curtius, 3.7.5; Arrian, 3.6.1; Arrian, *Indica*, 42.7–8. See also Diodorus, 19.17.2; 19.18.4.
134. Briant, 2002, 362–364; Herodotus, 7.34; 25.1: for Egyptian and Phoenician technicians; Hammond & Roseman, 1996: for bridges across the Bosphorus, Danube, Strymon and Hellespont (513–480 BC) with investigation into techniques used.

Chapter 14: Little-Known Combat Units in Asia

1. Polyaenus, 3.9.22.
2. Demosthenes, *Third Philippic*, 47–50. See also Euripides, *Rhesus*, 311–313.
3. Arrian, 2.7.1.
4. Dio Chrysostom, 4.8.
5. Strauss, 2004, 142; Jones, 2001, 29; Feugere, 2002, 18.
6. Corrigan, 2004, 381–388.
7. Curtius, 5.13.8.
8. Arrian, 3.21.7; 4.23.2; 4.24.5; Diodorus, 5.33.5.
9. Hesychius, s.v. *dimachai*; Herodotus, 7.160; Thucydides, 5.57.2; Xenophon, *Cavalry Commander*, 5.13.

10. Pollux, 1.10.132, trans. by D.M. Corrigan. See Corrigan, 2004, 385: the so-called 'Lydian misfortune' remains a mystery. Grooms probably kept up by holding on to the horse's tail – see Calcani, 1989, 131; fig. 76–77; Hamiaux, 1992, 223; no. 236.
11. Corrigan, 2004, 385; 387; 389.
12. Heckel, 1992, 300, n. 1.
13. Corrigan, 2004, 388.
14. Sekunda, 1984a, 30.
15. Gaebel, 2002, 195, n. 143.
16. Plutarch, *Eumenes*, 11.3–4.
17. Hyland, 2003, 159; 192; n. 60; Hammond, 1978.
18. Probably taking inspiration from Onomarchus of Phocis and his ingenious use of mobile stone-throwing catapults in the defeat of Philip in Thessaly (353 BC) – see Polyaenus, 2.38.2; Diodorus, 16.35.2; Keyser, 1994; Marsden, 1969, 165.
19. Pollux, 1.139: 'Macedonian catapults'.
20. Marsden, 1969, 68.
21. Diodorus, 20.93.5.
22. Hornblower, 1981, 30–31 and n. 46. See also *Suda* s.v. *katapeltes*. For more discussion on the use of military terms in Diodorus see Milns, 1983.
23. Arrian, 4.2.1; 4.2.3; Hammond & Griffith, 1979, 448.
24. Marsden, 1969, 67–68;106. 'Managers' of *katapeltai* also survive in epigraphic texts.
25. Morrison, 1984, 56, n. 21: citing an Italian article from 1936 detailing an inscription on the 'mess-mates of Alexidamas'. Two *katapeltaphetai* are noted.
26. Billows, 1990, 317: citing Diodorus, 20.49.4.

Epilogue: Macedonian Militarism and the Impact of Pydna

1. Chaniotis, 2005, 1–3; Lévêque, 1968, 279; Lendon, 2005, 140.
2. Chaniotis, 2005, 5–6; Baker, 2005, 386; Eckstein, 2008, 16; Livy, 42.52.1–2; 42.52.9–10.
3. Eckstein, 2010, 227–30.
4. Polybius, 16.22.4–5; Fraser, 1972, 80; Griffith, 1935, 139.
5. Cornelius Nepos, *Eumenes*, 3.4.
6. Diodorus, 21.2.2. See also Justin, 24.4.9–10: for Ptolemy Ceraunus' overconfidence and lack of fear in the face of the Galatian threat.
7. Polybius, 4.69.4–6; Walbank, 1957, 523.
8. Pausanias, 4.29.2–5.
9. Polybius, 5.2.5; For a modern summary, see Asirvatham, 2010, 109–10; Eckstein, 2000, 871; Eckstein, 2005, 487; Eckstein, 2008, 16–17.
10. Polybius, 4.8.10–12; Lendon, 2005, 170.
11. Livy, 9.17.10; 9.19.1–11.
12. Morello, 2002, 78–79.
13. Gabriel, 2005b 444; Livy, 42.52.
14. Lendering, 2009, 12.
15. Walbank, 1957, 8; Polybius, 1.14.1–9; 2.56.10. See also Marsden, 1974; Walbank, 1970.
16. As is actually implied under Plutarch, *Flamininus*, 5.5. See also Livy, 42.52.9–10: boasting Macedonian superiority.
17. Eckstein, 2008, 78–79; Walbank, 2002, 127–136; Polybius, 5.101.10; 102.1; 104.7; 108.5; 15.24.6.
18. Polybius, 7.9.10–14; Livy, 23.33.10–12.
19. Eckstein, 2009.
20. Billows, 1995, 219.
21. Arrian, 2.10.6–7.
22. Liampi, 1998, 43–47.
23. Liampi, 1990; Kosmetatou & Waelkens, 1998; Kosmetatou, 2005.
24. Billows, 1995, 55; 24–33; 172.

25. Polyaenus, Preface.
26. Bieber, 1964, pl. XV.25; Furtwängler, 1964, taf. XXXII.11. See also Sekunda, 2012b, 42, for the winged thunderbolt on a shield from the Agios Athanasios tomb as possibly the first appearance of this emblem 'as a shield-device in ancient iconography'.
27. For a commendable discussion of this, see Bennett & Roberts, 2009, 23–26.
28. Livy, 42.29.1; 45.7.3; 45.9.2.
29. Justin, 33.1.
30. Dionysius, 1.2.2–4; 1.3.5.
31. Plutarch, *Flamininus*, 7.3–4; Petrochilos, 1974, 98–100.
32. Plutarch, *Flamininus*, 7.4.
33. Livy, 32.17.4–18.1
34. Livy, 27.17.4–5; Aulus Gellius, 5.5.1–6; Plutarch, *Aemilius Paullus*, 13.3. That the Romans continued to be well aware of the Macedonian military reputation is discerned in the nature of their own preparations for the third war – see Lendon, 2005, 193.
35. Based on a lost work by Posidonius or possibly the eyewitness *testimonia* of the Roman officer Scipio Nasica.
36. Plutarch, *Aemilius Paullus*, 18.4. For modern analyses of the battle reconciling the discursive accounts and reconstructing events, see Hammond, 1984b; Taylor, 2009. See also Montagu, 2006, 220–227.
37. Onasander, 29.1–2.
38. Plutarch, *Aemilius Paullus*, 19.1–2; 20.1–3; 28.5; Polybius, 29.17.1.
39. Diodorus, 17.4.4; Heckel, 2006a, 13.
40. Arrian, 1.6.1–4.
41. Plutarch, *Aemilius Paullus*, 22.1.
42. Pritchett, 1985, 240; Polybius, 18.26.10–12; Livy, 44.42.1–4; Plutarch, *Aemilius Paullus*, 21.6. For an incisive analysis of the psychology of pike armed troops under the stress of a successful enemy assault, see Du Picq, 1921. For another ancient die hard account, this time of the Celtiberian contingent at the battle of the 'Great Plains' in 203 BC, see Polybius, 14.8.11–14; Livy, 30.8.8–9.
43. Polybius, 29.17.3.
44. Plutarch, *Aemilius Paullus*, 23.1.
45. Plutarch, *Demetrius*, 44.5.
46. Polybius, 18.32.13.
47. Wheatcroft, 2009, 25.
48. Appian, *Syrian Wars*, 7.37.
49. Connolly, 1998, 37; Gabriel, 2005b, 438–439; 448.
50. Polybius, 18.28.1–5; Brizzi, 2004, 131–144; Kitsos, 1999 (available at www.anistor.gr/english/enback/e991.htm, last accessed June 2012); Skarmintzos, 2008; Sekunda, 2001b, 96–98.
51. Cicero, *On the Orator*, 2.75.
52. Polybius, 31.25.6.
53. Pausanias, 1.13.2; Josephus, *Jewish War*, 2.387.
54. Polybius, 29.21.1–6. See also Pausanias, 7.8.8–9; Velleius Paterculus, 1.6.5–6; Appian, *History*, Preface 9; Claudian, 3.159–166; Swain, 1940; Mendels, 1998, 314–323.
55. Jesse, 2007, 30; Lasswell, 1941.
56. Plutarch, *Pyrrhus*, 19.5; 21.10; Appian, *Samnite Wars*, 3.24.
57. Morgan, 1969; Rubinsohn, 1988.
58. Sarikakis, 1977.
59. Livy, 45.30.2.
60. Echols, 1949.
61. Eckstein, 2008, 364; Livy, 42.52.15: the Romans sought 'to enslave Macedonia ... so that no people renowned in war [and geographically close to Rome] should retain its armaments'.
62. Dio Chrysostom, 33.27.
63. Tarn, 1952, 105–106. These values are in clear contradiction to the dismissive comments of Demosthenes, *Third Philippic*, 31.
64. Josephus, *Against Apion*, 2.70.

65. Josephus, *Jewish War*, 5.465.
66. Dio Chrysostom, 25.6; 32.65. For a sword being especially valued for promoting valour because it was Macedonian, see Gow & Page, 1968, 53.
67. Montgomery, 1968, 73; Fuller, 1958, 39.

Appendix I: Some Military Figures in Macedonian Funerary Art

1. Andronikos, 1988b, 56; 83–84.
2. Plutarch, *Pyrrhus*, 26.6.
3. From e-mail correspondence with the author, 29 November, 2002.
4. Saatsoglou-Paliadeli, 1984, 65–70; pl. 14: no. 5a–b; Prestianni Giallombardo, 1991, 270–272; 300, fig. 8.
5. Saatsoglou-Paliadeli, 1984, 120–122; pl. 27–28: no. 11a–b.
6. Saatsoglou-Paliadeli, 1984, 123–126: for text; pl. 29–30: no. 12a–b.
7. Saatsoglou-Paliadeli, 1984, 127–130; pl. 31–32: no. 13a–b.
8. Perdikatis, *et al.*, 2002.
9. Discussed in Andronikos, 1988b, 36–37; Drougou & Saatsoglou-Paliadeli, 1999, 65–70, fig. 92, fig. 94; Touchais, 1982, 573, fig. 76.
10. Mehl, 1980/81; Muller, 1973, 116–21; Stewart, 1993, 158–90.
11. This artistic evidence complements Xenophon, *Horsemanship*, 12.10.
12. In correspondence with Duncan Head.
13. Shefton, 1978, 13.
14. Arrian, 1.16.4–5; Curtius, 9.1.6.
15. Andronikos, 1988b, 36; Miller, 1993b, 56, n. 124. For Alexander and Macedonian cavalry being very occasionally equipped with shields, see Arrian 1.6.5–6; Plutarch, *Alexander*, 16.7; Diodorus, 17.20.3.
16. Liampi, 1998, 77–78: S32 and taf. 14.2.
17. Perhaps Cleitus, son of Dropides, whom Alexander murdered in 328 BC.
18. Pliny, 35.90–97. For discussion of the possible form of armour worn by Antigonus in this painting, see Cadario, 2004, 45.
19. Pliny, 35.106.
20. Chaniotis, 1988, 301; Chaniotis, 2005, 220.
21. Rice, 1993. See also Tracy, 2003; Karakas, 2002.
22. Philostratus, *Apollonius*, 2.20.62.
23. Pausanias, 2.21.4.
24. Diodorus, 17.115.3–4.
25. Diodorus, 18.27.1.
26. Plato, *Gorgias*, 524a.
27. Petsas, 1966, 153; Rhomiopoulou, 1997, 24–29, fig. 19–23.

Appendix II: Evidence for Cavalry Clothing Colours

1. Curtius, 6.6.10.
2. Curtius, 6.6.7–8.
3. Plutarch, *Eumenes*, 6.2.
4. Plutarch, *Moralia*, 330a–b; Pollux, 10.42; Athenaeus, 6.245a. See also Herman, 1980/81.
5. *Suda* s.v. *Krateros*; Arrian, *Successors*, F19; Plutarch, *Pyrrhus*, 8.1; Themistios, *Logoi Politikoi*, 13, 175b in Stewart, 1993, 349: T24; Reinhold, 1970, 31.
6. Curtius, 3.2.13; 3.3.26–27: 5.4.31.
7. Plutarch, *Moralia*, 177c.
8. Post, 2010, 17.
9. Rouveret, 2004, 37, 53–54, no. 7, Ma3636, 61–62, no. 11, Ma3640, 65–66, no. 13, Ma3642, 70–71, no. 15, Ma3644, 71–72, no. 16, Ma3645, 77–78, no. 19, Ma3648, 81–82, no. 21, Ma3650; Walter, *et al.*, 1998, 54–55; Rouveret, 1998a.
10. Griffith, 1935, 322, n. 2.
11. Snodgrass, 1967, 129.

12. Post, 2010, 16–17.
13. For more on colour terminology in literary sources, see Forbes, 1956, 119; Jensen, 1963, 111. See also Chenciner, 2000, 36–37. For an explanation as to how ancient peoples generally viewed colour, see Platnauer, 1921; Deutscher, 2010, 25–95.
14. E.g. see Brown, 1977, 79–87.
15. Pliny, 35.50; 35.92; Moreno, 2001, 34–35; Brown, 1977, 53–59.
16. For more on the study of paint fragments from Greece and ancient Macedonia, see Miller, 1998, 75–88; Brecoulaki, 1997; 2000; relevant papers in Tiverios & Tsiafakis, 2002, 1–8; 97–105; 147–154; 211–220.
17. Source for catalogue no. 5: Andronikos, 1988b, 85, no. 45; Saatsoglou-Paliadeli, 1984, 152–159, pl. 43: no. 20.
18. Source for catalogue no. 6: Saatsoglou-Paliadeli, 1984, 55–64; pl. 12–13: no. 4a–b.
19. Source for catalogue no. 7: Andronikos, 1988b, 111–113, fig. 68, 69. For a full bibliography of Philip's tomb, the efficacy of its identification and contents, see especially Hatzopoulos, 2008; O'Brien, 1992, 297–298.
20. Source for catalogue nos. 8 and 9: Winter, 1912, taf. 1–2. For appropriate colour reconstructions see also Sekunda, 1984a, pl. A2 and B1; 17–18; 37–38; Sekunda, 1984b; Heckel, 2006b.
21. Source for catalogue no. 10: Moreno, 2001, 18–19; 46–47: pl. III; Andreae, 1977, 44; Abb. 3.
22. Source for catalogue no. 11: Hatzopoulos and Loukopoulos, 1991, 70, col. pl. 48; Calcani, 1989, 67; 131: fig. 76, cf. 77; Brown, 1957, 26, cat. no. 21, pl. xi. See also Reinach, 1910, 8, fig. 10; Breccia, 1912, 10, no. 9, pl. 22–23; Brown, 1977, 79–87. Fragmentary inscription reads '... xenos a Macedonian'.
23. Source for catalogue no. 11: Sekunda, 2003, pl. x–xi; Picard, 1970, pl. XLI. See also Curtius, 1929, 240–241, fig. 140–141; Richter, 1955, 78 and fig. 231–232; Richardson, 2000, 68–71.
24. Source for catalogue no. 13: Wooton, 2002, 269; fig. 10.
25. Source for catalogue nos. 14 and 15: Zhivkova, 1975, 45–50; Verdiani, 1945, fig. 1–6.
26. Source for catalogue nos. 16–21: Head, 1999; Tsibidou-Avloniti, 2002b, tav. vi.2; vii.2. See also Tsibidou-Avloniti, 2002a; Tsibidou-Avloniti, 2004.
27. Source for catalogue nos. 22–24: Hatzopoulos and Loukopoulos, 1991, 67, fig. 46; Ginouvcs, 1993, 178–181; Petsas, 1966, 193: fold out.
28. Source for catalogue no. 25: Andronikos, 1988b, 35–37.
29. Source item no. 26: Rouveret, 2004, 45–46, fig. 3.
30. Source for catalogue nos. 27–29: Brown, 1957, 52–53, cat. no. 34, pl. xxiv. 1; Rostovtzeff, 1941a, 408, pl. 49; Sekunda, 1995, 75–76, fig. 108–109; Sekunda, 2011, 18–19. For a colour reconstruction of these figures set within the tomb itself, see also Grimm, 1998, 95.
31. Source for catalogue no. 30: Kinch, 1920; Couissin, 1932, planches I–II.
32. Source for catalogue no. 31: Soteriades, 1930, 43–49, fig. 4–5; Pandermalis, 1999, 13. There is evidence that the original painting showed twelve cavalrymen in combat.
33. Source for catalogue no. 34: Brown, 1957, 18, cat. no. 10, pl. iv.3; Rouveret, 2004, 30–31.
34. Source for catalogue no. 35: Brown, 1957, 18, cat. no. 11, pl. iv.4.
35. Source for catalogue no. 36: Brown, 1957, 24–25, cat. no. 16, pl. x; Sekunda, 2011, 18.
36. Source for catalogue no. 37: Brown, 1957, 24, cat. no. 15, pl. ix.1; Swindler, 1929, 344–345, fig. 552; Bernand, 1995, 42. For other evidence of Ptolemaic soldiers in similarly brightly coloured corselets, see Hanfmann, 1983, 247 and tav. LIV.4, XLV.1.
37. Source for catalogue no. 38: Fraser, 1964, pl. 10.1. Fragmentary inscription reads 'Zenodotos, a Colophonian'.
38. Source for catalogue no. 39: Brown, 1957, 26, cat. no. 22, pl. xiv. Fragmentary inscription reads '...os, son of Leontiskos, an Epeiretan'.
39. Source for catalogue no. 40: Brown, 1957, 28, cat. no. 26, pl. xix.1–2; Dintsis, 1986b, taf. 81.1. Fragmentary inscription reads 'Nikanor, a Macedonian'.
40. Source for catalogue no. 41: Brown, 1957, 16, cat. no. 4, pl. v; Rostovtzeff, 1941a, pl. 37.2; Swindler, 1929, 344–345, fig. 551.
41. Source for catalogue no. 42: Hourmouziadis, 1982, 67: fig. 37; Arvanitopoulos, 1928, 132: fig. 159.

Bibliography

Early Sources

Various translations were used for ancient sources, including the Loeb Classical Library series (Harvard University Press) in hardcopy, Perseus Digital Library at www.perseus.tufts.edu, and LacusCurtius at http://penelope.uchicago.edu/Thayer/E/Roman/home.html.

Aelian (Claudius), *Varia Historia*; *On Animals*.
Aelian, *Tactics*.
Aelius Donatus, *Comments on Terence ('Eunuchus')*.
Aeneas the Tactician, *How to Survive Under Siege*.
Aeschines, *Against Ctesiphon*.
Aeschylus, *Persians*.
Aetius, *Libri Medicinales*.
Alciphron, *Epistles*.
Ammianus Marcellinus, *Roman History*.
Appian, Preface to *History*; *Civil Wars*; *Samnite Wars*; *Macedonian Wars*.
Apuleius, *Golden Ass*.
Aristophanes, *Frogs*; *Acharnians*.
Aristotle, *Politics*; *Athenian Constitution*; *History of Animals*; *Rhetoric*; *Eudemian Ethics*.
Arrian, *Anabasis*; *Indica*; *Tactics*; *Cynegetica*; *Successors* (fragments).
Asclepiodotus, *Tactics*.
Athenaeus, *The Deipnosophists*.
Athenaeus Mechanicus, *On Machines*.
Aulus Gellius, *Attic Nights*.
Cassius Dio, *Roman History*.
Cicero, *Letters to his Brother Quintus*; *Tusculan Disputations*; *Letters to Friends*; *Lucullus*; *For Cornelius Balbus*; *On the Orator*; *Against Piso*; *On the Nature of the Gods*.
Claudian, *On Stilicho's Consulship*.
Clement of Alexandria, *Stromata*.
Cornelius Nepos, Lives of *Hannibal*; *Epaminondas*; *Iphicrates*; *Datames*; *Eumenes*; *Miltiades*.
Curtius, *History of Alexander*.
Demetrius, *On Style*.
Demosthenes, *On the Chersonese*; *On the Crown*; *Against Aphobus*; *For Phormio*; *First Olynthiac*; *Second Olynthiac*; *Second Philippic*; *Third Philippic*.
Dio Chrysostom, *Orationes*.
Diodorus Siculus, *Library of History*.
Diogenes Laertius, *Lives of Eminent Philosophers*.
Dionysius of Halicarnassus, *Roman Antiquities*.
Euripides, *Rhesus*; *Bacchae*.
Florus, *Epitome of Roman History*.
Frontinus, *Stratagems*.
Grattius, *Cynegeticon*.
Herodian, *History of the Empire*.
Herodotus, *Histories*.

Hesychius, *Lexicon*.
Homer, *Iliad*; *Odyssey*.
Isidore of Seville, *Etymologiae*.
Josephus, *Jewish Antiquities*; *Jewish War*; *Against Apion*.
Justin, *Epitome of the Philippic History of Pompeius Trogus*.
Livy, *History*.
Lucian, *Hippias or the Bath*; *Dialogues of the Dead*; *How to Write History*; *Octogenerians*; *Slip of the Tongue*; *Professor of Public Speaking*.
Lysias, *Speeches*.
Menander, *Perikeiromene*; *The Shield*; *The Flatterer*; *The Fisherman*.
Onasander, *The General*.
Oppian, *Cynegetica*.
Pausanias, *Description of Greece*.
Philon of Byzantium, *Mechanike syntaxis*.
Philostratus, *Imagines*; *Apollonius*.
Photius, *Bibliotheca*.
Pindar, *Pythian Odes*.
Plato, *Laws*; *Critias*; *Laches*; *Republic*; *Gorgias*.
Plautus, *Curculio*; *Miles Gloriosus*; *Pseudolus*; *Truculentus*.
Pliny the Elder, *Natural History*.
Plutarch, Lives of *Pyrrhus*, *Eumenes*, *Phocion*, *Pelopidas*, *Alexander*, *Agesilaus*, *Flamininus*, *Aemilius Paullus*, *Brutus*, *Mark Antony*, *Julius Caesar*, *Alcibiades*, *Philopoemen*, *Demetrius*, *Timoleon*, *Sulla*, *Solon*, *Aratus*, *Theseus*, *Marius*, *Cleomenes*, *Demosthenes*; *Moralia*.
Pollux, *Onomasticon*.
Polyaenus, *Stratagems*.
Polybius, *Histories*.
Pseudo-Aristotle, *On the Universe*.
Sallust, *Jugurthine War*; *Catiline War*.
Stephanus of Byzantium.
Stobaeus, *Eclogues*.
Strabo, *Geography*.
Suetonius, Lives of *Julius Caesar*, *Gaius*, *Nero*.
Tacitus, *Annals*.
Theocritus, *Idylls*.
Theophrastus, *Characters*; *Enquiry into Plants*; *Causes of Plants*.
Thucydides, *History of the Peloponnesian War*.
Valerius Maximus, *Memorable Doings and Sayings*.
Vegetius, *Military Science*.
Velleius Paterculus, *Roman History*.
Virgil, *Aeneid*; *Eclogues*.
Vitruvius, *On Architecture*.
Xenophon, *Memorabilia*; *Anabasis*; *Hellenica*; *Cyropaedia*; *Cavalry Commander*; *On Horsemanship*.

Other sources used

Suda.
Historia Augusta.
Alexander's Letter to Aristotle about India, trans. L.L. Gunderson, Meisenheim, 1980.
Alexander Romance (Greek version), trans. R. Stoneman, London, 1991.
Alexander Romance (Armenian version), trans. A.M. Wolohojian, New York, 1969.
Itinerarium Alexandri, trans. I. Davies, *Ancient History Bulletin*, 12, 1998, 29–54.
Maurice's Strategikon, trans. G.T. Dennis, Philadelphia, 1984.
The Holy Bible (*Book of Genesis*, *Book of Esther*, *First Book of Maccabees*, *Book of Acts*), King James Version.

Modern References

Adams, G.W., 'The Representation of Heroic Episodes in Plutarch's *Life of Pyrrhus*', *Anistoriton* 12, 2011 (online journal).

Adamson, P.B., 'The Influence of Alexander the Great on the Practice of Medicine', *Episteme: Rivista critica di storia delle scienze, mediche e biologiche* 7, 1973, 222–230.

Adam-Veleni, P., *Petres of Florina* (Thessaloniki, 1998).

Adam-Veleni, P., 'Arms and Warfare Techniques of the Macedonians' in Pandermalis (ed.), *Alexander the Great. Treasures from an Epic Era of Hellenism*, 2004, 47–63.

Agostinetti Simonetti, A., 'I mercenarii nell'esercito di Alexandro Magno', *Atti centro ricerche e documentazione sull'antichità classica* 9, 1977/78, 1–17.

Akamatis, I.M., *et al.*, *Pella and its Environs* (Athens, 2004).

Alexinsky, D., 'Early Hellenistic Helmets with Tiara-shaped Crown from Eastern Mediterranean', *Para Bellum* 25, 2005, 21–40.

Ali, T., *et al.*, 'South Asia's Oldest Incendiary Missile?', *Archeometry* 48, 2006, 641–655.

Allemandi, U., *Pompeii. Picta fragmenta* (Turin, 1998).

Alonso-Nunez, J.M., 'The Universal State of Alexander the Great' in W. Heckel & L.A. Tritle (eds), *Crossroads of History: The Age of Alexander*, 2003, 175–182.

Alvarez-Rico, M.G., 'The Greek Military Camp in the Ten Thousand's Army', *Gladius XXII*, 2002, 29–56.

Amberger, J.C., *The Secret History of the Sword: Adventures in Ancient Martial Arts* (Burbank, 1998).

Anderson, A.R., 'Heracles and his Successors: A Study of a Heroic Ideal and the Recurrence of a Heroic Type', *Harvard Studies in Classical Philology* 39, 1928, 7–58.

Anderson, A.R., 'Bucephalas and his Legend', *American Journal of Philology* 51, 1930, 1–21.

Anderson, J.K., *Ancient Greek Horsemanship* (Berkeley, 1961).

Anderson, J.K., 'Cleon's Orders at Amphipolis', *Journal of Hellenic Studies* 85, 1965, 1–14.

Anderson, J.K., *Military Theory and Practice in the Age of Xenophon* (Berkeley, 1970).

Anderson, J.K., 'Sickle and Xyele', *Journal of Hellenic Studies* 94, 1974, 158–166.

Anderson, J.K., 'Shields of Eight Palms' Width', *California Studies in Classical Antiquity* 9, 1976, 1–6.

Andreae, B., *Das Alexandermosaik aus Pompeji* (Recklinghausen, 1977).

Andreae, B., 'Seleukos Nikator als Pezhetairos im Alexandermosaik', *Mitteilungen des Deutschen Archäologischen Instituts. Römische Abteilung* 111, 2004, 69–82.

Andreotti, R., 'Die Weltmonarchie Alexanders des Grossen', *Saeculum* 8, 1957, 120–166.

Andrianou, D., 'Chairs, Beds, and Tables: Evidence for Furnished Interiors in Hellenistic Greece', *Hesperia* 75, 2006, 219–266.

Andronikos, M., 'Sarissa', *Bulletin de correspondance hellénique* 94, 1970/71, 91–107.

Andronikos, M., 'Vergina: The Royal Graves in the Great Tumulus', *Athens Annals of Archaeology* 10, 1977, 40–72.

Andronikos, M., 'The Finds from the Royal Tombs at Vergina', *Proceedings of the British Academy* 65, 1981, 355–367.

Andronikos, M., 'Some Reflections on the Macedonian Tombs', *British School at Athens* 82, 1987, 1–16.

Andronikos, M., 'Vergina 1988: Excavations in the Cemetery', *To archaiologiko ergo ste Makedonia kai Thrake* 2, 1988a, 1–3.

Andronikos, M., *Vergina: The Royal Tombs and the Ancient City* (Athens, 1988b).

Andronikos, M., 'Vergina 1989: The Excavations at the Cemetery', *To archaiologiko ergo ste Makedonia kai Thrake* 3, 1989, 1–11.

Ansari, M.A., 'The Encampment of the Great Mughals', *Islamic Culture* 37, 1963, 15–24.

Anson, E.M., 'Alexander's Hypaspists and Argyraspids', *Historia* 30, 1981, 117–120.

Anson, E.M., 'The Hypaspists: Macedonia's Professional Citizen-Soldiers', *Historia* 34, 1985, 246–248.

Anson, E.M., 'Hypaspists and Argyraspids after 323', *The Ancient History Bulletin* 2, 1988, 131–133.

Anson, E.M., 'Philip II and the Transformation of Macedonia: a Reappraisal' in T. Howe & J. Reames (eds), *Macedonian legacies: Studies in Ancient Macedonian History and Culture in Honor of Eugene N. Borza*, 2008, 17–30.

Anson, E.M., 'Asthetairoi: Macedonia's Hoplites', in E. Carney & D. Ogden (eds), *Philip II and Alexander the Great: Father and Son, Lives and Afterlives*, 2010a, 81–90.
Anson, E.M., 'The Introduction of the 'Sarisa' in Macedonian warfare', *Ancient Society* 40, 2010b, 51–68.
Aperghis, G.G., 'Alexander's Hipparchies', *Ancient World* 28, 1997, 133–148.
Aperghis, G.G., *The Seleukid Royal Economy: The Finances and Financial Administration of the Seleukid Empire* (Cambridge, 2004).
Archibald, Z.H., 'The Gold Pectoral from Vergina and its Connections', *Oxford Journal of Archaeology* 4, 1985, 165–185.
Archibald, Z.H., *The Odrysian Kingdom of Thrace: Orpheus Unmasked* (Oxford, 1998).
Arena, E., 'La lettera di Oleveni: Fra Filippo II et Filippo V di Macedonia', *Revue des études anciennes* 105, 2003, 49–82.
Arnott, W.G., 'New Evidence for the Opening of Menander's Perikeiromene', *Zeitschrift für Papyrologie und Ephigraphik* 71, 1988, 11–15.
Arvanitopoulos, A.S., *Graptai Stelai Demetriados-Pagason* (Athens, 1928).
Asirvatham, S.R., 'Perspectives on the Macedonians from Greece, Rome and Beyond', in J. Roisman & I. Worthington (eds), *A Companion to Ancient Macedonia*, 2010, 99–124.
Atkinson, J.E., 'Honour in the Ranks of Alexander the Great's Army', *Acta Classica* 53, 2010, 1–20.
Atkinson, J.E., *A Commentary on Q. Curtius Rufus' Historiae Alexandri Magni: Books 3 and 4* (Amsterdam, 1980).
Atkinson, J.E., 'The Infantry Commissions Awarded by Alexander at the End of 331BC', in W. Will & H. Heinrichs (eds), *Zu Alexander dem Grosse. Festschrift G. Wirth*, 1987, 413–436.
Atkinson, J.E., *A Commentary on Q. Curtius Rufus' Historiae Alexandri Magni: Books 5 to 7.2* (Amsterdam, 1994).
Austin, M., *The Hellenistic World from Alexander to the Roman Conquest* (Cambridge, 1981).
Austin, M., 'Alexander and the Macedonian Invasion of Asia: Aspects of the Historiography of War and Empire in Antiquity', in J. Rich & G. Shipley (eds), *War and Society in the Greek World*, 1995, 197–223.
Axworthy, M., *The Sword of Persia* (London, 2006).
Azzaroli, A., *An Early History of Horsemanship* (Leiden, 1985).
Badian, E., *Studies in Greek and Roman History* (Oxford, 1963).
Badian, E., 'The Administration of the Empire', *Greece & Rome* 12, 1965, 166–182.
Badian, E., 'Nearchus the Cretan', *Yale Classical Studies* 24, 1975, 147–170.
Badian, E., 'Alexander's Mules', *New York Review of Books* 20, 1979, 54–56.
Badian, E., 'Philip II and Thrace', *Pulpudeva* 4, 1983, 51–71.
Badian, E., 'The King's Indians', in W. Will (ed.), *Alexander der Grosse: Eine Welteroberung und ihr Hintergrund*, 1998, 205–224.
Bagnall, R.S. & Derow, P., *Historical Sources in Translation: The Hellenistic Period* (Oxford, 2004).
Baker, P., 'Warfare', in A. Erskine (ed.), *A Companion to the Hellenistic World*, 2005, 373–388.
Bakhuizen, S.C., *Salganeus and the Fortifications on its Mountains* (Groningen, 1970).
Bakhuizen, S.C., 'Renewed Investigations of Goritsa, the Military Base of the Macedonian Kings near Demetrias', *Athens Annals of Archaeology* 5, 1972, 485–495.
Baldassarre, I., 'Documenti di pittura ellenistica da Napoli', in *L'Italie méridionale et les premières expériences de la peinture hellénistique*, 1998, 95–160.
Baldus, H.R., 'Zum Siegel des Königs Lysimachos von Thrakien', *Chiron* 8, 1978, 195–199.
Baldus, H.R., 'Die Siegel Alexanders des Grossen: Versuch einer Rekonstruction auf literarischer und numismatischer Grunlage', *Chiron* 17, 1987, 395–449.
Baldwin, B., 'Medical Grounds for Exemptions from Military Service at Athens', *Classical Philology* 62, 1967, 42–43.
Bardunias, P., 'Don't Stick to Glued Linen: The *Linothorax* Debate', *Ancient Warfare* IV/3, 2010, 48–53.
Bardunias, P., response in 'The *Linothorax* Debate', *Ancient Warfare* V/1, 2011, 4–5.
Bar-Kochva, B., *The Seleucid Army: Organization and Tactics in the Great Campaigns* (Cambridge, 1979).
Bar-Kochva, B., *Judas Maccabaeus: The Jewish Struggle Against the Seleucids* (Cambridge, 1989).

Barnett, R.D., 'From Ivriz to Constantinople: A Study in Bird-Headed Swords', in R.M. Boehmer & H. Hauptmann (eds), *Beiträge zur Altertumskunde Kleinasiens*, 1983, 59–74.
Barr-Sharrar, B., *The Derveni Krater: Masterpiece of Classical Greek Metalwork* (Athens, 2008).
Baumann, H., *The Greek Plant World in Myth, Art and Literature* (Portland, 1993).
Bayliss, A.J., *After Demosthenes: The Politics of Early Hellenistic Athens* (London, 2011).
Bekker-Nielsen, T., 'Academic Science and Warfare in the Classical World', in T. Bekker-Nielsen & L. Hannestad (eds), *War as a Cultural and Social Force: Essays on Warfare in Antiquity*, 2001, 120–129.
Bellinger, A.R., *Essays on the Coinage of Alexander the Great* (New York, 1963).
Beloch, K.J., *Griechische Geschichte* III, 1 (Berlin & Leipzig, 1922).
Bengtson, H., 'Aus der Lebensgeschichte eines griechischen Distanzlaufers', *Symbolae Osloenses* 32, 1956, 35–39.
Bengston, H., *Die Staatsverträge des Altertums: Zweiter Band* (Munich, 1975).
Bennett, B. & Roberts, M., *The Wars of Alexander's Successors 323–281BC Volume 2: Armies, Tactics and Battles* (Barnsley, 2009).
Bennett, M.J., *Belted Heroes and Bound Women: the Myth of the Homeric Warrior-King* (Oxford, 1997).
Berg, B., 'An Early Source of the Alexander Romance', *Greek, Roman and Byzantine Studies* 14, 1973, 381–387.
Bergemann, J., *Romische Reiterstatuen: Ehrendenkmaler im offentlichen Bereich* (Mainz, 1990).
Berger, F. (ed.), *Antike Kunstwerke aus der Sammlung Ludwig III: Skulpturen* (Mainz, 1990).
Bergmann, B., 'The Roman House as Memory Theater: The House of the Tragic Poet in Pompeii', *The Art Bulletin* 76, 1994, 225–256.
Bernand, A., *Alexandrie des Ptolémées* (Paris, 1995).
Bernard, E., *Inscriptions métriques de l'Egypte gréco-romaine* (Paris, 1969).
Bernard, P., 'Ai Khanum: A Hellenistic City in Central Asia', *Proceedings of the British Academy* 53, 1967, 71–95.
Bernard, P., 'Campagne de fouille 1978 à Ai Khanoum (Afghanistan)', *Bulletin de l'École française d'Extrême-Orient* 68, 1980, 1–103.
Bertosa, B., 'The Supply of Hoplite Equipment by the Athenian State Down to the Lamian War', *Journal of Military History* 67, 2003, 361–369.
Bertrand, J.M., 'Sur les hyparques de l'empire d'Alexandre', in *Mélanges d'histoire ancienne offerts à William Seston*, 1974, 24–35.
Berve, H., *Das Alexanderreich Auf Prosopographischer Grundlage*, vol. I–II (Munich, 1926).
Besios, M., *Pieridon Stefanos: Pydna, Methoni ke i archeotites tis vorias pierias* (Katerini, 2010).
Bikerman, E.J., *Institutions de Séleucides* (Paris, 1938).
Bieber, M. & von Bothmer, D., 'Notes on the Mural Paintings from Boscoreale', *American Journal of Archaeology* 1956, 60, 171–172.
Bieber, M., 'The Portraits of Alexander the Great', *Proceedings of the American Philosophical Society* 93, 1949, 373–427.
Bieber, M., *History of the Greek and Roman Theater* (Princeton, 1961).
Bieber, M., *Alexander the Great in Greek and Roman Art* (Chicago, 1964).
Billows, R.A., *Antigonos the One-Eyed and the Creation of the Hellenistic State* (Berkeley, 1990).
Billows, R.A., *Kings and Colonists: Aspects of Macedonian Imperialism* (Leiden, 1995).
Bing, J.D., 'A Further Note on Cyinda/Kundi', *Historia* 22, 1973, 346–350.
Blackman, D.J., 'Archaeology in Greece 1996–97', *Archaeological Reports 1996–1997* (Society for the Promotion of Hellenic Studies), 1997, 1–125.
Blackman, D.J., 'Archaeology in Greece, 1999–2000', *Archaeological Reports 1999–2000* (Society for the Promotion of Hellenic Studies), 2000, 3–151.
Blackman, D.J., 'Archaeology in Greece 2001–2002', *Archaeological Reports 2001–2002* (Society for the Promotion of Hellenic Studies), 2002, 1–115.
Blasius, A., 'Army and Society in Ptolemaic Egypt – a Question of Loyalty', *Archiv für Papyrusforschung und verwandte Gebiete* 47, 2001, 81–98.
Blinski, B., 'L'Hemerodrome Philonides, son record et la nouvelle inscription d'Aigion', *Eos* 5, 1959/60, 69–80.

Bloedow, E.F., 'On 'Wagons' and 'Shields': Alexander's Crossing of Mount Haemus in 335BC', *Ancient History Bulletin* 10, 1996, 119–130.
Bloedow, E.F., 'On the Crossing of Rivers: Alexander's *Diptherai*', *Klio* 84, 2002, 57–75.
Bloedow, E.F., 'Why did Philip and Alexander Launch a War Against the Persian Empire?' *L'Antiquite Classique* 72, 2003, 261–274.
Blum, H., *Purpur als Statussymbol in der Griechischen Welt* (Bonn, 1998).
Blumberg, A., 'Inspired by the Bard: Philip II, Alexander the Great and the Homeric Ethos', *Ancient Warfare* III/3, 2009, 18–22.
Blyth, P.H., 'The Effectiveness of Greek Armour Against Arrows in the Persian Wars (490–479BC): An Interdisciplinary Study,' PhD Thesis, University of Reading, 1977.
Boardman, A.P., 'An Analysis of the Generalship of Alexander III of Macedon: Undermining or Underlining Greatness?' MA Thesis, University of Durham, 1999.
Boardman, J., 'Sickles and Strigils', *Journal of Hellenic Studies* 91, 1971, 136–137.
Boardman, J., 'Heroic Haircuts', *Classical Quarterly* 23, 1973, 196–197.
Boardman, J. (ed.), *The Cambridge Ancient History. Plates to Volumes V and VI* (Cambridge, 1994).
Bokonyi, S., *Data on Iron Age Horses of Central and Eastern Europe* (Harvard, 1968).
Bonfante, L., 'Nudity as a Costume in Classical Art', *American Journal of Archaeology* 93, 1989, 543–570.
Bonfante, L., 'Classical Nudity in Italy and Greece', in D. Ridgway (ed.), *Ancient Italy in its Mediterranean Setting: Studies in Honour of Ellen Macnamara*, 2000, 271–293.
Born, H. & Junkelmann, M., *Römische Kampf- und Turnierrüstungen* (Mainz, 1997).
Borza, E.N. & Palagia, O., 'The Chronology of the Macedonian Royal Tombs at Vergina', *Jahrbuch des Deutschen Archäologischen Instituts* 122, 2007, 81–125.
Borza, E.N., 'Alexander's Communications', *Ancient Macedonia II*, 1977, 295–303.
Borza, E.N., 'Some Observations on Malaria and the Ecology of Central Macedonia in Antiquity', *American Journal of Ancient History* 4, 1979, 102–124.
Borza, E.N., 'The Natural Resources of Early Macedonia', in W. Lindsay Adams & E.N. Borza (eds), *Philip II, Alexander the Great and the Macedonian Heritage*, 1982, 1–20.
Borza, E.N., 'Macedonian Tombs and the Paraphernalia of Alexander the Great', *Phoenix* 42, 1987a, 105–121.
Borza, E.N., 'Malaria in Alexander's Army', *Ancient History Bulletin* 1, 1987b, 36–38.
Borza, E.N., 'Timber and Politics in the Ancient World: Macedon and the Greeks', *Proceedings of the American Philological Society* 131, 1987c, 32–52.
Borza, E.N., *In the Shadow of Olympus: The Emergence of Macedon* (Princeton, 1990).
Borza, E.N., 'What Philip Wrought', *MHQ: The Quarterly Journal of Military History* 5, 1999, 104–109.
Bose, P., *Alexander the Great's Art of Strategy* (London, 2004).
Bosworth, A.B., 'The Government of Syria under Alexander the Great', *Classical Quarterly* 1974, 46–64.
Bosworth, A.B., 'Eumenes, Neoptolemus and *PSI* XII 1284', *Greek, Roman and Byzantine Studies* 19, 1978, 227–237.
Bosworth, A.B., *A Historical Commentary on Arrian's History of Alexander*, vol. I (Oxford, 1980a).
Bosworth, A.B., 'Alexander and the Iranians', *Journal of Hellenic Studies* 100, 1980b, 1–21.
Bosworth, A.B., 'Alexander the Great and the Decline of Macedon', *Journal of Hellenic Studies* 106, 1986a, 1–12.
Bosworth, A.B., 'Macedonian Manpower under Alexander the Great', *Ancient Macedonia IV*, 1986b, 115–122.
Bosworth, A.B., *From Arrian to Alexander: Studies in Historical Interpretation* (Oxford, 1988).
Bosworth, A.B., *A Historical Commentary on Arrian's History of Alexander*, vol. II (Oxford, 1995).
Bosworth, A.B., *Alexander and the East: The Tragedy of Triumph* (Oxford, 1996).
Bosworth, A.B., 'A Tale of Two Empires: Hernan Cortes and Alexander the Great', in A.B. Bosworth & E.J. Baynham (eds), *Alexander the Great in Fact and Fiction*, 2000, 23–49.
Bosworth, A.B., *The Legacy of Alexander: Politics, Warfare, and Propaganda under the Successors* (Oxford, 2002).

Bosworth, A.B., 'The Argeads and the Phalanx', in E. Carney & D. Ogden (eds), *Philip II and Alexander the Great: Father and Son, Lives and Afterlives*, 2010, 91–102.
Bowman, A., *Life and Letters on the Roman Frontier* (New York, 1998).
Braun, K., 'Der Dipylon-Brunnen B1: Die Funde', *Athenische Mitteilungen* 85, 1970, 126–269.
Breccia, E., *La necropoli di Sciatbi* (Cairo, 1912).
Brecoulaki, H., 'La couleur dans la peinture grecque antique de Macédoine', *Histoire de l'art* 39, 1997, 11–21.
Brecoulaki, H., 'Sur la techné de la peinture grecque ancienne d'après les monuments funéraires de Macédoine', *Bulletin de correspondance hellénique* 124, 2000, 189–216.
Brecoulaki, H., *La peinture funeraire de Macedoine II: Planches & tableaux* (Athens, 2006).
Breitenbach, H., *Historiographische Anschauungsformen Xenophons* (Freiburg, 1950).
Bresciani, E., *Kom Madi 1977 e 1978: Le pitture murali del cenotafio di Alessandro Magno* (Pisa, 1980).
Briant, P., *Rois, tributs et paysans: études sur les formations tributaires du Moyen-Orient ancien* (Paris, 1982).
Briant, P., 'Chasses d'Alexandre', *Ancient Macedonia IV*, 1986, 267–277.
Briant, P., 'Le nomadisme du grand roi', *Iranica Antiqua* 23, 1988, 253–273.
Briant, P., 'Chasses royales macédoniennes et chasses royales perses: le thème de la chasse au lion sur la chasse de Vergina', *Dialogues d'histoire ancienne* 17, 1991, 211–256.
Briant, P., *From Cyrus to Alexander: A History of the Persian Empire* (Winon Lake, 2002).
Brinkmann, V., 'Die blauen Augen der Perser: Die farbige Skulptur der Alexanderzeit und des Hellenismus', in V. Brinkmann & E. Wunsche (eds), *Bunte Gotter: Die Farbigkeit aintiker Skulptur*, 2003, 166–179.
Briscoe, J., *A Commentary on Livy Books XXXI–XXXXIII* (Oxford, 1973).
Briscoe, J., 'The Antigonids and the Greek States 276–196BC', in P.D.A. Garnsey & C.R. Whittaker (eds), *Imperialism in the Ancient World*, 1978, 145–157.
British Musem, *Guide to Exhibition Illustrating Greek and Roman Life* (London, 1920).
Brizzi, G., 'Hannibal: Punier und Hellenist', *Das Altertum* 37, 1991, 201–210.
Brizzi, G., *Le guerrier de l'Antiquité classique* (Monaco, 2004).
Brogan, T.M., 'Liberation Honors: Athenian Monuments from Antigonid Victories in their Immediate and Broader Contexts', in O. Palagia & S.V. Tracy (eds), *The Macedonians in Athens 322–229BC*, 2003, 194–205.
Brosius, M., 'Why Persia became the Enemy of Macedon', in W. Henkelman & A. Kuhrt (eds), *A Persian Perspective: Essays in Memory of Heleen Sancisi-Weerdenburg*, 2003, 227–238.
Brown, B.R., *Ptolemaic Paintings and Mosaics and the Alexandrian Style* (Cambridge, Mass., 1957).
Brown, B.R., *Royal Portraits in Sculpture and Coins: Pyrrhus and the Successors of Alexander the Great* (New York, 1995).
Brown, T.S., 'Alexander's Book Order (Plut. *Alex.* 8)', *Historia* 16, 1967, 359–368.
Brown, T.S., 'Alexander and Greek Athletics, in Fact and Fiction', in K. Kinzl (ed.), *Greece and the Eastern Mediterranean in Ancient History and Prehistory*, 1977, 76–88.
Brown, V.J., *Form and Colour in Greek Painting* (London, 1977).
Brunt, P.A., 'Alexander's Macedonian Cavalry', *Journal of Hellenic Studies* 83, 1963, 27–46.
Buchholz, E., *Die Homerischen Realien*, vol. II (Leipzig, 1881).
Buckler, J., 'Philip II's Designs on Greece', in R.W. Wallace & E.M. Harris (eds), *Transitions to Empire: Essays in Greco-Roman History 360–146BC*, 1996, 77–97.
Bugh, G.R., 'Introduction of the *Katalogeis* of the Athenian Cavalry', *Transactions of the American Philological Association* 112, 1982, 23–32.
Bull, S., *An Historical Guide to Arms and Armour* (New York, 1991).
Burford, A., *Craftsmen in Greek and Roman Society* (London, 1972).
Burliga, B., 'Aeneas Tacticus between History and Sophistry: the Emergence of the Military Handbook', in J. Pigon (ed.), *The Children of Herodotus: Greek and Roman Historiography and Related Genres*, 2008, 92–101.
Burnett, A., *et al.*, 'New Light on the Origins of Orichalcum', *Proceedings of the 9th International Congress of Numismatics*, 1982, 263–268.
Burnett, A., *et al.* (eds), *Coins of Macedonia and Rome: Essays in Honour of Charles Hersh* (London, 1998).

Burstein, S.M., 'IGII 1485a and Athenian Relations with Lysimachus', *Zeitschrift für Papyrologie und Epigraphik* 31, 1978, 181–185.
Burstein, S.M., *The Hellenistic Age from the Battle of Ipsos to the Death of Kleopatra VII* (Cambridge, 1985).
Bury, J.B, *et al.*, *The Hellenistic Age* (Cambridge, 1923).
Buschor, E., *Medusa Rondanini* (Stuttgart, 1958).
Busolt, G. & Swoboda, H., *Griechische Staatskunde*, vol. II (Munich, 1926).
Butler, M.E., 'Of Swords and Strigils: Social Change in Ancient Macedon,' PhD Thesis, Stanford University, 2008.
Cadario, M., *La corazza di Alessandro: Loricati di tipo ellenistico dal IV secolo a. C. al II d. C.* (Milan, 2004).
Cahill, N., 'The Treasury at Persepolis: Gift Giving at the City of the Persians', *American Journal of Archaeology* 89, 1985, 373–389.
Calcani, G., *Cavalieri di bronzo: La torma di Alessandro opera di Lisippo* (Rome, 1989).
Calcani, G., 'L'immagine di Alessandro Magno nel gruppo equestre del Granico', in J. Carlsen (ed.), *Alexander the Great, Reality and Myth*, 1993, 29–39.
Callaghan, P.J., 'On the Date of the Great Altar of Zeus at Pergamon', *Bulletin of the Institute of Classical Studies* 28, 1981, 115–121.
Cambitoglou, A., 'Military, Domestic and Religious Architecture at Torone in Chalkidike', in M. Stamatopoulou & M. Yeraoulanou (eds), *Excavating Classical Culture: Recent Archaeological Discoveries in Greece*, 2002, 21–56.
Camp, J.M., *Horses and Horsemanship in the Athenian Agora* (Athens, 1998).
Campbell, D.B., 'Theatre of War: The Hellenistic Mercenary in Contemporary Drama', *Ancient Warfare* III/1, 2009, 10–12.
Carlton, E., *Militarism: Rule without Law* (Aldershot, 2001).
Carney, E.D., 'Macedonians and Mutiny: Discipline and Indiscipline in the Army of Philip and Alexander', *Classical Philology* 91, 1996, 19–44.
Carney, E.D., 'The Trouble with Philip Arrhidaeus', *Ancient History Bulletin* 15, 2001, 63–89.
Carney, E.D., 'Hunting and the Macedonian Elite: Sharing the Rivalry of the Chase', in D. Ogden (ed.), *The Hellenistic World. New Perspectives*, 2002, 59–80.
Carney, E.D., 'Elite Education and High Culture in Macedonia', in W. Heckel & L.A. Tritle (eds), *Crossroads of History: The Age of Alexander*, 2003, 47–63.
Carney, E.D., 'The Role of the *Basilikoi Paides* at the Argead Court', in T. Howe & J. Reames (eds), *Macedonian Legacies: Studies in Ancient Macedonian History and Culture in Honor of Eugene N. Borza*, 2008a, 145–164.
Carney, E.D., 'Symposia and the Macedonian Elite: the Unmixed Life', *Syllecta Classica* 18, 2008b, 129–180.
Caro, A.L., 'Alessandro e Rossane come Ares ed Afrodite in un dipinto della casa Regio VI, Insula Occidentalis', in R.I. Curtius (ed.), *Studia Pompeiana & Classica in Honor of Wilhelmina F. Jashemski*, 1988, 75–80.
Carter, J.C., 'Relief Sculptures from the Necropolis of Taranto', *American Journal of Archaeology*, 74, 1970, 125–137.
Carter, M.J., 'Buttons and Wooden Swords: Polybius 10.20.3, Livy 26.51, and the Rudis', *Classical Philology* 101, 2006, 153–160.
Cartledge, P., 'Hoplites and Heroes: Sparta's Contribution to the Technique of Ancient Warfare', *Journal of Hellenic Studies* 97, 1977, 11–27.
Catling, H.W., 'Archaeology in Greece 1988–89', *Archaeological Reports 1988–89* (Society for the Promotion of Hellenic Studies), 1989, 3–116.
Caven, P., *Dionysus I: War-Lord of Sicily* (Newhaven, 1990).
Cawkwell, G., *Philip of Macedon* (London, 1978).
Ceka, N., *Iliret* (Tirana, 2001).
Chamay, J., 'Autour d'un vase phlyaque – un instrument de portage', *Antike Kunst* 20, 1977, 57–60.
Champion, J., 'Seleucus and the Availability of the Silver Shields', *Slingshot* 216, 2001, 42.
Champion, J., *Pyrrhus of Epirus* (Barnsley, 2009).
Chaniotis, A., *Historie und Historiker in den griechischen Inschriften* (Stuttgart, 1988).

Chaniotis, A., 'Mobility of Persons during the Hellenistic Wars: State Control and Personal Relations', in C. Moatti (ed.), *La mobilité des personnes en Méditerranée de l'Antiquité à l'époque moderne II*, 2004a, 481–500.

Chaniotis, A., 'New Inscriptions of Aphrodisias (1995–2001)', *American Journal of Archaeology* 108, 2004b, 377–409.

Chaniotis, A., *War in the Hellenistic World: A Social and Cultural History* (Oxford, 2005).

Chankowski, A.S., 'L'entraînement militaire des éphèbes dans les cités grecques d'Asie Mineure à l'époque hellénistique: Nécessité pratique ou tradition atrophée?' in J.C. Couvenhes & H.L. Fernoux (eds), *Les Cités grecques et la guerre en Asie Mineure à l'époque hellénistique*, 2004, 55–76.

Chenciner, R., *Madder Red: A History of Luxury and Trade* (Richmond, 2000).

Cheesman, G.L., *The Auxilia of the Roman Imperial Army* (Oxford, 1914).

Choremis, A., 'Metallic Armour from a Tomb at Prodromi in Thesprotia', *Athens Annals in Archaeology* 13, 1980, 3–20.

Christesen, P. & Murray, S.C., 'Macedonian Religion', in J. Roisman & I. Worthington (eds), *A Companion to Ancient Macedonia*, 2010, 428–445.

Christesen, P., 'Xenophon's *Cyropaedia* and Military Reform in Sparta', *Journal of Hellenic Studies* 126, 2006, 47–65.

Chroust, A.H., 'Aristotle and the Foreign Policy of Macedonia', *Review of Politics* 34, 1972, 367–394.

Chroust, A.H., *Aristotle: New Light on his Life and on Some of his Lost Works*, vol. I (London, 1973).

Chrysostomou, A. & Chrysostomou, P., 'Anaskaphe ste dytike nekropole tou Pellas kata to 2001', *To archaiologiko ergo ste Makedonia kai Thrake* 15, 2001, 477–488.

Chrysostomou, A. & Chrysostomou, P., 'Excavations in the west cemetery of Archontiko near Pella in 2002', *To archaiologiko ergo ste Makedonia kai Thrake* 16, 2002, 465–478.

Chrysostomou, P., 'Burial Mounds in the Area of Pella', *To archaiologiko ergo ste Makedonia kai Thrake* 1, 1987, 147–159.

Chrysostomou, P., 'To anaktoro tis Pellas', *To archaiologiko ergo ste Makedonia kai Thrake* 10, 1996, 105–142.

Chrysostomou, P., 'Two Early Hellenistic Relief Grave Stelai from Central Macedonia', *Archaiologikon deltion* 53A, 1998, 301–312.

Chugg, A.M., *Alexander's Lovers* (Morrisville, 2006).

Chugg, A.M., *The Quest for the Tomb of Alexander the Great* (2007).

Chugg, A.M., *Alexander the Great in India: A Reconstruction of Cleitarchus* (2009).

Cilliers, L. & Retief, F.P., 'Alexander die Grote se leer en die oorlogstres-sindroom, 326 v.C.', *Akroterion* 45, 2000, 27–35.

Cimok, F., *Pergamum* (Istanbul, 2001).

Clarysse, W. & Schepens, G., 'A Ptolemaic Fragment of Alexander History', *Chronique d'Égypte* 60, 1985, 30–47.

Clutton-Brock, J., *Horse Power: A History of the Horse and the Donkey in Human Societies* (Cambridge, Mass., 1992).

Codellas, P.S., 'The Epimonidion of Philon the Byzantine, 'The Hunger and Thirst Checking Pill' and other Emergency Foods', *Bulletin of the History of Medicine* 22, 1948, 630–634.

Cohen, A., 'Alexander and Achilles – Macedonians and 'Mycenaeans'', in J.B. Carter & S.P. Morris (eds), *The Age of Homer: A Tribute to Emily Townsend Vermeule* 1995, 483–505.

Cohen, A., *The Alexander Mosaic: Stories of Victory and Defeat* (Cambridge, 1997).

Cohen, G.M., *The Seleucid Colonies: Studies in Founding, Administration and Organization* (Wiesbaden, 1978).

Cohen, G.M., *The Hellenistic Settlements in Syria, the Red Sea Basin, and North Africa* (Berkeley, 2006).

Cohen, R., *By the Sword* (New York, 2003).

Cohn-Haft, L., *The Public Physicians of Ancient Greece* (Northampton, Mass., 1956).

Coleman Carter, E., 'The Sculptures of Taras', *Transactions of the American Philosophical Society* 65, 1975, 1–196.

Connolly, P., *Greece and Rome at War* (London, 1988).

Connolly, P., 'Legion versus Phalanx', *Military Illustrated* 124, 1998, 36–41.

Connolly, P., 'Experiments with the *Sarissa* – the Macedonian Pike and Cavalry Lance: a Functional View', *Journal of Roman Military Equipment Studies* 11, 2000, 103–112.
Cook, B.F., 'Footwork in Ancient Greek Swordsmanship', *Metropolitan Museum Journal* 24, 1989, 57–64.
Cook, J.M., *The Persian Empire* (London, 1983).
Cooksey, C.J., 'Making Tyrian Purple', *Dyes in History and Archaeology* 14, 1995, 70–77.
Cooper, J., *The Heart and the Rose* (Leigh-on-Sea, 2004).
Corrigan, D.M., 'Riders on High: an Interdisciplinary Study of the Macedonian Cavalry of Alexander the Great', PhD Thesis, University of Texas at Austin, 2004.
Corvisier, J.N., *Guerres et sociétés dans les mondes grecs (490–322 av. J.-C.)* (Paris, 1999).
Couissin, P., *Les institutions militaires et navales* (Paris, 1932).
Couvenhes, J.-C., 'La fourniture d'armes aux citoyens athéniens du IVe au IIIe siecle avant J.-C.', in P. Sauzeau & T. van Compernolle (eds), *Les armes dans l'Antiquité*, 2007, 521–540.
Cowan, R., 'The Battle of Nisibis, AD217', *Ancient Warfare*, III/5, 2009, 29–35.
Crawford, J.S., 'A Portrait of Alexander the Great at the University of Delaware', *American Journal of Archaeology* 83, 1979, 477–481.
Crawford, M.H., *Roman Republican Coinage*, vol. I (Cambridge, 1974).
Crowther, N.B., 'Male "Beauty" Contests in Greece: the *Euandria* and *Euexia*', *L'Antiquité classique* 54, 1985, 285–291.
Crowther, N.B., '*Euexia, Eutaxia, Philoponia*: Three Contests of the Greek Gymnasium', *Zeitschrift für Papyrologie und Epigraphik* 85, 1991, 301–304.
Culham, P., 'Chance, Command and Chaos in Ancient Military Engagements', *World Futures* 27, 1989, 191–205.
Cunningham, A., *Coins of Alexander's Successors in the East: Part 1 – The Greeks of Bactria, Ariana and India* (Delhi, 1971).
Curtis, J.W., 'Coinage of Pharaonic Egypt', *Journal of Egyptian Archaeology* 43, 1957, 71–76.
Curtius, L., *Die Wandmalerei Pompejis* (Leipzig, 1929).
Dahm, M., 'Make Camp!' *Ancient Warfare*, I/2, 2007a, 41–45.
Dahm, M., 'Survive a Siege', *Ancient Warfare*, I/3, 2007b, 40–44.
Dahm, M., 'By Word of Mouth: Oral Traditions of Teaching Warfare', *Ancient Warfare*, II/5, 2008, 48–51.
Dakaris, S., *Dodona* (Athens, 1993).
Dalby, A., *Siren Feasts: A History of Food and Gastronomy in Greece* (London, 1996).
Dalby, A., *Food in the Ancient World from A to Z* (London, 2003).
Dalloy, A., 'Greeks Abroad: Social Organisation and Food among the Ten Thousand', *Journal of Hellenic Studies* 112, 1992, 16–30.
Daniel, T., 'The Taxeis of Alexander and the Change to Chiliarch, the Companion Cavalry and the Change to Hipparchies: a Brief Assessment', *Ancient World* 23, 1992, 43–57.
Darnell, J.C. & Manassa, C., *Tutankhamun's Armies. Battle and Conquest during Ancient Egypt's Late Eighteenth Dynasty* (Hoboke, 2007).
Daux, G., 'Chronique des fouilles et découvertes archéologiques en Grèce', *Bulletin de correspondance hellénique* 85, 1961, 601–953.
Dàvidson, J., *Courtesans and Fishcakes* (London, 1997).
Davies, H.E.H., 'Designing Roman Roads', *Britannia* 29, 1998, 1–16.
Davies, J., 'Mines, Miners and Macedon', in N Sekunda (ed.), *Ergasteria: Works presented to John Ellis Jones*, 2010, 94–99.
Davies, O., 'Ancient Mines in Southern Macedonia', *Journal of the Royal Anthropological Institute of Great Britain and Ireland* 62, 1932, 145–162.
Davies, R.W., 'Roman Wales and Roman Military Practice-Camps', *Archaeologia Cambrensis* 117, 1968, 103–120.
Davis, B.L., *British Army Uniforms and Insignia of World War Two* (London, 1983).
De Callatay, F., 'Les trésors achéménides et les monnayages d'Alexandre: espèces immobilisées et espèces circulantes?' *Revue des études anciennes* 91, 1989, 25–74.
De Carolis, E., *Gods and Heroes in Pompeii* (Los Angeles, 2001).

De Franciscis, A., 'La Villa Romana di Oplontis', in B. Andraeae & H. Kyrieleis (eds), *Neue Forschungen in Pompeji*, 1975a, 9–18.
De Franciscis, A., *The Pompeian Wall Paintings in the Roman Villa of Oplontis* (Recklinghausen, 1975b).
De Grummond, N.T., 'The Real Gonzaga Cameo', *American Journal of Archaeology* 78, 1974, 427–429.
De Madariaga, I., *Ivan the Terrible: First Tsar of Russia* (New Haven, 2005).
De Sanctis, G., 'Il regolamento militare dei Macedoni', *Rivista di Filologia* 12, 1934, 515–521.
Deacy, S. & Villing, A., 'What was the Colour of Athena's Aegis?' *Journal of Hellenic Studies* 129, 2009, 111–119.
Delebecque, E., *Essai sur la vie de Xénophon* (Paris, 1957).
Deleman, I., 'An Unplundered Chamber Tomb on Ganos Mountain in Southeastern Thrace', *American Journal of Archaeology* 110, 2006, 251–273.
Delia, D., 'All Army Boots and Uniforms? Ethnicity in Ptolemaic Egypt', in M. True & K. Hamma (eds), *Alexandria and Alexandrianism*, 1996, 41–52.
Delrieux, F., 'La monnaie et la guerre dans l'antiquité classique: le cas des émissions d'Alexandre le Grand entre 333 et 323 av. J.-C.', in P. Brun (ed.), *Questions d'histoire: Guerres et sociétés dans les mondes grecs (490–322)*, 1999, 306–310.
Despini, A., 'Ho taphos tas Katerinas', *Athens Annals of Archaeology* 13, 1980, 190–209.
Despini, A., 'Ho taphos tas Katerinas', *He Makedonia apo ta mykenaika chronia hos ton Mega Alexandro*, 1988, 141–143.
Deszo, T. & Curtis, J., 'Assyrian Iron Helmets from Nimrud now in the British Museum', *Iraq* 53, 1991, 105–126.
Deszo, T., 'The Reconstruction of the Neo-Assyrian Army: As Depicted on the Assyrian Palace Reliefs, 745–612 BC', *Acta Archaeologica Academiae Scientiarum Hungaricae* 57, 2006, 87–130.
Detienne, M., *Dionysos Slain* (Baltimore, 1979).
Deutscher, G., *Through the Language Glass: How Words Colour your World* (London, 2010).
Devine, A.M., book review in *Phoenix* 33, 1979, 272–276.
Devine, A.M., 'Alexander's Propaganda Machine: Callisthenes as the Ultimate Source for Arrian, Anabasis 1–3', in I. Worthington (ed.), *Ventures into Greek History*, 1994, 89–102.
Devine, A.M., 'Polybius' Lost *Tactica*', *Ancient History Bulletin* 9, 1995, 40–44.
Devine, A.M., Book Review in *Classical Review* 47, 1997, 353–356.
Di Vita, A. & Alfano, C. (eds), *Alessandro Magno: Storia e mito* (Milan, 1995).
Dickinson, R.E., 'Length Isn't Everything – Use of the Macedonian Sarissa in the Time of Alexander the Great', *Journal of Battlefield Technology* 3, 2000, 51–62.
Hill, G.F., *A Catalogue of Greek Coins in the British Museum*, vol. 28 (London, 1922).
Dieulafoy, M., *L'Art antique de la Perse*, vol. I (Paris, 1884).
Dihle, A., 'The Conception of India in Hellenistic and Roman Literature', *Proceedings of the Cambridge Philological Society* 10, 1964, 15–23.
Dintsis, P., *Hellenistische Helme I: Text* (Rome, 1986a).
Dintsis, P., *Hellenistische Helme II: Tafeln* (Rome, 1986b).
Dintsis, P., 'Über die Bezeichnung *Konos* im Reglement von Amphipolis', *Ancient Macedonia IV*, 1986c, 171–182.
Dodge, T.A., *Alexander* (London,1994).
Doherty, P., *Alexander the Great: The Death of a God* (London, 2004).
Dohrn, T., 'Die Marmor-Standbilder des Daochos-Weihgeschenks in Delphi', *Antike Plastik* 8, 1968, 33–53.
Dorjahn, A.P., 'Smoke-Screens in Ancient Warfare', *Classical Bulletin* 35, 1959, 33.
Draganov, D., *The Coins of the Macedonian Kings. Pt. I: From Alexander I to Alexander the Great* (Sofia, 2000).
Drew, P., *Tensile Architecture* (Boulder, 1979).
Drougou, S. & Saatsoglou-Paliadeli, C., *Vergina: Wandering through the Archaeological Site* (Athens, 1999).
Droysen, H., *Untersuchungen über Alexander des Grossen Heerwesen und Kriegführung* (Freiburg, 1885).
Du Picq, A., *Battle Studies: Ancient and Modern Battle* (New York, 1921).

Dubs, H.H., 'Ancient Military Contact between Romans and Chinese', *American Journal of Philology* 62, 1941, 322–330.
Ducrey, P., *Guerre et guerriers dans la Grèce Antique* (Paris, 1985).
Durham, M.E., *Albania and the Albanians: Selected Articles and Letters 1903–1944* (London, 2001).
Echols, E.C., 'Macedon and Germany', *Classical Weekly* 43, 1949, 74–76.
Echols, E.C., 'Crossing a Classical River', *Classical Journal* 48, 1953, 215–224.
Eckstein, A.M., 'Review: Brigands, Emperors and Anarchy', *International History Review* 22, 2000, 862–879.
Eckstein, A.M., 'Bellicosity and Anarchy: Soldiers, Warriors and Combat in Antiquity', *International History Review* 27, 2005, 481–497.
Eckstein, A.M., *Rome Enters the Greek East* (Oxford, 2008).
Eckstein, A.M., 'What's in an Empire? Rome and the Greeks after 188B.C.', *South Central Review* 26, 2009, 20–37.
Eckstein, A.M., 'Macedonia and Rome, 221–146 BC', in J. Roisman & I. Worthington (eds), *A Companion to Ancient Macedonia*, 2010, 225–50.
Eddy, S.K., *The King is Dead: Studies in the Near Eastern Resistance to Hellenism 334–31 BC* (Lincoln, 1961).
Edmonds, J.M., *The Fragments of Attic Comedy*, vol. IIIA–B (Leiden, 1961).
Edson, C.F., 'The Antigonids, Heracles and Beroea', *Harvard Studies in Classical Philology* 45, 1934, 213–246.
Edson, C.F., 'The Location of Cellae and the Route of the Via Egnatia', *Classical Philology* 46, 1951, 1–16.
Edson, C.F., 'Early Macedonia', *Ancient Macedonia I*, 1970, 17–44.
Edwards, E.H., *Horses: Their Role in the History of Man* (London, 1987).
Eitrem, S., 'A Purificatory Rite and some Allied *Rites du Passage*', *Symbolae Osloenses* 25, 1947, 36–53.
Ellis, J.R., 'Amyntas III, Illyria and Olynthos 393/2 – 380/79', *Makedonika* 9, 1968, 1–7.
Ellis, J.R., 'Population Transplants under Philip II', *Makedonika* 9, 1969, 9–17.
Ellis, J.R., *Philip II and Macedonian Imperialism* (London, 1976).
Ellis, J.R., 'The Dynamics of Fourth Century Macedonian Imperialism', *Ancient Macedonia II*, 1977, 103–114.
Empereur, J.-Y., 'Collection Paul Canellopoulos (XVII): Petits objets inscrits', *Bulletin de correspondance hellénique* 105, 1981, 537–568.
Empereur, J.-Y., *Alexandria: Rediscovered* (London, 1998).
Enciclopedia dell'Arte Antica, vol. IV (Rome, 1961).
Engels, D.W., *Alexander the Great and the Logistics of the Macedonian Army* (Berkeley, 1978).
Engels, D.W., 'Alexander's Intelligence System', *Classical Quarterly* 30, 1980, 327–340.
English, S., 'The Army of Alexander the Great', MA Thesis, University of Durham, 2003.
English, S., *The Sieges of Alexander the Great* (Barnsley, 2010).
Errington, E. & Cribb, J. (eds), *The Crossroads of Asia: Transformation in Image and Symbol in the Art of Ancient Afghanistan and Pakistan* (Cambridge, 1992).
Errington, R.M., *A History of Macedonia* (Berkeley, 1993).
Errington, R.M., 'Recent Research on Ancient Macedonia', *Analele Univ. Galati, s. Istorie* 1, 2002, 10–23.
Erskine, A., 'The Pezhetairoi of Philip II and Alexander III', *Historia* 38, 1989, 385–394.
Etienne, R., 'Jason de Phères et Philippe II: Stratégies de deux condottieri', in F. Prost (ed.), *Armées et sociétés de la Grèce classique – Aspects sociaux et politiques de la guerre aux Ve et IVe s. av. J.-C.*, 1999, 276–286.
Evelyn-White, H., *Hesiod: The Homeric Hymns and Homerica* (Cambridge, 1970).
Everson, T., *Warfare in Ancient Greece: Arms and Armour from the Heroes of Homer to Alexander the Great* (Stroud, 2004).
Fabricius, E., 'Ein bemaltes Grab aus Tanagara', *Mitteilungen des Deutschen Archäologischen Instituts. Athenische Abteilung* 10, 1885, 158–164.
Faklaris, P., 'Peritrachelion', *Archaiologikon deltion* 40, 1985, 1–16; 283–284.
Faklaris, P., 'Harnesses from Vergina', *Archaiologikon deltion* 41, 1986, 1–58.

Faraone, C.A., 'Molten Wax, Spilt Wine and Mutilated Animals: Sympathetic Magic in Near Eastern and Early Greek Oath Ceremonies', *Journal of Hellenic Studies* 113, 1993, 60–80.
Farber, J.J., 'The Cyropaedia and Hellenistic Kingship', *American Journal of Philology* 100, 1979, 497–574.
Faridis, K., *Vergina: History – Archaeology* (Thessaloniki, 2001).
Faure, P., *La vie quotidienne des armées d'Alexandre* (Paris, 1982).
Fellman, M., *The Making of Robert E. Lee* (Baltimore, 2003).
Felten, F., 'Themen makedonischer Grabdenkmäler klassischer Zeit', *Ancient Macedonia V*, vol. I, 1989, 405–431.
Fermor, P.L., *Roumeli: Travels in Northern Greece* (London, 1966).
Ferrill, A., *The Origins of War: From the Stone Age to Alexander the Great* (London, 1985).
Feugere, M., *Les casques antiques* (Paris, 1994).
Feugere, M., *Weapons of the Romans* (Stroud, 2002).
Feyel, M., 'Un nouveau fragment du règlement militaire trouvé à Amphipolis', *Revue archaeologique* 5, 1935, 28–68.
Firasat, S., *et al.*, 'Y-Chromosomal Evidence for a Limited Greek Contribution to the Pathan Population of Pakistan', *European Journal of Human Genetics* 15, 2006, 121–126.
Fischer, J., *Griechisch-römische Terrakotten aus Ägypten* (Berlin, 1994).
Fol, A. & Marazov, I., *Thrace and the Thracians* (London, 1977).
Forbes, C.A., 'Expanded Uses of the Greek Gymnasium', *Classical Philology* 40, 1945, 32–42.
Forbes, R.J., *Studies in Ancient Technology*, vol. II (Leiden, 1955).
Forbes, R.J., *Studies in Ancient Technology*, vol. IV (Leiden, 1956).
Forbes, R.J., *Studies in Ancient Technology*, vol. VII (Leiden,1963).
Foxhall, L. & Forbes H.A., '*Sitometreia*: The Role of Grain as a Staple Food in Classical Antiquity', *Chiron* 12, 1982, 80–81.
Francfort, H.P., *Fouilles d'AikHanoum III: Le sanctuaire du temple à niches indentées. 2. Les trouvailles* (Paris, 1984).
Fraser, P.M., 'Inscriptions from Graeco-Roman Egypt', *Berytus* 15, 1964, 71–93.
Fraser, P.M., *Ptolemaic Alexandria*, vol. I (Oxford, 1972).
Fraser, P.M., *Cities of Alexander the Great* (Oxford 1996).
Frazer, A., 'Xenophon and the Boeotian Helmet', *Art Bulletin* 4, 1922, 99–108.
Fredricksmeyer, E., 'Alexander the Great and the Macedonian Kausia', *Transactions of the American Philological Society* 116, 1986, 215–227.
Fredricksmeyer, E., 'The Kausia: Macedonian or Indian?' in I. Worthington (ed.), *Ventures into Greek History*, 1994, 135–158.
Fuller, J.F.C., *A Military History of the Western World*, vol. I (New York, 1954).
Fuller, J.F.C., *Generalship of Alexander the Great* (London, 1958).
Furtwängler, A., *Die antiken Gemmen: Geschichte der Steinschneidekunst im Klassischen Altertum. Erster Band. Tafeln* (Amsterdam, 1964).
Gabbert, J., 'The Greek Hegemony of Antigonus II Gonatas (r. 283–239B.C.)', PhD Thesis, University of Cincinnati, 1982.
Gabbert, J., 'The Grand Strategy of Antigonus II Gonatas and the Chremonidian War', *The Ancient World* 8, 1983, 129–136.
Gabbert, J., 'The Language of Citizenship in Antigonid Macedonia', *Ancient History Bulletin* 2, 1988, 10–14.
Gabelmann, H., 'Pantherfellschabracken', *Bonner Jahrbücher des Rheinischen Landesmuseums in Bonn* 196, 1996, 11–39.
Gabriel, R.A. & Metz, K., *From Sumer to Rome: The Military Capabilities of Ancient Armies* (New York, 1991).
Gabriel, R.A. & Metz, K., *A History of Military Medicine: Volume I* (New York, 1992).
Gabriel, R.A., *The Culture of War* (New York, 1990).
Gabriel, R.A., *Great Captains of Antiquity* (Westport, 2000).
Gabriel, R.A., *The Great Armies of Antiquity* (Westport, 2002).
Gabriel, R.A., *The Military History of Ancient Israel* (Westport, 2003).

Gabriel, R.A., *Empires at War: Volume I* (Westport, 2005a).
Gabriel, R.A., *Empires at War: Volume II* (Westport, 2005b).
Gabriel, R.A., *Soldiers' Lives through History – the Ancient World* (Westport, 2007).
Gabriel, R.A., *Philip II of Macedonia: Greater than Alexander* (Washington DC, 2010).
Gabrielli, M., 'Transport et logistique militaire dans l'Anabase', *Pallas* 43, 1995, 109–122.
Gaebel, R., *Cavalry Operations in the Ancient Greek World* (Norman, 2002).
Gagsteiger, G., *Die ptolemaischen Waffenmodelle aus Memphis* (Hildesheim, 1993).
Gardiner-Garden, J., 'Ateas and Theopompus', *Journal of Hellenic Studies* 99, 1989, 29–40.
Gardner, P., *The Coins of the Greek and Scythic Kings of Bactria and India in the British Museum* (London, 1886).
Garlan, Y., 'Cités, armées et stratégie à l'époque hellénistique d'après l'oeuvre de Philon de Byzance', *Historia* 22, 1973, 16–33.
Garlan, Y., *Recherches de poliorcétique grecque* (Paris, 1974).
Garlan, Y., *War in the Ancient World: a Social History* (London, 1975).
Garoufalias, P., *Pyrrhus, King of Epirus* (London, 1979).
Gaukowsky, P., *Essai sur les origines du mythe d'Alexandre (336–270 av. J.-C.), II: Alexandre et Dionysos* (Nancy, 1981).
Gauthier, P. & Hatzopoulos, M.B., *La loi gymnasiarchique de Beroia* (Athens, 1993).
Geller, M.J., 'Babylonian Astronomical Diaries', *Bulletin of the School of Oriental and African Studies* 53, 1990, 1–7.
Gilley, D., 'Alexander and the Carmanian March of 324 BC', *Ancient History Bulletin* 20, 2004, 9–14.
Gilliver, C.M., *The Roman Art of War* (Stroud, 1999).
Gilliver, C.M., 'Display in Roman Warfare', *War in History* 14, 2007, 1–21.
Ginouves, R. (ed.), *Macedonia: From Philip II to the Roman Conquest* (Athens, 1993).
Girtzy, M., *Historical Topography of Ancient Macedonia* (Thessaloniki, 2001).
Gissel, J.A., 'Germanicus as an Alexander Figure', *Classica et Mediaevalia* 52, 2001, 277–301.
Giuliani, L., 'Alexander in Ruvo, Eretria und Sidon', *Antike Kunst* 20, 1977, 26–42.
Goralski, W.J., 'Arrian's *Events After Alexander*: Summary of Photius and Selected Fragments', *The Ancient World* 19, 1989, 81–108.
Gordon, D.H., 'Scimitars, Sabres and Falchions', *Man* 58, 1958, 22–27.
Gordon-Childe, V., 'Rotary Querns on the Continent and in the Mediterranean Basin', *Antiquity* 17, 1943, 19–26.
Gow, A.S.F. & Page, D.L. (eds), *The Greek Anthology: Hellenistic Epigrams* (Cambridge, 1965).
Gow, A.S.F. & Page, D.L. (eds), *The Greek Anthology: The Garland of Philip and Some Contemporary Epigrams. II: Commentary and Indexes* (Cambridge, 1968).
Grabsky, P.I., *Caesar: Ruling the Roman Empire* (London, 1997).
Graf, D.F., 'The Persian Royal Road System', in H. Sancisi-Weerdenburg, *et al.* (eds), *Achaemenid History VIII: Continuity and Change*, 1994, 167–189.
Grainger, J.D., *Seleukos Nikator* (London, 1990).
Grainger, J.D., *Alexander the Great Failure: The Collapse of the Macedonian Empire* (London, 2007).
Graninger, D., 'Macedonia and Thessaly', in J. Roisman & I. Worthington (eds), *A Companion to Ancient Macedonia*, 2010, 306–325.
Greek Ministry of Culture, *Ancient Macedonia* (Athens, 1988).
Green, P., 'Caesar and Alexander: *Aemulatio, imitatio, comparatio*', *American Journal of Ancient History* 3, 1978, 1–26.
Green, P., *Alexander of Macedon* (Berkeley, 1991).
Green, P., *Classical Bearings: Interpreting Ancient History and Culture* (Berkeley, 1998).
Greenewalt, C.H. & Heywood, A.M., 'A Helmet of the Sixth Century B.C. from Sardis', *Bulletin of the American Schools of Oriental Research* 285, 1992, 1–31.
Greenwalt, W., 'Macedonia's Kings and the Political Usefulness of the Medical Arts', *Ancient Macedonia IV*, 1986, 213–222.
Greenwalt, W., 'Why Pella?' *Historia* 48, 1999, 158–183.
Grentez, F., 'Old Samarkand: Nexus of the Ancient World', *Archaeology Odyssey* 6, 2003, 26–37.
Griffith, G.T., *The Mercenaries of the Hellenistic World* (Cambridge, 1935).

Griffith, G.T., 'Makedonika: Notes on the Macedonians of Philip and Alexander', *Proceedings of the Cambridge Philological Society* 4, 1956/7, 3–10.
Griffith, G.T., 'The Macedonian Background', *Greece and Rome* 12, 1965, 125–139.
Griffith, G.T., book review in *Gnomon* 48, 1976, 617–619.
Griffith, M., '"Public" and "Private" in Early Greek Institutions of Education', in Y.L. Too (ed.), *Education in Greek and Roman Antiquity*, 2001, 23–84.
Griffith, M., 'Horsepower and Donkeywork: Equids and the Ancient Greek Imagination', *Classical Philology* 101, 2006, 185–246.
Griffiths, W.B., 'The Sling and its Place in the Roman Imperial Army', in C. Van Driel-Murray (ed.), *5th Roman Military Equipment Conference*, 1989, 255–280.
Griffiths, W.B., 'The Hand-Thrown Stone', *Arbeia Journal* 1, 1992, 1–11.
Grimm, G., *Alexandria: Die erste Königsstadt der hellenistischen Welt* (Mainz, 1998).
Grose, S.W., *Fitzwilliam Museum: Catalogue of the McClean Collection of Greek Coins. II* (Cambridge, 1926).
Grote, K., 'Das griechische Soldnerwesen der hellenistischen Zeit', Dissertation, University of Jena, 1913.
Grozdanova, B., 'Macedonian Shield from Bonce', in M. Blecic, *et al.* (eds), *Scripta Praehistorica in Honorem Biba Terzan*, 2007, 863–874.
Gunderson, L.L., 'Early Elements in the Alexander Romance', *Ancient Macedonia I*, 1970, 353–375.
Gunn, B. & Gardiner, A.H., 'New Renderings of Egyptian Texts: II. The Expulsion of the Hyksos', *Journal of Egyptian Archaeology* 5, 1918, 36–56.
Guzzo, P.G. & Fergola, L., *Oplontis: La villa di Poppaea* (Milan, 2000).
Haagsma, M.J., 'Domestic Economy and Social Organization in New Halos,' Dissertation, University of Groningen, 2010.
Hackim, J., *Nouvelles recherches archéologiques à Begram: Planches* (Paris, 1954).
Hadley, R.A., 'Seleucus, Dionysus or Alexander?' *Numismatic Chronicle* 14, 1974, 9–13.
Hagemann, A., *Griechische Panzerung. I. Teil: der Metallharnisch* (Leipzig, 1919).
Hainsworth, B., *The Iliad: A Commentary. Volume III: Books 9–12* (Cambridge, 1993).
Hall Sternberg, R., 'The Transport of Sick and Wounded Soldiers in Classical Greece', *Phoenix* 53, 1999, 55–72.
Hamiaux, M., *Les sculptures grecques I: Des origines à la fin du IVe siècle avant J.-C.* (Paris, 1992).
Hamiaux, M., *Les sculptures grecques II: La période hellénistique (IIIe-Ier siècles avant J.-C.)* (Paris, 1998).
Hamilton, C.D., 'From Archidamus to Alexander: The Revolution in Greek Warfare', *Naval War College Review* 48, 1995, 84–95.
Hamilton, C.D., 'The Hellenistic World', in K. Raaflaub & N. Rosenstein (eds), *War and Society in the Ancient and Medieval Worlds*, 1999a, 170–171.
Hamilton, C.D., 'Macedonian Migration and its Effects', *Ancient Macedonia VI*, vol. I, 1999b, 839–850.
Hamilton, J.K., *Plutarch: Alexander. A Commentary* (Oxford, 1968).
Hammond, N.G.L. & Griffith, G.T., *A History of Macedonia. Vol. II: 550–336 BC* (Oxford, 1979).
Hammond, N.G.L. & Roseman, L.J., 'The Construction of Xerxes' Bridge over the Hellespont', *Journal of Hellenic Studies* 116, 1996, 88–107.
Hammond, N.G.L. & Walbank, F.W., *A History of Macedonia. Vol. III: 336–167 BC* (Oxford, 1988).
Hammond, N.G.L., *Epirus* (Oxford, 1967).
Hammond, N.G.L., *A History of Macedonia* (Oxford, 1972).
Hammond, N.G.L., 'A Note on 'Pursuit' in Arrian', *Classical Quarterly* 28, 1978, 136–140.
Hammond, N.G.L., book review in *Journal of Hellenic Studies* 100, 1980, 256–257.
Hammond, N.G.L., *Alexander the Great: King, Commander and Statesman* (London, 1981a).
Hammond, N.G.L., 'The Western Frontier of Macedonia in the Reign of Philip II', *Ancient Macedonian Studies in Honor of Charles F. Edson*, 1981b, 199–217.
Hammond, N.G.L., 'Army Transport in the Fifth and Fourth Centuries', *Greek, Roman & Byzantine Studies* 24, 1983a, 27–31.
Hammond, N.G.L., *Three Historians of Alexander the Great* (Cambridge, 1983b).

Hammond, N.G.L., 'Alexander's Veterans After his Death', *Greek, Roman and Byzantine Studies* 25, 1984a, 51–61.

Hammond, N.G.L., 'The Battle of Pydna', *Journal of Hellenic Studies* 104, 1984b, 31–47.

Hammond, N.G.L., 'A Papyrus Commentary on Alexander's Balkan Campaign', *Greek, Roman and Byzantine Studies* 28, 1987a, 331–347.

Hammond, N.G.L., 'Arms and the King: the Insignia of Alexander the Great', *Phoenix* 43, 1989a, 217–224.

Hammond, N.G.L., 'Casualties and Reinforcements of Citizen Soldiers in Greece and Macedonia', *Journal of Hellenic Studies* 109, 1989b, 56–68.

Hammond, N.G.L., *The Macedonian State: Origins, Institutions and History* (Oxford, 1989c).

Hammond, N.G.L., 'Royal Pages, Personal Pages and Boys Trained in the Macedonian Manner during the Period of the Temenid Monarchy', *Historia* 39, 1990, 261–290.

Hammond, N.G.L., *The Miracle that was Macedonia* (London, 1991a).

Hammond, N.G.L., 'The Various Guards of Philip II and Alexander III', *Historia* 40, 1991b, 396–418.

Hammond, N.G.L., 'The Archaeological and Literary Evidence for the Burning of the Persepolis Palace', *Classical Quarterly* 42, 1992, 358–364.

Hammond, N.G.L., 'The Macedonian Imprint on the Hellenistic World', in P. Green (ed.), *Hellenistic History and Culture*, 1993a, 12–23.

Hammond, N.G.L., *Sources for Alexander the Great: an Analysis of Plutarch's Life and Arrian's Anabasis Alexandrou* (Cambridge, 1993b).

Hammond, N.G.L., *Philip of Macedon* (London, 1994).

Hammond, N.G.L., 'Philip's Innovations in the Macedonian Economy', *Symbolae Osloenses* 70, 1995a, 22–29.

Hammond, N.G.L., 'Did Alexander Use One or Two Seals?' *Chiron* 25, 1995b, 199–203.

Hammond, N.G.L., 'Alexander and Armenia', *Phoenix* 50, 1996a, 130–137.

Hammond, N.G.L., 'A Macedonian Shield and Macedonian Measures', *British School at Athens* 91, 1996b, 365–367.

Hammond, N.G.L., 'Some Passages in Polyaenus' Stratagems Concerning Alexander', *Greek, Roman and Byzantine Studies* 37, 1996c, 23–53.

Hammond, N.G.L., 'The Frontiers of Philip II's Macedonia', *Ziva Antika* 47, 1997a, 43–50.

Hammond, N.G.L., *The Genius of Alexander the Great* (London, 1997b).

Hammond, N.G.L., 'What May Philip have Learnt as a Hostage in Thebes?' *Greek, Roman and Byyzantine Studies* 38, 1997c, 355–372.

Hammond, N.G.L., 'Alexander's Newly-Founded Cities', *Greek, Roman and Byzantine Studies* 39, 1998a, 243–269.

Hammond, N.G.L., 'Cavalry Recruited in Macedonia down to 322 BC', *Historia* 47, 1998b, 404–425.

Hammond, N.G.L., 'Roles of the Epistates in Macedonian Contexts', *Annual of the British School at Athens* 94, 1999a, 369–375.

Hammond, N.G.L., 'The Speeches in Arrian's *Indica* and *Anabasis*', *Classical Quarterly* 49, 1999b, 238–253.

Hammond, N.G.L., 'The Continuity of the Macedonian Institutions and the Macedonian Kingdoms of the Hellenistic Era', *Historia* 49, 2000, 141–159.

Hanfmann, G.M.A. & Vermeule, C., 'A New Trajan', *American Journal of Archaeology* 61, 1957, 223–253.

Hanfmann, G.M.A., 'New Fragments of Alexandrian Wall Painting', in G. Barone, *et al.* (eds), *Alessandria e il mondo ellenistico-romano*, 1983, 247–255.

Hansen, M.H., 'The Battle Exhortation in Ancient Historiography', *Historia* 42, 1993, 161–180.

Hansen, M.H., 'The Little Grey Horse – Henry V's Speech at Agincourt and the Battle Exhortation in Ancient Historiography', *Classica et Mediaevalia* 52, 2001, 95–115.

Hanson, J.A., 'The Glorious Military', in T.A. Dorey & D.R. Dudley (eds), *Roman Drama*, 1965, 51–85.

Hanson, V.D., *The Western Way of War: Infantry Battle in Classical Greece* (New York, 1989).

Hanson, V.D. (ed.), *Hoplites: The Classical Greek Battle Experience* (London, 1991).

Hanson, V.D., 'From Phalanx to Legion 350–250 BC', in *The Cambridge Illustrated History of Warfare*, 1995, 32–49.
Hanson, V.D., 'The Status of Ancient Military History: Traditional Work, Recent Research and Ongoing Controversies', *Journal of Military History* 63, 1999a, 379–413.
Hanson, V.D., *The Wars of the Ancient Greeks and their Invention of Western Military Culture* (London, 1999b).
Hanson, V.D., *Why the West Has Won: Carnage & Culture from Salamis to Vietnam* (London, 2001).
Harding, P., 'Athenian Defensive Strategy in the Fourth Century', *Phoenix* 42, 1988, 61–71.
Harding, P., 'Athenian Defensive Strategy Again', *Phoenix* 44, 1990, 377–380.
Harding, P., *Didymos on Demosthenes* (Oxford, 2006).
Hartle, R.W., 'The Search for Alexander's Portrait', in E.N. Borza (ed.), *Philip II, Alexander the Great and the Macedonian Heritage*, 1982, 153–176.
Hartmann, S.S., 'Dionysius and Heracles in India', *Temenos* 1, 1965, 55–64.
Hartwig, P., 'Ein Tongefäss des C. Popilius mit Scenen der Alexanderschlacht', *Mitteilungen des Deutschen Archäologischen Instituts, Römische Abteilung* 13, 1898, 399–408.
Hatzopoulos, M.B. & Juhel, P., 'Four Hellenistic Funerary Stelae from Gephyra, Macedonia', *American Journal of Archaeology* 113, 2009, 423–437.
Hatzopoulos, M.B. & Loukopoulos, L.D, *Philip of Macedon* (Athens, 1991).
Hatzopoulos, M.B. & Vidal-Naquet, P., *Cultes et rites de passage en Macédoine* (Athens, 1994).
Hatzopoulos, M.B., 'Strepsa: A Reconsideration of the Road System of Lower Macedonia', in M.B. Hatzopoulos & L.D. Loukopoulou (eds), *Meletemata 2: Two Studies in Ancient Macedonian Topography*, 1985, 21–60.
Hatzopoulos, M.B., *Macedonian Institutions under the Kings. I: A Historical and Epigraphic Study* (Athens, 1996).
Hatzopoulos, M.B., *L'organisation de l'armée macédonienne sous les antigonides: Problèmes anciens et documents nouveaux* (Athens, 2001).
Hatzopoulos, M.B., 'The Burial of the Dead (at Vergina) or the Unending Controversy on the Identity of the Occupant of Tomb II', *Tekmeria* 9, 2008, 91–118.
Haythornthwaite, P.J., *Napoleon's Military Machine* (London, 1988).
Head, B.V., *A Catalogue of the Greek Coins in the British Museum: Macedonia, etc* (London, 1879).
Head, D., *Armies of the Macedonian and Punic Wars 359 BC to 146 BC* (1982).
Head, D., 'Macedonian Military Costume on the Agios Athanasios Tomb Painting', *Slingshot* 201, 1999a, 35–41.
Head, D., 'Peritrachelion: the Derveni Gorget', *Slingshot* 202, 1999b, 15.
Heckel, W., 'Amyntas Son of Andromenes', *Greek, Roman & Byzantine Studies* 16, 1975, 393–398.
Heckel, W., 'The Somatophylakes of Alexander the Great: Some Thoughts', *Historia* 27, 1978, 224–228.
Heckel, W., 'The Somatophylax Attalos: Diodorus. 16.94.4', *Liverpool Classical Monthly* 4, 1979, 215–216.
Heckel, W., 'Marsyas of Pella, Historian of Macedon', *Hermes* 108, 1980, 444–462.
Heckel, W., 'Two Doctors from Kos', *Mnemosyne* 34, 1981, 396–398.
Heckel, W., 'The Career of Antigenes', *Symbolae Osloenses* 57, 1982, 57–67.
Heckel, W., 'The Boyhood Friends of Alexander the Great, *Emerita* 53, 1985, 285–289.
Heckel, W., 'Somatophylakia: A Macedonian *cursus honorum*', *Phoenix* 40, 1986, 279–294.
Heckel, W., *The Marshals of Alexander's Empire* (London, 1992).
Heckel, W., 'Kings and Companions: Observations on the Nature of Power in the Reign of Alexander', in J. Roisman (ed.), *Brill's Companion to Alexander the Great*, 2003, 197–225.
Heckel, W., '*Synaspismos*, Sarissas and Wagons', *Acta Classica* 48, 2005, 189–194.
Heckel, W., *Macedonian Warrior* (Oxford, 2006a).
Heckel, W., 'Mazaeus, Callisthenes and the Alexander Sarcophagus', *Historia* 55, 2006b, 385–396.
Heckel, W., *Who's Who in the Age of Alexander the Great: Prosopography of Alexander's Empire* (Oxford, 2006c).
Heinl, R.D., *Dictionary of Military and Naval Quotations* (Annapolis, 1966).
Heisserer, A.J., *Alexander the Great and the Greeks: The Epigraphic Evidence* (Norman, 1980).

Hekler, A., 'Beiträge zur Geschichte der antiken Panzerstatuen', *Jahreshefte des Österreichischen Archäologischen Institutes in Wien* 19/20, 1919, 190–241.
Hellmann, F., 'Zur Lustration des makedonischen Heeres', *Archiv für Religionswissenschaft* 29, 1931, 202–203.
Herman, G., 'The 'Friends' of the Early Hellenistic Rulers: Servants or Officials?' *Talanta* 12/13, 1980/81, 103–127.
Herman, G., *Ritualised Friendship and the Greek City* (Cambridge, 1987).
Hernandez, J.P.S., 'Procles the Carthaginian: a North African Sophist in Pausanias' *Pereigesis*', *Greek, Roman and Byzantine Studies* 50, 2010, 119–132.
Heskel, A., 'Macedon and the North, 400–336', in L.A. Tritle (ed.), *The Greek World in the Fourth Century*, 1997, 167–188.
Heuzy, L. & Daumet, H., *Mission archéologique de Macédoine* (Paris, 1876).
Hill, G.F., 'Alexander the Great and the Persian Lion-Gryphon', *Journal of Hellenic Studies* 43, 1923, 156–161.
Hobhouse, J.C., *Travels in Albania and other Provinces of Turkey in 1809 and 1810*, vol. I (London, 1858).
Hoddinott, R.F., *The Thracians* (London, 1981).
Hodges, A.L., 'The Mechanics of Marching: A Few Points for the Weary Foot-Soldier', *Scientific American*, 2 June 1917, 551.
Hoffmann, O., *Die Makedonen, ihre Sprache und ihr Volkstum* (Gottingen, 1906).
Holleaux, M., *Études d'épigraphie et d'histoire grecques*, vol. III (Paris, 1942).
Holt, F.L., 'Imperium Macedonicum and the East: The Problem of Logistics', *Ancient Macedonia V*, vol. I, 1993, 585–592.
Holt, F.L., 'Alexander of Macedon and the so-called Porus Medallions', *Military Archaeology – Weaponry and Warfare in the Historical and Social Perspective*, 1998, 125–127.
Holt, F.L., 'Alexander the Great and the Spoils of War', *Ancient Macedonia VI*, 1999a, 499–506.
Holt, F.L., *Thundering Zeus: The Making of Hellenistic Bactria* (Berkeley, 1999b).
Holt, F.L., 'The Death of Coenus: Another Study in Method', *Ancient History Bulletin* 14, 2000, 49–55.
Holt, F.L., *Alexander the Great and the Mystery of the Elephant Medallions* (Berkeley, 2003).
Holt, F.L., *Into the Land of Bones: Alexander the Great in Afghanistan* (Berkeley, 2005a).
Holt, F.L., 'Stealing Zeus's Thunder', *Saudi Aramaco World*, May/June 2005b, 10–19.
Holt, F.L., 'Ptolemy's Alexandrian Postscript', *Saudi Aramaco World*, Nov./Dec. 2006, 4–9.
Holt, F.L., 'Money and Finances in the Campaigns of Alexander', in J. Romm & R.B. Strassler (eds), *The Landmark Arrian: The Campaigns of Alexander*, 2010, 358–360.
Homolle, T., 'Comptes des hiéropes du temple d'Apollon délien', *Bulletin de correspondance hellénique* 6, 1882, 1–167.
Homolle, T., 'Comptes et inventaires des temples déliens en l'année 279', *Bulletin de correspondance hellénique* 14, 1890, 389–511 & 15, 1891, 113–168.
Hornblower, J., *Hieronymus of Cardia* (Oxford, 1981).
Hornell, J., 'Floats and Buoyed Rafts in Military Operations', *Antiquity* 19, 1945, 72–79.
Hornell, J., *Water Transport: Origins & Early Evolution* (Cambridge, 1946).
Houghton, A. & Stewart, A., 'The Equestrian Portrait of Alexander the Great on a New Tetradrachm of Seleucus I', *Schweizerische Numismatische Rundschau* 78, 1999, 27–35.
Houghton, A., 'Notes on the Early Seleucid Coinage of Persepolis', *Schweizer Numismatische Rundschau* 59, 1980, 5–14.
Hourmouziadis, G., *Ancient Magnesia from the Paleolithic Camp to the Palace of Demetrias* (Athens, 1982).
Houser, C., 'Alexander's Influences on Greek sculpture: As seen in a Portrait in Athens', in *Studies in the History of Art*, vol. 10, 1980, 229–238.
Houser, C., *Greek Monumental Sculpture of the Fifth and Fourth Centuries BC* (New York, 1987).
Howe, T., 'Alexander in India', in T. Howe & J. Reames (eds), *Macedonian Legacies: Studies in ancient Macedonian history and culture in honor of Eugene N. Borza*, 2008, 215–233.
Howell, E., *Escape to Live* (London, 1952).
Hrouda, B., 'Bits and Pieces', *Antiquity* 43, 1969, 289–300.

Huguenot, C., *La Tombe aux Erotes et la Tombe d'Amarynthos: Architecture funéraire et présence macédonienne en Grèce centrale. Eretria, vol. XIX* (Gollion, 2008).
Humphrey, J.W., *et al.*, *Greek and Roman Technology* (London, 1998).
Hunt, A.R., *et al.* (eds), *The Tebtunis Papyri*, vol. III, pt. II. (London, 1938).
Hurwit, J.M., *The Athenian Acropolis: History, Mythology and Archaeology from the Neolithic Era to the Present* (Cambridge, 1999).
Hyland, A., *Equus: The Horse in the Roman World* (London, 1990).
Hyland, A., *The Horse in the Ancient World* (Stroud, 2003).
Ignatiadou, D., 'Hellenistic Board Game with Glass Pillars', *Ancient Macedonia VI*, vol. I, 1999, 507–522.
Irby-Massie, G.L. & Keyser, P.T., *Greek Science of the Hellenistic Era: A Sourcebook* (London, 2002).
Jacoby, F. (ed.), *Die Fragmente der grieschischen Historiker* (Berlin, 1923–).
Jaeckel, P., 'Pergamenische Waffenreliefs', *Zeitschrift der Gesellschaft für Historische Waffen- und Kostümkunde* 7, 1965, 94–122.
Jaeger, W., *Paideia: the Ideals of Greek Culture*, vol. II (Oxford, 1944).
Jameson, M., 'The Asexuality of Dionysus', in T.H. Carpenter & C.A. Faraone (eds), *Masks of Dionysos*, 1993, 44–64.
Janssen, E., 'Die Kausia: Symbolik und Funktion der makedonischen Kleidung', DPhil Thesis, George August University, Gottingen, 2007.
Jarcho, S., 'A Roman Experience with Heat Stroke in 24 BC', *Bulletin of the New York Academy of Medicine* 43, 1967, 767–768.
Jarva, E., 'In Search of Xenophon's Ideal Helmet: an Alternative View of the *Kranos Boiotiourges*', *Faravid* 15, 1991, 33–83.
Jarva, E., *Archaiologia on Archaic Greek Body Armour* (Rovaniemi, 1995).
Jensen, L.B., 'Royal Purple of Tyre', *Journal of Near Eastern Studies*, Apr., 1963, 104–118.
Jesse, C., 'Rapid Conquest Societies: A Comparative Study', MSSc Thesis, University of Calgary, 2007.
Johnson, A.C., 'Ancient Forests and Navies', *Transactions of the American Philosophical Society* 58, 1927, 199–209.
Johnstone, S., 'Virtuous Toil, Vicious Work: Xenophon on Aristocratic Style', *Classical Philology* 89, 1994, 219–240.
Jones, A., *The Art of War in the Western World* (Chicago, 2001).
Jones, J.E., *et al.*, 'To Dema: a Survey of the Aigaleos-Parnes Wall', *British School at Athens* 42, 1957, 152–189.
Jost, M., 'Les divinités de la guerre', in F. Prost (ed.), *Armées et sociétés de la Grèce classique – Aspects sociaux et politiques de la guerre aux Ve et IVe s. av. J.-C.*, 1999, 163–178.
Juhel, P. & Temelkoski, D., 'Fragments de «boucliers macédoniens» au nom du roi Démétrios trouvés à Staro Bonce (République de Macédoine). Rapport préliminaire et présentation épigraphique', *Zeitschrift für Papyrologie und Epigraphik* 162, 2007b, 165–180.
Juhel, P., '"On Orderliness with Respect to the Prizes of War": The Amphipolis Regulation and the Management of Booty in the Army of the Last Antigonids', *Annual of the British School at Athens* 97, 2002, 401–412.
Juhel, P., 'The Regulation Helmet of the Phalanx and the Introduction of the Concept of Uniform in the Macedonian Army at the End of the Reign of Alexander the Great', *Klio* 91, 2009, 342–355.
Juhel, P., 'La stèle funéraire d'Amyntas, fils d'Alexandre, cavalier des confins macédoniens', *Acta Archaeologia* 81, 2010, 106–111.
Juhel, P., 'Un fantôme de l'histoire hellénistique: Le 'district' macédonien', *Greek, Roman and Byzantine Studies* 51, 2011, 579–612.
Junkelmann, M., *Die Legionen des Augustus: Der römische Soldat im archäologischen Experiment* (Mainz, 1986).
Kalita, S., '"Winged Soldiers" – a Special Unit in the Army of Alexander the Great', in *Sub Signis Aris et Martis*, 1995, 21–29.
Kalléris, J., *Les anciens Macédoniens: Étude linguistique et historique*, vol. I (Athens, 1954).

Kanatsoulis, D., 'Antipatros als Feldherr und Staatsmann in der Zeit Philipps und Alexanders des Grossen', *Hellenica* 16, 1958/59, 14–64.
Karageorghis, V., *Excavations in the Necropolis of Salamis III: Text* (Nicosia, 1973).
Karageorghis, V., *Excavations in the Necropolis of Salamis III: Plates* (Haarlem, 1974).
Karakas, S.L., 'Subject and Symbolism in Historical Battle Reliefs of the Late Classical and Hellenistic Periods', PhD Thesis, University of North Carolina, 2002.
Karamitrou-Mentessidi, G. (ed.), *Kozani, City of Elimiotis* (Thessaloniki, 1993).
Karunanithy, D., 'Of Ox-Hide Helmets and Three-Ply Armour: the Equipment of Macedonian Phalangites Described through a Roman Source', *Slingshot* 213, 2001, 33–40.
Karunanithy, D., 'Colour Clues and Considerations: Some Possible Terracotta Evidence for Hellenistic Military Dress and Shields', *Slingshot* 220, 2002, 34–41.
Karunanithy, D., 'Like an Actor on Stage: Demetrius the Besieger at Rhodes, 305BC', *Ancient Warfare*, V/3, 2011, 41–45.
Kasher, A., 'Further Revised Thoughts on Josephus' Report of Alexander's Campaign to Palestine (*ANT.* 11.304–347)', in L.L. Grabbe & O. Lipschits (eds), *Judah Between East and West: the Transition from Persian to Greek Rule (ca. 400–200 BCE)*, 2011, 131–157.
Kebric, R.B., 'Old Age, the Ancient Military, and Alexander's Army: Positive Examples for a Graying America', *The Gerontologist* 28, 1988, 298–302.
Kendrick Pritchett, W., *The Greek State at War* Part I (Berkeley, 1971).
Kendrick Pritchett, W., *The Greek State at War* Part II (Berkeley, 1974).
Kent, J.H., 'A Garrison Inscription from Rhamnous', *Hesperia* 10, 1941, 342–350.
Kerstein, M. & Hubbard, R., 'Heat Related Problems in the Desert: the Environment can be the Enemy', *Military Medicine* 149, 1984, 650–656.
Kertesz, I., 'Philip II the Sportsman', *Ancient Macedonia VII*, 2007, 327–332.
Kesner, L., 'Likeness of No One: (Re)presenting the First Emperor's Army', *Art Bulletin* 77, 1995, 115–132.
Keyser, P.T., 'The Use of Artillery by Philip II and Alexander the Great', *Ancient World* 25, 1994, 27–59.
Kinch, K.F., 'Le tombeau de Niausta: Tombeau Macedonien', *Mémoires de l'Académie Royale des Sciences et des Lettres de Danemark, Copenhague, 7me série, Section des lettres* 4, 1920, 283–288.
Kingsley, B.M., 'The Cap that Survived Alexander', *American Journal of Archaeology* 85, 1981a, 39–46.
Kingsley, B.M., 'The 'Chitrali': A Macedonian Import to the West', *Afghanistan Journal* 8, 1981b, 91–93.
Kingsley, B.M., 'The Kausia Diadematophoros', *American Journal of Archaeology* 88, 1984, 66–68.
Kingsley, B.M., 'Alexander's Kausia and the Macedonian Tradition', *Classical Antiquity* 10, 1991, 59–85.
Kitov, G., 'Mogilata Golyama Arsenalka', *Archeologija Sofia* 38, 1996, 31–42.
Kitov, G., 'Royal Insignia: Tombs and Temples in the Valley of the Thracian Rulers', *Archaeologia Bulgarica* 3, 1999, 1–20.
Kitov, G., 'A Newly Found Thracian Tomb with Frescoes', *Archaeologia Bulgarica* 5, 2001, 15–29.
Kitov, G., *Aleksandrovskata grobnitsa* (Sofia, 2009).
Kitsos, D., 'Gladius versus Sarissa: Roman Legion Against Greek Pike Phalanx', *Anistoriton* 3, 1999.
Kleiner, G., *Diadochen-Gräber* (Wiesbaden, 1963).
Kleinschmidt, H., 'Using the Gun: Manual Drill and the Proliferation of Portable Firearms', *Journal of Military History* 63, 1999, 601–629.
Koch, H.G., *Dachterrakotten aus Kampanien mit Ausschluss von Pompeji* (Berlin, 1912).
Koehl, R., 'Hairstyles, Rank and Status', *Journal of Hellenic Studies* 106, 1986, 99–110.
Kohl, M., 'Sièges et défense de Pergame Nouvelles réflexions sur sa topographie et son architecture militaires', in J.C. Couvenhes & H.L. Fernoux (eds), *Les Cités grecques et la guerre en Asie Mineure à l'époque hellénistique*, 2004, 177–198.
Kopcke, G. & Moore, M.B. (eds), *Studies in Classical Art and Archaeology: A tribute to Peter Heinrich von Blanckenhagen* (New York, 1979).
Kosmetatou, E. & Waelkens, M., 'The 'Macedonian' Shields at Sagalassos', in E. Kosmetatou, *et al.* (eds), *Sagalassos IV: Report on the Survey and Excavation Campaigns of 1994 and 1995*, 1998, 277–291.

Kosmetatou, E., 'Rhoxane's Dedications to Athena Polias', *Zeitschrift für Papyrologie und Epigraphik* 146, 2004, 75–80.
Kosmetatou, E., 'Macedonians in Pisidia', *Historia* 54, 2005, 216–221.
Kosmidou, E. & Malamidou, D., 'Arms and armour from Amphipolis, Northern Greece: Plotting the Military Life of an Ancient City', in M. Novotna, *et al.* (eds), *Proceedings of the International Symposium Arms and Armour through the Ages (from the Bronze Age to Late Antiquity)*, 4/5, 2004/05, 133–147.
Kottaridi, A. & Walker, S. (eds), *Heracles to Alexander the Great* (Oxford, 2011).
Koukouli-Chrysanthaki, C., 'À propos des voies de communication du royaume de Macédoine', in *Recherches récentes sur le monde hellénistique*, Lausanne, 1998, 53–64.
Kraft, K., 'Der behelmete Alexander der Grosse', *Jahrbuch für Numismatik und Geldgeschichte*15, 1965, 7–32.
Krasilnikoff, J.A., 'Aegean Mercenaries in the Fourth to Second Centuries BC: A Study in Payment, Plunder and Logistics of Ancient Greek Armies', *Classica et Mediaevalia* 43, 1992, 23–36.
Krasteva, M., 'Les Pectoraux de Thrace et de Macédoine du Ier mill. av. J.-C.', *Ancient Macedonia VII*, 2007, 445–453.
Kremydi-Sicilianou, S., 'Enas neos typos tetradrachmou tou Alexandrou', *Ancient Macedonia VI*, 1999, 643–654.
Krentz, P., 'The *Salpinx* in Greek Warfare', in V.D. Hanson (ed.), *Hoplites: The Greek Battle Experience*, 1991, 110–120.
Kroll, J.H., 'An Archive of the Athenian Cavalry', *Hesperia* 46, 1977a, 83–140.
Kroll, J.H., 'Some Athenian Armor Tokens', *Hesperia* 46, 1977b, 141–146.
Kron, G., 'Anthropometry, Physical Anthropology, and the Reconstruction of Ancient Health, Nutrition, and Living Standards', *Historia* 54, 2005, 68–83.
Kruglikova, I.T., *Dil'berdzin: Khram Dioskurov* (Moskva, 1986).
Kuhrt, A., *The Ancient Near East* (London, 1995).
Kuhrt, A., *The Persian Empire: A Corpus of Sources from the Achaemenid Period*, vol. II (London, 2007).
Kunze, E., *Beinschienen* (Berlin, 1991).
Kunzl, E., 'Waffendekor im Hellenismus', *Journal of Roman Military Equipment Studies* 8, 1997, 61–89.
Lainas, P., *et al.*, '"Most Brilliant in Judgement": Alexander the Great and Aristotle', *American Surgeon* 71, 2005, 275–280.
Lamb, H., *Alexander of Macedon* (London, 1946).
Landels, J.G., *Music in Ancient Greece and Rome* (London, 1999).
Landucci, G.F., 'Tra monarchia nazionale e monarchia militare: il caso della Macedonia', in *Gli stati territoriali nel mondo antico*, 2003, 199–214.
Lane-Fox, R., *Alexander the Great* (London, 1973).
Lane-Fox, R., 'Text and Image: Alexander the Great, Coins and Elephants', *Bulletin of the Institute of Classical Studies* 4, 1996, 87–108.
Lane-Fox, R., 'The Itinerary of Alexander: Constantius to Julian', *Classical Quarterly* 47, 1997, 239–252.
Lane-Fox, R., *The Long March: Xenophon and the Ten Thousand* (New Haven, 2004a).
Lane-Fox, R., *The Making of Alexander: The Official Guide to the Epic Film Alexander* (Oxford, 2004b).
Lape, S., *Reproducing Athens* (Princeton, 2004).
Lascaratos, J., *et al.*, 'The Opthalmic Wound of Philip II of Macedonia (360–336 BCE)' *Survey of Ophthalmology* 49, 2004, 256–261.
Lasswell, H.D., 'The Garrison State', *American Journal of Sociology* 46, 1941, 455–468.
Lattimore, S., 'The Chlamys of Daochos I', *American Journal of Archaeology* 79, 1975, 87–88.
Laubscher, H.P., 'Der 'Kameo Gonzaga'– Rom oder Alexandria?' *Mitteilungen des Deutschen Archäologischen Instituts: Athenische Abteilung* 100, 1995, 387–424.
Launey, M., *Recherches sur les armées hellénistiques*, vol. II (Paris, 1950).
Lawrence, A.W., *Greek Aims in Fortifications* (Oxford, 1979).
Lazar, S., 'Helmets of the Chalcidian Shape Found in the Lower Danube Area', *Dacia: revue d'archéologie et d'histoire ancienne* 53, 2009, 13–26.
Lazaridis, D., *Amphipolis* (Athens, 1997).

Le Bohec, S., *Antigone Doson, Roi de Macédoine* (Nancy, 1993).
Le Rider, G., *Le monnayage d'argent et d'or de Philippe II* (Paris, 1977).
Lee, J.W.I., 'Urban Combat at Olynthos, 348 BC', in P.W.M. Freeman & A. Pollard, *Fields of Conflict: Progress and Prospect in Battlefield Archaeology*, 2001, 11–22.
Lee, J.W.I., *A Greek Army on the March* (Cambridge, 2007).
Lefebvre des Noëttes, R. *L'attelage: Le cheval de selle à travers les âges* (Paris, 1931).
LeGrand, P.E., *The New Greek Comedy* (London, 1917).
Legras, B., *Néotês: recherches sur les jeunes Grecs dans l'Égypte ptolémaïque et romaine* (Geneva, 1999).
Leitao, D.D., 'The Perils of Leukippos: Initiatory Transvestism and Male Gender Ideology in the Ekdusia of Phaistos', *Classical Antiquity* 14, 1995, 130–163.
Lendering, J., *Alexander de Grote: De ondergang van het Perzische rijk* (Amsterdam, 2004).
Lendering, J., 'The Rise of a Hyperpower: Polybius, Rome, Carthage, and Capricious Fortune', *Ancient Warfare*, III/4, 2009, 10–12.
Lendering, J., 'Alexander the Invincible God: Uniting a Diverse Army through Deification', *Ancient Warfare*, IV/5, 2010, 38–41.
Lendon, J.E., 'The Rhetoric of Combat: Greek Military Theory and Roman Culture in Julius Caesar's Battle Descriptions', *Classical Antiquity* 18, 1999, 273–329.
Lendon, J.E., *Soldiers and Ghosts: A History of Battle in Classical Antiquity* (London, 2005).
Leon, C.F., 'Antigonus Gonatas Rediscovered', *The Ancient World* 20, 1989, 21–25.
Leslie Shear, T.L., 'The Athenian Agora: Excavations of 1971', *Hesperia* 42, 1973, 121–179.
Lévêque, P., *Pyrrhos* (Paris, 1957).
Lévêque, P., 'La guerre à l'époque hellénistique', in J.P. Vernant (ed.), *Problèmes de la guerre en Grèce ancienne*, 261–287.
Lewis, D.M., 'The King's Dinner (Polyaenus IV 3.32)', in H. Sancisi-Weerdenburg & A. Kuhrt (eds), *Achaemenid History II: The Greek Sources*, 1987, 79–87.
Lewis, M.J.T, *Surveying Instruments of Greece and Rome* (Cambridge, 2001).
Lewis, N., *Greeks in Ptolemaic Egypt* (Oxford, 1986).
Liampi, K., 'Zur Chronologie der sogenannten „anonymen" makedonischen Münzen des späten. 4. Jhs. v. Chr.', *Jahrbuch für Numismatik und Geldgesichte* 36, 1986, 41–66.
Liampi, K., 'Der makedonische Schild als politisch-propagandistisches Mittel in der hellenistischen Zeit', *Meletemata 10*, 1990, 157–171.
Liampi, K., *Der makedonische Schild* (Bonn, 1998).
Liapis, V., '*Rhesus* Revisited: the Case for a Fourth-Century Macedonian Context', *Journal of Hellenic Studies* 129, 2009, 71–88.
Liappas, J.A., *et al.*, 'Alexander the Great's Relationship with Alcohol', *Addiction* 98, 2003, 561–567.
Liddell, H.G. & Scott, R., *A Greek-English Lexicon* (Oxford, 1897).
Ligeti, D.A., 'The Role of Alexander the Great in Livy's Historiography', *Acta Antiqua Academiae Scientiarum Hungaricae* 48, 2008, 247–251.
Lilimpake-Akamate, M., *et al.* (eds), *To Archaiologiko Mouseio Pellas* (Athens, 2011).
Lindsay Adams, W., 'Dynamics of Internal Macedonian Politics in the Time of Cassander', *Ancient Macedonia III*, 1983, 17–30.
Lindsay Adams, W., 'Antipater and Cassander: Generalship on Restricted Resources in the Fourth Century', *Ancient World* 10, 1985, 79–88.
Lindsay Adams, W., 'In the Wake of Alexander the Great: The Impact of Conquest in the Aegean World', *Ancient World* 27, 1996, 29–37.
Lindsay Adams, W., 'Philip and the Thracian Frontier', *Actes 2e Symposium international des études thraciennes: Thrace ancienne*, 1, 1997, 81–89.
Lindsay Adams, W., 'The Frontier Policy of Philip II', *Ancient Macedonia VII*, 2007a, 283–291.
Lindsay Adams, W., 'The Games of Alexander the Great', in W. Heckel, *et al.* (eds), *Alexander's Empire: Formulation to Decay*, 2007b, 125–138.
Lindsay Adams, W., 'Sport and Ethnicity in Ancient Macedonia', in T. Howe & J. Reames (eds), *Macedonian Legacies: Studies in Ancient Macedonian History and Culture in Honor of Eugene N. Borza*, 2008, 57–78.

Lloyd, A.B., 'Philip II and Alexander the Great: The Moulding of Macedon's Army', in A.B. Lloyd (ed.), *Battle in Antiquity*, 1996, 169–198.
Litvinskij, B.A. & Picikyan, I.R., 'An Achaemenian Griffin-Handle from the Temple of the Oxus: The Machaira in Northern Bactria', in A. Invernizzi (ed.), *In the Land of the Gryphons: Papers on Central Asian Archaeology in Antiquity*, 1995, 107–128.
Litvinskij, B.A. & Picikyan, I.R., 'Handles and Ceremonial Scabbards of Greek Swords from the Temple of the Oxus in Northern Bactria', *East and West* 49, 1999, 47–104.
Loades, M., *Swords and Swordsmen* (Barnsley, 2010).
Lock, R.A., 'The Origins of the Argyraspids', *Historia* 26, 1977, 373–378.
Lonsdale, D.J., *Alexander: Killer of Men. Alexander the Great and the Macedonian Art of War* (London, 2004).
Lonsdale, D.J., *Alexander the Great: Lessons in Strategy* (London, 2007).
Lopez, J., 'The Rise and Fall of Macedonian Military Might', MLitt Thesis, Newcastle University, 1997.
Lorimer, H.L., *Homer and the Monuments* (London, 1950).
Lugli, N., *Il cavallo: origini, razze, attitudini* (Novara, 1972).
Lumpkin, H., 'The Weapons and Armour of the Macedonian Phalanx', *Journal of the Arms and Armour Society* VIII, 1975, 193–207.
Lund, H.S., *Lysimachus: A Study in Early Hellenistic Kingship* (London, 1992).
Lush, D., 'Body Armour in the Phalanx of Alexander's Army', *Ancient World* 38, 2007, 15–37.
Lyall, W.R., *History of Greece, Macedonia and Syria* (London, 1852).
Ma, J., 'Chaironeia 338: Topographies of Commemoration', *Journal of Hellenic Studies*, 2008, 128, 72–91.
Ma, J., 'Oversexed, Overpaid and Over Here: A Response to Angelos Chaniotis', in A. Chaniotis & P. Ducrey (eds), *Army and Power in the Ancient World*, 2002, 115–122.
Maass, M., 'Ein Terrakottaschmuckfund aus dem Zeitalter Alexanders d. Gr.', *Mitteilungen des Deutschen Archäologischen Instituts. Athenische Abteilung* 100, 1985, 309–326.
MacCary, W.T., 'Menander's Soldiers: their Names, Roles and Masks', *American Journal of Philology* 93, 1972, 279–298.
MacDonald, G., *Catalogue of Greek Coins in the Hunterian Collection, University of Glasgow*, vol. I (Glasgow, 1899).
MacMullen, R., 'Inscriptions on Armor and the Supply of Arms in the Roman Empire', *American Journal of Archaeology* 64, 1960, 23–40.
Macrory, P.A., *The Fierce Pawns* (London, 1966).
Mahaffy, J.P. & Smyly, J.G., *The Flinders Petrie Papyri* (Dublin, 1891).
Majno, G., *The Healing Hand: Man and Wound in the Ancient World* (Cambridge, Mass., 1975).
Makaronas, C.I. & Miller, S.G., 'The Tomb of Lyson and Kallikles', *Archaeology* 27, 1974, 248–259.
Manakidou, E., 'Heroic Overtones in Two Inscriptions from Ancient Lete', in E. Voutiras (ed.), *Epigraphes Makedonias*, 1996, 85–98.
Mann, M., *The Sources of Social Power*, vol. I (Cambridge, 1986).
Manning, S., 'Leading the Persian Spears: Persian Standards of the 5th and 4th centuries BC', *Ancient Warfare*, III/6, 2009, 22–26.
Mansel, A.M., 'Osttor und Waffenreliefs von Side', *Archäologischer Anzeiger*, 1968, 239–279.
Manti, P.A., 'The Cavalry Sarissa', *Ancient World* 8, 1983, 75–83.
Manti, P.A., 'The Sarissa of the Macedonian Infantry', *Ancient World* 23, 1992, 30–42.
Manti, P.A., 'The Macedonian Sarissa, again', *Ancient World* 25, 1994, 77–91.
Marinatos, S., 'Mycenaean Elements within the Royal House of Macedonia', *Ancient Macedonia I*, 1970, 45–52.
Mark, S., 'Alexander the Great, Seafaring, and the Spread of Leprosy', *Journal of the History of Medicine and Allied Sciences* 57/3, 2002, 302–305.
Markle, M.M., 'The Macedonian Sarissa, Spear and Related Armor', *American Journal of Archaeology* 81, 1977, 323–339.
Markle, M.M., 'Use of the Sarissa by Philip and Alexander of Macedon', *American Journal of Archaeology* 82, 1978, 483–497.

Markle, M.M., *Weapons from the Cemetery at Vergina and Alexander's Army* (Thessaloniki, 1980a).
Markle, M.M., 'Weapons from the Cemetery at Vergina and Alexander's Army', in *Alexander the Great: The 2300 Anniversary of his Death*, 1980b, 243–267.
Markle, M.M., 'Macedonian Arms and Tactics under Alexander the Great', in B. Barr-Sharrar & E. Borza (eds), *Studies in the History of Art Vol. 10: Macedonia and Greece in Late Classical and Hellenistic Times*, 1982, 87–111.
Markle, M.M., 'A Shield Monument from Veria', in *Ancient Macedonia. An Australian Symposium. Papers of the Second International Congress of Macedonian Studies* 7, 1994, 83–97.
Markle, M.M., 'A Shield Monument from Veria and the Chronology of Macedonian Shield Types', *Hesperia* 68, 1999, 219–254.
Markoulaki, P., 'In Search of the Linothorax Mystery: an Interview with Professor Gregory S. Aldrete', *Sparta* 6, 2010, 50–53.
Marrinan, R.D.M., 'The Ptolemaic Army: its Organization, Development and Settlement', PhD Thesis, University College London, 1998.
Marriner, N., *et al.*, 'Alexander the Great's Tombolos at Tyre and Alexandria, Eastern Mediterranean', *Geomorphology* 100, 2008, 377–400.
Marsden, E.W., *Greek and Roman Artillery: Historical Development* (Oxford, 1969).
Marsden, E.W., *Greek and Roman Artillery: Technical Treatises* (Oxford, 1971).
Marsden, E.W., 'Polybius as a Military Historian', in *Polybe, Entretiens sur l'Antiquité classique*, XX, 1974, 267–301.
Marsden, E.W., 'Macedonian Military Machinery and its Designers under Philip and Alexander', *Ancient Macedonia II* 1977, 211–223.
Martel, W.C., *Victory in War: Foundations of Modern Military Policy* (Cambridge, 2007).
Marx, P., 'The Introduction of the Gorgoneion to the Shield and Aegis of Athena and the Question of the Endoios', *Revue archaeologique*, 1993, 227–268.
Massar, N., 'Un savoir-faire à l'honneur: "Médecins" et "discours civique" en Grèce hellénistique', *Revue belge de philologie et d'histoire* 79, 2001, 175–201.
Mataranga, K., 'La ruse dans la guerre', in P. Brun (ed.), *Guerre et société dans le monde grec (490–322 av. J.-C.)*, 1999, 21–28.
Mathisen, R.W., 'The Shield/Helmet Bronze Coinage of Macedonia: a Preliminary Analysis', *Journal of the Society of Ancient Numismatics* 10, 1979, 2–6.
Matthews, R., *Alexander at the Battle of the Granicus: a Campaign in Context* (Stroud, 2008).
Matthews, V.J., 'The *Hemerodromoi*: Ultra Long-Distance Running in Antiquity', *Classical World* 68, 1974, 161–169.
Mattusch, C., 'Bronze and Ironworking in the Area of the Athenian Agora', *Hesperia* 46, 1977, 340–379.
Maxfield, V.A., *The Military Decorations of the Roman Army* (Berkeley, 1981).
Maxwell Edmonds, J., *The Fragments of Attic Comedy*, vol. III.A (Leiden, 1961).
Maxwell-Stuart, P.G., 'A Medical Recipe of Alexander the Great?' *Rivista di studi classici* 24, 1976, 321–323.
Mayor, A., *Greek Fire, Poison Arrows & Scorpion Bombs: Biological and Chemical Warfare in the Ancient World* (London, 2003).
Mayor, A., *The Poison King: The Life and Legends of Mithradates, Rome's Deadliest Enemy* (Princeton, 2010).
McCredie, J.R., *Fortified Military Camps in Attica* (Princeton, 1966).
McCrindle, J.W., *The Invasion of India by Alexander the Great* (London, 1896, repr. 1960).
McGeer, E., *Sowing the Dragons' Teeth: Byzantine Warfare in the Tenth Century* (Washington DC, 1995).
McGroarty, K., 'Did Alexander the Great Read Xenophon?' *Hermathena* 181, 2006, 105–124.
McKechnie, P., *Outsiders in the Greek Cities in the Fourth Century BC* (London, 1989).
McKechnie, P., 'Greek Mercenary Troops and their Equipment', *Historia* 43, 1994, 297–305.
McKechnie, P., 'Diodorus Siculus and Hephaistion's Pyre', *Classical Quarterly* 45, 1995, 418–432.
McNicoll, A.W., *Hellenistic Fortifications from the Aegean to the Euphrates* (Oxford, 1997).
Mehl, A., 'Doriktetos Chora', *Ancient Society* 11/12, 1980/81, 173–212.

Meiggs, R., *Trees and Timber in the Ancient Mediterranean World* (Oxford, 1982).
Meissner, B., *Historiker zwischen Polis und Konigshof* (Gottingen, 1992).
Mendels, D., *Identity, Religion and Historiography: Studies in Hellenistic History* (Sheffield, 1998).
Merker, I.L., 'The Ancient Kingdom of Paionia', *Balkan Studies* VI, 1965, 35–54.
Mieghem, T., 'Logistics Lessons from Alexander the Great', *Quality Progress*, January, 1998, 40–46.
Miesen, F.W., 'Reconstruyendo al pezhetairos', *Desperta ferro* 8, 2011, 24–27.
Miller, M.C.J., 'The 'Porus' Decadrachm of Alexander and the Founding of Bucephala', *Ancient World* 25, 1994, 109–120.
Miller, S.G., 'Menon's Cistern', *Hesperia* 43, 1974, 194–245.
Miller, S.G., *Arete: Greek Sports from Ancient Sources* (Berkeley, 1991).
Miller, S.G., 'The Iconography of Tomb Painting in Hellenistic Macedonia', *BABESCH supplemento 3*, 1993a, 115–118.
Miller, S.G., *The Tomb of Lyson and Kallikles* (Mainz, 1993b).
Miller, S.G., 'Macedonian Painting: Discovery and Research', in *Alexander the Great: From Macedonia to the Oikoumene*, 1998, 75–88.
Miller, S.G., book review in *American Journal of Archaeology* 105, 2001, 128–129.
Millett, P., 'The Political Economy of Macedonia', in J. Roisman & I. Worthington (eds), *A Companion to Ancient Macedonia*, 2010, 472–504.
Milns, R.D., 'Philip II and the Hypaspists', *Historia* 16, 1967, 509–512.
Milns, R.D., 'Asthippoi Again', *Classical Quarterly* 31, 1981, 347–354.
Milns, R.D., 'A Note on Diodorus and Macedonian Military Terminology in Book XVII', *Historia* 31, 1983, 123–126.
Milns, R.D., 'Army Pay and the Military Budget of Alexander the Great', in W. Will & H. Heinrichs (eds), *Zu Alexander der Grosse: Festschrift G. Wirth*, 1987, 232–256.
Miserachs Garcia, S. & Castillo Campill, L., 'La repercusión de la herida oftálmica de Filipo II de Macedonia', *Archivos de la Sociedad Española de Oftalmología* 85/2, 2010, 85–87.
Mitchell, S. & Waelkens, M., 'Sagalassus and Cremna, 1986', *Anatolian Studies* 37, 1987, 37–47.
Mitropoulou, E., 'The Origin and Significance of the Vergina Symbol', *Ancient Macedonia V*, vol. II, 1993, 843–958.
Montagu, J.D., *Greek and Roman Warfare: Battles, Tactics & Trickery* (London, 2006).
Montgomery of Alamein, A., *History of Warfare* (London, 1968).
Montgomery, H., 'The Economic Revolution of Philip II – Myth or Reality?' *Symbolae Osloenses* 60, 1985, 37–47.
Mooren, I., 'Kings and Courtiers: Political Decision making in the Hellenistic States', in W. Schuller (ed.), *Politische Theorie und Praxis im Altertum*, 1998, 122–133.
Moorman, E.M., *Ancient Sculpture in the Allard Pierson Museum Amsterdam* (Amsterdam, 2000).
Morello, R., 'Livy's Alexander Digression (9.17.9): Counterfactuals and Apologetics', *Journal of Roman Studies* 92, 2002, 62–85.
Moreno, P., *Vita e arte di Lisippo* (Milan, 1987).
Moreno, P., *Scultura ellenistica*, vol. I (Rome, 1994).
Moreno, P., *Lisippo: l'Arte e la fortuna* (Milan, 1995).
Moreno, P., *Apelles: The Alexander Mosaic* (Milan, 2001).
Morgan, C., 'Archaeology in Greece 2007–2008', *Archaeological Reports 2007–2008* (Society for the Promotion of Hellenic Studies), 2008, 1–113.
Morgan, J.D., 'Palaepharsalus – the Battle and the Town', *American Journal of Archaeology* 87, 1983, 23–54.
Morgan, M.G., 'Metellus Macedonicus and the Province Macedonia', *Historia* 18, 1969, 422–446.
Moritz, L.A., *Grain Mills and Flour in Classical Antiquity* (Oxford, 1958).
Morkholm, O., *Early Hellenistic Coinage: From the Accession of Alexander to the Peace of Apamea (336–188 BC)* (Cambridge, 1991).
Morris, D.R., *The Washing of the Spears: a History of the Rise of the Zulu Nation under Shaka and its Fall in the Zulu War of 1879* (New York, 1965).
Morrison, G., 'Alexander, Combat Psychology and Persepolis', *Antichthon* 35, 2001, 30–44.

Morrison, J.S., 'Hyperesia in Naval Contexts in the Fifth and Fourth Centuries BC', *Journal of Hellenic Studies* 104, 1984, 48–59.
Morrow, K.D., *Greek Footwear and the Dating of Sculpture* (London, 1985).
Muller, O., *Antigonos Monophthalmos und 'Das Jahr der Konige'* (Bonn, 1973).
Munn, M.H., *The Defense of Attica: The Dema Wall and the Boiotian War 378–375 BC* (Berkeley, 1993).
Murray, W.M., 'The Development of a Naval Siege Unit under Philip II and Alexander III', in T. Howe & J. Reames (eds), *Macedonian Legacies: Studies in Ancient Macedonian History and Culture in Honor of Eugene N. Borza*, 2008, 31–55.
Musgrave, J.H., *et al.*, 'The Skull from tomb II at Vergina: King Philip II of Macedon', *Journal of Hellenic Studies* 104, 1984, 60–78.
Nap, J.M., 'Ad Catonis librum de re militari', *Mnemosyne* 55, 1927, 79–87.
Napoli, M. & Pontrandolfo, A. (ed.), *La Pittura parietale in Macedonia e Magna Grecia* (Salerno, 2002).
Nash, C., *Word-games: the Tradition of Anti-realist Revolt* (London, 1987).
Naval Intelligence Division (Great Britain), *Persia* (London, 1945).
Nefedkin, A.K., 'Did Aristotle Teach Alexander Warfare?' *Para Bellum*, June 2000a.
Nefedkin, A.K., 'On the Origin of Greek Cavalry Shields in the Hellenistic Period', *Klio* 91, 2000b, 356–366.
Neill, D.A., 'Ancestral Voices: the Influence of the Ancients on the Military Thought of the Seventeenth Century', *Journal of Military History* 62, 1998, 487–520.
Németh, G., 'Der Preis einer Panoplie', *Acta Antique Academiae Scientiarum* 36, 1995, 5–13.
Neumann, C., 'A Note on Alexander's March Rates', *Historia* 20, 1971, 196–198.
Newell E.T., *The Coinages of Demetrius Poliorcetes* (London, 1927).
Nicgorski, A.M., 'The Iconography of the Herakles Knot and the Herakles-Knot Hairstyle of Apollo and Aphrodite', PhD Thesis, University of North Carolina, 1995.
Nielsen, A.M., 'Et upåagtet portræt af Alexander den Store', in *Meddelelser fra Ny Carlsberg Glyptotek* 40, 1984, 5–17.
Nilsson, M.P., *Griechische Feste von religiöser Bedeutung* (Leipzig, 1906).
Ninou, K., *Treasures of Ancient Macedonia* (Athens, 1978).
Noguera Borel, A., 'Le recrutement de l'armée macédonienne sous le royauté', in A.-M. Guimier-Sorbets & M.B. Hatzopoulos (eds), *Rois, cités, nécropoles: Institutions, rites et monuments en Macédoine*, 2006, 227–236.
Noguera Borel, A., 'L'armée macédonienne avant Philippe II', *Ancient Macedonia VII*, 2007, 97–111.
Nutt, S.W., 'Tactical Interaction and Integration: a Study in Warfare in the Hellenistic Period from Philip II to the Battle of Pydna', PhD Thesis, University of Newcastle, 1993.
Nutton, V., *Ancient Medicine* (London, 2004).
Nylander, C., 'The Standard of the Great King – a Problem in the Alexander Mosaic', *Opuscula Romana* 14, 1983, 19–37.
O'Brien, J.M., 'The Enigma of Alexander: the Alcohol Factor', *Annals of Scholarship* 1, 1980a, 31–46.
O'Brien, J.M., 'Alexander and Dionysus: the Invisible Enemy', *Annals of Scholarship* 1, 1980b, 83–105.
O'Brien, J.M., *Alexander the Great: the Invisible Enemy* (London, 1994).
O'Sullivan, L., 'Le Roi Soleil: Demetrius Poliorcetes and the Dawn of the Sun-King', *Antichthon* 42, 2008, 78–99.
Oakeshott, E.R., *The Archaeology of Weapons* (London, 1960).
Ober, J., *Fortress Attica: Defence of the Athenian Land Frontier 404–322 BC* (Leiden, 1985).
Ober, J., 'Defense of the Athenian Land Frontier 404–322 BC: a Reply', *Phoenix* 43, 1989, 294–301.
Ober, J., 'Mountains and Roads', in V.D. Hanson (ed.), *Hoplites: The Classical Greek Battle Experience*, 1999, 174–179.
Ognenova, L., 'Les cuirasses de bronze trouvées en Thrace', *Bulletin de correspondance hellénique* 85, 1961, 501–538.
Ognenova, L., 'Les motifs décoratifs des armures en Thrace', in *Actes du premier Congrès international des études balkaniques et sud-est européens*, 1970, vol. II, 397–418.
Ognenova-Marinova, L., 'L'armure des Thraces', *Archaeologica Bulgarica* 4, 2000, 11–24.
Oikonomides, A.N., 'On the Portraits of Alexander the Great used as Models for Portraits of Mithradates the Great', *Archeion Pontou* 22, 1958, 219–243.

Oikonomides, A.N., 'Macedonian Cavalry and Alexander the Great as Cavalry Commander in Ancient Art and Coinage', *Ancient World* 4, 1981, 81.

Oikonomides, A.N., 'The Portrait of Pyrrhos King of Epirus in Hellenistic and Roman Art', *The Ancient World* 8, 1983, 67–72.

Oikonomides, A.N., 'Coins, Archaeological Finds and the "Macedonian Shield"', *The Classical Bulletin* 64, 1988, 74–84.

Oikonomides, A.N., 'The Elusive Portrait of Antigonus I, the 'One-Eyed' King of Macedonia', *The Ancient World* 20, 1989, 17–20.

Olbrycht, M.J., *Aleksander Wielki i świat irański* (Rzeszów, 2004).

Olbrycht, M.J., 'Macedonia and Persia', in J. Roisman & I. Worthington (eds), *A Companion to Ancient Macedonia*, 2010, 342–369.

Olmstead, A.T., *History of the Persian Empire* (Chicago, 1948).

Olufsen, O., *The Emir of Bukhara and His Country* (Copenhagen, 1911).

Ottoni, C., *et al.*, 'Mitochondrial Analysis of a Byzantine Population Reveals the Differential Impact of Multiple Historical Events in South Anatolia', *European Journal of Human Genetics* 19, 2011, 571–576.

Pagenstecher, R., *Nekropolis: Untersuchungen über Gestalt und Entwicklung der alexandrinischen Grabanlagen und ihrer Malereien* (Leipzig, 1919).

Paipetis, S.A. & Kostopoulos, V., 'Defensive Weapons in Homer', in S.A. Paipetis (ed.), *Science and Technology in Homeric Epics*, 2008, 181–203.

Pais, E., 'Ritratti di re macedoni', *Rendiconti della Classe di scienze morali, storiche e filologiche* 6, 1926, 49–56.

Palagia, O., 'The Enemy Within: a Macedonian in Piraeus', O. Palagia & S.V. Tracy (eds), *Regional Schools in Hellenistic Sculpture*, 1998, 15–26.

Palagia, O., 'Impact of *Ares Macedon* on Athenian sculpture', in O. Palagia & S.V. Tracy (eds), *The Macedonians in Athens 322–229 BC*, 2003, 141–151.

Pandermalis, D., *Dion: The Archaeological Site and the Museum* (Athens, 1997).

Pandermalis, D., *Dion* (Athens, 1999).

Pandermalis, D., 'New Discoveries at Dion', in M. Stamatopoulou & M. Yeroulanou (eds), *Excavating Classical Culture: Recent Archaeological Discoveries in Greece*, 2002, 99–107.

Pandermalis, D., *Alexander the Great: Treasures from the Epic Era of Hellenism* (New York, 2004).

Papazoglu, F., *The Central Balkan Tribes in Pre-Roman Times* (Amsterdam, 1978).

Parke, H.W., *Greek Mercenary Soldiers* (Oxford, 1933).

Pasinli, A., *The Book of Alexander Sarcophagus* (Istanbul, 2000).

Pattenden, P., 'When did Guard Duty End? The Regulation of the Night Watch in Ancient Armies', *Rheinisches Museum für Philologie* 130, 1987, 164–174.

Paunov, E. & Dimitrov, D.Y., 'New Data on the Use of the War Sling in Thrace (4th–1st century BC)', *Archaeologia Bulgarica* 4, 2000, 44–57.

Peachey, S., *The Military History of Pontus: Volume 1* (Bristol, 2005).

Pearson, L., *The Lost Historians of Alexander the Great* (New York, 1960).

Pease, S., 'Xenophon's Cyropaedia: the Complete General', *Classical Journal* 29, 1933/34, 436–440.

Peatfield, A., 'Reliving Greek Personal Combat-Boxing and Pankration', in B. Molloy (ed.), *The Cutting Edge. Studies in Ancient and Medieval Combat*, 2007, 20–33.

Pédech, P., 'L'expédition d'Alexandre et la science grecque', *Megas Alexandros* 1980, 135–156.

Pekridou, A., *Das Alketas-Grab in Termessos* (Tübingen, 1986).

Pélékidis, C., *Histoire de l'éphébie attique des origines à 31 avant Jésus-Christ* (Paris, 1962).

Perdikatis, V., *et al.*, 'Characterisation of the Pigments and the Painting Technique used on the Vergina Stelae', in M.A. Tiverios & D.S. Tsiafakis (eds), *Color in Ancient Greece: The Role of Color in Ancient Greek Art and Architecture*, 2002, 245–257.

Perrin, Y., 'D'Alexandre à Néron : le motif de la tente d'apparat. La salle 29 de la Domus Aurea', in J.M. Croisille, (ed.), *Neronia IV: Alejandro Magna, modelo des los emperadores romanos*, 1990, 213–229.

Peters, J.P. & Thiersch, H., *Painted Tombs for the Necropolis of Marissa* (London, 1905).

Petrochilos, N., *Roman Attitudes to the Greeks* (Athens, 1974).

Petsas, P., 'Mosaics from Pella', in M.G. Picard & H. Stern (eds), *Colloque international sur la mosaïque gréco-romaine, Paris, 29 août–3 septembre 1963*, 1965, 41–56.
Petsas, P., *Ho taphos ton Leukadion* (Athens, 1966).
Petsas, P., *Pella: Alexander the Great's Capital* (Thessaloniki, 1979).
Pflug, H., 'Illyrische Helme', in A. Bottini, *et al.*, *Antike Helme: Sammlung Lipperheide und andere Bestande des Antikenmuseums Berlin*, 1988a, 42–64.
Pflug, H., 'Chalkidische Helme', in A. Bottini, *et al.*, *Antike Helme: Sammlung Lipperheide und andere Bestande des Antikenmuseums Berlin*, 1988b, 137–150.
Pflug, H., *Schutz und Zier: Helme aus dem Antikenmuseum Berlin und Waffen anderer Sammlungen* (Basel, 1989).
Pfrommer, M., *Alexander der Grosse: Auf den Spuren eines Mythos* (Mainz, 2001).
Pfuhl, E., *Malerei und Zeichnung Der Griechen*, vol. 3 (Munich, 1923).
Pfuhl, E., & Möbius, H., *Die ostgriechischen Grabreliefs: Textband II* (Mainz, 1979a).
Pfuhl, E., & Möbius, H., *Die ostgriechischen Grabreliefs: Tafelband II* (Mainz, 1979b).
Phang, S.E., *The Marriage of Roman Soldiers (13 BC–AD 235)* (Leiden, 2001).
Phillips, E.D., *Greek Medicine* (London, 1973).
Phul, R.K., *Armies of the Great Mughals (1526–1707)* (New Delhi, 1978).
Picard, C., 'Sépultures des compagnons de guerre ou successeurs macédoniens d'Alexandre le Grand', *Journal des Savants*, 1964, 215–228.
Picard, G., *Roman Painting* (London, 1970).
Pimouguet-Pedarros, I., 'Le siège de Rhodes par Démétrios et l'apogée de la poliorcétique grecque', *Revue des études anciennes* 105, 2003, 371–392.
Pinkwart, D., *Das Relief des Archelaos von Priene und die 'Musen des Philiskos'* (Kallmünz, 1965).
Plantzos, D., 'Hellenistic Cameos: Problems of Classification and Chronology', *Bulletin of the Institute of Classical Studies* 41, 1996, 115–137.
Platnauer, M., 'Greek Colour-Perception', *Classical Quarterly* 15, 1921, 153–162.
Podhajsky, A., *Die spanische Hofreitschule* (Vienna, 1948).
Polastron, L.X., *Books on Fire: The Tumultuous Story of the World's Great Libraries* (London, 2007).
Poliakoff, M.B., *Combat Sports in the Ancient World* (New Haven, 1987).
Pomeroy, S.B., *Spartan Women* (Oxford, 2002).
Pompei: pitture e mosaici, vol. I, regio I, parte prima (Rome, 1990a).
Popazoglou, F., *The Central Balkan Tribes in Pre-Roman Times* (Amsterdam, 1978).
Pope, A., *A Survey of Persian Art*, vol. IV (London, 1939).
Post, R., 'Bright Colours and Uniformity', *Ancient Warfare*, IV/6, 2010, 14–19.
Pottier, E., 'Sur le bronze du musée de Naples dit "Alexandre à cheval"', *Mélange Genève*, 1905, 427–443.
Pouilloux, J., *L'Histoire et les cultes de Thasos* (Paris, 1954).
Poulsen, B., 'Alexander the Great in Italy during the Hellenistic period', in J. Carlsen (ed.), *Alexander the Great, Reality and Myth*, 1993, 161–170.
Pounder, R.L., 'A Hellenistic Arsenal in Athens', *Hesperia* 52, 1983, 233–256.
Powell, B., 'Oeniadae I: History and Topography', *American Journal of Archaeology* 8, 1904, 137–173.
Prag, J.N.W., 'Reconstructing King Philip II: the "Nice" Version', *American Journal of Archaeology* 94, 1990, 237–247.
Praschniker, C., *et al.*, *Forschungen in Ephesos VI: Das Mausoleum von Belevi* (Vienna, 1979).
Prestianni Giallombardo, A.M., 'Recenti testimonianze iconografiche sulla *kausia* in Macedonia e la datazione del fregio della *caccia* della II tomba reale di Vergina', *Dialogues d'histoire ancienne* 17, 1, 1991, 257–304.
Prestianni Giallombardo, A.M., 'Un copricapo dell'equipaggiamento militare macedone: la kausia', *Numismatica e antichita classiche* 22, 1993, 61–90.
Prestianni Giallombardo, A.M. & Tripodi, B., 'Iconografia monetale e ideologia reale macedone: i tipi del cavaliere nella monetazione di Alessandro I e di Filippo II', *Revue des études anciennes* 98, 1996, 311–355.
Price, D.J., 'Medieval Land Surveying and Topographical Maps', *Geographical Journal* 121, 1955, 1–10.

Price, M., *Coins of the Macedonians* (London, 1974).
Price, M., 'Circulation at Babylon in 323 BC', in W. Metcalf (ed.), *Mnemata: Papers in Memory of Nancy M. Waggoner*, 1991a, 63–72.
Price, M., *The Coinage in the Name of Alexander the Great and Philip Arrhidaeus* (London, 1991b).
Primo, A., 'Il termine epigonoi nella storiografia sull ellenismo', *Klio* 91, 2009, 367–377.
Pritchett, W.K., *The Greek State at War: Part I* (Berkeley, 1971).
Pritchett, W.K., *The Greek State at War: Part II* (Berkeley, 1974).
Pritchett, W.K., *The Greek State at War: Part III* (Berkeley, 1979).
Pritchett, W.K., *The Greek State at War: Part IV* (Berkeley, 1985).
Psoma, S., 'Entre l'armée et l'oikos: l'éducation dans le royaume de Macédoine', in A.-M. Guimier-Sorbets & M.B. Hatzopoulos (eds), *Rois, cités, nécropoles: Institutions, rites et monuments en Macédoine*, 2006, 285–300.
Rakatsanis, D. & Otto, B., 'Das Grab von Prodromi', *Antike Welt* 11, 1980, 55–57.
Ramsay, W.M., 'Military Operations on the North Front of Mount Tarsus', *Journal of Hellenic Studies* 40, 1920, 89–112.
Rawlings, L., 'Celts, Spaniards, and Samnites: Warriors in a Soldier's War', in T. Cornell, *et al.* (eds), *The Second Punic War: A Reappraisal*, 1996, 81–92.
Raymond, D., *Macedonian Regal Coinage to 413 BC* (New York, 1953).
Rebuffat, F., 'Alexandre le Grand et les problèmes financiers au début de son règne (été 336 – printemps 334)', *Revue numismatique* 25, 1983, 43–52.
Reed, N., *More than a Game: The Military Nature of Greek Athletic Contests* (Chicago, 1998).
Reinach, A.J., 'Les Galates dans l'art alexandrine', *Foundation Eugène Piot, Monuments et memoires* 18, 1910, 1–115.
Reinders, H.R., *New Halos* (Utrecht, 1988).
Reinhold, M., *History of Purple as a Status Symbol in Antiquity* (Brussels, 1970).
Renault, M., *Nature of Alexander* (London, 1975).
Renault, M., *Funeral Games* (Harmondsworth, 1982).
Retief, F.P. & Cilliers, L., 'The Army of Alexander the Great and Combat Stress Syndrome (326 BC)', *Acta Theologica (Supplementum)* 7, 2005, 29–43.
Rey, C.F., *In the Country of the Blue Nile* (London, 1927).
Reyes, J.M., 'The 'Orientalism' of Alexander the Great. His Persian Clothes,' MA Thesis, University of California, Los Angeles, 1996.
Rhomiopoulou, K., *Lefkadia: Ancient Mieza* (Athens, 1997).
Rice, E., *The Grand Procession of Ptolemy Philadelphus* (Oxford, 1983).
Rice, E., 'The Glorious Dead: Commemoration of the Fallen and Portrayal of Victory in the Late Classical and Hellenistic World', in J. Rich & G. Shipley (eds), *War and Society in the Greek World*, 1993, 224–257.
Rice, T., *The Scythians* (London, 1957).
Richardson, A., 'The Order of Battle in the Roman Army: Evidence from Marching Camps', *Oxford Journal of Archaeology* 20, 2001, 171–185.
Richardson, E.H., 'The Muscle Cuirass in Etruria and Southern Italy: Votive Bronzes', *American Journal of Archaeology* 100, 1996, 91–120.
Richardson, L., *A Catalog of Identifiable Figure Painters of Ancient Pompeii, Herculaneum, and Stabiae* (Baltimore, 2000).
Richter, G.M.A., *Catalogue of Greek Sculptures* (Cambridge, Mass., 1954).
Richter, G.M.A., *Ancient Italy: A Study of the Interrelations of its Peoples as Shown in their Arts* (Ann Arbor, 1955).
Ridgeway, W., *Origin and Influence of the Thoroughbred Horse* (Cambridge, 1905).
Ridgway, B.S., *Hellenistic Sculpture I: The Styles of ca.331–200 BC* (Bristol, 1990).
Ridgway, B.S., *Hellenistic Sculpture II: The Styles of ca.200–100 BC* (Madison, 2000).
Riginos, A.S., 'The Wounding of Philip II of Macedon: Fact and Fabrication', *Journal of Hellenic Studies* 114, 1994, 103–119.
Rizakis, A., 'Une forteresse macédonienne dans l'Olympe', *Bulletin de correspondance hellénique* 110, 1986, 331–346.

Rizzo, G.E., 'La battaglia di Alessandro nell'arte italica e romana', *Bollettino d'Arte* II/5, 1925/26, 529–546.

Robert, J. & Robert, L., *Fouilles d'Amyzon en Carie. I. Exploration, histoire, monnaies, et inscriptions* (Paris, 1983).

Robert, L., *Noms indigènes dans l'Asie-Mineure gréco-romaine* (Paris, 1963).

Robertson, M., 'The Boscoreale Figure-Paintings', *Journal of Roman Studies* 45, 1955, 58–67.

Robertson, M., 'Early Greek Mosaics', in B. Barr-Sharrar & E.N. Borza (eds), *Studies in the History of Art vol.10: Macedonia and Greece in Late Classical and Hellenistic Times*, 1982, 241–249.

Robinson, C.A., *The History of Alexander the Great*, vol. I (Providence, 1953).

Robinson, D.M., 'Note: the Great Chlamys not Rectangular', *Greece & Rome* 8, 1939, 190.

Robinson, D.M., *Excavations at Olynthus. X. Metal and Minor Miscellaneous Finds* (Baltimore, 1941).

Robinson, J.P., 'Tyrian Purple', *Sea Frontiers* 17, 1971, 76–82.

Roesch, P., 'Une loi fédérale béotienne sur la préparation militaire', *Études béotiennes*, 1982, 307–354.

Roisman, J., 'Honour in Alexander's Campaign', in J. Roisman (ed.), *Brill's Companion to Alexander the Great*, 2003, 294–321.

Roisman, J., *Alexander's Veterans and the Early Wars of the Successors* (Austin, 2012).

Rolle, R., *The World of the Scythians* (London, 1989).

Romey, K.M., 'The Forgotten Realm of Alexander', *Archaeology* 57, 2004, 18–25.

Ross, R.S., 'Immigration and Integration: The Augmentation of Macedonia's Military Population under Philip II and Alexander III, 360–323 BC'. MA Thesis, University of Victoria, Canada, 1990.

Rossi, G.H., 'The Communications of Alexander the Great', PhD Thesis, Pennsylvania State University, 1973.

Rossi, L., *Trajan's Column and the Dacian Wars* (London, 1971).

Rostovtzeff, M., *A Large Estate in Egypt in the Third Century BC* (Madison, 1922).

Rostovtzeff, M., 'The Foreign Commerce of Ptolemaic Egypt', *Journal of Economic and Business History* 4, 1932, 728–769.

Rostovtzeff, M., *The Social and Economic History of the Hellenistic World*, vol. I (Oxford, 1941a).

Rostovtzeff, M., *The Social and Economic History of the Hellenistic World*, vol. II (Oxford, 1941b).

Rostovtzeff, M., *The Social and Economic History of the Hellenistic World*, vol. III (Oxford, 1941c).

Rotroff, S.I., 'Minima Macedonica', in O. Palagia & S.V. Tracy (eds), *The Macedonians in Athens 322–229 BC*, 2003, 213–225.

Roueche, C. & Sherwin-White, S., 'Some Aspects of the Seleucid Empire: the Greek inscriptions from Failaka in the Arabian Gulf', *Chiron* 15, 1985, 1–39.

Rouveret, A., 'Les stèles alexandrines du Musée du Louvre: Rapport des analyses techniques à l'histoire de la couleur dans la peinture hellénistique', *Bulletin de la SFAC, RA*, 1998a, 216–220.

Rouveret, A., 'Un exemple de diffusion des techniques de la peinture hellénistique: les stèles alexandrines du musée du Louvre', in *L'Italie méridionale et les premières expériences de la peinture hellénistique*, 1998b, 175–190.

Rouveret, A., *Peintures grecques antiques: la collection hellénistique du musée du Louvre* (Paris, 2004).

Rubinsohn, Z., 'Macedonian Resistance to Roman Occupation in the Second Half of the Second Century BC', in T. Yuge (ed.), *Forms of Control and Subordination in Antiquity*, 1988, 141–158.

Rubio, A.P., 'El agotamiento de Macedonia', *Desperta ferro* 8, 2011, 38–41.

Ruffin, J.R., 'The Efficacy of Medicine during the Campaigns of Alexander the Great', *Military Medicine* 157, 1992, 467–475.

Ruiz, A., *Spirit of Ancient Egypt* (New York, 2001).

Rumpf, A., 'Kranos boiōtiourges', *Abhandlungen der Königlich Preussischen Akademie der Wissenschaften* 8, 1943, 3–17.

Rumpf, A., 'Zum Alexander-Mosaik', *Mitteilungen des Deutschen Archäologischen Instituts. Athenische Abteilung* 77, 1962, 229–241.

Russell, F.S., *Information Gathering in Classical Greece* (Ann Arbor, 1999).

Rzepka, J., 'The Units of Alexander's Army and the District Divisions of Late Argead Macedonia', *Greek, Roman & Byzantine Studies* 48, 2008, 39–56.

Saatsoglou-Paliadeli, C., *Ta epitaphia mnemia apo te megale toumpa tes Verginas* (Thessaloniki, 1984).

Saatsoglou-Paliadeli, C., 'Aspects of Ancient Macedonian Costume', *Journal of Hellenic Studies* 113, 1993, 122–147.
Sahin, S., *Die Inschriften von Perge. Teil I. Vorrömische Zeit, frühe und hohe Kaiserzeit* (Bonn, 1999).
Sakellariou, M.B. (ed.), *Greek Lands in History. Macedonia: 4,000 years of Greek History and Civilization* (Athens, 1988).
Salazar, C.F., *The Treatment of War Wounds in Graeco-Roman Antiquity* (Leiden, 2000).
Sancisi-Weerdenburg, H., *et al.* (eds), *Achaemenid History VIII: Continuity and Change*, 1994, 362– 364.
Sandars, H., 'Weapons of the Iberians', *Archaeologia* 64, 1913, 205–294.
Sarikakis, C.T., 'Des soldats macédoniens dans l'armée romaine', *Ancient Macedonia II*, 1977, 431–464.
Schafer, T., *Andreas Agathoi* (Munich, 1997).
Schefold, K., *Der Alexander-Sarkophag* (Berlin, 1968).
Schmidt, E.F., *Persepolis II: Contents of the Treasury and other Discoveries* (Chicago, 1957).
Schmidt, S., *Hellenistische Grabreliefs. Typologische und chronologische Beobachtungen* (Cologne, 1991).
Scholl, R., 'Alexander der Grosse und die Sklaverei am Hofe', *Klio* 69, 1987, 108–121.
Schröder, B., 'Thrakische Helme', *Jahrbuch des Deutschen Archäologischen Instituts* 27, 1912, 317–344.
Schuchhardt, W.H., 'Relief mit Pferd und Negerknaben im Nationalmuseum in Athen N.M. 4464', *Antike Plastik* 17, 1978, 75–99.
Schulz, R., 'Militärische Revolution und politischer Wandel. Das Schicksal Griechenlands im 4. Jahrhundert v. Chr.', *Historische Zeitschrift* 268, 1999, 281–310.
Scranton, R.L., *Greek Walls* (Cambridge, Mass., 1941).
Sears, J.M., 'Oeniadae VI: The Ship Sheds', *American Journal of Archaeology* 8, 1904, 227–237.
Sekunda, N., *The Army of Alexander the Great* (London, 1984a).
Sekunda, N., 'Regimental Uniforms: a Look at the Army of Alexander the Great', *Popular Archaeology* 5, 1984b, 15–16.
Sekunda, N., 'Hellenistic Warfare', in J. Hackett (ed.), *Warfare in the Ancient World*, 1989, 130–135.
Sekunda, N., "The Macedonian White Shields' Regiment 3rd–2nd Century BC', *Military Illustrated*, Oct., 1990, 19–24.
Sekunda, N., 'Gallery: Philip II of Macedon', *Military Illustrated*, Nov., 1992a, 46–47; 50.
Sekunda, N., *The Persian Army 560–330 BC* (London, 1992b).
Sekunda, N., *Seleucid and Ptolemaic Reformed Armies 168–145 BC. Volume 1: The Seleucid Army* (Stockport, 1994a).
Sekunda, N., 'The Tomb of Lyson and Kallikles', *Military Illustrated*, May, 1994b, 29–33.
Sekunda, N., *Seleucid and Ptolemaic Reformed Armies 168–145 BC. Volume 2: The Ptolemaic Army under Ptolemy Philometer* (Stockport, 1995).
Sekunda, N., *Greek Hoplite 480–323 BC* (Oxford, 2000).
Sekunda, N., 'Greek Swords and Swordsmanship', *Osprey Military Journal* 3, 2001a, 34–42.
Sekunda, N., *Hellenistic Infantry Reform in the 160s BC* (Lodz, 2001b).
Sekunda, N., 'The Sarissa', *Acta Universitatis Lodziensis Folia Archeologica* 23, 2001c, 13–41.
Sekunda, N., 'Greek Military Cuisine', *Osprey Military Journal* 4, 2002, 3–6.
Sekunda, N., 'A Macedonian Companion in a Pompeian Fresco', *Archaeologia* 54, 2003, 29–33.
Sekunda, N., 'The Macedonian Army', in J. Roisman & I. Worthington (eds), *A Companion to Ancient Macedonia*, 2010, 446–471.
Sekunda, N., 'Cavalry about the Court: The Ptolemaic Horse Guard', *Ancient Warfare*, V/2, 2011a, 14–19.
Sekunda, N., 'La infantería de los sucesores: Macedonios e iranios al servicio de los Diádocos', *Desperta ferro* 8, 2011b, 16–22.
Sekunda, N., 'Intimate Accounts of the Great General. Pyrrhus and other Writers at his Court', *Ancient Warfare* VI/4, 2012a, 11–13.
Sekunda, N., *Macedonian Armies After Alexander 323–168 BC* (Oxford, 2012b).
Semple, E.C., 'The Influence of Geographic Conditions Upon Ancient Mediterranean Stock-Raising', *Annals of the Association of American Geography* 12, 1922, 3–38.
Shefton, B.B., *Greek Arms and Armour: the Greek Museum* (Newcastle upon Tyne, 1978).
Shepherd, R., *Ancient Mining* (London, 1993).

Sherk, R.K., 'Roman Geographical Exploration and Military Maps', *Aufstieg und Niedergang der römischen Welt* 2, 1974, 534–562.
Sherwin-White, S. & Kuhrt, A., *From Samarkand to Sardis: A New Approach to the Seleucid Empire* (Berkeley, 1993).
Sidnell, P., *Warhorse: Cavalry in Ancient Warfare* (London, 2006).
Siganidou, M. & Lilimpaki-Akamati, M., *Pella: Capital of Macedonians* (Athens, 2003).
Simon, A., *Comicae Tabellae* (Amsterdam, 1938).
Simpson, R.H., 'A Note on Cyinda', *Historia* 6, 1957, 503–504.
Singer, A., 'The Herbal in Antiquity and its Transmission to Later Ages', *Journal of Hellenic Studies* 47, 1927, 1–52.
Sinn, V., *Die homerischen Becher: Hellenistische Reliefkeramik aus Makedonien* (Berlin, 1979).
Six, J., 'Un ritratto del re Pirro d'Epiro', *Mitteilungen des Kaiserlich Deutschen Archaeologischen Instituts. Roemische Abteilung*, 6, 1891, 279–284.
Skarmintzos, S., 'Phalanx versus Legion: Greco-Roman Conflict in the Hellenistic Era', *Ancient Warfare*, II/2, 2008, 30–34.
Slowikowski, S., 'Alexander the Great and Sport History: a Commentary on Scholarship', *Journal of Sport History* 16, 1989, 70–78.
Smith, R.R.R., *Hellenistic Royal Portraits* (Oxford, 1988).
Smith, R.R.R., 'Spear-Won Land at Boscoreale: on the Royal Paintings of a Roman Villa', *Journal of Roman Archaeology* 7, 1994, 100–128.
Snodgrass, A.M., *Arms and Armour of the Greeks* (London, 1967).
Sofman, A.S. & Tsibukidis, D.I., 'Nearchus and Alexander', *The Ancient World* 16, 1987, 71–77.
Sokolovska, V., *Isar-Marvinci i povardarjeto vo antičko vreme* (Skopje, 1986).
Sokolovska, V., *The Ancient Town of Isar-Marvinci, Valandovo* (Skopje, 2011).
Soteriades, G., 'Das Schlachtfeld von Chäronea und der Grabhügel der Makedonen', *Mitteilungen des Deutschen Archäologischen Instituts. Athenische Abteilung* 28, 1903, 301–330.
Soteriades, G., 'Anaskaphai Diou Makedonias', *Praktika*, 1930, 36–51.
Sparkes, R.A., 'Illustrating Aristophanes', *Journal of Hellenic Studies* 95, 1975, 122–135.
Spaulding, O.L., 'Ancient Military Writers', *Classical Journal* 28, 1933, 657–669.
Spawforth, T., 'The Court of Alexander the Great between Europe and Asia', in A.J.S. Spawforth (ed.), *The Court and Court Society in Ancient Monarchies*, 2007, 82–120.
Speidel, M.P., 'Late Roman Military Decorations II: Gold-Embroidered Capes and Tunics', *Antiquite Tardive* 5, 1997, 231–237.
Spence, I.G., *The Cavalry of Classical Greece* (Oxford, 1993).
Spencer, D., *The Roman Alexander: Reading a Cultural Myth* (Exeter, 2002).
Spinazzola, V., *Pompei alla luce degli Scavi Nuovi di Via dell'Abbondanza (anni 1910–1923)*, vol. II (Rome, 1953).
Sprague de Camp, L., *Ancient Engineers* (London, 1977).
Sprawski, S., *Jason of Pherae* (Krakow, 1999).
Spyridakis, S.V., 'Cretan Soldiers Overseas: a Prosopography', *Kretologia* 12, 1981, 49–83.
Stadter, P.A., 'The Ars Tactica of Arrian: Tradition and Originality', *Classical Philology* 73, 1978, 117–128.
Stagakis, G., 'Observations on the *Hetairoi* of Alexander the Great', *Ancient Macedonia I*, 1970, 86–102.
Stein, A., 'Notes on Alexander's Crossing of the Tigris and the Battle of Arbela', *Geographical Journal* 100, 1942, 155–164.
Steinman, A., 'Adverse Effects of Heat and Cold on Military Operations', *Military Medicine* 152, 1987, 382–390.
Stemmer, K., *Untersuchungen zur Typologie, Chronologie und Ikonographie der Panzerstatuen* (Berlin, 1978).
Stephan, R.P. & Verhoogt, A., 'Text and Context in the Archive of Tiberianus', *Bulletin of the American Society of Papyrologists* 42, 2005, 189–201.
Stewart, A., *Faces of Power: Alexander's Image and Hellenistic Politics* (Berkeley, 1993).
Stojcev, V., 'Alexander the Great – the most Talented Strategist in the History of Mankind', *Defence* 63, 2001 (online).

Storcke, J. & Teague, W.D., *Flour for Man's Bread* (Minneapolis, 1952).
Strauss, B.S., 'Philip II of Macedon, Athens and Silver Mining', *Hermes* 112, 1984, 418–427.
Strauss, B.S., 'Alexander: the Military Campaign', in J. Roisman (ed.), *Brill's Companion to Alexander the Great*, 2004, 133–158.
Stronach, D., *Pasargadae: a Report on the Excavations Conducted by the British Institute of Persian Studies from 1961 to 1963* (Oxford, 1978).
Strootman, R., 'The Hellenistic Royal Court: Court Culture, Ceremonial and Ideology in Greece, Egypt and the Near East, 336–30 BCE', Dissertation, University of Utrecht, 2007.
Stroud, R.S., 'Greek Inscriptions: Theozotides and the Athenian Orphans', *Hesperia* 40, 1971, 280–301.
Stupka, D., 'Der Gürtel in der griechischen Kunst', PhD Thesis, University of Vienna, 1972.
Stutz, H., *Die Farbe Purpur im frühen Griechentum Beobachtet in der Literatur und in der bildenden Kunst* (Stuttgart, 1990).
Sumner, G., *Roman Military Dress* (Stroud, 2009).
Svoronos, J.N., *To en Athnais Ethnikon Mouseion* (Athens, 1903).
Swain, J.W., 'The Theory of the Four Monarchies: Opposition History under the Roman Empire', *Classical Philology* 35, 1940, 1–21.
Swindler, M.H., *Ancient Painting: From the Earliest Times to the Period of Christian Art* (New Haven, 1929).
Tarbell, F.B., 'The Form of the Chlamys', *Classical Philology* 1, 1906, 283–289.
Tarn, W.W., *Hellenistic Military and Naval Developments* (Cambridge, 1930).
Tarn, W.W., *Alexander the Great. Vol. II: Sources and Studies* (Cambridge, 1948).
Tarn, W.W., *Hellenistic Civilisation* (London, 1952).
Taylor, R., 'The Battle of Pydna', *Slingshot* 267, 2009, 13–22.
Tejada, J.V., 'Warfare, History and Literature in the Archaic and Classical Periods: the Development of Greek Military Treatises', *Historia* 53, 2004, 129–146.
Temple, R., *The Crystal Sun: Rediscovering a Lost Technology of the Ancient World* (London, 2000).
Thapliyal, U.P., 'Weapons, Fortifications and Military Training in Ancient India', in S.N. Prasad (ed.), *Historical Perspectives of Warfare in India*, 2002, 104–135.
Themelis, P.G. & Touratsoglou, G.P., *Hoi taphoi tou Derveniou* (Athens, 1997).
Themelis, P.G., 'Macedonian Dedications on the Akropolis', in O. Palagio & S.V. Tracy (eds), *The Macedonians in Athens 322–229 BC*, 2003, 162–172.
Thomas, C.G., 'Alexander's Garrisons: a Clue to his Administrative Plans?' *Antichthon* 8, 1974, 11–20.
Thomas, C.G., *Alexander the Great in His World* (Oxford, 2007).
Thompson, D.B., 'Mater Caelaturae', *Hesperia* 8, 1939, 289–293.
Thompson, D.J., *Memphis under the Ptolemies* (Princeton, 1988).
Thompson, E.A., *A Roman Reformer and Inventor being a new Text of the Treatise De Rebus Bellicis with a Translation and Introduction* (Oxford, 1952).
Thompson, G., 'Iranian Dress in the Achaemenid Period: Problems concerning the Kandys and other Garments', *Iran* 3, 1965, 121–126.
Thompson, H.A., 'Architecture as a Medium of Public Relations among the Successors of Alexander', in B. Barr-Sharrar & E.N. Borza (eds), *Macedonia and Greece in Late Classical and Early Hellenistic Times*, 1982, 173–189.
Thonemann, P.J., 'Hellenistic Inscriptions from Lydia', *Epigraphica Anatolica* 36, 2003, 95–108.
Tiverios, M.A. & Tsiafakis, D. (eds), *Color in Ancient Greece: The Role of Color in Ancient Greek Art and Architecture 700–31 BC* (Thessaloniki, 2002).
Tod, M.N., *A Selection of Greek Historical Inscriptions. Vol. II: From 403 to 323 BC* (Oxford, 1948).
Tornkvist, S., 'Notes on Linen Corselets', *Opuscula Romana* 7, 1969, 81–82.
Touchais, G., 'Chronique des fouilles et découvertes archéologiques en Grèce en 1981', *Bulletin de correspondance hellénique* 106, 1982, 529–635.
Touchais, G., *et al.*, 'Chronique des fouilles et découvertes archéologiques en Grèce en 1999', *Bulletin de correspondance hellénique* 124, 2000, 753–1006.
Touloupa, E., 'Kattymata Tyrrenika – Krepides Attikai', *Archaiologikon deltion* 28, 1973, 116–137.

Touratsoglou, I., 'To Xiphos tes Verginas Xumbole ste Makedonike hoplopia ton usteron klassikon chronon', *Ancient Macedonia IV*, 1986, 611–628.
Tracy, S.V., 'Antigonos Gonatas, King of Athens', in O. Palagia & S.V. Tracy (eds), *The Macedonians in Athens 322–229 BC*, 2003, 56–60.
Treister, M.Y., *Hammering Techniques in Greek and Roman Jewellery and Toreutics* (Leiden, 2001).
Tritle, W., 'Two Armies in Iraq: Tommy Franks in the Footsteps of Alexander the Great', in M.B. Cosmopoulos (ed.), *Experiencing War: Trauma and Society in Ancient Greece and Today*, 2007, 177–178.
Troncoso, V.A., 'The Bearded King and the Beardless Hero: From Philip II to Alexander the Great', in E. Carney & D. Ogden (ed.), *Philip II and Alexander the Great: Father and Son, Lives and Afterlives*, 2010, 13–24.
Trundle, M., *Greek Mercenaries* (London, 2004).
Tsibidou-Avloniti, M., 'Excavating a Painted Macedonian Tomb near Thessaloniki: An Astonishing Discovery', in M. Stamatopoulou & M. Yeroulanou (eds), *Excavating Classical Culture: Recent Archaeological Discoveries in Greece*, 2002a, 91–97.
Tsibidou-Avloniti, M., 'Revealing a Painted Macedonian Tomb near Thessaloniki', in M. Napoli & A. Pontrandolfo (eds), *La pittura parietale in Macedonia e Magna Grecia*, 2002a, 37–42.
Tsibidou-Avloniti, M., 'The Macedonian Tomb at Aghios Athanasios, Thessaloniki', in D. Pandermalis (ed.), *Alexander the Great: Treasure from an Epic Era of Hellenism*, 2004, 149–151.
Tsigarida, B. & Ignatiadou, D., *The Gold of Macedon* (Athens, 2000).
Turchin, T.D.H. & Adams, J.M., 'East-West Orientation of Historical Empires', *Journal of World-Systems Research* 12, 2006, 219–229.
Turner, E.G. (ed.), *The Hibeh Papyri Part II* (London, 1955).
Tzifopoulos, Y.Z., '"Hemerodromoi" and Cretan "Dromeis": Athletes or Military Personnel? The Case of the Cretan Philonides', *Nikephoros* 11, 1998, 137–170.
Ueda-Sarson, L., 'The Evolution of Hellenistic Infantry, Part 1: The Reforms of Iphikrates', *Slingshot* 222, 2000, 30–36.
Uhlenbrock, J.P., *The Coroplast's Art: Greek Terracottas of the Hellenistic World* (New Paltz, 1990).
van der Spek, R.J., 'Darius III, Alexander the Great and Babylonian Scholarship', in W. Henkelman & A. Kuhrt (eds), *A Persian Perspective: Essays in Memory of Heleen Sancisi-Weerdenburg*, 2003, 289–342.
van Wees, H., 'Kings in Combat: Battles and Heroes in the *Iliad*', *Classical Quarterly* 38, 1988, 1–24.
van Wees, H., 'Heroes, Knights and Nutters: Warrior Mentality in Homer', in A.B. Lloyd (ed.), *Battle in Antiquity*, 1996, 1–86.
van Wees, H., *Greek Warfare: Myths and Realities* (London, 2004).
Vandorpe, K., 'When a Man has Found a Horse in his Mind (Xen., De equitandi ratione, IV.1). On Greek Horsemanship in Ptolemaic Egypt', in B. Kramer (ed.), *Akten des 21. Internationalen Papyrologenkongresses: Berlin, 13–19.8.1995, 1997, 984–990.*
Vasic, R., 'Reflecting on Illyrian Helmets', *Starinar* 60, 2010, 37–55.
Vasilev, V.P., 'Production de casques thraces dans les Rhodopes', *Archeologia Sofia* 22, 1984, 1–18.
Vatin, C., *Recherches sur le mariage et la condition de la femme mariée à l'époque hellénistique* (Paris, 1970).
Vaughan, P.B., 'Local Cold Injury: Menace to Military Operations', *Military Medicine* 145, 1980, 306–307.
Venedikov, I. & Gerassimov, T., *Thrakische Kunst* (Vienna, 1973).
Verdiani, C., 'Original Hellenistic Paintings in a Thracian Tomb', *American Journal of Archaeology* 49, 1945, 402–415.
Vermeule, C., 'Hellenistic and Roman Cuirassed Statues: the Evidence of Paintings and Reliefs in the Chronological Development of Cuirass Types', *Berytus* 13, 1959/60, 3–82.
Vermeule, C., 'Hellenistic and Roman Cuirassed Statues: a Supplement', *Berytus* 15, 1964, 95–110.
Vermeule, C., 'A Greek Theme and its Survivals: the Ruler's Shield (Tondo Image) in Tombs and Temples', *Proceedings of the American Philosophical Society* 109, 1965, 361–397.
Vermeule, C., 'Hellenistic and Roman Cuirassed Statues: Second Supplement', *Berytus* 16, 1966, 49–59.
Vermeule, C., 'Cuirassed Statues – 1974 Supplement', *Berytus* 23, 1974, 5–26.

Vermeule, C., 'The Horse and Groom Relief in Athens', *Greek, Roman and Byzantine Monographs* 10, 1984, 297–304.
Vickers, M., 'Hellenistic Thessaloniki', *Journal of Hellenic Studies* 92, 1972, 156–170.
Vidal-Naquet, P. & Lévêque, P., 'Epaminondas the Pythagorean, or the Tactical Problem of Right and Left', in P. Vidal-Naquet (ed.), *The Black Hunter*, 1986, 61–82.
Vidal-Naquet, P., 'La tradition de l'hoplite athénien', in J.P. Vernant (ed.), *Problèmes de la guerre en Grèce ancienne*, 1968, 161–181.
Vieillefond, J.R. (ed.), *Jules Africain: Fragments des Vestes, provenant de la collection des tacticiens grecs* (Paris, 1932).
Vigneron, P., *Le cheval dans l'Antiquité gréco-romain*, vol. II (Nancy, 1968).
Virgilio, B., 'Eumene I ei mercenari di Filetereia e di Attaleia', *Studi Classici e Orientali* 32, 1983, 97–140.
Vokotopoulou, I., *et al.* (eds), *Sindos Exhibition Catalogue* (Thessaloniki, 1985).
Vollmoeller, K.G., 'Über zwei euboische Kammergraber mit Totenbetten', *Mitteilungen des Kaiserlich Deutschen Archaologischen Instituts. Athenische Abteilung* 26, 1901, 333–376.
von Graeve, V., *Der Alexandersarkophag und seine Werkstatt* (Berlin, 1970).
von Merhart, G., 'Panzer-studie', in *Origines. Raccolta di scritti in onore di Mons. Giovanni Baserga*, 1954, 33–61.
von Oppenheim, M.F., *Vom Mittelmeer zum persischen Golf*, vol. II (Berlin, 1900).
von Rohden, H., *Die Terracotten von Pompeji* (Stuttgart, 1880).
von Schwartz, F., *Alexanders des Grossen Feldzüge in Turkestan* (Stuttgart, 1906).
von Szalay, A. & Boehringer, E., *Altertümer von Pergamon X. Die Hellenistischen Arsenale* (Berlin, 1937).
Wace, A.J.B., 'The Cloaks of Zeuxis and Demetrius', *Jahreshefte des Österreichischen Archäologischen Institutes in Wien* 38, 1950, 111–118.
Walbank, F.W., *Philip V of Macedon* (Cambridge, 1940).
Walbank, F.W., *A Historical Commentary on Polybius*, vol. I (Oxford, 1957).
Walbank, F.W., *A Historical Commentary on Polybius*, vol. II (Oxford, 1967).
Walbank, F.W., 'Polybius and Macedonia', *Ancient Macedonia I*, 1970, 291–307.
Walbank, F.W., *Polybius, Rome and the Hellenistic World* (Cambridge, 2002).
Wallace, M., 'Sutor Supra Crepidam', *American Journal of Archaeology* 44, 1940, 213–221.
Wallinga, H., 'Naval Installations in Cilicia Pedias: the Defence of the Parathalassia in Achaemenid Times and After', *Anatolia Antiqua* 32, 1991, 277–281.
Walter, P., *et al.*, 'Peinture hellénistique, les stèles alexandrines', *Techne* 7, 1998, 53–56.
Watanabe, E., 'Floating Bridges: Past and Present', *Structural Engineering International* 13, 2003, 128–132.
Waterfield, R., *Dividing the Spoils: The War for Alexander the Great's Empire* (Oxford, 2011).
Watson, G.R., *The Roman Soldier* (Ithaca, 1985).
Waurick, G., 'Helme der hellenistischen Zeit und ihre Vorläufer', in A. Bottini, *et al.*, *Antike Helme: Sammlung Lipperheide und andere Bestande des Antikenmuseums Berlin*, 1988, 151–180.
Webber, C., *The Gods of Battle* (Barnsley, 2011).
Weber, G., 'The Court of Alexander the Great as Social System', in W. Heckel & L.A. Tritle (eds), *Alexander the Great: A New History*, 2009, 83–98.
Webster, T.B.L., *Studies in Later Greek Comedy* (Manchester, 1953).
Webster, T.B.L., *Monuments Illustrating New Comedy: Bulletin of the Institute of Classical Studies* (London, 1995).
Welles, C.B., 'New Texts from the Chancery of Philip V of Macedonia and the Problems of the "Diagramma"', *American Journal of Archaeology* 42, 1938, 245–260.
Westholm, A., 'Cypro-Archaic Splint Armour', *Acta Archaeologica* 9, 1938, 163–173.
Westlake, H.D., *Thessaly in the Fourth Century BC* (Groningen, 1969).
Wheatcroft, A., *The Enemy at the Gate: Habsburgs, Ottomans and the Battle for Europe* (London, 2009).
Wheeler, E.L., 'The Origins of Military Theory in Ancient Greece and China', *International Commission of Military History* 5, 1980/81, 74–79.
Wheeler, E.L., '*Hoplomachia* and Greek Dances in Arms', *Greek, Roman and Byzantine Studies* 23, 1983a, 223–233.

Wheeler, E.L., '"The Hoplomachoi and Vegetius" Spartan Drillmasters', *Chiron* 13, 1983b, 1–20.
Wheeler, E.L., '*Polla Kena tou Polemou*: the History of a Greek Proverb', *Greek, Roman and Byzantine Studies* 21, 1988, 153–184.
Wheeler, E.L., 'Why the Romans can't Defeat the Parthians: Julius Africanus and the Strategy of Magic', in W. Groenman-van Waateringe, *et al.* (eds), *Roman Frontier Studies 1995*, 1997, 575–579.
Whitby, M., 'The Spinning of Alexander', *Omnibus* 40, 2000, 15–17.
White, D.A., 'A Survey of Millstones from Morgantina', *American Journal of Archaeology* 67, 1963, 199–206.
White, J.R., *The Prussian Army* (Lanham: Maryland, 1996).
Whitehead, D., 'Who Equipped Mercenary Troops in Classical Greece?' *Historia* 40, 1991, 105–113.
Whitehead, D., *Aineias the Tactician: How to Survive under Siege* (London, 2001).
Whitehead, D., 'Fact and Fantasy in Greek Military Writers', *Acta Antiqua Academiae Scientiarum Hungaricae* 48, 2008, 139–155.
Whitehead, R.B., 'The Eastern Satrap Sophytes', *Numismatic Chronicle* 2, 1943, 60–72.
Whitley, J., 'Archaeology in Greece 2004–2005', *Archaeological Reports 2004–2005* (Society for the Promotion of Hellenic Studies), 2005, 1–118.
Whitley, J., *et al.*, 'Archaeology in Greece 2005–2006', *Archaeological Reports 2005–2006* (Society for the Promotion of Hellenic Studies), 2006, 1–112.
Whitley, J., *et al.*, 'Archaeology in Greece 2006–2007', *Archaeological Reports 2006–2007* (Society for the Promotion of Hellenic Studies), 2007, 1–121.
Widengren, G., 'Some Remarks on Riding Costume and Articles of Dress among Iranian Peoples in Antiquity', *Studia Ethnographica Upsaliensia* 11, 1956, 228–276.
Wiesehöfer, J., *Die 'dunklen Jahrhunderte' der Persis* (Munich, 1994).
Wilcken, U., 'Ein Sosylos-Fragment in der Würzburger Papyrussammlung', *Hermes* 41, 1906, 141.
Wilkins, R.H., *Neurosurgical Classics* (New York, 1992).
Wilkinson, F., *Arms and Armour* (London, 1978).
Willems, E., *A Way of Life and Death: Three Centuries of Prussian-German Militarism* (Nashville, 1986).
Williams, M.F., 'Philopoemen's Special Forces: Peltasts and a New Kind of Greek Light-Armed Warfare', *Historia* 53, 2004, 257–277.
Winkes, R., 'Pliny's Chapter on Roman Funeral Customs in the Light of the *Clipeatae Imagines*', *American Journal of Archaeology* 83, 1979, 481–484.
Winkes, R., 'The Pyrrhus Portrait' in T. Hackens, *et al.* (eds), *The Age of Pyrrhus: Archaeology, History and Culture in Early Hellenistic Greece and Italy*, 1992, 175–188.
Winter, F., 'Die Sarkophage von Sidon', *Archäologischer Anzeiger* 19, 1894, 1–23.
Winter, F., *Der Alexandersarkophagaus Sidon* (Strasbourg, 1912).
Winter, F.E., *Greek Fortifications* (Toronto, 1971).
Wood, M., *In the Footsteps of Alexander the Great* (London, 1997).
Wood, N., 'Xenophon's Theory of Leadership', *Classica et Medievalia* 25, 1964, 33–66.
Wooton, W., 'Another Alexander Mosaic: reconstructing the Hunt Mosaic from Palermo', *Journal of Roman Archaeology* 15, 2002, 264–274.
Worthington, I., *Alexander the Great: Man and God* (Harlow, 2004).
Worthington, I., *Philip II of Macedonia* (New Haven, 2008).
Wright, J.H. (ed.), 'Archaeological News', *American Journal of Archaeology* 2, 1898, 95–158.
Wright, N.L., 'Who's Killing whom on the Coinage of Patraos of Paionia', *Journal of the Numismatic Association of Australia* 22, 2012, 19–49.
Wrightson, G., 'The Nature of Command in the Macedonian Sarissa Phalanx', *The Ancient History Bulletin* 24, 2010, 73–94.
Yalouris, N., *et al.*, *The Search for Alexander: an Exhibition* (Boston, 1980).
Zhivkova, L., *The Kazanluk Tomb* (Recklinghausen, 1975).
Zimmerman, J.L., 'Une Cuirasse de Grand Grece', *Museum Helveticum* 36, 1979, 177–184.
Zuckerman, C., 'The Military Compendium of Syrianus Magister', *Jahrbuch der Österreichischen Byzantinistik* 40, 1990, 209–224.

Index

Note: References such as '178–9' indicate (not necessarily continuous) discussion of a topic across a range of pages. Wherever possible in the case of topics with many references, these have either been divided into sub-topics or only the most significant discussions of the topic are listed. Because the entire work is about 'Macedonia' and the 'Macedonian army', the use of these terms (and certain others which occur constantly throughout the book) as an entry point has been minimized. Information will be found under the corresponding detailed topics.